nation about this series at http://www.springer.com/series/11987

More infor

Cave and Karst Systems of t

Series Editor

James W. LaMoreaux, P.E.LaMoreaux and Associates,

Augusto S. Auler · Paulo Pessoa
Editors

Lagoa Santa Karst: Brazil's Iconic Karst Region

Springer

Editors
Augusto S. Auler
Instituto do Carste
Belo Horizonte, Minas Gerais, Brazil

Paulo Pessoa
Instituto do Carste
Belo Horizonte, Minas Gerais, Brazil

ISSN 2364-4591 ISSN 2364-4605 (electronic)
Cave and Karst Systems of the World
ISBN 978-3-030-35939-3 ISBN 978-3-030-35940-9 (eBook)
https://doi.org/10.1007/978-3-030-35940-9

This Springer imprint is published by the registered company Springer Nature Switzerland AG
The registered company address is: Gewerbestrasse 11, 6330 Cham, Switzerland

Preface

If not for a fortuitous moment in 1835, the surroundings of the town of Lagoa Santa, in southeastern Brazil, would have been just another karst area among the extensive karst terrains in the state of Minas Gerais, Brazil. Fascinated by the caves and their fossil content, Danish naturalist Peter Wilhelm Lund decided to establish himself in the small town and start a detailed study of the abundant fossil remains in the caves. For about 10 years, Peter Lund researched approximately 1000 caves and brought to life a remarkable collection of, until then, unknown extinct species. As a naturalist, Lund maintained an interest in several other areas, and invited colleagues and welcomed visitors to join him in the solitude of Lagoa Santa. In the mid-nineteenth century, several lines of research were being pursued in Lagoa Santa, mostly related to paleontology, the living fauna, and botany. Lund's influence continued after his retirement from research and, ultimately, his death in 1880 in Lagoa Santa, without him ever having returned to his native Copenhagen. The bait provided by Lund worked as a magnet attracting a continuous stream of scientists until the present day. The Lagoa Santa Karst is by far the best-researched karst area in Brazil, with hundreds of publications in many fields of science. Lagoa Santa's proximity to the new state capital, Belo Horizonte (inaugurated in 1897), also facilitated access and made it a convenient study site for many generations of scientists.

The Lagoa Santa Karst is the cradle of Brazilian paleontology, archaeology, speleology, and karst research. The iconic saber-toothed tiger (*Smilodon populator*) was first described by Lund in the area. In the late nineteenth century, much of the contentious debate about the contemporaneity of humans and the Pleistocene megafauna centered on Lagoa Santa. The largest collection of early Holocene human skulls also comes from the Lagoa Santa Karst. The area contains Brazil's largest number of caves—in excess of 1500—as well as hundreds of archaeological sites. Although the area presents several scientific milestones, it is also experiencing challenging times. Encroachment of the suburban areas of Belo Horizonte and its satellite cities is threatening the integrity of the landscape as well as the groundwater quality and quantity. Approximately 500,000 people live at or near the karst, and several cement and lime plants have active quarries in the area. Sinkhole collapse, groundwater pollution, landscape degradation, deforestation, and cave vandalism are among the many ongoing impacts.

This book aims to contribute to knowledge of the Lagoa Santa Karst, while also aiding its conservation. It is an attempt to summarize in one comprehensive volume the vast amount of research, in several disciplines, performed over nearly 200 years. Each chapter is designed as a comprehensive review of its theme, and the book intends to integrate the many fields of research performed in the region.

This book starts with a historical review of research performed in the Lagoa Santa Karst. In chapter "History of Research in the Lagoa Santa Karst", Augusto S. Auler provides insights into the initial human involvement with caves in pre-history and the economic exploitation of caves for saltpeter mining since the eighteenth century. The arrival of Peter Lund is highlighted alongside the studies performed by several of his visitors. In the twentieth century, the area continued to attract the attention of many scientists, which resulted in continuing interest in the

Lagoa Santa Karst up to the present day. In chapter "The Geology of Lagoa Santa Karst", Paulo Galvão and Jorge Peñaranda present a synthesis of the geological knowledge of the Lagoa Santa Karst, emphasizing the main characteristics of the two carbonate units (Lagoa Santa and Pedro Leopoldo members of the Sete Lagoas Formation) where the karst is developed. The tectonic framework is also discussed, providing a basis on which to understand the geomorphology of the Lagoa Santa Karst.

In chapter "The Vegetation of Lagoa Santa Karst", Ana Elisa Brina reviews the vegetation of the area, first studied over 100 years ago by Eugen Warming. The vegetation is, in many ways, peculiar, being in a transition zone between two biomes (Atlantic Forest and Cerrado) and displaying species especially adapted to living in karst settings, such as limestone outcrops, under a highly seasonal climate. Over 2,000 species have been recorded in the area. The chapter points out the need for conservation in order to preserve the significant biodiversity of the area. In chapter "The Lagoa Santa Fauna: Historical Records", Gisele Lessa, Flávia Henriques e Souza, and Natália Lima Boroni present an updated review of the fauna in the Lagoa Santa Karst. This fauna is notable for presenting species related to both biomes and is among the best known in the country, with 70 species of fish, 25 species of amphibians, 41 species of reptiles, 240 species of birds, and 107 species of mammals recorded. A reduction in the number of species due to environmental impacts has been demonstrated. In chapter "The Waterbirds and Dynamics of Lagoa Santa Karst Temporary Lakes", José Eugênio Figueira and collaborators discuss one of the most remarkable features of the Lagoa Santa Karst: its shallow lakes, which represent a habitat for several bird species. The dynamics of the lakes are discussed, together with variations in the richness, abundance, and foraging guilds of species, depending on a lake's characteristics and the season. The list of waterbirds is updated, and initiatives for protecting the lakes and their fauna are highlighted.

The Lagoa Santa Karst is mostly soil covered. In chapter "Soils of the Lagoa Santa Karst", Luís B. Piló and Selma Simões de Castro discuss the karst's important soil cover. Red latosols predominate and can reach considerable thickness, masking the limestone. The soil has its origin in the upper metapelite cover, and two soil horizons (an upper red and a lower yellow) can be recognized throughout the region. Much of the initial development of karst occurs under soil at the epikarst zone. A detailed characterization of the soil and its importance in shaping the karst landscape is presented. In chapter "Karst Landforms in the Lagoa Santa Area", Augusto S. Auler presents an overview of the karst landforms in the Lagoa Santa area, starting with a new proposal of geomorphological domains. The area displays a full suite of karst landforms, but with some peculiar features, especially the ubiquitous lakes and limestone cliffs, which are discussed in detail. Karst forms are described, from the smaller ones initially generated under soil to the macroforms that evolve after exposure. The chapter closes with a schematic model for the evolution of the Lagoa Santa Karst. In chapter "Karst Hydrogeology of the Lagoa Santa Area", Paulo Pessoa discusses the hydrogeology of the area, showing how the distinct lithological units behave in terms of groundwater flow. Infiltrating waters percolating through complex vadose pathways allow for a conspicuous recharge distribution, interconnecting various portions of the karst aquifer system and flowing through deep epikarst zones. More than 500 pumping wells and dozens of karst springs serve as the main source of potable water supply. The chapter ends by highlighting that the Lagoa Santa Karst aquifer lacks spatial and seasonal hydrogeological monitoring plans. In chapter "Caves and Speleogenesis in the Lagoa Santa Karst", Augusto S. Auler describes the main characteristics of the large number of caves in the area. The area displays considerable speleogenetic variability, with both syngenetic and paragenetic caves, of which a significant number were formed by lakes or due to dissolution over a sediment cover. The morphological features of the caves and their chemical, clastic, and organic sedimentation are also discussed. The chapter closes with remarks on the caves' age and evolution.

The caves in the Lagoa Santa Karst harbour a remarkable cave fauna, which is discussed in chapter "Biospeleology of the Lagoa Santa Karst" by Rodrigo Lopes Ferreira, Marcus Paulo Alves de Oliveira, and Marconi Souza-Silva. The cave fauna is structurally diverse due to the variety of habitats. It has 41 species of obligate cave dwellers, but only four have so far been formally described. The number of troglobitic species is likely to increase, as new research is being performed. The chapter highlights the many impacts taking place, leading to a highly vulnerable situation, which threatens the cave ecosystem. The fossil vertebrate fauna, first revealed by Peter Lund, is described by Cástor Cartelle in chapter "Cave Paleontology in the Lagoa Santa Karst". The chapter opens with a review of the life and works of Lund, his most iconic findings and the research of those who followed in his footsteps, such as the Danish zoologist Herluf Winge and other scientists during the twentieth century. This historical review provides a timely update on the nearly 200 years of paleontological studies in the area. Human occupation in the Lagoa Santa Karst is discussed by André Strauss and collaborators in chapter "The Archaeological Record of Lagoa Santa (East-Central Brazil): From the Late Pleistocene to Historical Times". The area is known for its archaeological sites and evidence of ancient occupation (since approximately 12,500 years ago). The chapter presents a comprehensive survey of several archaeological aspects, such as chronology, rock art, pottery, stone and bone technology, diet, funerary rituals, site formation, and so forth, providing the reader with an updated review of the Lagoa Santa Karst archaeology.

The Lagoa Santa Karst is presently being subjected to environmental impacts that must be addressed urgently. The two last chapters aim to describe these impacts and the conservation efforts. In chapter "Environmental Problems in the Lagoa Santa Karst", Paulo Pessoa, Dora Atman, and Gisele Kimura focus on the misuse of land and the inadequate occupation in the Lagoa Santa Karst aquifer system, showing that several environmental problems still persist. Besides the spread of urban centers and allotments, severe secondary issues related to inappropriate sanitation infrastructure, aquifer vulnerability, mining activity hazards, and deforestation in sensitive areas emerge. However, as happens at the root of most environmental problems, recent studies demonstrated how important it is to break away from current educational and social patterns, avoiding an unrestricted focus on basic water cycle reconnaissance studies. The chapter points out that stakeholders need to realize how important they are in order to preserve water resources. In chapter "Protection of the Environmental and Cultural Heritage in the Lagoa Santa Karst: History and Challenges", which brings the book to a close, Luciana Alt and Vitor Moura summarize the threats and conservation efforts in the Lagoa Santa Karst, and stress the gaps and challenges for adequate protection of the area. Although there are significant conservation units, the reality shows the need for more effective measures; otherwise, the environmental quality of the area may be jeopardized.

We would like to thank all the authors, who contributed their time to producing a synthesis of their field of knowledge. The chapters benefited enormously from the reviews provided by authorities, especially in areas beyond our field of expertise. We acknowledge the critical review of Dr. Renata Andrade (biospeleology), Dr. Ulysses Pardiñas (fauna), Dr. André Prous (archaeology), Dr. Alex Hubbe (paleontology), Dr. Luiz P. Travassos (history), Dr. Esther Sebastian (waterbirds), Pablo Hendrigo Melo (vegetation), Dr. Luis Piló (impacts), Dr. Paulo Galvão (hydrogeology), and Tatiana Souza (karst geomorphology and speleogenesis). The staff at Carste Ciência e Meio Ambiente and Hidrovia Hidrogeologia e Meio Ambiente provided continuous support during the long gestation of this book.

The Lagoa Santa Karst is where we started our scientific careers—first, as enthusiastic youngsters, looking for caves and being surprised by the intriguing karst landforms, and later, as young professionals learning how science should be done. After nearly 40 years, as we look back, we still see the Lagoa Santa Karst as mysterious and full of secrets, but also crying out for support. We are grateful for the continuous support provided by several local farmers, as well as industries and the staff at the conservation units. Friends from Grupo Bambuí de Pesquisas Espeleológicas, a caving club, helped in many ways.

The Lagoa Santa Karst is an emblematic karst area in Brazil, a classical site where much of what is known about Brazil's past humans and fauna was first brought to light. In its thousands of caves, cliffs, dolines, and lakes lies a significant portion of Brazil's pre-European history. We hope this book will help disseminate knowledge about this outstanding area, fostering new research, and prompting the enactment of effective protection measures in order to preserve the Lagoa Santa Karst.

Belo Horizonte, Brazil

Augusto S. Auler
Paulo Pessoa

Contents

History of Research in the Lagoa Santa Karst

Augusto S. Auler

Abstract

The Lagoa Santa Karst is the most researched and best-known karst area in Brazil; it is an emblematic karst area within the context of Brazilian karst. Its modern history spans over 200 years, since the arrival of the first European settlers, although prehistoric cultures had occupied the area since at least the early Holocene. The arrival of Danish naturalist Peter Wilhelm Lund in Lagoa Santa in 1835 started a period of intense research that continues to this day. Lund was responsible for bringing international recognition to the area through outstanding paleontological work that has formed, up to this day, the basis of Brazilian palaeontology. Numerous other European scientists, notably Johannes Reinhardt, Hermann Burmeister, Eugen Warming and Herluf Winge, took advantage of Lund's residence in Lagoa Santa (and later, of Lund's collection in Copenhagen) to perform additional research and publish original data in several disciplines. In the early twentieth century, members of the Academy of Sciences of Minas Gerais (mainly Aníbal Mattos and Harold Walter) and the National Museum of Rio de Janeiro resumed Lund's work in the caves and revived old controversies related to the antiquity of human remains and their contemporaneity with the extinct megafauna. In the second half of the twentieth century, major research continued to be performed by foreign scientists, including archaeologists and geomorphologists (respectively, led by Annette Laming-Emperaire and Heinz Charles Kohler), which helped create local research groups based in the Minas Gerais's state capital, Belo Horizonte. The area is still a hot spot of scientific research in various fields, mostly led by Brazilian groups, and continues to enjoy international scientific relevance.

A. S. Auler (✉)
Instituto do Carste, Rua Barcelona 240/302,
Belo Horizonte, MG 30360-260, Brazil
e-mail: aauler@gmail.com

Keywords

Lagoa Santa Karst • History • Saltpetre • Peter Lund • Cave science

1 Introduction

The Lagoa Santa Karst was the first karst area in Brazil to be the object of scientific studies in the early nineteenth century. Since then, research there has been continuous, including up through the present day. The total amount of research in the Lagoa Santa Karst far exceeds what has been produced in other karst regions in Brazil. The research has been interdisciplinary, from physical aspects such as geomorphology (including speleology and hydrogeology), to biological (fauna, flora, palaeontology) and social/cultural (archaeology) fields. However, it is unequivocal that the caves and their fossil content provided the initial impetus that later evolved into wide-ranging studies. The Lagoa Santa Karst owes much of its international recognition to the initial work of Danish naturalist Peter Wilhelm Lund. The Lagoa Santa Karst, although unique in some aspects, has some similarities with other sites throughout the vast limestone areas that occur throughout much of the states of Minas Gerais, Bahia and Goiás (Auler 2019). What made the Lagoa Santa Karst the focus of so much early attention was arguably the outstanding scientific contributions of Peter Lund's work.

In addition to Lund's own pioneering research in the caves, he also attracted to Lagoa Santa a considerable number of colleagues and visitors who expanded the scope of research in the area. The publications by Lund and his collections, which were sent to Copenhagen, also provided a rich field for discussion and reinterpretation by other scientists, many of them would never visit Lagoa Santa. This stream of research continued throughout the twentieth and twenty-first centuries.

A. S. Auler and P. Pessoa (eds.), *Lagoa Santa Karst: Brazil's Iconic Karst Region*, Cave and Karst Systems of the World, https://doi.org/10.1007/978-3-030-35940-9_1

In addition to Lund's role in raising the scientific visibility of the Lagoa Santa Karst, the establishment of the new capital of the state of Minas Gerais, Belo Horizonte, in the surroundings of the karst, also played a major role. From an isolated area in the backlands of Brazil, suddenly the Lagoa Santa Karst was very close (approximately 40 km) from a fast-growing metropolis. The population living in Belo Horizonte and the adjoining areas grew rapidly, reaching 2.5 million in 2014. Previously small towns such as Lagoa Santa, Matozinhos, Pedro Leopoldo and Sete Lagoas also expanded over the karst. Presently, it is estimated that approximately 500,000 people live within or in the immediate surroundings of the Lagoa Santa Karst (Auler 2016). Although this massive population growth had major impacts in the region, it also facilitated access to the area, as being close to important universities and research centers made the Lagoa Santa Karst a natural target for research. Presently, a significant portion of the cement produced in Brazil comes from the Lagoa Santa Karst. Several other industries and the state's only international airport are also located in the area. Environmental assessment studies performed for the installation or due to compensation measures for these industries have also considerably expanded scientific knowledge of the karst. No other karst area in Brazil is located so close to such a sprawling urban metropolis, a dubious honour that carries positive and negative consequences.

This review will detail human involvement with the Lagoa Santa Karst, the main scientific landmarks and the people who worked in (or based their research on data from) the Lagoa Santa Karst, from pre-colonial times to the late twentieth century, excluding ongoing research.

2 Initial Human Occupation

Abundant archaeological data point to the start of human occupation over 12,000 years ago (Da-Gloria et al. 2017). These early Americans had a close relationship with the karst area, as attested by a large number of cave entrances and shelters associated with archaeological sites (see chapter "The Archaeological Record of Lagoa Santa (East-Central Brazil): From the Late Pleistocene to Historical Times", this volume). In contrast to pre-colonial humans in North and Central America, there is scarce evidence that these Paleo-Indians entered caves. The rock art and archaeological findings are mostly limited to the entrance zone. No major environmental impact can be attributed to these early populations. Unlike in other areas in Brazil, limited indigenous geographical denominations are known to be related to the Lagoa Santa Karst, with the exception of the word 'anhanhonhacanhuva', which was applied to the swallet at Sumidouro Lake (Silveira 1929).

The first experience Europeans had with the Lagoa Santa Karst took place during 'bandeiras', expeditions led by Portuguese frontiersmen with the goal of finding precious stones. There are records of one of these expeditions, led by Fernão Dias in 1674, having reached Sumidouro Lake (Lima Junior 1962) and founded the small hamlets at Fidalgo and Quinta do Sumidouro.

Gradually, over the eighteenth century, the area became occupied by large farms. The significant discoveries of gold in the surroundings of the town of Ouro Preto and diamond in the Espinhaço Ridge, near the town of Diamantina, made the Lagoa Santa area (located approximately midway between these two mining centers) a natural stopover. This time marks the initial settlement of most of the towns in the area. The town of Lagoa Santa is associated with the reportedly miraculous waters of its local lake. The name 'Lagoa Santa' means 'Holy Lake', and the healing properties of its waters were publicized in Lisbon in the eighteenth century (Anonymous 1749).

2.1 Saltpetre Mining

Saltpetre is a nitrate-rich soil that was an essential ingredient for producing gunpowder until the early twentieth century. Saltpetre occurs in many natural settings but is especially abundant in caves, where it forms fine-grained sediment that can be leached in order to concentrate the nitrates. The Lagoa Santa Karst was one of the areas where saltpetre was extracted from caves. Mentions of saltpetre-rich caves in the surroundings of Matozinhos, Fidalgo and Sete Lagoas are recorded in early saltpetre records, such as a series of documents dated from 1813 (Anonymous 1899), with some of the caves still retaining their names to this day.

The extent of the economic use of Lagoa Santa Karst caves cannot be assessed with precision. However, evidence of the extraction, normally pickaxe marks, can be observed in several caves, and entire passages have had their sediment completely or partially removed. Knowledge about the caves already existed when Peter Lund arrived in the area, and he had little problem finding guides to lead him into them, although he lamented that saltpetre extraction had damaged the paleontological content of several caves (Lund 1841a, b, c). Although saltpetre extraction peaked during the late eighteenth and early nineteenth centuries (Gomes and Piló 1992; Ferraz 2000), there has been little research performed on this subject, despite the abundant documentation available.

3 Peter Wilhelm Lund

The arrival of Danish naturalist Peter Wilhelm Lund (1801–1880) in Lagoa Santa in 1835 forever changed the scientific destiny of the then sleepy village. Lund's paleontological

Fig. 1 **a** Photograph of Peter Wilhelm Lund by Eugen Warming. Archives of the Natural History Museum, Copenhagen. **b** Portrait of Peter Andreas Brandt by an unknown artist

cave work lasted until 1845, but his influence endured for far longer. Not only did Lund produce highly original work—mostly related to palaeontology (see chapter "Cave Paleontology in the Lagoa Santa Karst", this volume), though with insights on archaeology and geomorphology as well (see chapters "Karst Landforms in the Lagoa Santa Area, Caves and Speleogenesis in the Lagoa Santa Karst and The Archaeological Record of Lagoa Santa (East-Central Brazil): From the Late Pleistocene to Historical Times", this volume) but he also welcomed numerous other scientists who performed complementary research or delved into other disciplines.

Peter Lund (Fig. 1a) came to Lagoa Santa with an open mind. A botanist with a comprehensive interest in and knowledge of natural history, he intended to perform botanical studies in his travels in the interior of Brazil with his German colleague, Ludwig Riedel (see Holten and Sterll 2011 for details). However, a fortuitous meeting with Dane Peter Claussen and his first visit to caves containing fossil bones almost immediately changed the focus of his research. His conversion to palaeontology was instantaneous, showing how prepared he was to delve into an area where he had never done any previous practical field or laboratory work. At the same period, an equally fortuitous meeting with the Norwegian artist Peter Andreas Brandt (1792–1862, Fig. 1b) provided Lund with a capable assistant in charge of drawing the caves (Fig. 2) and producing essential scientific drawings of fossil bones (Holten et al. 2012). In particular, Brandt produced several maps of caves in the Lagoa Santa Karst, showing considerable skill, despite the difficult conditions and his lack of experience with this highly specialized task (Auler and Piló 2017). Lund's work on palaeontology, archaeology,

Fig. 2 **a** Map of Cerca Grande Cave. **b** A scene of Lapa Vermelha Cave. Both drawings by P.A. Brandt

Fig. 3 **a** Hermann Burmeister and **b** Emmanuel Liais visited Lund and performed original research on the Lagoa Santa Karst

karst geomorphology, and speleology were initially published in Danish in a series of memoirs (Lund 1837, 1841a, b, c, 1845, 1846), which were later translated into Portuguese (Lund 1950), English (Lund 1840) and French (Lund 1845). Among Lund's main discoveries, one can list the scientific description of 22 fossil species and nine living species (see chapter "Cave Paleontology in the Lagoa Santa Karst", this volume), the proposition of the contemporaneity between Man and the extinct megafauna (Neves and Piló 2003) and the recognition of the old age of human remains (Holten and Sterll 2011). Although initially a proponent of the catastrophism view of his mentor, Georges Cuvier, Lund slowly abandoned these ideas to formulate his own original concepts. Lund's life has been portrayed in several biographical essays. Reinhardt (1880), Lund (1885) and Holten and Sterll (2011) are among the most comprehensive, while Marchesotti (2011) and Luna (2007) represent additional recent studies.

Peter Lund received several visitors in Lagoa Santa. Some of them were scientists who, attracted by Lund's research, came to Lagoa Santa to perform scientific studies—the Danes Johannes Reinhardt and Eugen Warming, for example, in addition to the German Hermann Burmeister and the French Emmanuel Liais. Others were in Lagoa Santa for short periods, and, although they were scientists, they did not contribute any significant new data about the Lagoa Santa area. Among this group were Joel Allen, Orestes St John, George Sceva, Virgil von Helmreichen, Richard Burton, Jakob Heusser, George Claraz and Marianne North. Among the non-scientists who visited were members of the royal entourage of the Duke of Saxe, including photographer Augusto Riedel, who provided important early photographs of Lagoa Santa (Riedel 1868).

Some of these scientists deserve special mention because they had the chance to visit karst sites and make original observations. The German Hermann Burmeister (1807–1892) (Fig. 3a) stayed in Lagoa Santa with Lund and Reinhardt for approximately three months (Burmeister 1853). Burmeister published work about the fossils in Lagoa Santa (Burmeister 1871, 1885). Frenchman Emmanuel Liais (1826–1900) (Fig. 3b) also visited caves and provided accounts of cave genesis, sedimentation and fossiliferous beds (Liais 1872).

There were major scientific contributions from colleagues who were among the continuous stream of visitors, particularly from Danish scientists somehow associated with Lund. Johannes Reinhardt and Eugen Warming were two such colleagues who stayed in Lagoa Santa for extended periods and organized collections that were later shipped to Denmark, enabling more detailed work to be performed later by other colleagues.

3.1 Reinhardt and Warming

Although isolated in a remote village in Brazil, Lund kept constant contact with his Danish colleagues. Johannes Reinhardt (1816–1882, Fig. 4a) was the son of one of Lund's mentors. Reinhardt remained in Lagoa Santa for an extended period during his three visits to Lagoa Santa—enough time to acquaint himself with the local fauna. Reinhardt's contributions were his own, mainly in the field of palaeontology (Reinhardt 1867, 1875a, b—see additional references in Mones 1986) but also covering the topics of vegetation (Reinhardt 1856) and birds (Reinhardt 1870). In addition to being nominated as the curator of the portion of Lund's

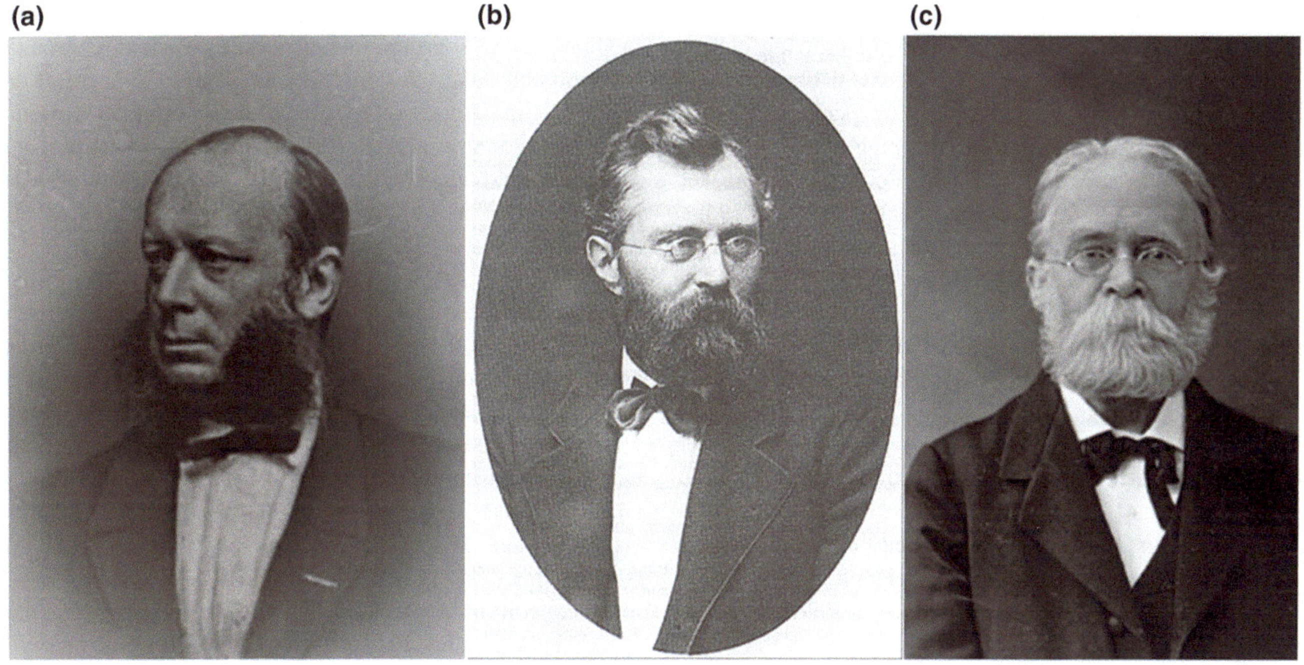

Fig. 4 **a** Johannes Reinhardt. **b** Eugen Warming. **c** Herluf Winge

collection that had been sent to Copenhagen, Reinhardt himself also made important collections that were used by other scientists to produce original work on fish, amphibians and reptiles (Lütken 1875; Reinhardt and Lütken 1861).

In addition to his own work, Reinhardt was able to reinterpret Lund's view on the formation of caves and the introduction of bones (Reinhardt 1867, 1888). This fact is important because it allows us to learn about a change in Lund's own interpretation that is only hinted at in his personal correspondence.

In need of a new secretary, Lund welcomed the assistance of young botanist, Eugen Warming, in 1863. Warming (1841–1924, Fig. 4b) was able to find considerable spare time away from his duties with Lund during his 3-year stay in Lagoa Santa and performed a pioneering and influential study on the ecology of plants in the Lagoa Santa area (Warming 1892).

3.2 Lund's Collection in Denmark

Lund's paleontological collection in Copenhagen remained in storage for many years, and proper studies on it were not performed until much later when its content was finally re-examined (Holten and Sterll 2011). The result of this re-evaluation was published as a set of five volumes entitled *E Museu Lundii* from 1888 to 1915. This massive and comprehensive work was coordinated by Herluf Winge (1857–1923—Fig. 4c) and contains detailed studies of the fossil bones, consisting of chapters on birds, rodents and mammals, including humans (see full references in Mones 1986). This re-evaluation added considerably to the knowledge of Peter Lund's findings in Lagoa Santa's caves. In particular, towards the end of the nineteenth century, Lund's concepts on the contemporaneity between man and the megafauna and on the antiquity of man, based on his findings at Sumidouro Cave (Lund 1845), attracted considerable international attention and generated a scientific debate that lasted many decades (see Piló et al. 2005; Neves et al. 2007; Hubbe and Neves 2017 and references therein). Some of these topics are subject to renewed interest being the focus of recent archaeological projects (Bernardo et al. 2017).

4 Early Twentieth Century

Limited original field research was performed in Lagoa Santa Karst in the years after Lund's death. In the first decade of the twentieth century, a local engineer named Cássio Umberto Lanari, whose family owned the large Mocambo Farm at the heart of the karst, resumed archaeological work in the area, making new findings of human remains, mainly at Caetano Cliff (Lanari 1909).

In approximately the same period, Álvaro Astolpho da Silveira, also a local engineer and naturalist (with an emphasis on botany), described a number of caves and the karst landscape in the area (Silveira 1921, 1929, 1931). Some of the Lagoa Santa Karst caves are mentioned in the first overview of Brazilian caves, by Antonio Olinto dos Santos Pires on the occasion of the centenary of the Brazilian independence (Pires 1929).

In 1937, students at the Federal University of Ouro Preto, led by Victor Dequech, established the first caving club in the Americas, the Sociedade Excursionista e Espeleológica (SEE). The very first field studies were performed at the Lagoa Santa Karst, including the mapping of many important caves, such as Morro Redondo, Poções, Lavoura and Estudantes (Dequech 1940; Bittencourt 1945; Parada 1949a, b). The SEE continues to have a strong link with the Lagoa Santa Karst to the present day. A few years later, the state government of Minas Gerais commissioned the creation of a catalogue of caves in the state (IBGE 1939), and caves in the Lagoa Santa Karst were briefly described.

4.1 The Academy of Sciences of Minas Gerais

During the first half of the twentieth century, amateur archaeologists and palaeontologists, belonging to the Academy of Sciences of Minas Gerais, based in Belo Horizonte, explored several caves and archaeological sites in the Lagoa Santa Karst. Among those, the most influential were artist and painter Aníbal Mattos (1886–1969) and the English vice-consul, Harold Walter (1897–1976) (Fig. 5). Other members of this group included medical doctor Arnaldo Cathoud and dentist Josaphat Penna. These members of the Academy, most notably Harold Walter, performed original excavations at several sites, such as Confins, Eucalipto, and Mãe Rosa, and published extensively about archaeology and palaeontology in the Lagoa Santa Karst (Cathoud 1935; Mattos 1941, 1961; Walter 1948, 1958; Walter et al. 1939). Harold Walter, the most scientifically inclined of the group, attempted to determine the first absolute ages of the ancient human skulls of the area (Stewart and Walter 1955). Anibal Mattos was responsible for bringing much of the Lund's work in Lagoa Santa back into the limelight and promoting Lund's legacy through a series of books (Mattos 1934, 1935). A recent review of the work performed by the Academy of Sciences of Minas Gerais has been presented by Costa (2017).

Fig. 5 The three main organizers of the Academy of Sciences of Minas Gerais at Lapa Vermelha Cave. From left to right, Aníbal Mattos, Arnaldo Cathoud and Harold Walter. From Walter (1948)

In the late 1920s, German Jorge Padberg-Drenkpohl, working for the National Museum at Rio de Janeiro, organized expeditions to the Lagoa Santa Karst (Gaspar Neto and Rodrigues-Carvalho 2017), during which general observations about the landscape and caves were performed. The expeditions focused on archaeology, and Padberg-Drenkpohl performed pioneering excavations at Confins Cave, where he unearthed human skulls that reignited the controversial debates with the members of the Academy of Sciences of Minas Gerais regarding the antiquity and contemporaneity of the extinct fauna (see Keuller 2017; Gaspar Neto and Rodrigues-Carvalho 2017). The expedition results remain in the form of unpublished reports in the National Museum's archives. J. Bastos Ávila, also from the National Museum, directed excavations at Carrancas site in 1936, but the results, mostly unpublished (but see Mattos (1941)), are short and descriptive in nature (Gaspar Neto and Rodrigues-Carvalho 2017).

5 Late Twentieth Century

The Lagoa Santa Karst continued to receive a steady stream of scientists during the second half of the twentieth century. Unlike the members of the Academy of Sciences of Minas Gerais, most of these were professional scientists working through official projects. Several nationalities were represented, and some of these scientists had a lifelong involvement with the Lagoa Santa Karst.

5.1 Palaeontology

Carlos de Paula Couto was Brazil's foremost palaeontologist in the late twentieth Century, and he organized a new complete annotated translation of Lund's writings (Lund 1950), which allowed Brazilians to be able to fully analyse

Fig. 6 **a** Annette Laming-Emperaire. **b** Heinz Charles Kohler (Travassos and Kranjc 2011)

Lund's ideas and findings. Paula Couto also performed new excavations in the area and described a new species of rodent (Paula Couto 1951). Paula Couto benefited from Harold Walter's paleontological collection, studying the fossil bear found in Lagoa Funda Cave (Paula Couto 1960) and re-describing some of the species initially found by Lund (Paula Couto 1947). Hoffstetter (1954), a French palaeontologist, studied material on Xenarthra from Lund's collection in Denmark. Some of the archaeological expeditions described in the following section by Hurt and Blasi and the French mission led by A. Laming-Emperaire also yielded significant paleontological material. For an updated review of paleontological research in the area see chapter "Cave Paleontology in the Lagoa Santa Karst", this volume.

5.2 Archaeology

To test Lund's hypothesis on the age of human remains, American archaeologist Wesley Hurt teamed up with Brazilian archaeologist Ondemar Blasi and palaeontologist Paula Couto to excavate a number of Lagoa Santa sites in 1954/1955, such as Boleiras (Escadas), Ballet and, mainly, the Cerca Grande site. This research resulted in the first reliable radiocarbon dating for the Lagoa Santa archaeological sites—over 9,000 years BP old (Hurt 1960; Hurt and Blasi 1969)—which demonstrated the considerable antiquity of human occupation in the area (see review in Prous 2017). A new impetus was provided by the French-Brazilian archaeological mission, which took place in Lagoa Santa between 1971 and 1977 (Laming-Emperaire et al. 1974). The mission was led by the French Annette Laming-Emperaire (1917–1977) (Fig. 6a) and had among its members a number of archaeologists who would later become important mentors in the field, including André Prous and Maria Beltrão, among many others (see Prous 2017 for a detailed account of the mission). Among the most publicized findings of the French-Brazilian mission is the skull belonging to 'Luzia'. Indirectly dated between 11,200 and 11,680 years BP (Bernardo et al. 2017), the skull would become an icon, representing the antiquity of South American human occupation. In the late 1970s, Hungarian amateur archaeologist Mihaly Banyai performed numerous excavations (Banyai 1997), mostly at the Lapinha outcrop, where he would later establish a private museum that would eventually be incorporated by the state of Minas Gerais.

5.3 Geomorphology and Speleology

Studies on karst geomorphology were largely dormant during the first half of the twentieth century. They were slowly resumed in the 1950s and intensified during the 1970s–1980s. Frenchman Jean Tricart produced a detailed description of karren types at the now largely destroyed Nova Granja outcrop in São José da Lapa (Tricart 1956). At about the same time, Elzio Dolabela, a native from the town of Lagoa Santa, published his monograph on karst landforms containing information on the Lagoa Santa area (Dolabela 1958). Brief descriptions of the karst were provided by archaeologists (Hurt and Blasi 1969; Laming-Emperaire et al. 1974) but merely as an introduction to the main archaeological focus of the research.

In the 1970s, due to an initiative of A. Laming-Emperaire, a cooperative project was established between French geomorphologists André Journaux and Jean-Pierre Coutard and the Federal University of Minas Gerais (Prous 2017), represented by the Swiss-Brazilian Heinz Charles Kohler; this

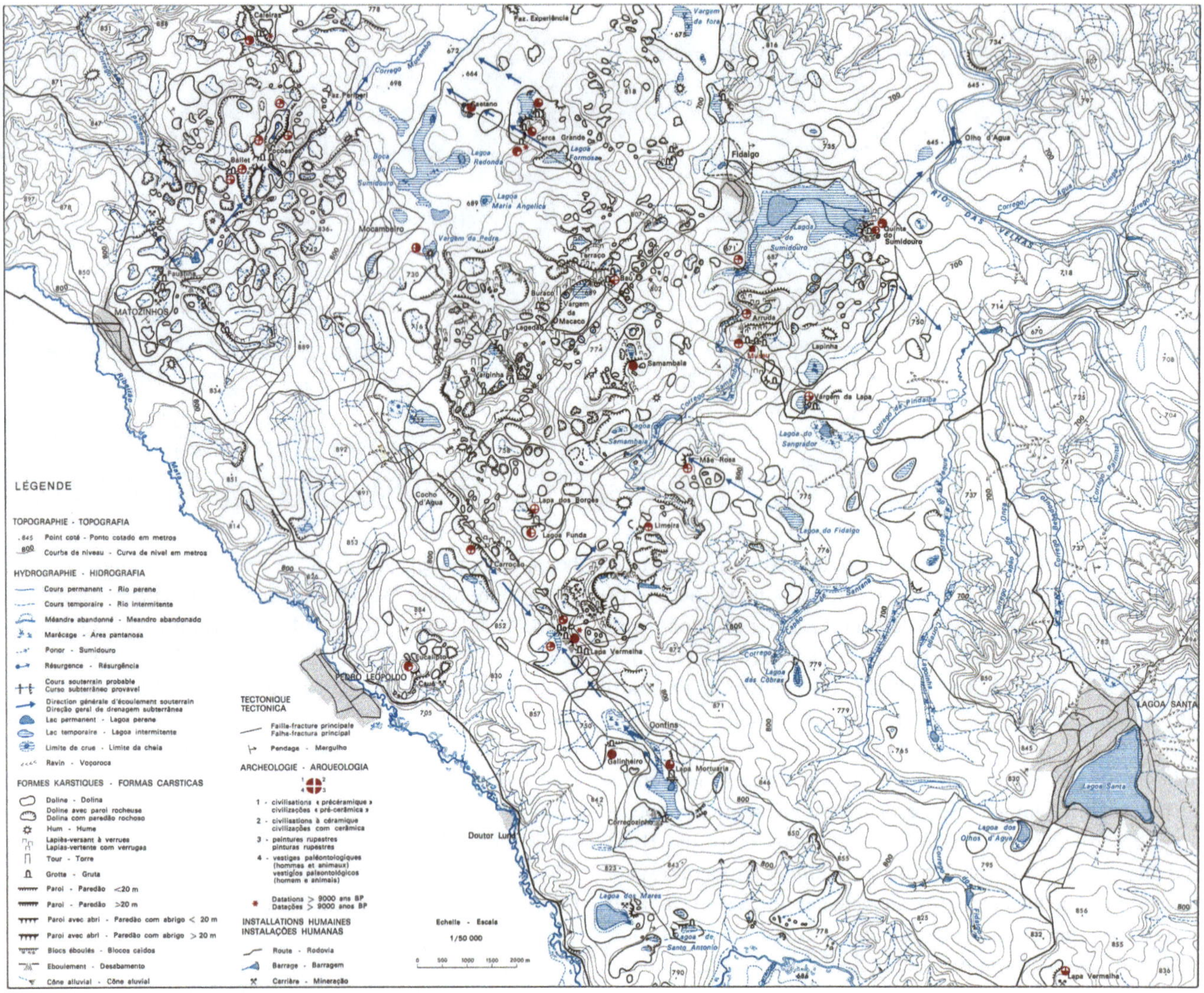

Fig. 7 The Lagoa Santa Karst map by Kohler et al. (1978)

project led to a marked advance in the understanding of the karst landforms (including caves) and their dynamics. The map of the Lagoa Santa Karst (Kohler et al. 1978) was the first of its kind in Brazil and showed the main features in the central portion of the karst, including caves and archaeological sites, on a 1:50,000 scale (Fig. 7). This map was accompanied by detailed text that described the main landforms and delved into topics such as cave genesis and karst evolution (Coutard et al. 1978). An original interpretation of the genesis of cliffs was published by Journaux (1977). Heinz Charles Kohler (1945–2010) (Fig. 6b) was instrumental in later years, as he performed additional research in the area (Kohler 1989; Kohler and Karfunkel 2002). An isolated initiative was provided by the geomorphological work of Hungarian Denes Balazs in the Lapinha outcrop (Balasz 1984). Descriptions of species sampled in caves in the Lagoa Santa Karst were also published during this period (Wygodzinsky 1950; Schubart 1956).

Cave exploration and research was limited to a few practitioners until the appearance of additional speleological groups in the late 1970s and particularly in the 1980s, most of which were based in Belo Horizonte. Exploration and mapping of caves became frequent, resulting in important discoveries, including some of the longest and deepest caves in the area. Among the groups, the Grupo Bambuí de Pesquisas Espeleológicas was the most productive. Two governmental cave and karst-related projects deserve mention. The 'Projeto Grutas' (CETEC 1981) produced a basic inventory of several caves in the state, including a number of important caves in the Lagoa Santa Karst. Starting in the early 1990s, the 'Projeto Vida' aimed to produce basic information to help achieve sustainable development for the area. This project performed original work in several areas, including a new geological mapping effort. Several cave maps and detailed reports were produced that were later applied to the management plan of the Carste of Lagoa Santa Environmental Protection Area (IBAMA/CPRM 1998).

6 Present Times and Final Remarks

Starting in the late 1990s, research in Lagoa Santa Karst became more quantitative, with measurements taking the place of descriptions. Dating techniques (U-series, radiocarbon, thermoluminescence) were routinely applied to the archaeological sites and caves. Chemical analysis was used to characterize the origin and evolution of the soils. Palynological analyses were performed to better understand lake evolution in the area. The first hydrogeological inventories took place, with the application of tracing techniques, hydrochemical analysis and applied research related to sinkhole collapse. Additional archaeological sites have been excavated, and new paleontological research has improved our knowledge of the late Pleistocene fossil fauna. Exciting new research has been performed in areas such as fauna and biospeleology regarding ecology, parasitology and taxonomy, among others. Academic work in all areas is continuously being performed in the Lagoa Santa Karst, resulting in a significant number of theses and dissertations. The several chapters in this volume bring an updated review of the present status of research in Lagoa Santa Karst.

The implementation of Federal Decree 6640 (Brasil 2008) brought about new procedures for the classification of caves. Mining companies are now required to produce a comprehensive inventory of caves during the licensing process, including detailed biological and geological assessments, in order to classify caves according to their levels of significance (Auler and Piló 2015). This procedure has resulted in studies in hundreds of new caves, expanding knowledge about the speleological sites in the Lagoa Santa Karst considerably. The database of caves is ever increasing, reaching 1,400 in 2017. New conservation areas have been implemented, although increased urbanization, together with other threats (see chapters "Environmental Problems in the Lagoa Santa Karst and Protection of the Environmental and Cultural Heritage in the Lagoa Santa Karst: History and Challenges", this volume), has challenged attempts to protect the remarkable karst heritage in the area. As a karst area rich in history, the Lagoa Santa Karst and its future depend on a successful science-based approach to tackle existing threats in order to sustainably guarantee the karst's continued existence.

References

Anonymous (1749) Prodigiosa lagoa descoberta nas Congonhas das Minas do Sabará, que tem curado a varias pessoas dos achaques, que nesta relação se expõem. Officina de Miguel Manescal da Costa, Lisboa

Anonymous (1899) Salitreiras. Revista do Archivo Publico Mineiro 4:299–316

Auler AS (2016) Cave protection as a karst conservation tool in the environmentally sensitive Lagoa Santa Karst, southeastern Brazil. Acta Carsologica 45:131–145

Auler AS (2019) Histórico, ocorrência e potencial de cavernas no Brasil. In: Auler AS, Menin DS, Brandi R (eds) Rubbioli EL. Cavernas. Atlas do Brasil Subterrâneo. Editora IABS, Brasília

Auler AS, Piló LB (2015) Lagoa Santa Karst: Cradle of Brazilian cave studies. In: Vieira BC et al (eds) Landscapes and landforms of Brazil. Springer, Dordrecht

Auler AS, Piló LB (2017) Peter Wilhelm Lund's scientific contributions. In: Da-Gloria P, Neves WA, Hubbe M (eds) Archaeological and paleontological research in Lagoa Santa. Springer, Cham, pp 27–41

Balasz D (1984) Exhumált Trópusi Oskarszt Lapinha Vidéken (Minas Gerais, Brazília). Karst és Barlang 2:87–92

Banyai M (1997) Minhas pesquisas arqueológicas na região de Lagoa Santa. Symbiose, Budapest

Bernardo DV, Neves WA, Kipnis R (2017) The origins project and the first Americans' controversy. In: Da-Gloria P, Neves WA, Hubbe M (eds) Archaeological and paleontological research in Lagoa Santa. Springer, Cham

Bittencourt AU (1945) Gruta dos Estudantes. Revista Brasileira de Geografia 7:486–489

Brasil (2008) Decreto Federal 6.640. Presidência da República. http://www.planalto.gov.br/ccvil_03/_Ato2007-2010/2008/Decreto/D6640.htm. Accessed 11 July 2019

Burmeister H (1853) Reise nach brasilien durch die provinzen von Rio de Janeiro und Minas Geraes. Georg Reimer, Berlin

Burmeister HG (1871) Ueber *Hoplophorus euphracthus*. Archiv. für Anat, Phys und Wissensc Med 164–179

Burmeister HG (1885) Berichtigung zu Coelodon. Sitzungsberichte der königlich preussischen. Akad der Wissensc zu Berlin 2:567–573

Cathoud A (1935) A raça de Lagoa Santa e o Pleistoceno Americano. Biblioteca Mineira de Cultura, Belo Horizonte

CETEC (1981) Cadastramento de grutas nas regiões metalúrgica e Alto Jequitinhonha. Fundação Centro Tecnológico de Minas Gerais, Belo Horizonte

Costa FWS (2017) The Minas Gerais Academy of Sciences: Lund's inheritors. In: Da-Gloria et al (eds) Archaeological and paleontological research in Lagoa Santa. Springer, Cham

Coutard JP et al (1978) Map of Lagoa Santa Karst, description. Caen

Da-Gloria P, Neves WA, Hubbe M (2017) Archaeological and paleontological research in Lagoa Santa. Springer, Cham

Dequech V (1940) Atividades Speleologicas no Brasil. Revista Mineira de Engenharia 2:54–62

Dolabela E (1958) Estudo das regiões cársicas. UFMG, Belo Horizonte

Ferraz MHM (2000) A produção do salitre no Brasil colonial. Quim Nova 23:845–850

Gaspar Neto VV, Rodrigues-Carvalho C (2017) The physical anthropology archives of the National Museum of Rio de Janeiro: Lagoa Santa in the first half of the twentieth century. In: Da-Gloria et al (eds) Archaeological and paleontological research in Lagoa Santa. Springer, Cham

Gomes MCA, Piló LB (1992) As minas de salitre: a exploração econômica das cavernas em Minas Gerais nos fins do período colonial. Espeleo-Tema 16:83–93
Hoffstetter R (1954) Les gravigrades (Edentés Xénarthres) des cavernes de Lagoa Santa (Minas Gerais, Brasil). Annales des Sciences Naturelles, Zoologie et Biologie Animale 16:741–761
Holten B, Sterll M (2011) Peter Wilhelm Lund e as grutas com ossos em Lagoa Santa. Editora UFMG, Belo Horizonte
Holten B, Sterll M, Fjeldså J (2012) O artista desaparecido. P.W. Lund e P.A. Brandt no Brasil. Editora UFMG, Belo Horizonte
Hubbe M, Neves WA (2017) The repercussions of the human skeletons from Lagoa Santa in the international scenario. In: Da-Gloria P, Neves WA, Hubbe M (eds) Archaeological and paleontological research in Lagoa Santa. Springer, Cham
Hurt WR (1960) The cultural complexes from the Lagoa Santa region. Am Anthropol 62:569–585
Hurt WR, Blasi O (1969) O Projeto Arqueológico "Lagoa Santa" – Minas Gerais, Brasil. Arquivos do Museu Paranaense 4:1–63
IBAMA/CPRM (1998) APA Carste de Lagoa Santa. Relatórios vol I, II, III. CPRM, Belo Horizonte
IBGE (1939) As grutas em Minas Gerais. Oficinas Gráficas da Estatística, Belo Horizonte
Journaux A (1977) Geomorphologie et préhistoire. Méthodologie pour une cartographie de l'environement des sites préhistoriques; l'exemple de Lagoa Santa (Minas Gerais, Brésil). Norois 95: 319–335
Keuller ATAM (2017) The anthropological studies of Lagoa Santa in the National Museum of Rio de Janeiro: insertion, debates and scientific controversies at the turn of the nineteenth to the twentieth century. In: Da-Gloria et al (eds) Archaeological and paleontological research in Lagoa Santa. Springer, Cham
Kohler HC (1989) Geomorfologia cárstica da região de Lagoa Santa, MG. Dissertation, Universidade de São Paulo
Kohler HC, Karfunkel J (2002) The quaternary morphogenesis of the Lagoa Santa tropical karst, Minas Gerais state, SE Brazil. Theor Appl Karstology 15:93–99
Kohler HC, Cançado AM, Gomes D, Macieira FL, Nascimento N (1978) Carte du Karst. Laboratoire d'Analyse et Cartographie des Formations Superficielles de l'Université de Caen et le Centre de Geomorphologie, CNRS, Caen
Laming-Emperaire A, Prous A, Moraes AV, Beltrão MC (1974) Grottes et abris de la région de Lagoa Santa, Minas Gerais, Brésil. Cahiers d'Archéologie d'Amerique du Sud 1. EPHE, Paris
Lanari CU (1909) Ossadas humanas fósseis encontradas numa caverna calcária das vizinhanças do Mocambo. Annaes da Escola de Minas 11:15–35
Liais E (1872) Climats, geologie, faune et geographie botanique du Brésil. Garnier Freres, Paris
Lima Junior A (1962) As primeiras vilas do ouro. Estabelecimentos gráficos Santa Maria, Belo Horizonte
Luna PE (2007) Peter Wilhelm Lund: O auge das suas investigações e a razão para o término de suas pesquisas. PhD thesis, Universidade de São Paulo
Lund H (1885) Naturforskeren Peter Wilhelm Lund. En biografisk Skizze. C.A. Reitzels Forlag, Copenhagen
Lund PW (1837) Om huler i kalksteen i det indre af Brasilien, der tildeels indeholde fossile knokler. Anden Afhandling. Det Kongelige danske Videnskabernes Selskabs Skrifter 6:309–330
Lund PW (1840) View of the fauna of Brazil previous to the last geological revolution. Mag Nat Hist 4:1–8, 49–57, 105–112, 153–161, 207–213, 251–259, 307–317, 373–389ww
Lund PW (1841a) Blik paa Brasiliens dyreverden för sidste jordomvæltning. Første Afhandling. Det Kongelige Danske Videnskabernes Selskabs Naturvidenskabelige og Mathematiske Afhandlinger 8:29–60
Lund PW (1841b) Blik paa Brasiliens dyreverden för sidste jordomvæltning. Anden Afhandling: Fortsættelse af Pattedyrene. Det Kongelige Danske Videnskabernes Selskabs Naturvidenskabelige of Mathematiske Afhandlinger 8:61–144
Lund PW (1841c) Blik paa Brasiliens dyreverden för sidste jordomvæltning. Tredje Afhandling: Fortsættelse af Pattedyrene. Det Kongelige Danske Videnskabernes Selskabs Naturvidenskabelige of Mathematiske Afhandlinger 8:217–272
Lund PW (1845) Notice sur des ossements humains fossiles, trouvés dans une caverne du Brésil. Mémoires de la Societé Royale des Antiquaires du Nord 49–77
Lund PW (1846) Meddelelse af det Udbytte de I 1844 undersogte Knoglehuler have afgivet til Kundskaben om Brasiliens Dyreverden for sidste Jordomvaeltning. Det Kongelige Danske Videnskabernes Selskabs Naturvidenskabelige og Mathematiske Afhandlinger 12:60–92
Lund PW (1950) Memórias sobre a paleontologia brasileira. Instituto Nacional do Livro, Rio de Janeiro
Lütken CF (1875) Velhas-flodens fisk. Bianco Lunos, Copenhagen
Marchesotti APA (2011) Peter Wilhelm Lund. O naturalista que revelou ao mundo a pré-história brasileira. E-papers, Rio de Janeiro
Mattos A (1934) O sábio Dr. Lund e estudos sobre a pré-história brasileira. Edições Apollo, Belo Horizonte
Mattos (1935) Peter Wilhelm Lund no Brasil. Companhia Editora Nacional, São Paulo
Mattos (1941) A raça de Lagoa Santa. Companhia Editora Nacional, São Paulo
Mattos (1961) O homem das cavernas de Minas Gerais. Editora Itatiaia, Belo Horizonte
Mones A (1986) Palaeovertebrata sudamericana. Courier Forschungsinstitut Senckenberg 82:1–625
Neves WA, Piló LB (2003) Solving Lund's dilemma: new AMS dates confirm that humans and megafauna coexisted at Lagoa Santa. Curr Res Pleistocene 20:57–60
Neves WA, Hubbe M, Piló LB (2007) Early Holocene human skeletal remains from Sumidouro Cave, Lagoa Santa, Brazil: History of discoveries, geological and chronological context, and comparative cranial morphology. J Hum Evol 52:16–30
Parada JM (1949a) Gruta do Morro Redondo. Revista da Escola de Minas 14(1):29–35
Parada JM (1949b) Gruta dos Poções, da Lavoura e das Cacimbas. Revista da Escola de Minas 14(3):17–29
Paula Couto C (1947) Contribuição para o estudo de *Hoplophorus euphractus* Lund, 1839. Summa Brasiliensis Geologiae 1:33–53
Paula Couto C (1951) Uma espécie do gênero *Tetrastylus* Ameghino, 1886, em Lagoa Santa. Boletim do Museu Nacional 15:1–16
Paula Couto (1960) Um urso fóssil de Minas Gerais. An Acad Bras Ciênc 29(2):3
Piló LB, Auler AS, Neves WA, Wang X, Cheng H, Edwards RL (2005) Geochronology, sediment provenance, and fossil emplacement at Sumidouro Cave, a classic Late Pleistocene/Early Holocene paleoanthropological site in eastern Brazil. Geoarchaeology 20:751–764
Pires AOS (1929) Speleologia. Revista do Archivo Público Mineiro 23:105–167
Prous A (2017) Archaeological missions to the Lagoa Santa region in the second half of the twentieth century. In: Da-Gloria et al (eds) Archaeological and paleontological research in Lagoa Santa. Springer, Cham
Reinhardt JT (1856) Nogle Bemærkninger om den Indflydelse, de idelige Markbrande har udøvet paa Vegetationen in de brasilianiske Campos. Videnskabelige Meddelelser fra den naturhistorike Forening I Kjobenhavn 63–87

Reinhardt JT (1867) De brasilianske Knoglehuler og de i dem forekommende Dyrelevninger. Tidsskrift for populaere Fremstillingen af Naturvidenskaben 4(193–207):291–353
Reinhardt JT (1870) Bidrag til Kundskab om Fuglefaunaen i Brasiliens Campos. Bianco Lunos, Copenhagen
Reinhardt JT (1875a) Bidrag til kundskab om klaemedovendyret Lestodon armatus. Det Kongelige Danske Videnskabernes Selskabs 11:1–39
Reinhardt JT (1875b) De i Brasiliens knogelhuler fundne Glyptodont-levninger of en ny, til de gravigrade edentater hörende slaegt. Videnskabelige Meddelelser fra den naturhistorike Forening I Kjobenhavn 165–236
Reinhardt JT (1880) Naturforskeren Peter Wilhelm Lund, hans Liv of hans Virksonhed. Det Kongelige Danske Videnskabernes Selskabs 147–210
Reinhardt JT (1888) De Brasilianske knogelhuler og de i dem forekommende dyrelevninger. E Museo Lundii 1:1–56
Reinhardt JT, Lutken CF (1861) Bidrag til Kundskab om Brasiliens Padder og Krybdyr. Videnskabelige Meddelelser fra den naturhistorike Forening I Kjobenhavn 3:143–242
Riedel A (1868) Viagem de S.S.A.A. Reaes Duque de Saxe e seu augusto irmão D. Luis Philippe ao interior do Brasil no anno 1868. Fundação Biblioteca Nacional
Schubart O (1956) Leptodesmidae brasileiras II. O gênero *Obiricodesmus* (Diplopoda, Proterospermophora). Revta Bras Biol 16 (3):341–348
Silveira AA (1921) Memórias Chorographicas. Imprensa Oficial, Belo Horizonte
Silveira AA (1929) Geographia do estado de Minas. Oliveira, Costa & Cia, Belo Horizonte
Silveira AA (1931) Floralia Montium. Imprensa Oficial, Belo Horizonte
Stewart TD, Walter HV (1955) Fluorine analysis of putatively ancient human and animal bones from Confins Cave, Minas Gerais, Brazil. Proc 31st Int Congr Am 2:925–927
Travassos LEP, Kranjc A (2011) Prof. Dr. Heinz Charles Kohler (1945–2010). Acta Carsologica 40:219
Tricart J (1956) O karst das vizinhanças setentrionais de Belo Horizonte. Revista Brasileira de Geografia 18:451–469
Walter HV (1948) The Pre-history of the Lagoa Santa Region (Minas Gerais). Tipografia Brasil, Belo Horizonte
Walter HV (1958) Archaeology of the Lagoa Santa region. Sedegra, Rio de Janeiro
Walter HV, Cathoud A, Mattos A (1939) The Confins Man. Grafica Queiroz Breymer, Belo Horizonte
Warming E (1892) Lagoa Santa. Et Bidrag til den biologiske Plantegeografi. Bianco Lunos, Copenhagen
Wygodzinsky P (1950) Sobre alguns Emesinae (Reduviidae, Hemiptera). Revta bras Biol 10(1):73–78

The Geology of Lagoa Santa Karst

Paulo Galvão and Jorge Peñaranda

Abstract

This chapter provides information about the geological evolution and structural geometry of the Lagoa Santa Karst region, located in the São Francisco sedimentary basin, where Neoproterozoic rocks from the Bambuí Group are deposited. This group is constituted of karstified limestones from the Sete Lagoas and metapelites rocks from the Serra de Santa Helena Formation, covered by unconsolidated sediments of Cenozoic age, in a context of basin border. The area presents some detachment faults, displacing the whole stratigraphic succession, allowing formations of distinct ages to stand alongside. The Sete Lagoas Formation, most frequent in the region, is divided by the lower Pedro Leopoldo Member with karst dissolution features developed mostly in impure limestone bedding planes; and the upper and better exposed Lagoa Santa Member, with dissolutions in pure limestones in the form of caves, small cavities and sinkholes.

Keywords

Lagoa Santa Karst • Geology • Stratigraphy • Structural geology • Sete Lagoas Formation

P. Galvão (✉)
Institute of Geosciences, Federal University of Minas Gerais, Pampulha Campus, Belo Horizonte, MG 31270-901, Brazil
e-mail: hidropaulo@gmail.com

J. Peñaranda
Institute of Geosciences, University of São Paulo, Rua do Lago 562, São Paulo, SP 05508-080, Brazil
e-mail: jorgeologist@gmail.com

1 Introduction

The Lagoa Santa Karst region is situated around 30 km north from Belo Horizonte, capital of the Minas Gerais state, Brazil, and it is an important example of the karst environment in South America. The region presents a dense set of karst geomorphological features associated with subaerial fluvial hydrography and underground karst components developed in the Sete Lagoas Formation's carbonate rocks. In some places, the rocks are covered by metapelites of the Serra de Santa Helena Formation and Cenozoic unconsolidated sediments (Berbert-Born 2000). Due to its geological particularities, the region provides significant economic resources for many surrounding cities, boosting the economy and nearby population growth.

The Bambuí Group, where the Lagoa Santa Karst is emplaced over, was first studied by Rimann (1917)—being further detailed by Freyberg (1932), who divided the previously known as Bambuí series into eastern and western facies, and Branco and Costa (1961), where the first stratigraphic column was proposed, modified in sequence by Oliveira (1967), Schöll and Fogaça (1979), Dardenne (1978) and Ribeiro et al. (2003). Moreover, other significant studies were developed to understand the region, such as structural and geological cartography (Magalhães 1989; Danderfer Filho 1991; Peñaranda 2016), geochronology (Babinski and Kaufman 2003; Babinski et al. 2007), hydrogeology (Pessoa 1996, 2005; Galvão et al. 2015a, b; Tayer and Velásques 2017; de Paula and Velásquez 2019), and geological evolution (Nobre-Lopes 1995; Uhlein et al. 2004; Vieira et al. 2007; Tuller et al. 2010; Galvão et al. 2016).

This chapter provides relevant information about the Lagoa Santa Karst geology and structural geometry. Geological mappings, aerial photography interpretations, lithological and image well profiles analyses, and a compilation of the most relevant studies were carried out to set up a geological framework of the region.

A. S. Auler and P. Pessoa (eds.), *Lagoa Santa Karst: Brazil's Iconic Karst Region*, Cave and Karst Systems of the World, https://doi.org/10.1007/978-3-030-35940-9_2

2 Site Description

The Lagoa Santa Karst area is in the south-central of the Minas Gerais state, Brazil, around 30 km north from Belo Horizonte, the capital city (Fig. 1). The area covers more than 1,000 km^2, where the cities of Vespasiano, Confins, Funilândia, Pedro Leopoldo, Sete Lagoas, Matozinhos, Capim Branco and Prudente de Morais are located (red contour line, Fig. 1b). This karst area has the highest density of caves in Brazil, with more than 1,000 known caves, many with significant archaeological and paleontological values (Auler and Farrant 1996) (see chapter "Caves and Speleogenesis in the Lagoa Santa Karst", this volume). The Lagoa Santa Karst Environmental Protection Area (green contour line, Fig. 1b) is in the east-central area of the region, covering 39,957 hectares (98,740 acres). This area is known for containing hundreds of Early Holocene/Late Pleistocene archaeological sites, giving important anthropological pieces and information about the origin and age of human colonization in the Americas (Auler and Piló 2015).

Geologically, the area overlies the southeastern region of the Paleoproterozoic São Francisco Craton (Almeida 1977) (Fig. 1a), composed of crystalline rocks, being a result of an amalgamation of distinct Archean craton blocks. Between 900 and 600 Myr ago (Condie 2002) extension events occurred, resulting in the deposition of Neoproterozoic silicate–carbonate sediments, giving origin to the Bambuí Group.

The Bambuí Group occurs over an area of about 300,000 km^2 (Almeida 1977) in the states of Minas Gerais, Goiás and Bahia. Its genesis is related to subsidence due to the overloading of Brasiliano orogenic belts in a foreland basin context within the São Francisco Craton (Castro and Dardenne 2000; Alkmim and Martins-Neto 2001; Sial et al. 2010; Uhlein 2013; Perrella Júnior et al. 2017). The basement structure is controlled by faults responsible for sediment thickness variations (Misi et al. 2007). Seismic surveys made by Petrobras S. A. has indicated that the central area of this group may reach thicknesses of about 1,000 m (Pedrosa-Soares et al. 1994). Due to a shallow epicontinental sea with a low gradient ramp, the Bambuí Group sediments are regionally distributed, enabling the recognition of the same lithofacies sequences in large areas (Schobbenhaus 1984).

Regarding stratigraphy, three transgressive–regressive sedimentary cycles occurred: (1) transgressive pelite–carbonate cycle (Pedro Leopoldo Member of Sete Lagoas Formation); (2) regressive carbonate cycle (Lagoa Santa Member of Sete Lagoas Formation); and (3) transgressive pelite–arenaceous cycle (Serra de Santa Helena Formation) (Fig. 1c and Table 1).

Rested over the gneiss–migmatitic undifferentiated complex (the basement), the Bambuí Group is constituted, from the bottom to the top, by the Carrancas Formation (rhythmites, sandstones and diamictites); Sete Lagoas Formation (carbonaceous sequences); Serra de Santa Helena Formation (metapelites, and marlstones), Lagoa do Jacaré Formation (oolitic limestones); Serra da Saudade Formation (siltstones and sandstones); and Três Marias Formation (sandstones from fluvial platforms). Unconsolidated sediments of Cenozoic age overlay in some places those formations (Tuller et al. 2010).

Considering just the limits of the Lagoa Santa Karst region, the Bambuí Group displays only the Sete Lagoas and Serra de Santa Helena formations, being locally covered by Cenozoic unconsolidated sediments (Fig. 1c and Table 1), which will be further discussed in detail.

In terms of structural geology, normal and thrust faults trending NNW-SSE are common, where fractures and foliations are the main planar fabrics (Danderfer Filho 1991; Tuller et al. 2010; Galvão et al. 2016). Mineral lineations and stretching kinematically indicate tectonic transport towards the west, where interstratal ramp thrust belts were formed being accommodated in argillaceous upper planes. According to structural geological surveys, the Bambuí Group can be divided in four structural domains, considering its deformation magnitude (Tuller et al. 2010): (I) high intensity—the Espinhaço Supergroup rocks outcrop with mass movement signals towards the west; (II) intermediate—more deformed in the east than in the west, extending from the Velhas river to the extreme western limit of the city of Sete Lagoas. This is the domain where the Lagoa Santa Karst region is situated; (III) rocks with less intense deformation, which in some places no deformations can be found; and (IV) in the extreme west area, incipient deformation affects only the Serra da Saudade Formation's rocks, where sedimentary structures of the other formations are more preserved.

The Lagoa Santa Karst region has an important aquifer, named Sete Lagoas Karst Aquifer (Pessoa 1996; Galvão et al. 2015a), which consists of Neoproterozoic karstified limestones composed of the Pedro Leopoldo Member (bottom) and Lagoa Santa Member (top) (see chapter "Karst Hydrogeology of the Lagoa Santa Area", this volume). The primary porosity and matrix permeability are very low (Galvão et al. 2015b; Peñaranda 2016) and the secondary porosity (micro-fractures) is mostly filled by calcite and quartz (Tonietto 2010; Galvão et al. 2015b, 2016).

Most of the water flows through karst dissolution conduits classified as tertiary porosity. Image well loggings data in the Sete Lagoas' urban area indicated that two dominant karstified bedding planes have high permeability and significant storage capacity, being the most relevant discontinuities for groundwater circulation. In the same area, the overall aquifer is about 75 m thick (Galvão et al. 2015a).

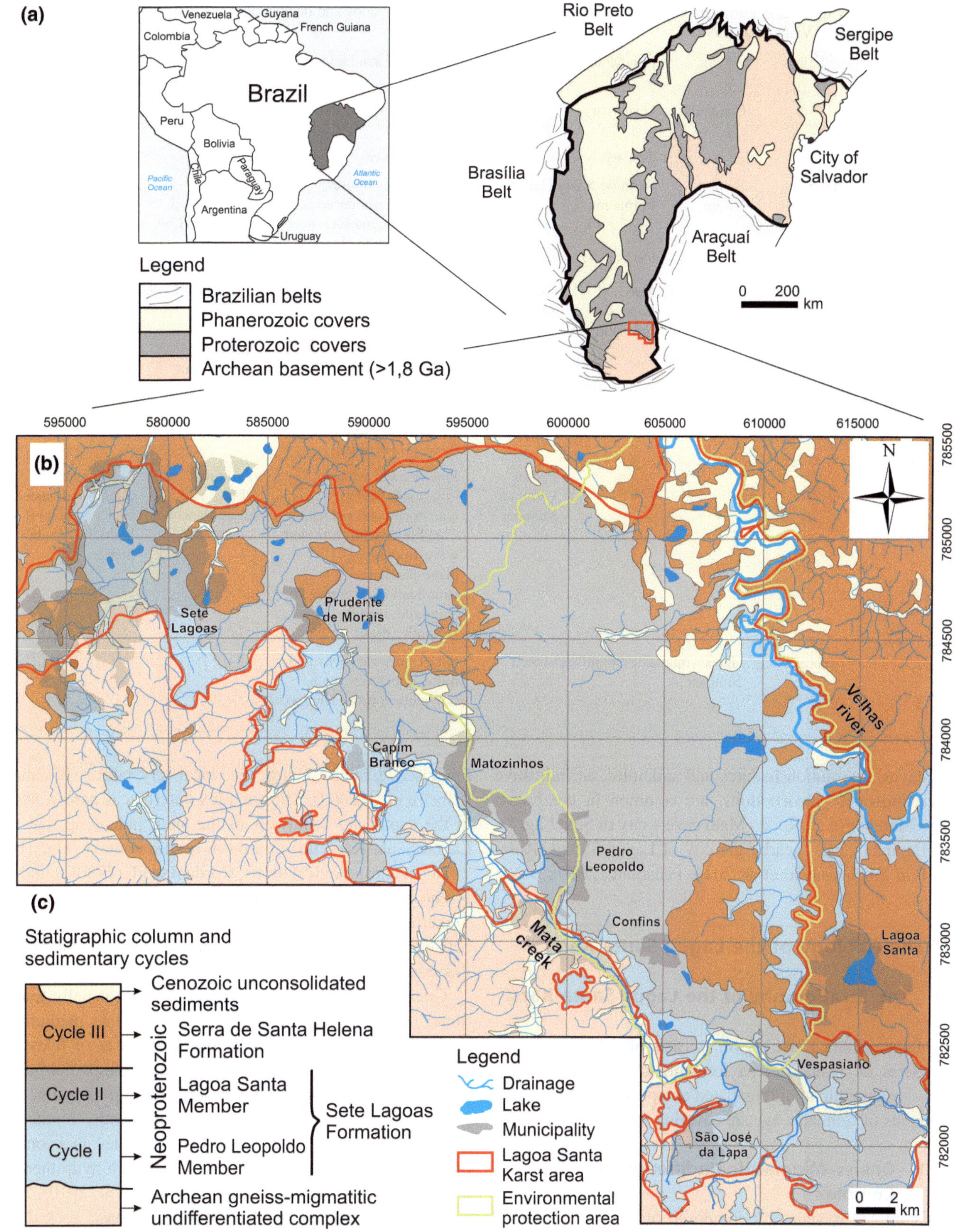

Fig. 1 **a** The Lagoa Santa Karst area in the São Francisco Craton (modified from Almeida and Hasui 1984; Alkmim et al. 1993); **b** geology of the Lagoa Santa Karst (simplified from Viana 1998); **c** stratigraphic representation and the sedimentary cycles of the Lagoa Santa Karst (modified from Ribeiro et al. 2003). Figure "**c**" can be used as the legend of the lithologies seen in figure "**b**"

Table 1 Summary of main lithostratigraphic units and their respective contact relationships and depositional systems of the Lagoa Santa Karst region (modified from Viana et al. 1998)

Chronostratigraphic unit			Lithostratigraphic unit		Lithology/Lithofacies	Contact relationship	Depositional system
Phanerozoic		Cenozoic	Quaternary		Alluvial terraces and alluviums	Discordant	Continental
			Tertiary		Detritic cover	Discordant	Continental
Precambrian	Upper Proterozoic	Bambuí Group	Serra de Santa Helena Formation		Silt-clay rhythmites, generally carbonaceous, dark shales, marble, siltstones, slate cleavage	Abrupt, tectonic and gradational contacts	Transgressive argillo-arenaceous cycle. Epicontinental marine
			Sete Lagoas Formation	Lagoa Santa Member	Medium-grained grey limestones. Grey calcarenites and calcisiltites, plane-parallel lamination, wave truncation, medium-sized crossed stratifications. Presence of stromatolites (type *gyminosolens*)	Gradational, or abrupt contact through detachment faults	Regressive carbonated cycle. Subtidal and supratidal marine
				Pedro Leopoldo Member	Fine-grained white limestones, dolomites, marlstones and pelites. Calcilutite, with truncated wave-ripple or plane-parallel stratification. Presence of aragonite crystals. Dolomite rocks	Faulting, abrupt and discordant contacts	Transgressive argillo-carbonated cycle. Structural highs and basin border
	Archean	Gneiss–migmatitic undifferentiated complex			Set of gneissic rocks and migmatite zones, with polymetamorphic features	–	–

Dry caves, dissolution features and sinkholes, an indicative of groundwater paleoconduits, are common in the Lagoa Santa Member (e.g., Rei do Mato cave—city of Sete Lagoas, and Lapinha cave—nearby the city of Lagoa Santa) (Auler and Piló 2015, Galvão et al. 2016, Peñaranda 2016).

3 The Geology of the Lagoa Santa Karst

3.1 The Stratigraphy of the Lagoa Santa Karst

The Lagoa Santa Karst region consists of the Sete Lagoas and Serra de Santa Helena formations, which overly the gneiss–migmatitic undifferentiated complex. Unconsolidated sediments of Cenozoic age are covering some of these rocks.

3.1.1 Gneiss–Migmatitic Undifferentiated Complex

This complex is generally composed of a set of migmatite zones, gneiss and granitoid rocks, corresponding to the São Francisco Craton, being the local crystalline basement (Ribeiro et al. 2003). The complex outcrops as narrow strips in the south and southwest of the area, indicating that the Lagoa Santa Karst lies in a basin border context (Fig. 1b).

In geochronological terms, Cordani et al. (1980) found an isochron age of 1 Ga for gneisses in the city of Sete Lagoas; Teixeira et al. (2000) recorded T_{DM} model ages varying between 3.4 and 3.1 Gyr, indicating continuous evolution to this complex.

Gneisses, migmatites and granitoids have usually transitional contacts (Fig. 2). Light grey to whitish, medium to coarse-grained granitoids with porphyroblasts of feldspars and xenoliths are common, where restites from partial fusion are also noted in those granitoids. Petrographically, the gneisses vary in composition with hornblende, biotite, or amphibole, indicating that granites and tonalites, in a context of metamorphism at the amphibole facies with hydrothermal alteration, would be the protolytes (Tuller et al. 2010). This crystalline basement was also latter cut by basic rock dikes (Ribeiro et al. 2003).

Fig. 2 Example of a gneiss–migmatite rock from the crystalline basement (left), characterized by mesoscopic migmatite features (right)—city of Sete Lagoas, northwest of the Lagoa Santa Karst region (modified from Galvão et al. 2016)

3.1.2 Sete Lagoas Formation: Pedro Leopoldo and Lagoa Santa Members

According to Dardenne (1978), the Sete Lagoas Formation regionally is over the Carrancas Formation or it is in contact with the gneiss–migmatitic undifferentiated complex. Geochronological data suggests that this formation ranges from 740 ± 22 Myr (Pb–Pb—Babinski et al. 2007) to 630 Myr (C-O-Sr—Caxito et al. 2012), or 540 Myr, via fossil investigations (Warren et al. 2014). The sedimentary depositional environment is characterized by a shallowing upward carbonate cycle, with three deposit systems being recognized, according to Lima (1997): internal, intermediary and external ramps. The Sete Lagoas Formation is separated by the members Pedro Leopoldo (bottom) and the Lagoa Santa (top), covering about 80% of the area (Fig. 1b). The lithological contact between the Pedro Leopoldo Member and the crystalline basement is abrupt and/or discordant (tectonic faults) (Table 1). The contact of the Lagoa Santa Member with the Serra de Santa Helena Formation is abrupt (Galvão et al. 2016).

Pedro Leopoldo Member

This member has a significant exposure in the region (about 20% of the total area), occurring especially on the western margin of the Velhas River, bordering the Mata Creek and the cities of Vespasiano, São José da Lapa, Pedro Leopoldo, Prudente de Morais, Capim Branco and Sete Lagoas (Fig. 1b). The Pedro Leopoldo Member overlies the crystalline basement usually by faulting, showing discordant contacts. Dolomites, marlstones, and pelites (less frequent) and fine-grained white limestones (more common) are the rocks from this member (Table 1). This limestone is composed of 85% micrite, 8% sparite and 7% quartz (microscopy analysis), being classified as micrite (Folk 1959) or mudstone (Dunham 1962) (Galvão et al. 2015b, 2016).

In the Sete Lagoas Formation, six lithological facies were described by Vieira et al. (2007), four being recognized in the Pedro Leopoldo Member: Facies A—light grey calcilutite intercalating yellowish pelite displaying hummocky and swaley cross-stratifications, indicating a tidal flat deposits with local storm influence; Facies B—tabular beds and layers of deep-sea sedimentation presenting aragonite crystals; Facies C—dolomites with metric to sub-metric dimensions; and Facies D—light grey calcilutites displaying plane-parallel stratifications and rhythmic terrigenous indicating zones of minor to major agitated marine sedimentation.

Lagoa Santa Member

With the largest exposure in the area (approximately 60%), this member occurs as NW-SE wide strip starting from the Mata Creek valley in the southwest, through the cities of Vespasiano, Confins, Matozinhos and bordering Prudente de Morais and Sete Lagoas (Fig. 1b). The Lagoa Santa Member superimposes the Pedro Leopoldo Member through gradational or abrupt contacts through detachment faults (Fig. 3). Near the contact, an intense occurrence of concordant/discordant calcite veins is observed.

Fig. 3 Exposure of the abrupt contact (dashed white line) between the Pedro Leopoldo Member (impure white limestones at the bottom) and the Lagoa Santa Member (pure grey limestones on the top) in a quarry situated within the city of Sete Lagoas. The yellowish-orange areas are weathered rocks (modified from Galvão et al. 2016)

Overlying the member, metasedimentary rocks from the Serra de Santa Helena Formation occur, usually in tectonic contact (detachment fault). The Lagoa Santa Member is featured by medium-grained grey limestones (Table 1), being classified as grainstone (Dunham 1962), or oobiosparite (Folk 1959) because, in general, the limestones are constituted of 55% sparite, 30% ooids and 15% micrite (microscopy analysis) (Galvão et al. 2015b, 2016). The two remaining facies (Vieira et al. 2007) are: Facies E—grey calcarenites displaying plane-parallel laminations, hummocky and swaley cross-stratifications; and Facies F—well-preserved stromatolites mounds (type *gyminosolens*, Marchese 1974), typical of tidal flat sedimentation.

3.1.3 Serra de Santa Helena Formation

This formation occurs in the SE, NE and NW areas occupying approximately 15% of the region, bordering the cities of Prudente de Morais and Sete Lagoas (Fig. 1b). Regionally, the metapelite rocks overlie the Sete Lagoas Formation with gradational and abrupt contacts (Table 1), being recognized three stratigraphic members, according to Grossi-Sad et al. (1998): (1) lower—carbonaceous dark shales and silt-clay rhythmites (90 m thick); (2) medium—marble and shales (50 m thick); and (3) upper—shales and siltstones with bearing slate cleavage (140 m thick) (Galvão et al. 2016).

Extensional events generated subvertical fractures and relief zones that were filled by quartz veins with hexagonal prisms and, locally, by pegmatite veins composed of feldspar and calcite grains (Viana 1998; Tuller et al. 2010; Galvão et al. 2016). Schistosity (a type of secondary foliation) discordant and concordant with the bedding is present in the rocks, where intercalated slate with well-developed rock cleavage at the bottom of the formation are also noted (Fig. 4).

3.1.4 Cenozoic Unconsolidated Sediments

This Cenozoic unit can be separated into detrital cover, alluvial terraces and alluviums, which can be grouped as unconsolidated sediments (Table 1). They occupy about 5% of the Lagoa Santa Karst area, along the Velhas River and Mata Creek, in some depressions and bordering the cities of Vespasiano, Pedro Leopoldo, Matozinhos, Capim Branco and Sete Lagoas (Fig. 1b).

The detrital cover is characterized predominantly by red color, clayey and sandy sediments with levels of gravel, occurring over all the formations, mainly covering the rocks from the Serra de Santa Helena Formation. Alluvial terraces are common along main rivers (e.g. Velhas River) and streams (e.g. Mata Creek) and their tributaries, as well as in the central urban area of the cities of Sete Lagoas and Capim Branco due to their lower topographies. These terraces can reach 5 m in thickness and consist of yellowish to reddish clayey-sandy semi-consolidated sediments (Fig. 5). The alluviums are common along meanders of significant watercourses, being composed of fine to coarse-grained sediments, with levels of rounded and poorly sorted quartz pebbles.

Fig. 4 Well-developed cleavage slate rocks coinciding with the bedding planes at the Santa Helena ridge (city of Sete Lagoas) (modified from Galvão et al. 2016)

Fig. 5 Cenozoic clayey-sandy sediments with a thickness of about 5 m, with yellowish to reddish color. These sediments are mostly situated along the Velhas River

4 Geological Framework of the Lagoa Santa Karst

The crystalline basement outcrops in the extreme SW of the region and dips to NE direction in a basin border setting context. Because the framework is controlled by faulting, the basement may have different depths over short distances.

According to geological cross sections (Danderfer Filho 1994; Galvão et al. 2016), some detachment faults are observed placing side by side rocks from different formations (Figs. 6, 7 and 8). Rock layers from the Sete Lagoas Formation are horizontally continuous dipping and getting thicker towards the NE being covered in the SE and NE regions by rocks from the Serra de Santa Helena (Figs. 6 and 7). Both formations are thinner towards the SW where the basement outcrops. The thicknesses of the Sete Lagoas and Serra de Santa Helena formations within the Lagoa Santa Karst region can reach, in some places, 300 m and 160 m, respectively.

The Cenozoic unconsolidated sediment thickness is thin compared to other geological formations. According to Viana (1998), these sediments may vary from a few centimeters to several meters. Good examples are seen along the Velhas River, in some depressions, in the central urban area of the city of Sete Lagoas (Fig. 1b), where sediments may reach 40 m in thickness (Galvão et al. 2016), or in the vicinity of the city of Lagoa Santa, where the thickness can reach 100 m (Pessoa 2005).

5 Structural Geology of the Lagoa Santa Karst

The Lagoa Santa Karst region is in the structural domain II, extending from the Velhas River to the extreme western portion of the city of Sete Lagoas (Tuller et al. 2010), being more deformed towards the east and less deformed towards the west. The domain II is characterized by less intense deformation compared to the domain I, located right in the eastern side of the area. It is typical to the domain II the minor presence of isoclinal folds, being more common asymmetric west-verging folds influenced by interstratal sliding.

The structures mapped are foliations, lineations, fractures and faults. Fractures and faults are often associated with drainage valleys and surface alignments of karstified features, where numerous sinkholes and caves can be found. Structural features such as normal faults and low-angle detachment faults between layers were observed by Danderfer Filho (1994), Pessoa (2005), Galvão et al. (2016) and Peñaranda (2016), explaining lithological successions of different ages occurring side by side (Figs. 6, 7 and 8).

According to Danderfer Filho (1991), there were three stages of deformation associated with the Sete Lagoas and the Serra de Santa Helena formations and with the gneiss–migmatitic undifferentiated complex. In the crystalline basement, due to several stages of deformation making it difficult the

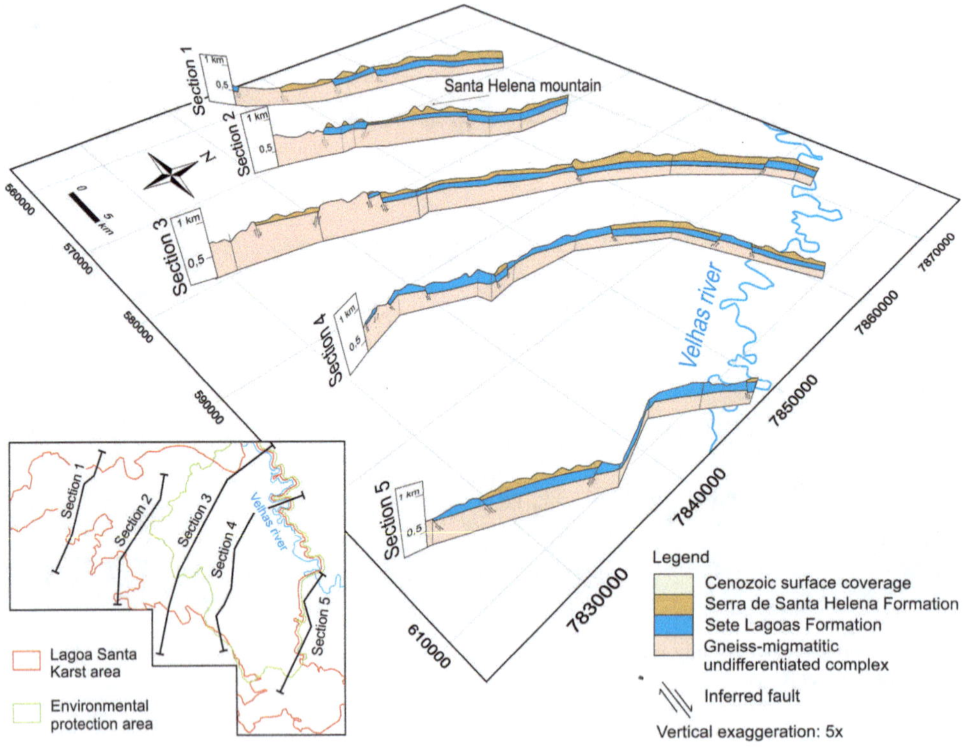

Fig. 6 Geologic SW-NE cross sections indicating the geometry of the basement border, where the gneiss–migmatitic undifferentiated complex outcrops in the SW region dipping to NE directions

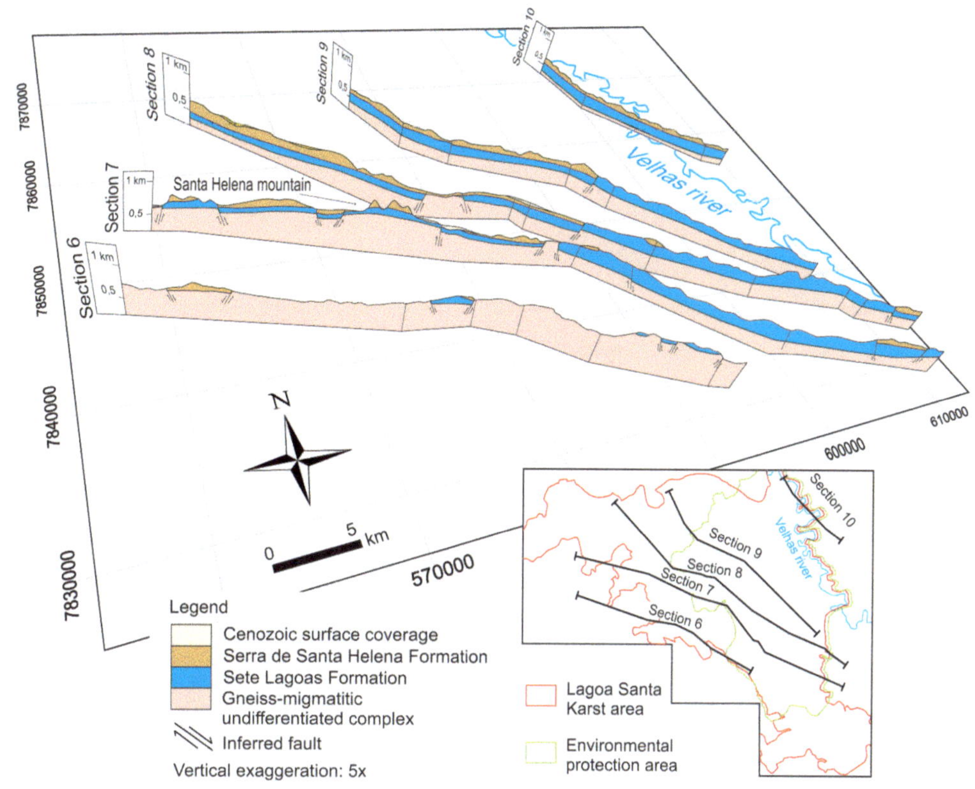

Fig. 7 Geologic NW-SE cross sections. The Sete Lagoas Formation dips becoming thicker in the NE, being covered by rocks from the Serra de Santa Helena Formation towards the SE and NE

understanding of the evolution of its rocks, it is possible to highlight the predominant gneissic foliation/banding with the direction NNW-SSE and quartz veins with the direction ENE-WSW as a result of the first stage. The second stage occurred in the rocks from the Sete Lagoas and the Serra de Santa Helena formations. East-verging high-angle faults with structural trend approximately NNW-SSE are common features of those rocks, revealing an initial phase of extension characteristic of the inception of the São Francisco Basin. The third stage of deformation, structurally more intense, is featured by low-angle compressive shear of E-W direction, with low-grade metamorphism. Subhorizontal detachment faults, shear zones and low-angle dip foliations associated with lineations are common.

In a study made by Peñaranda (2016), in the city of Sete Lagoas, the association between fracture systems measured in the Sete Lagoas Formation's limestones and karst tertiary porosity development was analyzed. Three sets of predominant fractures (all subverticals) intersecting the limestones with subhorizontal bedding planes of 6–10° dip were identified: (1) N10–20E; (2) N30–40E; (3) N80–90 W. These dominant intersecting lines (N10–20E/6–10°, N30–40E/6–10°, N80–90 W/6–10°) are coincident with the development of tertiary porosity represented by caves in the surroundings of the city of Sete Lagoas. Thus, similar behavior is expected in the Lagoa Santa Karst area in terms of karstification processes. However, the degree of deformation in the eastern side of the Lagoa Santa Karst region is more intense, showing different sets of fracture/fault in comparison to the northwestern side of the region, where the city of Sete Lagoas is located.

6 Geological Evolution of the Lagoa Santa Karst

Extensional geotectonic events between 900 and 600 Myr (Condie 2002) in the São Francisco Craton generated a low gradient epicontinental shelf giving the opportunity to exist three regressive sedimentary cycles, where Neoproterozoic silicate-carbonaceous sediments of the Bambuí Group were deposited originated the São Francisco Basin. These cycles resulted in the respective sedimentation of the Sete Lagoas and Serra de Santa Helena formations.

Around 740 ± 22 Myr (Babinski et al. 2007), 630 Myr (Caxito et al. 2012), or 540 Myr (Warren et al. 2014), firstly fine-grained white limestones of the Pedro Leopoldo Member and, after that, medium-grained grey limestones of the Lagoa Santa Member where deposited in intermediary internal and external ramps, originating the Sete Lagoas Formation. Some hummocky and swaley cross-stratifications noted in some outcrops indicate that the Pedro Leopold Member sedimentation occurred in a tidal flat environment with local storm influence, as well as in zones of minor to

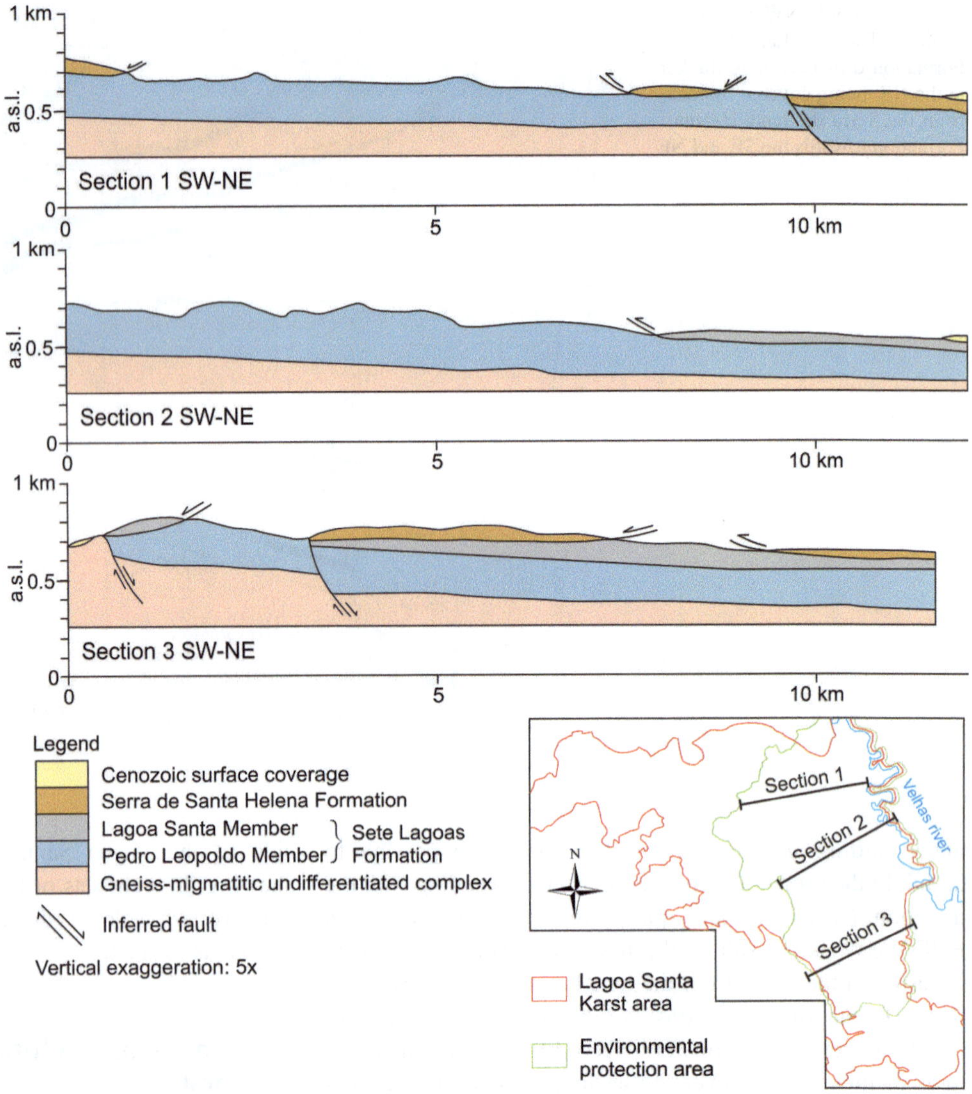

Fig. 8 Geologic SW-NE cross sections inside the Lagoa Santa Karst Environmental Protection Area (modified from Danderfer Filho 1994) showing, in more detail, detachment faults

major agitated marine areas, also indicated by plane-parallel stratifications and rhythmic terrigenous. In the Lagoa Santa Member, plane-parallel laminations and hummocky and swaley cross-stratifications with well-preserved stromatolites mounds (evidence of ancient life) indicating that, in that time, tidal flat sedimentation was occurring.

Over the Sete Lagoas Formation rocks, low flow coastal and marine environments in a transgressive cycle (Ribeiro et al. 2003) deposited pelite sediments characterized by plane-parallel and cross-stratifications, as well as ripple marks. This sedimentary depositional environment originated the Serra de Santa Helena Formation, achieving thickness values of about 160 m, observed in tubular well lithological profiles located in the city of Sete Lagoas (Galvão et al. 2016).

After that, the Bambuí Group was faulted and folded to the east because of the deformation of the Espinhaço Supergroup during the Brasiliano event ($\sim$600 Myr) (Trompette et al. 1993). This event occurred within the structural domain I (deformation with high intensity) close to the Lagoa Santa Karst area (belonging to the structural domain II of deformation with intermediate magnitude). It is important to mention that the deformations are more expressive in the east portion, near the stress from the domain I. In the west, sedimentary features with no significant deformations are more preserved (Alkmim and Martins-Neto 2001). This deformation event resulted in some detachment faults displacing the stratigraphy, which enabled lithologic successions of different ages to be placed side by side. Because of this, the limestones of the Pedro Leopoldo Member outcrop in some places (Figs. 1b, 6, 7 and 8).

The limestones present important dissolution features resulting in enlarged bedding planes and subvertical fractures, most frequent in the Pedro Leopoldo Member, especially in the city of Sete Lagoas (Fig. 9). Dry caves and small cavities, an indicative of former saturated paleoconduits, are

Fig. 9 Terrigenous sediments filling the karstified bedding planes from the Pedro Leopoldo Member (left) and inactive cave occurrence in the medium-grained limestone of the Lagoa Santa Member. Both of them are in the city of Sete Lagoas (modified from Galvão et al. 2016)

Fig. 10 General view at the Lagoa Santa Environmental Protection Area. In the foreground, the medium-grained grey limestones from the Lagoa Santa Member with typical vegetation of the region (photo by Ataliba Coelho)

also noted, being the most important dissolution features of the Lagoa Santa Member, easily noted in the cities of Matozinhos, Pedro Leopoldo and Lagoa Santa (Fig. 9). These processes originated the most important and beautiful landscapes in the area, represented by a set of caves, small cavities and dolines (Fig. 10) (see chapers "Karst Landforms in the Lagoa Santa Area" and "Caves and Speleogenesis in the Lagoa Santa Karst", this volume).

Finally, during the Cenozoic age, the rocks from Sete Lagoas and Serra de Santa Helena formations located right in the higher portions of the area were weathered and the erosion resulted in the deposition of unconsolidated sediments in low relief areas and along the Velhas riverbeds.

7 Conclusions

The nature of the Lagoa Santa Karst region provides such a great variety of information related to geological evolution, geochronology, hydrogeology, geomorphology and even anthropology resulting in a very rich area for multidisciplinary studies. The area is in a basement border of the São Francisco Craton where, after extensional events, silicate–carbonate sediments were deposited in the Neoproterozoic Era. These sedimentations originated the Bambuí Group, locally constituted by the Sete Lagoas and Serra de Santa Helena formations, with layer thicknesses reaching 300 m and 160 m, respectively. In the Brasiliano event, around 600 Myr ago, in the east of the Lagoa Santa Karst area, the Espinhaço Supergroup was deformed generating detachment faults in the Bambuí Group, displacing the whole stratigraphic succession allowing different formations to be set side by side. The region is in the intermediate magnitude of deformation, characterized by the presence of asymmetric folds. Fractures and faults are often associated with drainage valleys and surface alignments of karst features, where numerous dolines and caves can be found. Probably, the association between fractures/faults with subhorizontal bedding planes in the rocks of the Sete Lagoas Formation explains the direction trends of the development of karst tertiary porosity. These rocks, especially the limestones from the Lagoa Santa Member, present dissolution features in the form of caves, small cavities and dolines, indicating groundwater paleoconduits. Because of all those characteristics described here, the Lagoa Santa Karst region is known for having and being a remarkable karst landscape.

Acknowledgements Special thanks go to geologist Arnaldo Cordeiro for the fieldwork, to Ataliba Coelho for allowing the use of his photos, and to Professor André Marconato for revisions.

References

Alkmim FF, Brito Neves BB, Alves JAC (1993) Arcabouço tectônico do Cráton do São Francisco - uma revisão. In: Dominguez JML, Misi A (eds) O Cráton do São Francisco. Reunião Preparatória para o II Simpósio Sobre o Cráton do São Francisco. SBG/SGM/CNPq, Salvador, pp 45–62

Alkmim FF, Martins-Neto MA (2001) A bacia intracratônica do São Francisco: arcabouço estrutural e cenários evolutivos. In: Pinto CP, Martins-Neto MA (eds) Bacia do São Francisco: geologia e recursos minerais. Sociedade Brasileira de Geologia-MG, Belo Horizonte, pp 9–30

Almeida FFM (1977) O Cráton do São Francisco. Rev Bras Geoc 7:349–364

Almeida FFM, Hasui Y (1984) O Pré-Cambriano do Brasil. Edgard Blücher, São Paulo

Auler AS, Farrant AR (1996) A brief introduction to karst and caves in Brazil. Proc Univ Bristol Spelaeol Soc 20(3):187–200

Auler AS, Piló LB (2015) Lagoa Santa Karst: Cradle of Brazilian Cave Studies. In: Vieira BC et al (eds) Landscapes and landforms of Brazil. Springer, Dordrecht, pp 183–190. https://doi.org/10.1007/978-94-017-8023-0_16

Babinski M, Kaufman AJ (2003) First direct dating of a Neoproterozoic post-glacial cap carbonate. In: South American symposium on isotope Geology Salvador. Short Papers, pp 321–323

Babinski M, Vieira CV, Trindade RIF (2007) Direct dating of the Sete Lagoas cap carbonate (Bambuí Group, Brazil) and implications for the Neoproterozoic glacial events. Terra Nova 19(6):401–406

Berbert-Born M (2000) The Lagoa Santa karst. In: Schobbenhaus C, Campos DA, Queiroz ET, Winge M, Berbert-Born M (eds) Sítios Geológicos e Paleontológicos do Brasil. http://www.unb.br/ig/sigep/sitio015/sitio015.htm. Accessed 25 Sep 2016

Branco JJR, Costa MT (1961) Roteiro da excursão Belo Horizonte – Brasília. In: 14 Congresso Brasileiro de Geologia. Instituto de Pesquisas Radioativas, UFMG, Publicação 15, Belo Horizonte

Castro PTA, Dardenne MA (2000) The sedimentology, stratigraphy and tectonic context of the São Francisco Supergroup at the southern boundary of the São Francisco craton, Brazil. Revista Brasileira de Geociências 30:345–437

Caxito FA, Halverson GP, Uhlein A, Stevenson R, Dias TG, Uhlein GJ (2012) Marinoan glaciation in east central Brazil. Precambr Res 200–203:38–58

Condie KC (2002) The supercontinent cycle: are there two patterns of ciclicity? J Afr Earth Sci 35:179–183

Cordani UG, Kawashita K, Müller G, Quade H, Reimer V, Röser H (1980) Interpretação tectônica e petrológica de dados geocronológicos do embasamento no bordo sudeste do Quadrilátero Ferrífero, MG. Rio de Janeiro: Anais da Academia Brasileira de Ciências 52 (4):785–799

Dardenne MA (1978) Síntese sobre a estratigrafia do Grupo Bambuí no Brasil Central. In: Anais 30 Congresso Brasileiro de Geologia, Sociedade Brasileira de Geologia 2:597–610

Danderfer Filho A (1991) Cartografia geológica dos municípios Sete Lagoas, Funilândia, Lagoa Santa, Pedro Leopoldo, Vespasiano, Capim Branco, Matozinhos e Prudente de Morais – Capítulo 6 - Geologia Estrutural. Projeto Vida. MME, CPRM, Belo Horizonte

Danderfer Filho A (1994) Geologia estrutural da área do Projeto VIDA. CPRM, Belo Horizonte

De Paula RS, Velásquez LNM (2019) Balanço hídrico em sistema hidrogeológico cárstico, região de Lagoa Santa, Minas Gerais [Water balance in karstic hidrogeologic system, region of Lagoa Santa, Minas Gerais]. Águas Subterrâneas 33(2):119–133. https://doi.org/10.14295/ras.v33i2.29252

Dunham RJ (1962) Classification of carbonate rocks according to depositional texture. In: Ham WE (ed) Classification of carbonate rocks. Am. Assoc. Petrol. Geol. Mem. 1:108–121

Folk RL (1959) Practical petrographic classification of limestones. Bull Am Assoc Petrol Geologists 43:1–38

Freyberg BV (1932) Ergebnisse geologischer Forschungen in Minas Gerais, Brasilien: Stuttgart, Neus Jahrb, sonderband 2, 403 pp

Galvão P, Halihan T, Hirata R (2015a) The Karst permeability scale effect of Sete Lagoas, MG, Brazil. J Hydrol 531:85/15–105. https://doi.org/10.1016/j.jhydrol.2015.11.026

Galvão P, Halihan T, Hirata R (2015b) Evaluating karst geotechnical risk in the urbanized area of Sete Lagoas, Minas Gerais, Brazil. Hydrogeol J 23(7):1499–1513

Galvão P, Hirata R, Cordeiro A, Osório DB, Peñaranda J (2016) Geologic conceptual model of the municipality of Sete Lagoas (MG, Brazil) and the surroundings. An Acad Bras Ciênc 88(1):35–53

Grossi-Sad JH, Chiodi-Filho C, Chiodi DK (1998) Panorama do setor de ardósias Estado de Minas Gerais, Brasil. COMIG, Belo Horizonte, CD-ROM

Lima SAA (1997) Fácies, ambiente deposicional e aspectos diagenéticos das rochas carbonáticas da Formação Sete Lagoas na região norte de Minas Gerais, Bacia do São Francisco. Dissertation, Universidade Federal de Ouro Preto

Magalhães PM (1989) Análise estrutural qualitativa das rochas do Grupo Bambuí, na porção sudoeste da Bacia do São Francisco. Dissertation, Universidade Federal de Ouro Preto

Marchese AG (1974) Litoestratigráfia y petrología del Grupo Bambuí en los estados de Minas Gerais y Goiás. Brasil Rev Bras Geoc 3:172–190

Misi A, Kaufman AJ, Veizer J, Powis K, Azmy K, Boggiani PC, Gaucher C, Teixeira JBG, Sanches AL, Iyer SSS (2007) Chemostratigraphic correlation of Neoproterozoic successions in South America. Chem Geol 237:161–185

Nobre-Lopes J (1995) Faciologia e gênese dos carbonatos do Grupo Bambuí na região de Arcos, Estado de Minas Gerais. Dissertation, Universidade de São Paulo

Oliveira MAM (1967) Contribuição à geologia da parte sul da Bacia do São Francisco e áreas adjacentes. Coletânea de Relatórios de Exploração. Petrobrás, Rio de Janeiro 3:71–105

Pedrosa-Soares AC, Dardenne MA, Hasui Y, Castro FDC, Carvalho MVA, Reis AC (1994) Mapa Geológico do estado de Minas Gerais e Nota Explicativa. Companhia Mineradora de Minas Gerais, Belo Horizonte

Peñaranda JR (2016) Condicionamento estrutural e litológico da porosidade cárstica da Formação Sete Lagoas, Município de Sete Lagoas (MG). Dissertation, Universidade de São Paulo

Perrella Júnior P, Uhlein A, Uhlein GJ, Sial NA, Pedrosa-Soares AC, Lima ONB (2017) Facies analysis, sequence stratigraphy and chemostratigraphy of the Sete Lagoas Formation (Bambui Group), northern Minas Gerais State, Brazil: evidence of a cap carbonate deposited on the Januária basement high. Braz J Geol 47(1):59–77

Pessoa PFP (1996) Caracterização hidrogeológica da região cárstica de Sete Lagoas - MG: Potencialidades e Riscos. Dissertation, Universidade de São Paulo

Pessoa PFP (2005) Hidrogeologia dos aquíferos cársticos cobertos de Lagoa Santa, MG. Thesis, Universidade Federal de Minas Gerais

Ribeiro JH, Tuller MP, Danderfer Filho A (2003) Mapeamento geológico da região de Sete lagoas, Pedro Leopoldo, Matozinhos, Lagoa Santa, Vespasiano, Capim Branco, Prudente de Morais, Confins e Funilândia, Minas Gerais (1:50.000) – In: Projeto Vida, CPRM, Belo Horizonte (CD-ROM)

Rimann ET (1917) A kimberlita no Brasil. Anais da Escola de Minas, Ouro Preto 15:27–32

Schobbenhaus C (1984) Geologia do Brasil. Departamento Nacional de Produção Mineral, Brasília

Schöll WU, Fogaça ACC (1979) Estratigrafia da Serra do espinhaço na região de Diamantina. In: I Simpósio de Geologia de Minas Gerais Sociedade Brasileira de Geologia/MG, Belo Horizonte, pp 55–73

Sial AN, Dardenne MA, Misi A, Pedreira AJ, Gaucher C, Ferreira VP, Silva Filho MA, Uhlein A, Pedrosa-Soares AC, Santos RV, Egydio-Silva M, Babinski M, Alvarenga CJS, Fairchild TR, Pimentel MM (2010) The São Francisco Palaeocontinent. In: Gaucher C, Sial AN, Halverson GP, Frimmel HE (eds) Neoproterozoic-Cambrian tectonics, global change and evolution: a focus on South Western Gondwana. Developments in Precambrian Geology, vol 16. Elsevier, Amsterdan, pp 31–69

Tayer TC, Velásques LNM (2017) Assessment of intrinsic vulnerability to the contamination of karst aquifer using the COP method in the Carste Lagoa Santa Environmental Protection Unit, Brazil. Environ Earth Sci 76:445

Teixeira W, Sabaté P, Barbosa J, Noce CM, Carneiro MA (2000) Archean and Paleoproterozoic tectonic evolution of the São Francisco Craton, Brazil. In: Tectonic evolution of South America —International Geological Congress, 31, 2000, Rio de Janeiro. Sociedade Brasileira de Geologia, Anais, Rio de Janeiro, pp 101–137

Tonietto SN (2010) Diagenesis and hydrothermal rocks in Proterozoic carbonates: Bambuí and Vazante Groups, São Francisco Basin. Dissertation, Universidade de Brasília

Trompette RR, Uhlein A, Selva ME, Karmann I (1993) O Cráton brasiliano do São Francisco - Uma revisão. Rev Bras Geoc 22:481–486

Tuller MP, Ribeiro JH, Signorelli N, Féboli WL, Pinho JMM (2010) Projeto Sete Lagoas – Abaeté, Estado de Minas Gerais. CPRM, Belo Horizonte

Uhlein A, Lima ONB, Fantinel LM, Baptista MC (2004) Excursão 2. Estratigrafia e evolução geológica do Grupo Bambuí, Minas Gerais. In: 42 Congresso Brasileiro de Geologia, Sociedade Brasileira de Geologia, Araxá

Uhlein A (2013) Coberturas cratônicas proterozóicas do Brasil. Boletim de Geociências da Petrobras 21(1):9–38

Viana, HS, Tavares VP, Kohler HC (1998) Síntese da geologia, recursos minerais e geomorfologia. In: APA Carste de Lagoa Santa - Meio físico. CPRM/IBAMA, Belo Horizonte

Vieira LC, Trindade RIF, Nogueira ACR, Ader M (2007) Identification of a Sturtian cap carbonate in the Neoproterozoic Sete Lagoas carbonate platform, Bambuí Group, Brazil. Comptes Rendus, Geosciences 339:240–258

Warren LV, Quaglio F, Riccomini C, Simões MG, Poiré DG, Strikis NM, Anelli LE, Strikis PC (2014) The puzzle assembled: Ediacaran guide fossil Cloudina reveals an old proto-Gondwana seaway. Geology 42:391–394

The Vegetation of Lagoa Santa Karst

Ana Elisa Brina

Abstract

Lagoa Santa Karst is located in the transition zone of two Brazilian biomes: Atlantic Forest and Cerrado. On limestone outcrops can be found an open deciduous vegetation, conditioned by physical restrictions imposed by low soil thickness and low moisture retention. Around limestone outcrops, in dolines and at cave entrances, deciduous forests occur on soils that present carbonate rock outcrops. Semideciduous forest occurs associated with deeper soils with higher water retention. Cerrado vegetation, typical of Brazilian central highlands, occurs associated with deep poor soils with high aluminium rate. In some dolines among limestone outcrops, shallow seasonal ponds are formed and on these partially flooded eutrophic soils, grow aquatic herb species. Considering taxonomically precise botanical registers of these several habitats, floristic surveys revealed a considerable richness related to the Lagoa Santa Karst region, with more than two thousand species. Regional vegetation phenology reflects the climate dynamic: at the end of the rainy summer, when temperature and photoperiod decrease and wind speed increase, the main phenological event is leaf falling and forest canopy opening. The beginning of a new leaf sprouting phase, to most species, occurs at the transition from dry season to the rainy season. Leaf loss seems to represent the most important strategy of the water economy in forest communities. Community flowering and fruiting present a less marked seasonality but increase at the transition from dry to rainy season. The forests show a peak of litter production from June to September, under water deficit. Leaves are the major constituent of litter composition and are responsible for the return of nutrients to the soil. The knowledge of karst vegetation composition and dynamics can help in the process of decision making involved in the choice of biodiversity protection areas and the valuation of environmental services.

Keywords

Brazilian biomes • Karst vegetation • Floristic • Semideciduous and deciduous forest • Lagoa Santa Karst • Phenology • Litter production • Nutrient cycling • Water economy strategies • Biodiversity conservation

1 Introduction

The Lagoa Santa Karst is located in the transition zone of two Brazilian biomes: Atlantic Forest and Cerrado (Ab'Saber 1977a; Eiten 1983; IBGE 1993), both considered *Biodiversity Hotspots*—planet's most biologically rich and endangered regions, with high species richness and high level of endemic species with restrict range area (Mittermeier et al. 2004).

The description of the vegetation presented in this chapter was based on data obtained by several studies already developed in the Environmental Protection Area of Lagoa Santa Karst, particularly in Cauaia Farm, inside a zone considered a priority to biodiversity conservation. The study emphasized the physiognomic type of vegetation more strongly associated with exokarst, adapted to water restrictions imposed by low soil deepness and low moisture retention on limestone outcrops and on shallow soils around them. Taking into consideration the importance of the vegetation associated to limestone outcrops (Figs. 1 and 2) and its interactions with cave's biota, as well as the lack of information about its dynamics, the selected information focuses on basic composition and structure of vegetation, and also aspects linked to phenology and ecophysiology of

A. E. Brina (✉)
Universidade Federal de Minas Gerais, Rua Doutor Juvenal dos Santos 12/302, Belo Horizonte, MG 30380-530, Brazil
e-mail: anaelisa.brina@gmail.com

A. S. Auler and P. Pessoa (eds.), *Lagoa Santa Karst: Brazil's Iconic Karst Region*, Cave and Karst Systems of the World, https://doi.org/10.1007/978-3-030-35940-9_3

Fig. 1 Landscape of Lagoa Santa Karst: forests associated with limestone outcrops; pastures and remaining Cerrado on plain areas. *Photo* Ataliba Coelho

Fig. 2 Dry forest in the dry season: most of the trees without leaves. *Photo* Ataliba Coelho

plant species adapted to exokarst substrate and microclimate conditions.

Since the beginning of the twentieth century, there is evidence of deforestation and the use of fire in the Lagoa Santa region, in order to remove original vegetation and to give place to agriculture and pastures. In the past decades, farms were divided into smaller areas due to inheritance processes or rearranged and converted to agricultural enterprises or recreational or urban areas. Mineral exploitation—currently one of the region's most important sources of economic resources—has modified landscape over the years, eliminating several large limestone outcrops and their associated ecosystems. Besides, several forest remnants are still subject to illegal logging, hunting, and uncontrolled visitation.

More than ever, it is important to correctly manage environmental protection areas, and make efforts to build ecological corridors to maintain the ecological function of these preserved areas. Lagoa Santa Karst Environmental Protection Area was created in 1990 by the Brazilian Federal Government, aiming to conciliate sustainable use of natural resources and protection of natural and cultural heritage represented by speleological, archeological and paleontological sites, associated to limestone outcrops (Fundação Biodiversitas 1996 and chapter "Protection of the Environmental and Cultural Heritage in the Lagoa Santa Karst: History and Challenges", this volume).

The first and classical study about Lagoa Santa karst region vegetation was performed by Danish botanist Eugen Warming (1841–1924). Between 1863 and 1866, while working as secretary of the Danish paleontologist Peter Wilhelm Lund (1801–1880), Warming collected a great amount of plants and informations that came up to one of the first texts in Plant Ecology (Warming 1908) and where can be found the first floristic list of the Brazilian savanna (Cerrado), the *Florula Lagoensis* (Moreira 2015). He not only described floristic composition but also explored aspects of growing forms, morphological patterns, seasonal events—flowering, fruiting, senescence, and leaf sprouting. He classified vegetational types into natural or primitive (forests, fields, wetlands, and lakes) and secondary or introduced vegetation (agriculture and pastures). For Warming, fields (savannah physiognomies) were found on plain and less fertile surfaces; on low lands, wetlands and lakes occurred; forests occurred along rivers and associated with limestone outcrops.

In the past decades, several studies were carried out in this region, such as Pedersoli and Martins (1972) and Rizzini (1979, 1986), about phytogeographical aspects and karst flora; Domingos et al. (1994) and Vasconcelos et al. (1996) about Cerrado; Reis and Barbosa (1993) and Pelli (1994) about ecology and flora of karst lakes; Carvalho (1995) about human perception of changes of the Lagoa Santa Cerrado; Pedralli and Meyer (1994) and Brina and Silva (1996) about floristic and phytosociology of Lagoa Santa vegetation, as a working basis for the Environmental Protection Area of Lagoa Santa karst zoning (Fundação Biodiversitas 1995, 1996); Brina (1998), about dynamics of karst vegetation; Melo (2008) and Melo et al. (2013) contributed to the knowledge of plants in karst areas of the state of Minas Gerais; Moreira (2015) reviewed Warming's plant registers to Lagoa Santa region. Other recent studies, with restricted scientific dissemination, are those associated with environmental assessment studies or to support the selection of protection areas.

2 Vegetation in Lagoa Santa Karst Present Landscape

2.1 Floristic Aspects

Floristic surveys made by Melo (2008) revealed a considerable richness related to the karstic environment, mainly herbaceous plant species—839 species of vascular plants distributed into 95 families and 414 genres. According to Moreira (2015), only 539 (23.7%) from 2,270 of Warming's samples in Lagoa Santa have been collected by botanists after almost 150 years and new additions to the list were only 228 (9.1%); the author recognizes Eugen Warming as the father of floristic studies in Brazil due to the completeness of his work, and emphasizes the importance of collecting intensively; observing details in the field; and working in a collaborative way.

The result of a recent revision of herbarium registers related to Lagoa Santa Karst (Brina et al. 2019), considering species from different habitats with complete and valid taxonomy (genus and species), indicated 2,165 species of 158 botanical families. Among these, 47 species are included in Brazilian official protection list (MMA 2014); some are considered endemic according to Carmo et al. (2018) and REFLORA (2019); or rare (Oliveira Filho 2006; Giulietti 2009); other are included in the Convention on International Trade in Endangered Species of Wild Fauna and Flora (CITES). Besides, many species are important as food resources to native fauna. The distribution of species registers underestimates sampling already performed in the karst region and points to lack of sampling especially in some municipalities as Prudente de Morais, Capim Branco, Pedro Leopoldo, São José da Lapa, and Vespasiano.

According to Melo et al. (2013), the mosaic of different physiognomies including forests and open areas results in a noteworthy plant diversity. During the inventory of angiosperm species in areas of carbonatic rocks outcrops in the upper São Francisco River basin karst region, a total of 456 angiosperm species were registered, 161 of which were herbaceous.

2.2 Vegetation Associated with Limestone Outcrops

On limestone outcrops (Figs. 3, 4 and 5), an open deciduous vegetation is conditioned by low soil thickness and low moisture retention. On spots with rock cracks or soil pockets, vegetation is composed by tree species also present in forests in the surroundings, such as *Cedrela odorata* (cedro), *Myracrodruon urundeuva* (aroeira), *Astronium fraxinifolium* (gonçalo), *Aspidosperma cylindrocarpon* (peroba); on exposed rock areas, grow species whose roots support trees in very shallow soil as *Pseudobombax grandiflorum* (embiruçu) and *Ficus calyptroceras* (gameleira), conducting secondary roots and their assimilation zones even to places far from the base of the trunk (CETEC 1980). Among bushes, *Cnidosculus urens* (cansansão) and *Sapium hematospermum* (leiteiro) stand out. On the rocks, there are succulent plants such as cactuses *Cereus jamacaru* subsp. *calcirupicola* (mandacaru) and *Brasiliopuntia*

Fig. 3 Xerophytic vegetation on limestone outcrops in Cauaia Farm. *Photo* Daniel Souza

Fig. 4 Xerophytic vegetation on limestone outcrops in Experiência da Jaguara Farm. *Photo* Daniel Souza

brasiliensis (palma), and also herb species of genus *Encholirium*, *Sinningia,* and *Hippeastrum*. Present all over the karst region, limestone outcrops and caves associated with them shelter many native fauna species. Despite its low density, this vegetation is rich in species of high scientific interest because of their adaptability to unfavorable habitat conditions. Several species present on limestone outcrops do not occur in other areas, being endemic of these specific

Fig. 5 Vegetation in limestone outcrops: plants adapted to specific soil and microclimate conditions. *Photo* Ataliba Coelho

habitats. Furthermore, these vegetal communities have high scenic value.

2.3 Deciduous Forest—The 'Dry Forests'

Surrounding limestone outcrops, in soils rich of rock outcrops, in dolines and cave entrances (Figs. 6, 7, 8, and 9), dry forests develop, composed by species that loose completely their leaves on dry season. Dry forests are found over central areas of Brazilian Cerrado, under the same climate regime. They are common in the states of Minas Gerais, Goiás, Mato Grosso, and Bahia. However, they receive less attention from researchers when compared with other tropical forests.

Some authors state that dry forests are relicts of past Caatinga, from when dry climate alternated with more humid phases during the Quaternary Period (Bigarella et al. 1975; Ab'Saber 1977b; Rizzini 1986; Andrade-Lima 1989; Eiten 1990; Oliveira 1992; Parizzi 1993). Fernandes and

Fig. 6 Dry forests around limestone outcrops, in the rainy season. *Photo* Ataliba Coelho

Fig. 7 Dry forest around a limestone outcrop and xerophyte plant species on the rocks. *Photo* Ataliba Coelho

Fig. 8 Deciduous forest around limestone outcrops during the dry season. *Photo* Daniel Souza

Bezerra (1990) state that Caatinga would have been dispersed through xeric corridors during dry periods, maintaining itself in contact with other vegetation and promoting interpenetrations of some elements that would have fixed themselves since then in both areas. As the vegetation typical from dry areas retracted or the medical vegetation expanded, some spots of vegetation similar to caatinga would have remained in the form of disjunctions or relicts.

In deciduous forests of Lagoa Santa Karst, Shannon diversity index corresponds to 3.06–3.17. Trees range between four and ten meters high, some of them reaching 20 m; most diameters range between five and ten centimeters.

Fig. 9 Interior of the deciduous forest during the dry season. *Photo* Daniel Souza

Some trees of the forest canopy stand out: *Myracrodruon urundeuva* (aroeira), *Anadenanthera peregrina* (angico), *Astronium fraxinifolium* (gonçalo), *Cedrela odorata* (cedro), *Machaerium scleroxylon* (jacarandá), *Ceiba pubiflora* (paineira), *Jaracatia spinosa* (jaracatiá), *Aspidosperma polyneuron* (peroba), *Centrolobium tomentosum* (araribá), *Trichilia catigua* and *T. pallida* (catiguá), *Guettarda viburnoides* (veludo), *Randia armata* (espeto), *Senegalia polyphylla* (acácia), *Cereus jamacaru* (mandacaru), *Dilodendron bipinnatum* (maria-pobre), *Celtis brasiliensis* (grão-de-galo), *Sterculia striata* (chichá), *Ficus calyptroceras* (gameleira), *Guazuma ulmifolia* (mutamba). The cactus *Brasiliopuntia brasiliensis* (palma) is frequent at the understory. Storage structures as bulbs and rhizomes of some annual herbal species such as *Calathea* sp., *Costus* sp. and *Dorstenia* sp. allow them to survive under adverse habitat conditions (Crawley 1986; Seghieri et al. 1995).

Some species of the Lagoa Santa Karst region are also common in arid regions of the Brazilian Caatinga: *Astronium fraxinifolium*, *Handroanthus impetiginosus* Mattos, *Callisthene fasciculata* Mart., *Anadenanthera colubrina* var. *cebil*, *Platypodium elegans*, *Guazuma ulmifolia*, *Hymenaea courbaril*, *Peltophorum dubium*, *Machaerium acutifolium*, *Jatropha* sp., *Cereus* sp. and *Opuntia* sp. *Myracrodruon urundeuva*, considered endemic from caatinga (Fernandes and Bezerra 1990) and present in the list of Brazilian endangered species as 'Vulnerable' because of habitat destruction and declining populations, is dominant in dry forests in Lagoa Santa Karst (Brina and Silva 1996).

2.4 Semideciduous Forest

Semideciduous forest occurs along river margins or as a continuous transition to deciduous forest, associated with deeper soils with higher water retention (Figs. 10, 11 and 12). Leaf loss is only partial. Shannon diversity index corresponds to 3.05–3.72. Besides species present in deciduous forest, some other species stand out due to Significant Value Index, that expresses species frequency (spatial distribution), dominance (size) and density (spatial occupation): *Senegalia polyphylla* (monjoleiro), *Platycyamus regnellii* (pau-pereira), *Peltophorum dubium* (canafístula), *Cecropia pachystachya* (embaúba), *Randia armata*, *Cupania vernalis* and *Matayba guianensis*, *Nectandra lanceolata*, *Acosmium cardenasii*, *Holocalyx balansae*, *Machaerium lanceolatum*, *Psidium* sp., *Aspidosperma cylindrocarpon* (canafístula). Trees range between 4 and 20 m high and have medium diameters that range from 5 to 10 centimeters; some are emergent, reaching more than 20 m in height and 30 cm in diameter.

Different degrees of leaf loss in deciduous and semideciduous forests are due to different contributions of totally deciduous trees, especially at the high canopy. In the semideciduous forest, the trees possibly have access to more

Fig. 10 Deciduous forest on the upper part of limestone outcrops and semideciduous forest on the surroundings. *Photo* Ataliba Coelho

Fig. 11 Semideciduous forest in dry season *Photo* Daniel Souza

Fig. 12 Interior of semideciduous forest in dry season *Photo* Daniel Souza

available groundwater, through larger or deeper root systems (Jackson et al. 1995; Seghieri et al. 1995).

2.5 Herbal Vegetation Associated to Dolines and Ponds

In some dolines among limestone outcrops, shallow ponds are formed (Fig. 13, 14 and 15), generally restricted to the rainy season. The density of vegetal covering depends on the level of eutrophication and on the depth of the water body. On these partially flooded, eutrophic soils, herbal vegetation includes species that grow on wet soils like *Ludwigia* sp. (cru-de-malta), *Juncus* sp. (junco), *Typha* sp. (taboa), *Cyperus* sp. (tiririca), *Polygonum* spp. (erva-de-bicho) and *Echinodorus grandiflorus* (chapéu-de-couro); those floating species such as *Eichornia crassipes* (aguapé), *Lemna* sp. (lentilha-d'água), *Pistia stratiotes* (alface-d'água) and

Fig. 13 Herbal vegetation in the seasonally flooded doline

Fig. 14 Pond located in Confins. *Photo* Daniel Souza

Fig. 15 Sumidouro Lake pond in the dry season. *Photo* Daniel Souza

Salvinia auriculata (salvínia); the floating free or fixed on substrate like *Nymphaea ampla*, *Nymphoides indica* (ninfeias); and those fixed to substrate like *Ceratophyllum* sp. (cabelo-de-urso).

2.6 Cerrado

Cerrado vegetation (Figs. 16, 17, and 18), typical of Brazilian central highlands, occurs associated with deep and poor soils

Fig. 16 Cerrado in Confins. *Photo* Daniel Souza

Fig. 17 Cerrado in Confins. *Photo* Daniel Souza

Fig. 18 Remnants of Cerrado in Lagoa Santa. *Photo* Daniel Souza

with a high aluminum rate. Cerrado is the second largest Brazilian biome, with two million km^2 in area (Ratter et al. 1997). In the Lagoa Santa Karst region, cerrado remnants are three to eight meters high, and the rare trees that exceed this height are *Tapirira guianensis* (pau-pombo), *Platypodium elegans* (jacarandá-canzil), *Luehea grandiflora* (açoita-cavalo), and *Myracrodruon urundeuva* (aroeira). Shannon diversity index corresponds to 2.85–3.74. Richness and density vary between remnants, due to conservation degree or time of regeneration after disturbances, particularly fire or cattle entrance.

In the arboreal stratum of the vegetation, can be found species such as *Qualea grandiflora* and *Qualea parviflora* (pau-terra); *Kielmeyera coriacea* (pau-santo); *Erythroxylum tortuosum* (mercúrio); *Hymenaea stigonocarpa* (jatobá-do-cerrado); *Stryphnodendron adstringens* (barbatimão); *Eugenia dysenterica* (cagaiteira); *Salvertia convallariodora* (bate-caixa); *Schefflera macrocarpa* (mandiocão); *Bowdichia virgilioides* (sucupira-do-cerrado); *Vochysia elliptica* and *V. rufa* (pau-de-tucano).

3 Phenology, Litter Production, and Nutrient Cycling in Deciduous and Semideciduous Forests

Seasonal variation of some climatic events is illustrated in Fig. 19. At the end of the rainy summer (October–March), temperature decreases (A), photoperiod gets shorter and wind velocity increases, particularly in September (B). This is the period of forest canopy opening (C), due to leaf falling that reaches its maximum in July/August, peak of dry season (D). The decrease of leaf falling corresponds to the beginning of a new leaf sprouting phase that to most species occurs at the transition from dry season to rainy season. Temperatures rise again, the period of solar radiation increases and winds are weaker. Community flowering and fruiting present a less marked seasonality but increase at the transition from dry to rainy season.

Karst vegetation is subject to soil water deficit, due to shallow and rocky substrate; as a response to environmental conditions, plants usually show marked phenological events, high synchronicity inside and between species; concentrating productivity efforts on more favorable periods in terms of humidity, temperature, soil nutrients availability, higher dispersal agents activity, and low impact of herbivory.

Leaf sprouting in most species occurs immediately before or soon after the beginning of the rainy season, since the enhanced humidity helps to avoid water stress in young leaves and leaf expansion is promoted by high temperature and luminosity. A rapid leaf expansion reduces the time of vulnerability to herbivory (Coley and Kursar 1996). Leaf sprouting synchronicity represents a massive food resource offering to herbivores, satisfying their hunger and avoiding excessive damages into few individuals or populations (Janzen 1976; Wolda 1978; Aide 1988; Coley and Kursar 1996). In addition to this, the constitution of a dense treetop and a broad root system during the high productivity season

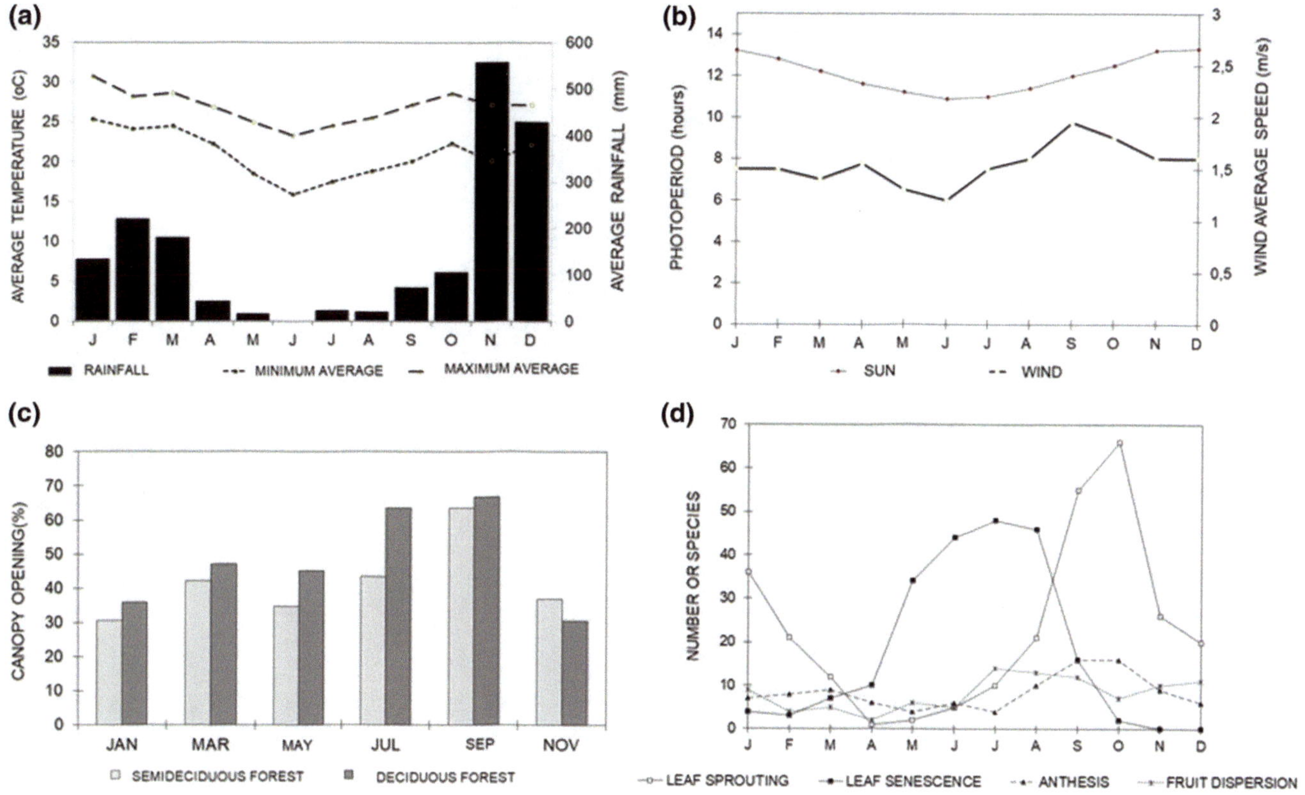

Fig. 19 Main climate and phenological events on the vegetation associated with Lagoa Santa Karst region. **a** Medium temperature and precipitation; **b** sun hours and wind velocity; **c** canopy opening at the deciduous and semideciduous forest; **d** range of senescence and leaf sprouting, flowering and fruiting at forests and limestone outcrops

is a quick response to resource availability variation inside the habitat; a competitive ability in plants (Grime 1986). Sprouting and flowering of herbal annual plants (Fig. 20) use accumulated nutritional reserves (Batalha et al. 1997), accumulated in the underground organs of these plants (Mantovani and Martins 1988). Dormancy break in these species can be related to temperature increase and variation of the dry weight of the underground organs (Figueiredo and Dietrich 1981).

Flowering is favored by conditions as long day length, high pollinator insects' activity, absence of strong rains that break flowers or dilute nectar, and leaf fall that makes the flower more evident (Janzen 1976; Bawa 1974). Most species flowering is short and intense, thus tending to attract many pollinators (Mosquin 1971; Gentry 1974; Rathcke and Lacey 1985; Machado et al. 1997). Flowering peak during the rainy season is a strategy for species subject to water restriction (Ivizi and Araújo 1995, Borchert 1996). Earlier flowering of some species in relation to the community as a whole helps to avoid competition for pollinators (Wheelwright 1985). Flowering during the dry season, while trees lack their leaves, is observed in species of *Pseudobombax* and *Handroanthus*, as well as in *Myracrodruon urundeuva*, dominant species in deciduous forests. This flowering pattern let flowers more visible to pollinators (Van Schaik et al. 1993; Borchert 1996; Batalha et al. 1997; Machado et al. 1997). Increased flowering in forest understory during the dry season is related to light distribution that can increase because of canopy opening due to leaf loss.

Dispersion of fruits by wind as occurs with *Ceiba speciosa*, *Centrolobium tomentosum*, *Cedrela fissilis* and *Senegalia polyphylla*, is more significant at higher stratum of forests, especially on dry season when most species loose leaves, opening canopy (Janzen 1976; Frankie et al. 1974; Augspurger 1986; Morellato et al. 1989; Morellato and Leitão Filho 1990; Araújo et al. 1995; Kageyama et al. 1995; Sun et al. 1996; Fischer 1997). Fleshy fruits, dispersed by animals (Fig. 23), are more frequent in the understory, and get ripe during rainy season, with more heat and humidity (Smythe 1970; Hilty 1980; Mantovani and Martins 1988; Willson et al. 1989; Murali and Sukumar 1994; Sun et al. 1996). Fruit dispersion during the rainy season minimizes the risk of seedling death (van Schaik et al. 1993). The different extension of the reproductive period is related

Fig. 20 Some annual herbs in dry forests have underground reserve organs and sprout on rainy season

mainly to the fruit type: species of long term reproduction are those whose dry fruits get ripe slowly, such as *Anadenanthera colubrina* var. *cebil*, *Platymiscium pubescens*, *Aspidosperma ramiflorum*, *Cedrela odorata,* and *Cedrela fissilis*; fleshy fruits as those of Myrtaceae (Fig. 21) show a shorter period of ripening.

Some species of karst vegetation have long reproductive time, offering important food resources during less favorable times of the year. It is the case of *Psychotria* sp., *Nectandra megapotamica*, *Ficus calyptroceras*, *Piper amalago*. Some of them are bat pollinated—*Luehea*, *Pseudobombax* e *Ceiba* (Sazima et al. 1982)—or bat dispersed—*Cereus* (Figs. 21 and 22), *Ficus* (Fig. 24) and *Piper* (Galetti and Morellato 1994; Figueiredo 1996; Valiente-Banuet et al. 1996). Possibly they represent a relevant source of energy to save life when carried underground by bats.

Leaf fall (Figs. 25 and 26) represents the reduction of assimilation surface and increased nutrient return to the soil. The main factors related to foliar senescence are decreased daylight period, low temperatures, water deficit, and exhaustion of immediately available soil nutrients (Proctor et al. 1983; Larcher 1986; Morellato 1992). The Lagoa Santa Karst forests show a peak of litter production from June to September, under water deficit (Fig. 27). Leaves are the major constituent of litter composition and are responsible for the return of nutrients to the soil (Table 1, Golley et al. 1978).

Litter production has an important biological meaning, because it reflects the environmental conditions of the whole ecosystem (Oliveira and Lacerda 1993). After the peak of litter production, on the transition from dry to rainy season, litter decomposition is accelerated (Alvarez-Sanchez and Guevara-Sada 1993), and nutrients are made available again to vegetation, favoring flowering, fruiting, and seed sprouting.

Nutrient transferring to plants (Table 1) is favored by elevated transpiration rates due to high temperatures, and mineral availability due to intense decomposition (Wolda 1978; Lieberman and Lieberman 1984; Morellato et al. 1989; van Schaik et al. 1993; Wright and van Schaik 1994).

The higher nitrogen return to soil through litter occurred at the end of the dry season and beginning of the rainy season, related to the contribution of leaves, branches, and reproductive parts (Fig. 28). The amount of N found in litter indicates that this element is not a limiting factor in studied forests (Vitousek 1982). Possibly, the abundance of potentially nitrogen-fixing plants as Leguminosae contributes to the high level of this element.

Phosphorus content of litter is one of the highest found in literature about Brazilian forests and showed relation to leaf and reproductive structures fractions of the litter. This element is one of those with a small reservoir and high cycling speed and has been considered limiting in tropical forests (Golley et al. 1978; Attiwill and Adams 1993). This does not

Fig. 21 Undersory fleshy fruits of Myrtaceae (Eugenia sp.) ripened in the rainy season

seem to be the case in the studied forests, which can be considered nutritionally rich. The content of Potassium in the litter is among the lowest found in the literature and was related to the leaves and reproductive structures litter fractions. Despite the fact that, in general, this element deposition is high, it cycles quickly and is leached by pluvial water (Golley et al. 1978; Barbosa and Fearnside 1996). Calcium amount in litter was higher than literature references and this suggests its accumulation in plants that are developed on soils very rich in this element, which was related to leaf and branch fractions of litter, the main structural material from trees (Oliveira and Lacerda 1993; Barbosa and Fearnside 1996; Songwe et al. 1997). Calcium belongs to the group of elements that show big reservoirs and low cycling rate (Golley et al. 1978) and is not translocated before leaf fall. Magnesium showed high values in litter, strongly related to the litter fractions of leaves and branches; this was an expected relationship since leaves and green branches have Mg in chlorophyll and so the soil reservoir of this element is, in general, large (Golley et al. 1978).

4 Water Economy in Plants of the Exokarst

In dry regions, periods of leaf sprouting represent a water stress event even when there is available soil water (Reich and Borchert 1984). Shallow soils and limestone outcrops in the Lagoa Santa Karst region make it difficult for plants to get water for their root systems, especially during the dry season, from April until October. So, plants associated with exokarst display different strategies to avoid water stress during the dry season.

Leaf area reduction during the dry season as a water economy strategy can happen as total or partial canopy leaf loss. In the case of partial leaf loss, the remaining leaves can diminish radiation effect reducing their angle in relation to sun explosion (Mulkey and Wright 1996). The high proportion of deciduous species gives vegetation associated with limestone outcrops a more intense seasonal variation if compared to other forests outside karst regions and makes them similar to the vegetation of arid zones as in the Brazilian Caatinga biome.

Fig. 22 Flowering *Cereus jamacaru* subsp *calcirupicola*

Fig. 23 Fruiting *Cereus jamacaru* subsp *calcirupicola*

Leaf loss seems to represent the most important strategy of the water economy in forest communities. At the beginning of the dry season, water steam pressure causes an increase of plant evapotranspiration and a more intense motion of water with nutrients from the soil, roots, and branches to the leaves (Golley et al. 1978). During the dry season, the availability of soil water hampers this process and so the tendency of trees is to lose leaves to decrease

Fig. 24 *Ficus calyptroceras* (gameleira), one of the most frequent trees of limestone outcrops, offers fruits all over the year

Fig. 25 Dry season: population of *Myracrodruon urundeuva* (aroeira) without their leaves

evapotranspiration. At the end of the dry season, despite water deficit and low availability of soil nutrients, plants have nutritional reserves accumulated at the beginning of the dry season to promote intense flowering and fruiting. Thus, leaf loss allows rehydration of tissues, and flowering or leaf sprouting before the end of the dry season (Borchert 1983; Reich and Borchert 1984; Lemos Filho 1992).

Leaf loss in species of limestone outcrops occurs at the beginning of the dry season, probably as a response to water deficit in rocky soils. In forests of the surroundings, understory plants are less exposed to factors such as insolation and winds, and this can delay or diminish the leaf loss period (Reich et al. 1992).

Among morphological strategies, besides leaf loss (Fig. 29), some plant species have extensive root systems (Fig. 30) and low-density wood that help to maintain tissues inside water, in trees such as *Ficus calyptroceras*, *Cedrela odorata,* and *Pseudobombax grandiflorum*. These features

Fig. 26 Dry season: accumulated litter on soil of the dry forest. *Photo* Ataliba Coelho

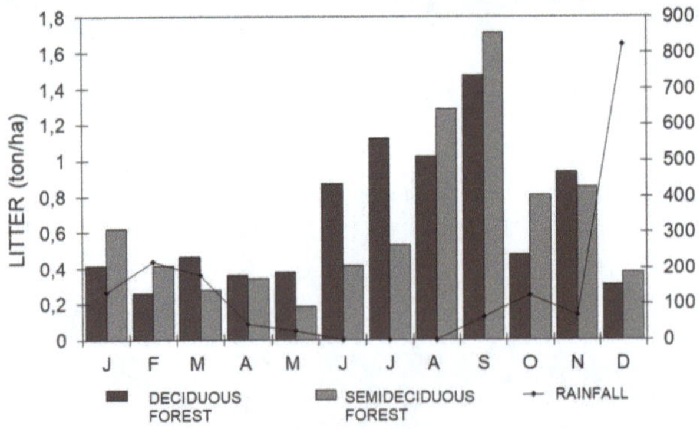

Fig. 27 Litter production in the deciduous and semideciduous forest in the Lagoa Santa Karst

contribute to less variations of hydric potential in both dry and rainy season (Holbrook et al. 1995; Medina 1995).

Other strategies include the presence of bulbs or tuberous roots that accumulate water. Xerophytic plants such as succulents of families Cactaceae and Bromeliaceae, frequent in limestone outcrops, also use water economy strategies, maintaining elevated water potential during the entire year, concentrating flowering in the dry season and growing restricted to the rainy season (Grime 1986).

Some species show important physiological strategies to avoid water stress (hydric relations stability, drought tolerance, periodic growing, etc.); stomatic control is similar to that of dry vegetation of Brazilian Caatinga (arid zones vegetation). For three studied species—*Ficus calyptroceras*, *Pseudobombax grandiflorum,* and *Myracrodruon urundeuva*, there is low seasonal or diary variation in water potential as well; low stomatic conductance was observed in *M. Urundeuva*; and preventive behavior in relation to transpiratory demand increase was observed in *P. grandiflorum* and *Ficus calyptroceras* (Fig. 31).

5 The Conservation of Lagoa Santa Karst Vegetation

Lagoa Santa Karst vegetal cover is gradually becoming more fragmented, making it difficult to maintain landscape permeability to native fauna fluxes. Thus, territory planning may consider degraded areas rehabilitation and the creation of protected areas to favor the connection between cerrado and forests remnants and biodiversity conservation. Planning actions to protect vegetation associated with the exokarst put into discussion the need of attaching an economic value comparable to that of the activities which take place in sites originally occupied by dry forests. Traditionally, arguments in favor of the protection of limestone outcrops are linked to the protection of speleological, archeological and paleontological sites. However, the importance of the vegetation to the exchange process between exokarst and trophic relations

Table 1 Monthly contribution of each litter fraction on the soil of deciduous and semideciduous forest associated with limestone outcrops in Cauaia Farm, Lagoa Santa Karst

Semideciduous forest											
kg/ha	Leaves		Branches		Flowers		Fruits		Other		Total
Month	M	SD	M	SD	M	SD	M	SD	M	SD	M
January	231.74	101.14	338.09	324.95	5.11	16.37	7.06	12.29	39.51	23.92	621.52
February	208.18	91.30	163.61	427.24	2.07	8.83	3.90	12.95	36.94	84.27	414.70
March	179.22	94.81	74.34	73.96	0.80	1.98	1.35	2.67	22.90	16.13	278.62
April	228.66	98.61	72.41	136.81	1.96	4.51	1.16	5.79	37.83	17.83	342.03
May	106.82	53.93	32.09	53.45	2.89	8.31	16.88	58.42	28.87	12.35	187.55
June	287.50	136.48	68.07	61.45	8.24	20.54	12.00	18.90	39.79	61.74	415.60
July	441.53	161.00	55.91	60.08	1.09	2.78	16.02	31.99	13.96	12.96	528.52
August	1014.35	343.03	204.50	213.63	14.41	58.25	39.57	53.09	15.44	11.68	1288.27
September	1202.84	382.67	452.70	748.88	0.24	0.92	40.95	131.94	13.08	12.18	1709.81
October	622.35	281.65	141.09	138.61	0.23	1.16	23.03	53.66	18.87	17.19	805.58
November	570.00	233.65	168.30	156.21	0.36	1.02	35.62	75.51	79.47	86.14	853.76
December	152.39	65.82	158.31	179.54	1.49	6.47	12.76	18.89	51.75	29.06	376.69
Total	5245.59		1929.44		38.90		210.30		398.42		7822.65
Deciduous forest											
January	162.70	102.56	168.31	213.74	0.76	2.41	18.02	69.65	64.21	67.16	414.00
February	157.56	76.42	59.43	58.50	2.01	5.17	8.99	20.08	29.75	18.35	257.74
March	207.95	107.40	208.53	196.43	0.16	0.57	5.05	19.38	42.24	32.07	463.92
April	195.21	102.85	105.23	264.79	2.51	4.54	4.56	9.89	50.41	32.90	357.91
May	265.49	268.53	75.05	196.73	1.31	4.32	0.82	1.99	34.08	36.95	376.75
June	589.61	431.59	220.75	320.92	22.13	102.35	27.57	94.17	12.32	6.83	872.38
July	680.91	554.84	357.63	988.35	31.43	90.66	46.26	189.99	5.60	6.78	1121.83
August	767.26	315.77	139.72	105.17	51.14	67.17	49.46	137.28	14.73	10.68	1022.31
September	1056.51	528.76	366.65	602.54	10.06	18.99	25.15	32.96	16.47	18.03	1474.84
October	265.20	188.69	128.96	139.46	2.13	5.32	52.35	82.55	25.67	20.86	474.30
November	364.87	335.22	499.87	1426.17	3.60	9.82	10.32	17.02	55.82	60.33	934.47
December	121.79	61.26	125.33	101.14	1.33	3.73	6.67	8.09	51.15	62.50	306.26
Total	4835.04		2455.45		128.57		255.21		402.45		8076.72

M mean, *SD* standard deviation

inside caves has already been recognized, as well as the importance of the vegetation shade to rock paintings protection and of providing knowledge about palynology and the evolution of vegetation and its use by extinct megafauna or primitive humans.

In addition to this, it is also important to consider the value of forest ecosystems themselves; the importance of geographically restricted species and their capability of surviving under adverse environmental conditions on limestone outcrops; the species richness and diversity of

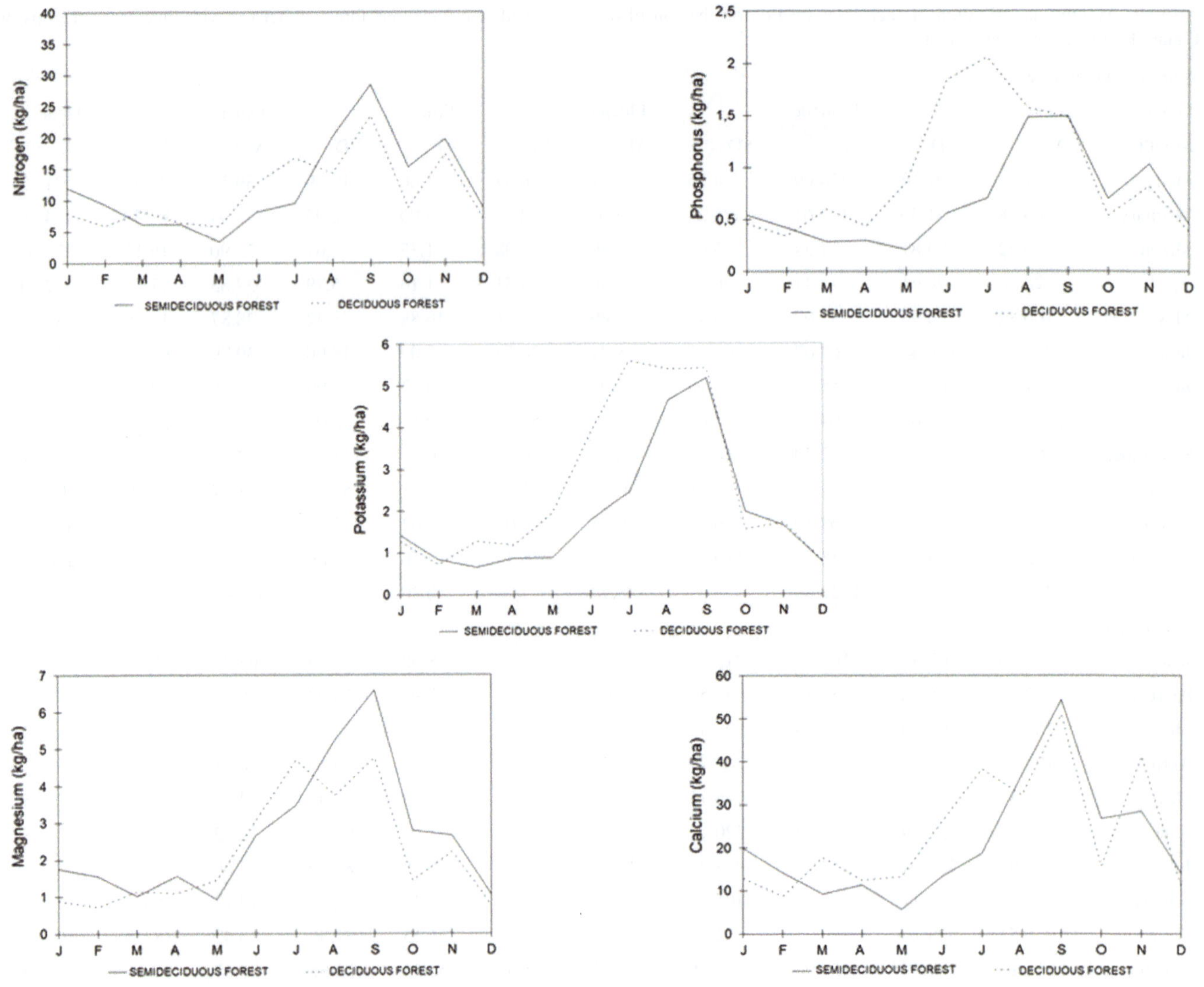

Fig. 28 Return of N, P, K, Ca and Mg to the soil through litter production in the deciduous and semideciduous forest in the Lagoa Santa Karst

interactions between flora and fauna; the importance of preserving sources of seeds and propagating materials of native species; and even scenic or scientific value.

In the context of sustainable development, several initiatives can be stimulated in Lagoa Santa Karst region such as agroforestry, ecological restoration of mining areas using plant species adapted to local conditions of soil and climate, environmental education of human communities and tourism oriented to protecting regional natural heritage.

Carbonate rocks outcrops are speciation places and vegetal diversity reservoirs that show floristic heterogeneity at local and regional scales of analysis, and demand protection in the different biomes where they occur.

Regarding the protection of the buffer areas of caves, maintaining native vegetation around their entrances may be positive to ensure organic input to the interior of caves, to maintain microclimatic conditions, minimize abiotic changes related to light intensity, air and soil temperature and humidity or erosion, and sediment carrying. To define

Fig. 29 Lack of leaves during the flowering of *Handroanthus* sp. in the dry season

Fig. 30 *Pseudobombax grandiflorum* large root system

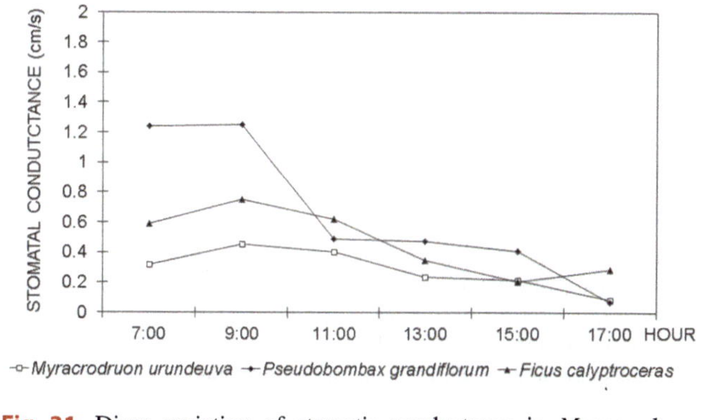

Fig. 31 Diary variation of stomatic conductance in *Myracrodruon urundeuva*, *Pseudobombax grandiflorum* e *Ficus calyptroceras*

protection zones some parameters of vegetation may be considered, such as structural complexity, preservation conditions, connectivity to other preserved areas, and floristic composition, especially related to the presence of species attractive to bats that carry organic material to their interior and generate trophic chains in the guano deposits.

Acknowledgements Many thanks to Ataliba Coelho and Daniel Souza for providing photos, to Pablo Hendrigo Melo for the revision and suggestions, to Leandro Scoss for the analysis of Kernel density of plant registers and to José Pires de Lemos Filho for the orientation during master course.

References

Ab'saber AN (1977a) Os domínios morfoclimáticos na América do Sul. Primeira aproximação. Geomorfologia 52:3–23

Ab'saber AN (1977b) Espaços ocupados pela expansão dos climas secos na América do Sul, por ocasião dos períodos glaciais quaternários. Paleoclimas 3:1–19

Aide TM (1988) Dry season leaf production: an escape from herbivory. Biotropica 24:532–537

Alvarez-Sanchez J, Guevara-Sada S (1993) Litterfall dynamics in a Mexican lowland tropical rain forest. Trop Ecol 34:127–142

Andrade-Lima D (1989) Plantas das Caatingas. Academia Brasileira de Ciências, Rio de Janeiro

Araújo GM, Rodrigues LA, Arantes AA (1995) Fenologia de dez espécies arbóreas de mata decídua em Uberlândia, MG. Resumos do XLVI Congresso Nacional de Botânica. Ribeirão Preto, pp 22–27

Attiwill PM, Adams MA (1993) Nutrient cycling in forests. New Phytol 124:561–582

Augspurger CK (1986) Morphology and dispersal potential of wind-dispersed diaspores of neotropical trees. Am J Bot 73:353–363

Barbosa RI, Fearnside PM (1996) Carbon and nutrient flows in an Amazonian forest: fine litter production and composition at Apiaú, Roraima, Brazil. Trop Ecol 37:115–125

Batalha MA, Aragaki S, Mantovani W (1997) Variações Fenológicas das espécies do cerrado em Emas (Pirassununga, SP). Acta Botanica Brasilica 11:61–78

Bawa KS (1974) Breeding systems of tree species of a low land tropical community and their evolutionary significance. Evolution 28:85–92

Bigarella JJ, Andrade-Lima D, Riehs PJ (1975) Considerações a respeito das mudanças paleoambientais na distribuição de algumas espécies vegetais e animais no Brasil. An Acad Bras Ciênc 47:411–464

Borchert R (1983) Phenology and control of flowering in tropical trees. Biotropica 15:81–89

Borchert R (1996) Phenology and flowering periodicity of Neotropical dry forest species: evidence from herbarium collections. J Trop Ecol 12:65–80

Brina AE (1998) Aspectos da dinâmica da vegetação associada a afloramentos calcários na APA Carste de Lagoa Santa. Dissertation, Universidade Federal de Minas Gerais

Brina AE, Silva LVCS (1996) Flora da APA Carste de Lagoa Santa. In: Fundação Biodiversitas. APA Carste de Lagoa Santa - Proposta de Zoneamento - Meio Biótico

Brina AE, Souza DT, Scoss LM (2019) Diagnóstico do território expandido do carste de Lagoa Santa - tema: vegetação e flora. Instituto PROAPA, Belo Horizonte

Carmo FF, Mota RC, Kamino LHY, Jacobi CM (2018) Check-list of vascular plant communities on ironstone ranges of south-eastern Brazil: dataset for conservation. Biodivers Data J 6:e27032

Carvalho PGS (1995) O Cerrado de Lagoa Santa: transformações e a percepção do homem. Dissertation, Universidade Federal de Minas Gerais

CETEC - Fundação Centro tecnológico de Minas Gerais. Superintendência de Ecologia e Engenharia Ambiental (1980) Estudos integrados do Vale do Jequitinhonha - Diagnóstico da situação ambiental do Vale do Jequitinhonha. CETEC, Belo Horizonte

Coley PD, Kursar TA (1996) Anti-herbivores defenses of young tropical leaves: physiological constraints and ecological trade-offs. In: Mulkey SS, Chazdon RL, Smith AP (eds) Tropical forest plant ecophysiology. Chapman & Hall, New York

Crawley MJ (1986) Life history and environment. In: Crawley MJ (ed) Plant ecology. Blackwell Scientific Publications, London

Domingos DS, Gontijo TA, Andrade MP, Vasconcelos, MF (1994) Efeito do fogo sobre a fauna de térmitas em cerrados de Lagoa Santa, MG. Resumos da 46a Reunião Anual da SBPC, Vitória, p 702

Eiten G (1983) Classificação da Vegetação do Brasil. CNPq - Conselho Nacional de Desenvolvimento Científico e Tecnológico, Brasília

Eiten G (1990) Vegetação do cerrado. In: Pinto MN (ed) Cerrado: caracterização, ocupação e perspectivas. Editora Universidade de Brasília, Brasília

Fernandes AG, Bezerra P (1990) Estudo fitogeográfico do Brasil. Stylus Comunicações, Fortaleza

Figueiredo RA (1996) Vertebrates at neotropical fig species in a tropical fragment. Trop Ecol 37:139–141

Figueiredo RCL, Dietrich SMC (1981) Variações estacionais nos compostos de reserva e no metabolismo do xilopódio de Ocimum nudicaule Benth. Var. anisifolia Giul. (Labiatae). Revista Brasileira de Botânica 4:73–82

Fischer EA (1997) The role of plumes in Eriotheca pentaphylla (Bombacaceae) seed survival in south-eastern Brazil. J Trop Ecol 13:133–138

Frankie GW, Baker HG, Opler PA (1974) Comparative phenological studies of trees in tropical wet and dry forests in the lowlands of Costa Rica. J Ecol 62:881–913

Fundação Biodiversitas (1995) Documento preliminar do Plano de Gestão da Área de Proteção Ambiental (APA) Carste de Lagoa Santa. Fundação Biodiversitas, Belo Horizonte

Fundação Biodiversitas (1996) Proposta de zoneamento para a Área de Proteção Ambiental (APA) Carste de Lagoa Santa. Fundação Biodiversitas, Belo Horizont

Galetti M, Morellato PC (1994) Diet of the large fruit-eating bat Artibeus lituratus in a forest fragment in Brasil. Mammalia 58:661–665

Gentry AH (1974) Flowering phenology and diversity in tropical Bignoniaceae. Biotropica 6:64–68

Giulietti AM (2009) Plantas raras do Brasil. Conservação Internacional, Belo Horizonte

Golley FB, Mc Ginnis JT, Clements RG, Child GI, Duever MJ (1978) Ciclagem de minerais em um ecossistema de floresta tropical úmida. EPU EDUSP, São Paulo

Grime JP (1986) Plant strategies and vegetation processes. Wiley, Chichester

Hilty SL (1980) Flowering and fruiting periodicity in a premontane rais forest in Pacific Colombia. Biotropica 12:292–306

Holbrook NM, Whitbeck JL, Mooney HA (1995) Drought responses of neotropical dry forest trees. In: Bullock SH, Mooney HA, Medina E (eds) Seasonally dry tropical forests. Cambridge University Press, Cambridge

IBGE - Instituto Brasileiro de Geografia e Estatística (1993) Mapa de Vegetação do Brasil. IBGE, Brasília

Ivizi L, Araújo GM (1995) Fenologia de treze espécies arbóreas de mata decídua em Uberlândia, MG. Resumos do XLVI Congresso Nacional de Botânica. Ribeirão Preto

Jackson PC, Cavelier J, Goldstein G, Meinzer FC, Holbrook NM (1995) Partitioning of water resources among plants of a lowland tropical forest. Oecologia 101:197–203

Janzen DH (1976) Seeding patterns of tropical trees. In: Tomlinson PB, Zimmermann MH (eds) Tropical trees as living systems. Cambridge University Press, Cambridge

Kageyama PY, Castilho CAR, Souza LMI, Gandara FB (1995) Dispersão anemocórica em Chorisia speciosa St. Hil. (Bombacaceae). Resumos do XLVI Congresso Nacional de Botânica. Ribeirão Preto

Larcher W (1986) Ecofisiologia vegetal. Editora Pedagógica e Universitária Ltda, São Paulo

Lemos Filho JP (1992) Crescimento de frutos e folhas de Tabebuia ochracia (Cham.) Standl. Anais do 8° Congresso da Sociedade Botânica de São Paulo, pp 27–31

Lieberman D, Lieberman M (1984) The causes and consequences of synchronous flushing in a tropical dry forest. Biotropica 16:193–201

Machado ICS, Barros LM, Sampaio EVSB (1997) Phenology of Caatinga species at Serra Talhada, PE, Northeastern Brazil. Biotropica 29:57–68

Mantovani W, Martins FR (1988) Variações fenológicas em espécies do cerrado da reserva Biológica de Mogi-Guaçu, Estado de São Paulo. Revista Brasileira de Botânica 11:101–112

Medina E (1995) Diversity of life forms of higher plants in neotropical dry forests. In: Bullock SH, Mooney HA, Medina E (eds) Seasonally dry tropical forests. Cambridge University Press, Cambridge

Melo PHA (2008) Flora vascular relacionada aos afloramentos de rocha carbonática no interior do Brasil. Dissertation, Universidade Federal de Lavras

Melo PHA, Lombardi JA, Salino A, Carvalho DA (2013) Composição florística de angiospermas no carste do Alto São Francisco, Minas Gerais, Brasil. Rodriguésia 64:29–36

Ministério do Meio Ambiente – MMA (2014) Portaria 443 de 17 de dezembro de 2014. Diário oficial da União – Seção 1:245

Mittermeier RA, Gil PR, Hoffmann M, Pilgrim J, Brooks T, Mittermeier CG, Lamoreux J, Fonseca GAB. (2004) Hotspots revisited. Earth's biologically richest and most endangered terrestrial ecoregions. CEMEX, Mexico

Moreira NC (2015) A Florula Lagoensis de Warming revisitada. Dissertation, Universidade Federal de Minas Gerais

Morellato LPC (1992) Nutrient cycling in two south-east Brazilian forests. I Litterfall and litter standing crop. J Trop Ecol 8:205–215

Morellato LPC, Leitão Filho HF (1990) Estratégias fenológicas de espécies arbóreas em floresta mesófila na Serra do Japi, Jundiaí, São Paulo. Rev Bras Biol 50:163–173

Morellato LPC, Rodrigues RR, Leitão Filho HF, Joly CA (1989) Estudo comparativo da fenologia de espécies arbóreas de floresta de altitude e floresta mesófila semidecídua na Serra do Japi, Jundiaí, São Paulo. Revista Brasileira de Botânica 12:85–98

Mosquin T (1971) Competition for pollinators as a stimulus for the evolution of flowering time. Oikos 22:398–402

Mulkey SS, Wright SJ (1996) Influence of seasonal drought on the carbon balance of tropical forest plants. In: Mulkey SS, Chazdon RL, Smith AP (eds) Tropical forest plant ecophysiology. Chapman & Hall, New York

Murali KS, Sukumar R (1994) Reproductive phenology of a tropical dry forest in Mudumalai, southern India. J Ecol 82:759–767

Oliveira PE (1992) A palynological record of the late quaternary vegetational and climatic change in Southeastern Brazil. Dissertation, Ohio State University

Oliveira Filho ATde (2006) Catálogo das árvores nativas de Minas Gerais: mapeamento e inventário da floranativa e dos reflorestamentos de Minas Gerais. Lavras: UFLA, 423p

Oliveira RR, Lacerda LD (1993) Produção e composição química da serrapilheira na Floresta da Tijuca (RJ). Revista Brasileira de Botânica 16:93–99

Parizzi MP (1993) A gênese e a dinâmica da Lagoa Santa com base em estudos palinológicos, geomorfológicos e geológicos de sua bacia. Dissertation, Universidade Federal de Minas Gerais

Pedersoli JL, Martins JL (1972) A vegetação dos afloramentos de calcário. Oreades 5:27–29

Pedralli G, Meyer ST (1994) Florística e principais fisionomias na região da área de proteção ambiental (APA) de Lagoa Santa, Minas Gerais, Brasil. Estudos de Biologia 3:91–98

Pelli A (1994) Taxa de crescimento de Salvinia molesta Michel e sua entomofauna associada em um lago do planalto de Lagoa Santa, MG. Dissertation, Universidade Federal de Minas Gerais

Proctor J, Anderson JM, Fogden SCL, Vallack HW (1983) Ecological studies in four contrasting lowland rain forests in Gunung Mulu National Park, Sarawak. II. Litterfall, litter standing crop and preliminary observations on herbivory. J Ecol 71:261–283

Rathcke B, Lacey EP (1985) Phenological patterns of terrestrial plants. Annu Rev Ecol Syst 16:179–214

Ratter JA, Ribeiro JF, Bridgewater S (1997) The Brazilian cerrado vegetation and threats to its biodiversity. Ann Bot 80:223–230

REFLORA (2019) Flora do Brasil 2020 em construção. Jardim Botânico do Rio de Janeiro. Accessedat http://floradobrasil.jbrj.gov.br/reflora/listaBrasil/ConsultaPublicaUC

Reich PB, Borchert R (1984) Water stress and tree phenology in a tropical dry forest in the lowlands of Costa Rica. J Ecol 72:61–74

Reich PB, Walters MB, Ellsworth DS (1992) Leaf life-span in relation to leaf, plant, and stand characteristics among diverse ecosystems. Ecol Monogr 62:365–392

Reis SPW, Barbosa FAR (1993) Estudo da composição de macrófitas aquáticas da Lagoa dos Mares, município de Lagoa Santa, com ênfase em aspectos ecológicos de Salvinia herzogii (Aubl.) Acta Limnol. Brasil. VI:196–208

Rizzini CT (1979) Tratado de Fitogeografia do Brasil, vol 2. HUCITEC & EDUSP, São Paulo

Rizzini CT (1986) Sobre a cactácea dendróide do calcário de Minas Gerais. Rev Bras Biol 46:781–784

Sazima M, Fabián ME, Sazima I (1982) Polinização de Luehea speciosa (Tiliaceae) por Glossophaga soricina (Chiroptera, Phyllostomidae). Rev Bras Biol 42:505–513

Seghieri J, Floret C, Pontanier R (1995) Plant phenology in relation to water availability: herbaceous and woody species in the savannas of northern Cameroon. J Trop Ecol 11:237–254

Smythe N (1970) Relationships between fruiting seasons and seed dispersal methods in a neotropical forest. Am Nat 104:25–35

Songwe NC, Fasehun FE, Okali DUU (1997) Leaf nutrient dynamics of two tree species and nutrient content in Southern Bakundu Forest Reserve, Cameroon. J Trop Ecol 4:25–37

Sun C, Kaplin BA, Kristensen KA, Munyaligoga V, Mvukiyumwami J, Kajondo KK, Moermond TC (1996) Tree phenology in a tropical montane forest in Rwanda. Biotropica 28:668–681

Valiente-Banuet A, Arizmendi MDC, Rojas-Martínez A, Dominguez-Canseco L (1996) Ecological relationships between columnar cacti and nectar-feeding bats in Mexico. J Trop Ecol 12:103–119

Van Schaik CP, Terborgh JW, Wright SJ (1993) The phenology of tropical forests: adaptive significance and consequences for primary consumers. Annu Rev Ecol Syst 24:353–377

Vasconcelos, MF, Gontijo TA, Domingos DJ (1996) A vegetação de cerrado tem hoje a mesma estrutura e composição apresentada há cinco anos atrás? Resumos do XLVII Congresso Nacional de Botânica, Nova Friburgo

Vitousek P (1982) Nutrient cycling and nutrient use efficiency. Am Nat 119:553–572

Warming E (1908) Lagoa Santa. Imprensa Oficial do Estado de Minas Gerais, Belo Horizonte

Wheelwright NT (1985) Competition for dispersers, and the timing of flowering and fruiting in a guild of tropical trees. Oikos 44:465–477

Willson MF, Irvine AK, Walsh NG (1989) Vertebrate dispersal syndromes in some Australian and New Zealand plant communities, with geographic comparisons. Biotropica 21:133–147

Wolda H (1978) Seasonal fluctuations in rainfall, food and abundance of tropical insects. J Anim Ecol 47:369–381

Wright SJ, Van Schaik CP (1994) Light and the phenology of tropical trees. Am Nat 143:192–199

The Lagoa Santa Fauna: Historical Records

Gisele Lessa, Flávia Henriques e Souza, and Natália Lima Boroni

Abstract

Lagoa Santa is one of the most important places in South America in the realm of natural history knowledge. Its renown as the birthplace of Brazilian paleontology is due to the research first undertaken there by Peter Lund in the 19th century. The fauna of the Lagoa Santa Karst is notable for the presence of typical Cerrado savannah species and of the Atlantic Forest domain. Zoologically, the region is extremely important insofar as, in addition to sheltering a peculiar fauna of its own, it is one of the few areas of Brazil for which meticulous inventories have been published. That historical data has made it possible to understand the composition of the groups and to evaluate natural and anthropic-induced changes that have occurred in the course of the past 200 years. Based on an analysis of recent studies, it has been shown that the vertebrate fauna in the region is composed of 70 species of fish distributed among 22 families; 25 species of amphibians belonging to 9 families, and 41 species of reptiles in 14 families. There are also 240 bird species belonging to 58 families and the mammal fauna consists of 107 species belonging to 28 families. In comparison with the historical records of species, there has been a serious reduction in numbers. Evidently, the most important explanatory factor for that reduction is Man's action in deteriorating the natural environment.

Keywords

Vertebrates • Lagoa Santa Karst • Diversity • Historical data

G. Lessa (✉) · F. H. e Souza
Animal Biology Department, Federal University of Viçosa, Viçosa, Brazil
e-mail: gislessa@gmail.com

F. H. e Souza
e-mail: flavia_hs@hotmail.com

N. L. Boroni
Federal University of Minas Gerais, Belo Horizonte, Brazil
e-mail: natalia_boroni@hotmail.com

1 Introduction

As far back as the beginning of the colonial period, the karstic areas in Brazil have been exploited especially for the extraction of saltpeter employed in the manufacture of gunpowder (Travassos et al. 2008). Occasionally, the bone remains of humans and animals were uncovered during the removal of the saltpeter and they caught the attention of the naturalist researchers of those times. Beginning in the nineteenth century, there are records of the first systematic explorations of the caves in Minas Gerais undertaken by Danish naturalist Peter Lund (Paula Couto 1953). The fossilized remains described in those works have formed the basis of paleontological research into Pleistocene vertebrates in Brazil (Paula Couto 1953 and chapter "Cave Paleontology in the Lagoa Santa Karst", this volume).

From the seventeenth century to the eighteenth century, the early colonial settlers in the Lagoa Santa region engaged in mining activities, especially in the quest for gold which was to be found in the Velhas River (Burton 1869). The nineteenth and twentieth centuries were noted for deforestation, clearing the way for pasture land to raise cattle and for the production of charcoal (Cavalcanti et al. 2011). Generally speaking, the karstic environments in Brazil are sites with environmental problems, mainly due to socioeconomic conflicts in the areas where the caves are to be found. The principal causes are ventures or activities that use and occupy the land and the subsoil, such as mining, agriculture, infrastructure, and energy-generating installations, deforestation, urban expansion, and tourism, (Cavalcanti et al. 2011).

The municipality of Lagoa Santa lies within Lagoa Santa Karst Environmental Protection Area (EPA) which in turn is located in the Greater Belo Horizonte Metropolitan Area. The EPA embraces an area of 35,600 ha and parts of the municipalities of Lagoa Santa, Pedro Leopoldo, Matozinhos, Confins and Funilândia (Schobbenhaus et al. 2002). A great quantity of vertebrate remains has been collected from caves

A. S. Auler and P. Pessoa (eds.), *Lagoa Santa Karst: Brazil's Iconic Karst Region*, Cave and Karst Systems of the World, https://doi.org/10.1007/978-3-030-35940-9_4

in the region, especially of mammals and birds (see chapter "Cave Paleontology in the Lagoa Santa Karst", this volume). Today, most of that material is in the Museum of Natural History in Denmark (Lund 1840; Liais 1872).

Some researchers have dedicated themselves to studying the environmental conditions that prevailed in the past in an effort to gain a better understanding of the processes that led to the formation of the present-day landscape (e.g. Lanari 1909; Ab'Sáber 1977; Auler 1999) and based on that to reconstitute the environment in which the animals lived and gain an understanding of the distribution of species among the ecosystems (e.g., Paula Couto 1975; Cartelle 1999). Such studies also make it possible to envisage how living beings responded to changes in the climate, whether they were of natural origin or anthropic origin (e.g. Barnosky et al. 2003). For some authors, geomorphic features and paleobiotic remains indicate important paleoenvironmental changes during the past periods (Auler et al. 2004). The increase in rainfall determined more humid conditions during much of the Pleistocene and late Holocene, indicating the existence of forested links between the Amazon and the Atlantic forests, rich in biodiversity (Auler et al. 2004; Wang et al. 2004; Stríkis et al. 2011). Some works dedicated to reconstructing the paleoenvironmental conditions in Brazil have been based on pollen finds in the sediments of the lakes, rivers and palm swamps (Soubiès et al. 1991; Ledru 1993; Servant et al. 1993; Bauermann et al. 2008) and they attribute the greatest climatic changes that occurred during the Quaternary period to the glacial–interglacial oscillations of the end of the Pleistocene epoch (Perez 2009). The relations animals establish with the environment occur as a function of their physiological, morphological, and behavioral characteristics (Dutra 2000). The presence or absence of suitable habitats frequently determines the distribution of a given population, albeit other factors like ecological interactions can also have an influence (Faunmap Working Group 1996; Ricklefs 2001). Thus, fossils can provide information on the kind of environment that existed in the past. That type of analysis is based on the principle that a species ecological needs such as diet and climate preferences do not change with the passage of time (Davis 1987; Reitz and Wing 2001). Also, knowledge of local fauna can serve as a paleoenvironmental marker to indicate the environmental conditions in which they lived in the past (Perez 2009).

The fossils of still-extant species are those that supply the most accurate paleoenvironmental information. By analyzing a species' behavior and its capacity to adapt to a given environment, it is possible to infer the existence of open grasslands, savannahs, or forests in the past and also the greater or lesser presence of watercourses (Dutra 2000; Bergqvist et al. 2000; Auler et al. 2006). Animals that live in very specific conditions are capable of characterizing a given type of environment that existed in the past merely by their presence in the fossil records (Reitz and Wing 2001). A typical example would be that of species that are restricted to polar regions whose fossilized remains would indicate the cold conditions that existed at that site at the time they were buried down (Yalden 1982).

About climate change in Lagoa Santa, there seems to have been great climatic stability in the region ever since the beginning of the Holocene epoch and up until about three thousand years ago when intense densification of vegetation in the local landscape occurred (Perez 2009). That expansion of denser vegetation can be readily perceived by the relative abundance of *Euphractus sexcinctus*, an armadillo that inhabits regions of open vegetation and the appearance in the fossil records of the tapir *Tapirus terrestris*, which is found in more humid areas with denser vegetation (Perez 2009).

A historical assessment shows that in the period from 12,000 to 10,000 years ago, the planet was living through the end of the last glacial period of the ice age and in Brazil both temperatures and rainfall were much lower than present-day levels (Ledru 1993; Ledru et al. 1996). In the central region of Brazil, where the Lagoa Santa Karst is located, there is even evidence that it may have been a semiarid region in the late Pleistocene epoch with considerably lower rainfall volumes (De Oliveira 1992). Palynological records from lakes and palm swamps indicate that conditions were unfavorable for the formation of permanent aquatic surfaces (De Oliveira 1992). Other palynological records for central Brazil reveal the presence of non-arboreal elements indicative of a prolonged dry season in the course of the year (Salgado-Labouriau et al. 1997). From 12,000 to 7,500 years ago, in the period known as the Mid-Holocene, the records confirm the presence of Pequi trees (*Caryocar brasiliensis*) which together with the gramineae and cyperaceae species confirm the long extension of the dry periods. Possibly, the vegetation cover at that time corresponded to the modern classification of Cerradão (transition between Cerrado savannah and forest vegetation) (Behling 1995; Passacantili 2008). From 7,500 to 4,000 years ago, in the mid-Holocene epoch, the Lagoa Santa region shows evidence of the expansion of forest formations (Kipnis 2002). Finally, from 4,000 years ago to the present day, the palynological records suggest the presence of semi-deciduous forests with open areas and an herbaceous stratum. Around 1,320 years ago, the vegetation was a veritable mosaic with a rich diversity of species (De Oliveira 1992).

Today the fauna in the Lagoa Santa Karst Environmental Protection Area (EPA) is typical of the Cerrado biome but there are also elements more typical of the Atlantic Forest formation. That is because the EPA is in a region of transition between the two biomes and therefore influenced by both (Fundação Biodiversitas 1996). The region is highly important in zoological terms and is a priority conservation area for the Cerrado biome (Figueira et al. 2004). In addition to sheltering its own specific fauna, it is one of the few areas of Brazil where a meticulous inventory was made in the past century, especially of bird and mammal species. The results of that inventory are highly valuable insofar as they can be compared with current data and provide important supporting information for delineating conservation and management strategies for the area (Fundação Biodiversitas 1996).

The portion of the Cerrado biome in this region of the state of Minas Gerais is home to 124 mammal species, which is 78% of the 157 species that occur in the biome as a whole, and of those species that are restricted to the Cerrado alone, 40% can be found in Minas Gerais (Costa et al. 1998).

2 Fish

The Velhas River basin which surrounds the Lagoa Santa region is located in the central part of Minas Gerais and the main source of the river's waters is in the municipality of Ouro Preto. It is the longest tributary in the São Francisco River basin and along its 761 km extension it passes through 51 of that state's municipalities (Polignano 2001). When Lund settled in the region in the first half of the nineteenth century, the region of the upper course of the Velhas River began to receive visits. At Lund's invitation, Johannes T. Reinhardt spent two seasons in the region in the period from 1850 to 1856 when he collected fish specimens that made it possible to scientifically describe around 20 species (Alves and Pompeu 2010) (Table 1). Today the Velhas River is in an advanced state of degradation resulting from the many anthropic activities carried out in its watershed. The most serious pollution occurs in its upper course which receives everything from mining waste to part of the domestic and industrial effluents of the greater Belo Horizonte metropolitan area (Alves and Pompeu 2010). The pollutants discharged into the upper course of the river affect all the downstream course of the river where the water is of very poor quality and there are frequent incidents of fish dying in great numbers (Alves and Pompeu 2010) (Fig. 1).

3 Amphibians

There are 209 known species of anuran amphibians for the Cerrado biome of which 108 are considered to be endemic (Valdujo et al. 2012). The rich variety of the Cerrado herpetofauna has been attributed to environmental heterogeneity; the existence of a mosaic of horizontally contrasting habitats such as open habitats and dry and moist forests, all of which enhance the resources available to the species (Colli et al. 2002; Nogueira et al. 2005, 2009).

In recent times, there has been an increase in the number of studies of anuran assemblages in the more interior parts of the biome (Uetanabaro et al. 2007; Giaretta et al. 2008; Oda et al. 2009; Kopp et al. 2010; Morais et al. 2011; Nomura et al. 2012; Morais et al. 2012; Campos et al. 2013). In the region of the Lagoa Santa Karst, however, there is still very little knowledge of the composition of anuran groups albeit Warming (1908) did conduct a large survey of the vertebrate communities but since his time little work has focused on the animals in question (Table 2 and Fig. 2).

4 Reptiles

In regard to reptiles, there is a greater amount of information available on the snake fauna in the Lagoa Santa region (Table 3). Some authors (Cunha and Nascimento 1978; Sazima 1989) report that in studies of such animals, encounters are generally merely fortuitous because they are continually on the move and hardly ever establish a defined area of use. Generally, the populations are neither large nor concentrated and the snakes are quick to hide from intruders or their persecutors, which makes them difficult to find (Fundação Biodiversitas 1996).

The species found in this region are those typical of the Cerrado biome with its open spaces but some of the species also occur in other biomes such as the Atlantic Forest biome and the Amazon Forest (Peters and Orejas-Miranda 1970; Cunha and Nascimento 1978; Vanzolini 1986; Campbel and Lamar 1989) (Fig. 3). Two species of vipers have been registered, namely *Crotalus durissus* and *Bothrops neuwiedi*. Both are essentially nocturnal in their habits but can sometimes be found in activity during the day. They are poisonous snakes feeding mainly on small mammals, birds, and lizards (Fundação Biodiversitas 1996). The rattlesnake (*Crotalus durissus*) has been registered as occurring throughout the area and is normally found in dry open areas, on stony land, and in pastures. It is the most frequently occurring and most well-known snake among the local people, probably because of its renowned rattle on the tip of the tail, typical of its species. The jararaca-de-rabo-branco, or jararaca pintada (*Bothrops neuwiedi*) (Neuwied's lancehead) is found in more closed environments and usually near to water (Fundação Biodiversitas 1996).

The true coral species (Elapidae) registered for the region are *Micrurus frontalis* and *Micrurus lemniscatus*. They are fossorial or semi-fossorial but come out on the surface, usually in search of food. Nocturnal or crepuscular by habit, they hunt and feed on other snakes and amphisbaenids (worm lizards) using venom to kill their prey (Fundação Biodiversitas 1996). In spite of living in the ground, which

Table 1 List of fish species currently occurring in the Lagoa Santa Karst region

Family	Species
Acestrorhynchidae	*Acestrorhynchus lacustris* (Lütken, 1875)
Curimatidae	*Cyphocharax gilbert* (Quoy and Gaimard, 1824)
	Steindachnerina elegans (Steindachner, 1874)
	Prochilodus costatus (Valenciennes, 1850)
Gymnotidae	*Gymnotus carapo* (Linnaeus, 1758)
Callichthyidae	*Callichthys callichthys* (Linnaeus, 1758)
	Hoplosternum littorale (Hancock, 1828)
Cichlidae	*Cichlasoma facetum* (Jenyns, 1842)
	Tilapia rendalli (Boulenger, 1897)
	Cichla monoculus (Spix and Agassiz, 1831)
	Oreochromis niloticus (Linnaeus, 1758)
Hemiodontidae	*Parodon hilarii* (Reinhardt, 1866)
Anostomidae	*Leporinus elongatus* (Valenciennes, 1850)
	Leporinus reinhardti (Lütken, 1875)
	Leporinus taeniatus (Lütken, 1875)
	Leporinus marcgravii (Lütken, 1875)
	Leporellus vittatus (Valenciennes, 1850)
Characidae	*Characidium lagosantensis* (Travassos, 1947)
	Serrapinnus heterodon (Eigenmann, 1915)
	Astyanax lacustris (Lütken, 1875)
	Serrasalmus brandti (Lütken, 1875a)
	Characidium fasciatum (Reinhardt, 1866)
	Psellogrammus kennedyi (Eigenmann, 1903)
	Hasemania nana (Lütken, 1875)
	Hemigrammus gracilis (Lütken, 1875)
	Deuterodon longirostris (Steindachner, 1907)
	Astyanax fasciatus (Cuvier, 1819)
	Serrapinnus piaba (Lütken, 1875)
	Brycon orthotaenia (Günther, 1864)
	Brycon nattereri (Günther, 1864)
	Roeboides xenodon (Reinhardt, 1851)
	Salminus brasiliensis (Cuvier, 1816)
	Salminus hilarii (Valenciennes, 1850)
	Pygocentrus piraya (Cuvier, 1819)
	Serrasalmus brandtii (Lütken, 1875)
	Myleus micans (Lütken, 1875)
	Hyphessobrycon santae (Eigenmann, 1907)
	Piabina argentea (Reinhardt, 1867)
Heptapteridae	*Imparfinis minutus* (Lütken, 1874)
	Rhamdia quelen (Quoy and Gaimard, 1824)
	Rhamdiopsis microcephala (Lütken, 1874)
	Imparfinis minutus (Lütken, 1874)
Pseudopimelodidae	*Pseudopimelodus charus* (Valenciennes, 1840)
Erythrinidae	*Hoplias malabaricus* (Bloch, 1794)

(continued)

Table 1 (continued)

Family	Species
Sternopygidae	*Eigenmannia microstoma* (Reinhardt, 1852)
	Sternopygus macrurus (Bloch and Schneider, 1801)
Apteronotidae	*Apteronotus brasiliensis* (Reinhardt, 1852)
Sciaenidae	*Pachyurus francisci* (Cuvier, 1830)
	Pachyurus squamipennis (Agassiz, 1831)
Aplocheilidae	*Neofundulus acutirostratus* (Costa, 1992)
Erythrinidae	*Hoplias malabaricus* (Bloch, 1794)
	Hoplias lacerdae (Miranda Ribeiro, 1908)
Poeciliidae	*Pamphorichthys hollandi* (Henn, 1916)
Trichomycteridae	*Stegophilus insidiosus* (Reinhardt, 1859)
	Trichomycterus brasiliensis (Lütken, 1874)
Loricariidae	*Rineloricaria lima* (Kner, 1853)
	Hypostomus lima (Lütken, 1874)
	Hypostomus alatus (Castelnau, 1855)
Doradidae	*Franciscodoras marmoratus* (Lütken, 1874)
Auchenipteridae	*Trachelyopterus lacustris* (Lütken, 1874)
	Glanidium albescens (Lütken, 1874)
Pimelodidae	*Duopalatinus emarginatus* (Valenciennes, 1840)
	Pseudoplatystoma corruscans (Spix and Agassiz, 1829)
	Bagropsis reinhardti (Lütken, 1874)
	Bergiaria westermanni (Lütken, 1874)
	Pimelodus maculatus (Lacepède, 1803)
	Pimelodus fur (Lütken, 1874)
	Pimelodella lateristriga (Lichtenstein, 1823)
	Pimelodella vittata (Lütken, 1874)
incertae sedis	*Conorhynchos conirostris* (Valenciennes, 1840)

Adapted from Warming (1908) and Alves and Pompeu (2010)

makes them difficult to find, they are seen quite often (Fundação Biodiversitas 1996).

Only one species of the Boidae family has been detected in the region of the EPA. The jibóia is a boa constrictor that grows to a large size inhabiting treetops and bushes and rarely coming down to the ground. It is most active at night feeding on small and medium-sized mammals, birds, and lizards that it kills by constriction only, as it is not a poisonous snake (Fundação Biodiversitas 1996).

The Colubridae family has the greatest representation in terms of species. *Drymoluber brazili* and *Phimophis iglesiasi* are both rare species. *D. brazili* has an arboreal habit and is oviparous and omnicarnivorous (Cunha and Nascimento 1978). Little is known about the biology of *P. iglesiasi*, even though it is a typical species of the Cerrado biome (Fundação Biodiversitas 1996).

5 Birds

In the field of ornithology, Johannes Theodor Reinhardt made an impressive collection of bird skins from the Lagoa Santa region and surrounding areas which today is preserved in the Copenhagen Zoological Museum (Warming 1896; Pinto 1952; Silva 1995; Christiansen and Pitter 1997; Rodrigues and Gomes 2004; Rodrigues and Michelin 2005). The number of species he collected at that time is far superior to the number found by recent studies. Reinhardt explored the region in the period from 1847 to 1855 and identified 343 bird species some of which are now considered to be locally extinct (Christiansen and Pitter 1997; Rodrigues and Goulart 2005).

For the Cerrado biome as a whole, Silva (1995) conducted a survey that registered 837 species of birds

Fig. 1 Fish species found in Lagoa Santa Karst. **a** *Acestrorhynchus lacustris*, 160 mm (photo by Ivan Sazima), **b** *Characidium fasciatum*, 45 mm (photo by Mucio de Biagio Rebelo), **c** *Tilapia rendalli*, 430 mm (photo by Neil McNicoll), **d** *Gymnotus carapo*, 240 mm (photo by Carvajal-Vallejos), **e** *Hoplosternum littorale*, 210 mm (photo by Planet Catfish), **f** *Cyphocharax gilbert*, 150 mm (photo by Erica Pauls)

Table 2 List of amphibian species currently found in the region of the Lagoa Santa karst

Family	Species
Bufonidae	*Rhinella marina* (Linnaeus, 1758)
	Rhinella schneideri (Werner, 1894)
Caeciliidae	*Siphonops annulatus* (Mikan, 1820)
Cycloramphidae	*Proceratophrys boiei* (Wied-Neuwied, 1824)
	Odontophrynus cultripes (Reinhardt and Lütken, 1862)
Ceratophryidae	*Ceratophrys cornuta* (Linnaeus, 1758)
Dicroglossidae	*Hoplobatrachus tigerinus* (Daudin, 1802)
Hylidae	*Hypsiboas faber* (Wied-Neuwied, 1821)
	Hypsiboas boans (Linnaeus, 1758)
	Hypsiboas lundii (Burmeister, 1856)
	Hypsiboas pardalis (Spix, 1824)
	Hypsiboas polytaenius (Cope, 1870)
	Hypsiboas rubracylus (Cochran and Goin, 1970)
	Scinax nasicus (Cope, 1862)
	Phyllomedusa burmeisteri (Boulenger, 1882)
	Dendropsophus bipunctatus (Spix, 1824)
	Dendropsophus seniculus (Cope, 1868)
	Dendropsophus rubicundulus (Reinhardt and Lütken, 1862)
Hylodidae	*Crossodactylus gaudichaudii* (Duméril and Bibron, 1841)
Leiuperidae	*Physalaemus marmoratus* (Reinhardt and Lütken, 1862)
	Physalaemus signifer (Girard, 1853)
	Physalaemus albifrons (Spix, 1824)
Leptodactylidae	*Leptodactylus ocellatus* (Linnaeus, 1758)
	Leptodactylus mystacinus (Burmeister, 1861)
	Leptodactylus pentadactylus (Laurenti, 1768)

Adapted from Fundação Biodiversitas (1996) and Warming (1908)

Fig. 2 Amphibian species found in Lagoa Santa Karst. **a** *Ceratophrys cornuta*, 120 mm (photo by Lucas Bustamante), **b** *Hoplobatrachus tigerinus*, 150 mm (photo by Geraldo Barros), **c** *Hypsiboas lundii*, 60.4 mm (photo by Fábio Maffei), **d** *Physalaemus albifrons*, 33.8 mm (photo by Renato Gaiga), **e** *Rhinella marina*, 13 mm (photo by Paulo S. Bernarde), **f** *Siphonops annulatus*, 500 mm (photo by Renato Gaiga)

belonging to 64 Families. According to that author, only 3.8% of those species can be considered endemic to the Cerrado. That low number is due to the fact that many of the species also occur in the neighboring biomes (Fig. 4). Sick (1965) reported that it is not very easy to determine what would be the typical avifauna for this biome because several of the species found in it occur in other open formations with different floristic compositions from the Cerrado. Furthermore, many species considered to be typical of the Cerrado biome have been expanding their geographic distribution, especially because of deforestation and the consequent expansion of open areas (Sick 1965).

An analysis of the studies conducted by Rodrigues (2008) gives a good idea of the species that occur in the Lagoa Santa region and, by comparing them to Reinhardt's work towards the end of the nineteenth century, enables us to arrive at some interesting conclusions (Table 4). In Rodrigues's work, species are mentioned that are considered to be rare or important in some way. Among them is the Jacupemba (*Penelope superciliaris*) (Rusty-margined Guan) which in the Brazilian southeast has been drastically reduced in numbers by hunting and deforestation (Sick 1993). Even in large protected areas like the Serra do Cipó National Park, around 40 km to the north of Lagoa Santa, the Jacupemba, which was once considered to be a common species, today has been practically extirpated (Rodrigues et al. 2005). On the other hand, the roseate spoonbill (*Platalea ajaja*) was still represented by a tiny population of 10–20 individuals at the Sumidouro Lake, the largest in the region (Rodrigues and Michelin 2005). Apparently it, too, was once quite common in the region according to Warming's (1896) records, but today it is represented by a scanty population and considered to be a vulnerable species in the state of Minas Gerais (COPAM 2010).

Three Maguary storks (*Ciconia maguari*) were observed on a single occasion at the Sumidouro Lake (Rodrigues et al. 2005). It is possible that the numbers of individuals are dropping because of the destruction of marshland to make way for agriculture (Rodrigues et al. 2005). That species was recorded by Reinhardt but not by Lins et al. (1998), Herrmann et al. (1998) and today it is considered to be a rare species in Minas Gerais (Rodrigues and Michelin 2005).

Only four individuals of the Wood Stork (*Mycteria americana*) have been registered at the Sumidouro Lake (Rodrigues et al. 2005). That species was registered by Reinhardt but today it is considered to be a rare species in Minas Gerais (Rodrigues et al. 2005; COPAM 2010).

Another species now considered to be vulnerable in the state of Minas Gerais (COPAM 2010) is the blue-and-yellow macaw. It is well known that the decline in the population of these birds is due to loss of habitat and illegal trafficking to be sold as pets (Machado et al. 1998; Birdlife International 2000).

A group of many individuals of the Blue-fronted Amazon Parrot (*Amazona aestiva*) was observed (Rodrigues et al. 2005), and the species was also formally registered by Reinhardt (Warming 1896) and by Lins et al. (1998). These birds nest in cavities in the limestone rock outcrops that are scattered around the region (Silva 1995) and are under threat, not only from illegal trafficking but also from the destruction of the rock formations which are raw material for the cement industry. This parrot species is the one most

Table 3 List of reptile species currently found in the Lagoa Santa Karst region

Family	Species
Boidae	*Boa constrictor* (Linnaeus, 1758)
Colubridae	*Apostolepis assimilis* (Reinhardt, 1861)
	Drymoluber brazili (Gomes, 1918)
	Leptodeira annulata (Linnaeus, 1758)
	Helicops modestus (Günther, 1861)
	Sibynomorphus ventrimaculatus (Boulenger, 1885)
	Liophis poecilogyrus (Wied-Neuwied, 1825)
	Mastigodryas bifossatus (Raddi, 1820)
	Oxyrhopus guibei (Hoge and Romano, 1977)
	Philodryas olfersii (Lichenstein, 1823)
	Philodryas patagoniensis (Girard, 1857)
	Phimophis iglesiasi (Gomes, 1915)
	Pseustes sexcarinatus (Wagler, 1824)
	Sibynomorphus mikanii (Schlegel, 1837)
	Simophis rhinostoma (Schlegel, 1837)
	Spilotes pullatus (Linnaeus, 1758)
	Thamnodynastes strigilis (Thunberg, 1787)
	Waglerophis merremii (Wagler, 1824)
	Xenodon rabdocephalus (Wied-Neuwied, 1824)
Elapidae	*Micrurus frontalis* (Duméril, Bibron and Duméril, 1854)
	Micrurus lemniscatus (Linnaeus, 1758)
Viperidae	*Bothrops neuwiedi* (Wagler, 1824)
	Crotalus durissus (Linnaeus, 1758)
Chelidae	*Hydromedusa maximiliani* (Mikan, 1825)
	Acanthochelys spixii (Duméril and Bibron, 1835)
	Phrynops hilarii (Duméril and Bibron, 1835)
Alligatoridae	*Caiman latirostris* (Daudin, 1802)
	Caiman crocodilus (Linnaeus, 1758)
Iguanidae	*Polychrus marmoratus* (Linnaeus, 1758)
	Enyalius bilineatus (Duméril and Bibron, 1837)
Tropiduridae	*Tropidurus torquatus* (Wied-Neuwied, 1820)
Leiosauridae	*Urostrophus vautieri* (Duméril and Bibron, 1837)
Diploglossidae	*Ophiodes striatus* (Spix, 1824)
Teiidae	*Tupinambis nigropunctatus* (Spix, 1825)
	Ameiva ameiva (Linnaeus, 1758)
Gymnophthalmidae	*Heterodactylus lundii* (Reinhardt and Lütken, 1862)
	Colobosaura modesta (Reinhardt and Lütken, 1862)
Amphisbaenidae	*Amphisbaena alba* (Linnaeus, 1758)
	Amphisbaena vermicularis (Wagler, 1824)
Scincidae	*Aspronema dorsivittatum* (Cope, 1862)
	Notomabuya frenata (Cope, 1862)

Adapted from Fundação Biodiversitas (1996) and Warming (1908)

Fig. 3 Reptile species found in the Lagoa Santa Karst. **a** *Acanthochelys spixii*, 153 mm (photo by Gabriel F. Horta), **b** *Aspronema dorsivittatum*, 70 mm (photo by Michel de Aguiar Passos), **c** *Caiman latirostris*, 2000 mm (photo by Marcos Wiese), **d** *Crotalus durissus*, 1500 mm (photo by Pedro Martins), **e** *Micrurus frontalis*, 85 mm (photo by Cristiano Nogueira), **f** *Polychrus marmoratus*, 135 mm (photo by Mauro Teixeira Jr)

Fig. 4 Bird species found in Lagoa Santa Karst. **a** *Platalea ajaja*, 685–865 mm (photo by Jeff Vanuga), **b** *Crypturellus tataupa*, 150 mm (photo by Klaus Rudloff), **c** *Dendrocygna autumnalis*, 480 mm (photo by Wyatt Berka), **d** *Ara ararauna*, 800 mm (photo by Florian Demmer), **e** *Caracara plancus*, 560 mm (photo by Hermann Redies), **f** *Hirundinea ferruginea*, 150 mm (photo by Bertrando Campos)

commonly found in the illegal trafficking of wild species which supplies the pet market in North America (Brien 1996).

6 Mammals

Recent studies (Grelle et al. 1996; Fundação Biodiversitas 1996; Trolle et al. 2007) have registered 42 species of mammals inside the limits of the Environmental Protection Area (Table 5 and Fig. 5). Small mammals (marsupials and rodents) and species of the Chiroptera order make up 28.6% and 23.0%, of the total number of species, respectively. The dominance in species numbers of those groups reflects the fact that the three orders are those that have the highest numbers of species among the Brazilian mammals as a whole (Fonseca et al. 1996).

Actually, much of what is known about small mammals in South America is due to the study of owl regurgitation pellets (*Tyto furcata*), especially in Argentina (e.g., Pardiñas

Table 4 List of bird species currently occurring in the region of the Lagoa Santa Karst

Family	Species
Tinamidae	*Crypturellus parvirostris* (Wagler, 1827)
	Crypturellus tataupa (Temminck, 1815)
	Rhynchotus rufescens (Temminck, 1815)
	Nothura maculosa (Temminck, 1815)
Anatidae	*Dendrocygna viduata* (Linnaeus, 1766)
	Dendrocygna autumnalis (Linnaeus, 1758)
	Cairina moschata (Linnaeus, 1758)
	Sarkidiornis sylvicola (Ihering and Ihering, 1907)
	Amazonetta brasiliensis (Gmelin, 1789)
	Netta erythrophthalma (Wied, 1832)
	Nomonyx dominica (Linnaeus, 1766)
Podicipedidae	*Tachybaptus dominicus* (Linnaeus, 1766)
	Podilymbus podiceps (Linnaeus, 1758)
Ciconiidae	*Ciconia maguari* (Gmelin, 1789)
	Mycteria americana (Linnaeus, 1758)
Phalacrocoracidae	*Nannopterum brasilianus* (Gmelin, 1789)
	Phalacrocorax brasilianus (Gmelin, 1789)
Anhingidae	*Anhinga anhinga* (Linnaeus, 1766)
Ardeidae	*Tigrisoma lineatum* (Boddaert, 1783)
	Ardea alba (Linnaeus, 1758)
	Cochlearius cochlearius (Linnaeus, 1766)
	Nycticorax nycticorax (Linnaeus, 1758)
	Butorides striata (Linnaeus, 1758)
	Bubulcus ibis (Linnaeus, 1758)
	Ardea cocoi (Linnaeus, 1766)
	Ardea alba (Linnaeus, 1758)
	Syrigma sibilatrix (Temminck, 1824)
	Pilherodius pileatus (Boddaert, 1783)
	Egretta thula (Molina, 1782)
Threskiornithidae	*Phimosus infuscatus* (Lichtenstein, 1823)
	Theristicus caudatus (Boddaert, 1783)
	Platalea ajaja (Linnaeus, 1758)
Cathartidae	*Coragyps atratus* (Bechstein, 1793)
	Sarcoramphus papa (Linnaeus, 1758)
Accipitridae	*Rupornis magnirostris* (Gmelin, 1788)
	Elanus leucurus (Vieillot, 1818)
	Leptodon cayanensis (Latham, 1790)
	Rostrhamus sociabilis (Vieillot, 1817)
	Heterospizias meridionalis (Latham, 1790)
Aramidae	*Aramus guarauna* (Linnaeus, 1766)
Rallidae	*Aramides cajaneus* (Statius Muller, 1776)
	Aramides saracura (Spix, 1825)
	Porzana flaviventer (Boddaert, 1783)
	Porzana albicollis (Vieillot, 1819)

(continued)

Table 4 (continued)

Family	Species
	Pardirallus nigricans (Vieillot, 1819)
	Gallinula galeata (Lichtenstei, 1818)
	Porphyrio martinicus (Linnaeus, 1766)
	Porphyrio flavirostris (Gmelin, 1789)
Charadriidae	*Vanellus cayanus* (Latham, 1790)
	Vanellus chilensis (Molina, 1782)
	Pluvialis dominica (Statius Muller, 1776)
	Charadrius semipalmatus (Bonaparte, 1825)
	Charadrius collaris (Vieillot, 1818)
Recurvirostridae	*Himantopus mexicanus* (Statius Muller, 1776)
	Himantopus melanurus (Vieillot, 1817)
Scolopacidae	*Gallinago paraguaiae* (Vieillot, 1816)
	Tringa solitaria (Wilson, 1813)
	Tringa melanoleuca (Gmelin, 1789)
	Tringa flavipes (Gmelin, 1789)
	Calidris melanotos (Vieillot, 1819)
Jacanidae	*Jacana jacana* (Linnaeus, 1766)
Sternidae	*Sternula superciliaris* (Vieillot, 1819)
	Gelochelidon nilotica (Gmelin, 1789)
Columbidae	*Columbina talpacoti* (Temminck, 1811)
	Columbina squammata (Lesson, 1831)
	Patagioenas picazuro (Temminck, 1813)
Cuculidae	*Piaya cayana* (Linnaeus, 1766)
	Tapera naevia (Linnaeus, 1766)
	Coccyzus americanus (Linnaeus, 1758)
	Crotophaga ani (Gmelin, 1788)
	Guira guira (Gmelin, 1788)
Tytonidae	*Tyto furcata* (Temminck, 1827)
Strigidae	*Athene cunicularia* (Molina, 1782)
	Pulsatrix perspicillata (Latham, 1790)
	Glaucidium brasilianum (Gmelin, 1788)
Caprimulgidae	*Hydropsalis albicollis* (Gmelin, 1789)
	Chordeiles sp.
	Antrostomus rufus (Boddaert, 1783)
	Lurocalis semitorquatus (Gmelin, 1789)
Apodidae	*Streptoprocne zonaris* (Shaw, 1796)
	Chaetura meridionalis (Hellmayr, 1907)
Trochilidae	*Phaethornis pretrei* (Lesson and Delattre, 1839)
	Phaethornis ruber (Linnaeus, 1758)
	Colibri serrirostris (Vieillot, 1816)
	Eupetomena macroura (Gmelin, 1788)
	Chlorostilbon lucidus (Shaw, 1812)
	Amazilia lactea (Lesson, 1832)

(continued)

Table 4 (continued)

Family	Species
	Heliomaster squamosus (Temminck, 1823)
Trogonidae	*Trogon surrucura* (Vieillot, 1817)
Momotidae	*Baryphthengus ruficapillus* (Vieillot, 1818)
Alcedinidae	*Megaceryle torquata* (Linnaeus, 1766)
	Chloroceryle amazona (Latham, 1790)
	Chloroceryle americana (Gmelin, 1788)
Galbulidae	*Galbula ruficauda* (Cuvier, 1816)
Bucconidae	*Malacoptila striata* (Spix, 1824)
	Nonnula rubecula (Spix, 1824)
Ramphastidae	*Ramphastos toco* (Statius Muller, 1776)
Picidae	*Picumnus cirratus* (Temminck, 1825)
	Melanerpes candidus (Otto, 1796)
	Picumnus cirratus (Temminck, 1825)
	Dryocopus lineatus (Linnaeus, 1766)
	Campephilus melanoleucos (Gmelin, 1788)
	Veniliornis passerinus (Linnaeus, 1766)
	Colaptes melanochloros (Gmelin, 1788)
	Colaptes campestris (Vieillot, 1818)
Cariamidae	*Cariama cristata* (Linnaeus, 1766)
Falconidae	*Caracara plancus* (Miller, 1777)
	Herpetotheres cachinnans (Linnaeus, 1758)
	Falco sparverius (Linnaeus, 1758)
	Milvago chimachima (Vieillot, 1816)
Cracidae	*Penelope superciliaris* (Temminck, 1815)
Psittacidae	*Psittacara leucophthalmus* (Statius Muller, 1776)
	Ara ararauna (Linnaeus, 1758)
	Forpus xanthopterygius (Spix, 1824)
	Eupsittula aurea (Gmelin, 1788)
	Pyrrhura frontalis (Vieillot, 1817)
	Brotogeris chiriri (Vieillot, 1818)
	Pionus maximiliani (Kuhl, 1820)
	Amazona aestiva (Linnaeus, 1758)
Furnariidae	*Furnarius rufus* (Gmelin, 1788)
	Furnarius figulus (Lichtenstein, 1823)
	Synallaxis frontalis (Pelzeln, 1859)
	Synallaxis albescens (Temminck, 1823)
	Certhiaxis cinnamomeus (Gmelin, 1788)
	Synallaxis spixi (Sclater, 1856)
	Lochmias nematura (Lichtenstein, 1823)
	Anumbius annumbi (Vieillot, 1817)
	Xenops rutilans (Temminck, 1821)
	Automolus leucophthalmus (Wied-Neuwied, 1821)
	Sittasomus griseicapillus (Vieillot, 1818)

(continued)

Table 4 (continued)

Family	Species
	Dendrocolaptes platyrostris (Spix, 1824)
	Lepidocolaptes angustirostris (Vieillot, 1818)
	Xiphorhynchus fuscus (Vieillot, 1818)
	Phacellodomus rufifrons (Wied, 1821)
Thamnopilidae	*Hypoedaleus guttatus* (Vieillot, 1816)
	Taraba major (Vieillot, 1816)
	Thamnophilus caerulescens (Vieillot, 1816)
	Dysithamnus mentalis (Temminck, 1823)
	Herpsilochmus atricapillus (Pelzeln, 1868)
	Formicivora serrana (Hellmayr, 1929)
	Pyriglena leucoptera (Vieillot, 1818)
Conopophagidae	*Conopophaga lineata* (Wied-Neuwied, 1831)
Tyrannidae	*Camptostoma obsoletum* (Temminck, 1824)
	Todirostrum poliocephalum (Wied, 1831)
	Elaenia flavogaster (Thunberg, 1822)
	Elaenia obscura (d'Orbigny and Lafresnaye, 1837)
	Griseotyrannus aurantioatrocristatus (d'Orbigny and Lafresnaye, 1837)
	Myiopagis viridicata (Vieillot, 1817)
	Myiopagis caniceps (Swainson, 1835)
	Phyllomyias fasciatus (Thunberg, 1822)
	Leptopogon amaurocephalus (Tschudi, 1846)
	Myiornis auricularis (Vieillot, 1818)
	Hemitriccus nidipendulus (Wied-Neuwied, 1831)
	Poecilotriccus plumbeiceps (Lafresnaye, 1846)
	Poecilotriccus latirostris (Pelzeln, 1868)
	Corythopis delalandi (Lesson, 1831)
	Platyrinchus mystaceus (Vieillot, 1818)
	Tolmomyias sulphurescens (Spix, 1825)
	Myiophobus fasciatus (Statius Muller, 1776)
	Lathrotriccus euleri (Cabanis, 1868)
	Cnemotriccus fuscatus (Wied-Neuwied, 1831)
	Hirundinea ferruginea (Gmelin, 1788)
	Xolmis cinereus (Vieillot, 1816)
	Xolmis velatus (Lichtenstein, 1823)
	Machetornis rixosa (Vieillot, 1819)
	Colonia colonus (Vieillot, 1818)
	Arundinicola leucocephala (Linnaeus, 1764)
	Casiornis rufus (Vieillot, 1816)
	Myiarchus tyrannulus (Statius Muller, 1776)
	Myiarchus ferox (Gmelin, 1789)
	Sirystes sibilator (Vieillot, 1818)
	Megarynchus pitangua (Linnaeus, 1766)
	Empidonomus varius (Vieillot, 1818)

(continued)

Table 4 (continued)

Family	Species
	Serpophaga subcristata (Vieillot, 1817)
	Pitangus sulphuratus (Linnaeus, 1766)
	Myiodynastes maculatus (Statius Muller, 1776)
	Myiozetetes similis (Spix, 1825)
	Tyrannus melancholicus Vieillot, 1819
	Tyrannus albogularis (Burmeister, 1856)
	Tyrannus savana (Vieillot, 1808)
	Fluvicola nengeta (Linnaeus, 1766)
	Satrapa icterophrys (Vieillot, 1818)
Tityridae	*Pachyramphus polychopterus* (Vieillot, 1818)
	Tityra cayana (Linnaeus, 1766)
	Pachyramphus validus (Lichtenstein, 1823)
Pipridae	*Neopelma pallescens* (Lafresnaye, 1853)
Corvidae	*Cyanocorax cristatellus* (Temminck, 1823)
Hirundinidae	*Pygochelidon cyanoleuca* (Vieillot, 1817)
	Tachycineta leucorrhoa (Vieillot, 1817)
	Alopochelidon fucata (Temminck, 1822)
	Hirundo rustica (Linnaeus, 1758)
	Stelgidopteryx ruficollis (Vieillot, 1817)
	Progne tapera (Vieillot, 1817)
Troglodytidae	*Troglodytes musculus* (Naumann, 1823)
Donacobiidae	*Donacobius atricapilla* (Linnaeus, 1766)
Turdidae	*Turdus leucomelas* (Vieillot, 1818)
	Turdus rufiventris (Vieillot, 1818)
	Turdus amaurochalinus (Cabanis, 1850)
Motacillidae	*Anthus lutescens* (Pucheran, 1855)
Mimidae	*Mimus saturninus* (Lichtenstein, 1823)
Vireonidae	*Cyclarhis gujanensis* (Gmelin, 1789)
	Hylophilus poicilotis (Temminck, 1822)
	Vireo chivi (Vieillot, 1817)
Parulidae	*Setophaga pitiayumi* (Vieillot, 1817)
	Geothlypis aequinoctialis (Gmelin, 1789)
	Basileuterus culicivorus (Deppe, 1830)
	Myiothlypis flaveola (S. F. Baird, 1865)
Cardinalidae	*Cyanoloxia brissonii* (Lichtenstein, 1823)
Emberizidae	*Arremon flavirostris* (Swainson, 1838)
	Ammodramus humeralis (Bosc, 1792)
	Zonotrichia capensis (Statius Müller, 1776)
Icteridae	*Chrysomus ruficapillus* (Vieillot, 1819)
	Sturnella superciliaris (Bonaparte, 1850)
	Cacicus haemorrhous (Linnaeus, 1766)
	Gnorimopsar chopi (Vieillot, 1819)
	Molothrus bonariensis (Gmelin, 1789)

(continued)

Table 4 (continued)

Family	Species
Thraupidae	*Tangara sayaca* (Linnaeus, 1766)
	Tangara palmarum (Wied, 1823)
	Sporophila lineola (Linnaeus, 1758)
	Sporophila bouvreuil (Statius Müller, 1776)
	Emberizoides herbicola (Vieillot, 1817)
	Sicalis luteola (Sparrman, 1789)
	Lanio pileatus (Wied, 1821)
	Tangara cayana (Linnaeus, 1766)
	Conirostrum speciosum (Temminck, 1824)
	Nemosia pileata (Boddaert, 1783)
	Hemithraupis ruficapilla (Vieillot, 1818)
	Tangara sayaca (Linnaeus, 1766)
	Saltator similis (d'Orbigny and Lafresnaye, 1837)
	Schistochlamys ruficapillus (Vieillot, 1817)
	Compsothraupis loricata (Lichtenstein, 1819)
	Sicalis flaveola (Linnaeus, 1766)
	Volatinia jacarina (Linnaeus, 1766)
	Tersina viridis (Illiger, 1811)
	Dacnis cayana (Linnaeus, 1766)
	Sporophila nigricollis (Vieillot, 1823)
Coerebidae	*Coereba flaveola* (Linnaeus, 1758)
Fringillidae	*Euphonia chlorotica* (Linnaeus, 1766)
Estrildidae	*Estrilda astrild* (Linnaeus, 1758)
Passeridae	*Passer domesticus* (Linnaeus, 1758)

Adapted from Fundação Biodiversitas (1996), Warming (1908) and Rodrigues (2008)

and Cirignoli 2002). In Brazil, the use of this method in studies is less common but it has gradually increased over the years (Souza et al. 2010; Bonvicino and Bezerra 2003; Scheibler and Christoff 2007; Roda 2006; Rocha et al. 2011; Motta-Junior and Talamoni 1996). The first registered survey of species via owl pellets in Brazil was in 1836 when Lund collected various fossil and contemporary specimens in limestone caves in Minas Gerais, and that collection was later reassessed by Herluf Winge (Winge 1887; Avila-Pires 1960). One of the most important aspects of Lund and Winge's work is that the information was collected over a period of 40 years in a single region in the form of hundreds of specimens of mammals existing at the time and of fossilized mammals, constituting a complete survey for the region at that time (Avila-Pires 1960). That makes the karstic region of central Minas Gerais one of the Brazilian regions for which there is the greatest body of knowledge of extant and extinct mammal species.

A recent survey in the Lagoa Santa Karst area (2009–2012) identified 13 and 16 species of small rodents in owl regurgitation pellets (Martins et al. 2018). Apart from Lund's pioneering work (1836), this last study was only the second investigation of owl pellets ever carried out in the karst region. Greater richness of species was found than has been reported by similar studies in the Cerrado and Atlantic Forest biomes but diversity was only moderate insofar as the species *Calomys tener* (Winge, 1888) and *Calomys expulsus* (Lund 1841) were by far the most abundant in all the samples (Martins et al. 2018). Some of the rodent species described by Lund as being common in the region in his day were no longer found by recent studies, namely, four species of the Echimyidae family and five of the Cricetidae family (Martins et al. 2018). Another interesting factor is in regard to the species that were most abundant in the pellets. Today species of the *Calomys* genus represent 64% of the individuals in the samples (Martins et al. 2018) whereas Lund (1837) reported that *Neromys lasiurus* (Lund 1840) was the commonest rodent species found in the pellets, responding for 80% of the individuals identified. Probably those differences in the rodent communities are related to the changes

Table 5 List of mammal species to be found today in the Lagoa Santa Karst region

Family	Species
Myrmecophagidae	*Myrmecophaga tridactyla* (Linnaeus, 1758)
	Tamandua tetradactyla (Linnaeus, 1758)
Dasypodidae	*Euphractus sexcinctus* (Linnaeus, 1758)
	Cabassous tatouay (Desmarest, 1804)
	Cabassous unicinctus (Linnaeus, 1758)
	Dasypus septemcinctus (Linnaeus, 1758)
	Dasypus novemcinctus (Linnaeus, 1758)
Tayassuidae	*Tayassu pecari* (Link, 1795)
	Pecari tajacu (Linnaeus, 1758)
Tapiridae	*Tapirus terrestris* (Linnaeus, 1758)
Cervidae	*Blastocerus dichotomus* (Illiger, 1815)
	Mazama gouazoubira (G. Fischer [von Waldheim], 1814)
	Mazama americana (Erxleben,1777)
Felidae	*Panthera onca* (Linnaeus, 1758)
	Puma concolor (Linnaeus, 1771)
	Leopardus pardalis (Linnaeus, 1758)
	Leopardus wiedii (Schinz, 1821)
	Leopardus tigrinus (Schreber, 1775)
	Puma yagouaroundi (É. Geoffroy Saint-Hilaire, 1803)
Canidae	*Lycalopex vetulus* (Lund, 1842)
	Cerdocyon thous (Linnaeus, 1766)
	Chrysocyon brachyurus (Illiger, 1815)
	Speothos venaticus (Lund, 1842)
Procyonidae	*Nasua nasua* (Linnaeus, 1766)
	Procyon cancrivorus (G. [Baron] Cuvier, 1798)
Mustelidae	*Eira barbara* (Linnaeus, 1758)
	Galictis vittata (Schreber, 1776)
	Lontra longicaudis (Olfers, 1818)
	Pteronura brasiliensis (Gmelin, 1788)
Mephitidae	*Conepatus semistriatus* (Boddaert, 1785)
Didelphidae	*Thylamys macrurus* (Olfers, 1818)
	Micoureus demerarae (Thomas, 1905)
	Marmosops incanus (Lund, 1840)
	Gracilinanus agilis (Burmeister, 1854)
	Gracilinanus microtarsus (Wagner, 1842)
	Thylamys velutinus (Wagner, 1842)
	Caluromys lanatus (Olfers, 1818)
	Didelphis albiventris (Lund, 1840)
	Didelphis marsupialis (Linnaeus, 1758)
	Chironectes minimus (Zimmermann, 1780)
	Monodelphis domestica (Wagner, 1842)
	Monodelphis americana (Müller, 1776)
Cricetidae	*Calomys tener* (Winge, 1887)
	Calomys expulsus (Lund, 1840)

(continued)

Table 5 (continued)

Family	Species
	Akodon cursor (Winge, 1887)
	Thaptomys nigrita (Lichtenstein, 1829)
	Thalpomys lasiotis (Thomas, 1916)
	Holochilus brasiliensis (Desmarest, 1819)
	Necromys lasiurus (Lund, 1840)
	Bibimys labiosus (Winge, 1887)
	Kunsia tomentosus (Litchtenstein, 1830)
	Oligoryzomys nigripes (Olfers, 1818)
	Oligoryzomys fornesi (Massoia, 1973)
	Hylaeamys laticeps (Lund, 1840)
	Pseudoryzomys simplex (Winge, 1887)
	Cerradomys subflavus (Wagner, 1842)
	Cerradomys scotti (Langguth and Bonvicino, 2002)
	Rhipidomys mastacalis (Lund, 1840)
	Nectomys squamipes (Brants, 1827)
Muridae	*Rattus rattus* (Linnaeus, 1758)
	Mus musculus (Linnaeus,1758)
Erethizontidae	*Sphiggurus insidiosus* (Olfers, 1818)
	Coendou prehensilis (Linnaeus, 1758)
Dasyproctidae	*Dasyprocta azarae* (Lichtenstein, 1823)
Cuniculidae	*Cuniculus paca* (Linnaeus, 1766)
Caviidae	*Cavia aperea* (Erxleben, 1777)
	Hydrochoerus hydrochaeris (Linnaeus, 1766)
Echimyidae	*Phyllomys brasiliensis* (Lund, 1840)
	Trinomys setosus (Desmarest, 1817)
	Thrichomys apereoides (Lund, 1839)
	Clyomys laticeps (Thomas, 1909)
	Carterodon sulcidens (Lund, 1841)
Sciuridae	*Guerlinguetus brasiliensis* (Gmelin, 1788)
Leporidae	*Sylvilagus brasiliensis* (Linnaeus, 1758)
Phyllostomidae	*Micronycteris megalotis* (Gray, 1842)
	Chrotopterus auritus (Peters, 1856)
	Phyllostomus hastatus (Pallas, 1767)
	Carollia perspicillata (Linnaeus, 1758)
	Vampyressa pusilla (Wagner, 1843)
	Molossus molossus (Pallas, 1766)
	Mesophylla macconnelli (Thomas, 1901)
	Glossophaga soricina (Pallas, 1766)
	Anoura caudifer (E. Geoffroy, 1818)
	Platyrrhinus lineatus (E. Geoffroy, 1810)
	Sturnira lilium (E.Geoffroy, 1810)
	Artibeus lituratus (Olfers, 1818)
	Pygoderma bilabiatum (Wagner, 1843)

(continued)

Table 5 (continued)

Family	Species
	Desmodus rotundus (E. Geoffroy, 1810)
Emballonuridae	*Peropteryx macrotis* (Wagner, 1843)
Natalidae	*Natalus stramineus* (Gray, 1838)
Vespertilionidae	*Myotis nigricans* (Schinz, 1821)
	Eptesicus brasiliensis (Desmarest, 1819)
	Histiotus velatus (I. Geoffroy, 1824)
	Lasiurus borealis (Müller, 1776)
	Lasiurus cinereus (Palisot de Beauvois, 1796)
	Lasiurus ega (Gervais, 1856)
Molossidae	*Eumops bonariensis* (Peters, 1874)
	Eumops perotis perotis (Schinz, 1821)
	Cynomops abrasus (Temminck, 1827)
	Molossops temminckii (Burmeister, 1854)
Callitrichidae	*Callithrix penicillata* (É. Geoffroy, 1812)
Cebidae	*Cebus apella* (Linnaeus, 1758)
Pitheciidae	*Callicebus personatus* (É. Geoffroy, 1812)
	Alouatta guariba (Humboldt, 1812)

Adapted from Paula Couto (1953) and Fundação Biodiversitas (1996)

Fig. 5 Mammal species found in Lagoa Santa Karst. **a** *Necromys lasiurus*, 118–128 mm (photo by Felipe Peters), **b** *Dasypus septemcinctus*, 550 mm (photo by Claudio Dias Timm), **c** *Lontra longicaudis*, 510 mm (photo by Mark Bowler), **d** *Puma concolor*, 1950 mm (photo by James Carmichael Jr), **e** *Alouatta guariba*, 300–750 mm (photo by Ronai Rocha), **f** *Sturnira lilium*, 60–73 mm (photo by Ilan Ejzykowicz)

in the environment that have taken place over the past two hundred years, especially those resulting from anthropic activities like deforestation, agriculture, and mining. Such impacts, in the course of time, have led to a homogenization of the habitats, a reduction in specialist species and an expansion, in the region, of the generalist species of the *Calomys* genus which commonly dominate anthropically degraded areas (e.g., Mills et al. 1991; Pardiñas et al. 2000). A similar pattern was observed in regard to the bird species with a drop in species richness in the present-day scenario (Christiansen and Pitter 1997).

Despite the importance of the Cerrado mammal fauna, it is the least studied and those studies that do exist have conflicting results. Some authors report that the mammal fauna of the Cerrado biome is impoverished, characteristic of open vegetation areas and poor in endemism (Avila-Pires 1996), while others assert that the mammal fauna is adapted to the savannahs and rich in endemism (Muller 1979).

The small mammals in the pellet samples reflect local wildlife in the landscape (Hadly 1999; Terry 2010a). Based on this, in this study, several species listed by Lund (Paula Couto 1950) and Winge (1887) as living in Lagoa Santa were not found in the modern owl pellet. Among these, we can cite *Callistomys* sp. and *Kannabateomys amblyonyx* and *Kunsia tomentosus*, rodents were only found in the "Fossil" material (Winge 1887), not occurring presently in the Lagoa Santa region, which may indicate a recent reduction in the geographic distribution of the genus (Emmons and Vucetich 1998).

7 Final Remarks

The Lagoa Santa region is an ecosystem extremely susceptible to environmental degradation because of the peculiarities of its karstic characteristics. Deforestation and habitat fragmentation are responsible for the loss of plant and animal species, for erosion and for the silting up of innumerable bodies and courses of water in the surrounding areas, causing irreversible changes to the cave-dwelling biota. Various fauna groups that appear on Lund's lists for the region have been suffering the effects of anthropic actions for a very long time. Since the earliest first fauna inventories were made, several species have become extinct or have come under threat of extinction (Christiansen and Pitter 1997, Trolle et al. 2007; Martins et al. 2018).

In spite of its high degree of biological diversity and the strong anthropic pressure it is under, the Cerrado has only recently received the attention its biological importance deserves and the number of protected areas that have been implanted in it is still quite small (CI 1999). Outstanding among those few is the Lagoa Santa Karst Environmental Protection Area (EPA). It is an extremely important protected area in biological terms because it is a migration point for rare bird species; it has a complex array of dry and mesophyll forests; and because it shelters endemic and threatened vertebrate species (Costa et al. 1998). Furthermore, in addition, to its great historical importance, the region is endowed with great scenic beauty in the form of cave shelters and grottos, temporary lakes, and sinkholes. The limestone substrate has preserved registers of the past; rock paintings and many fossils of fauna species of both the very distant and the recent past have been preserved there. In spite of its great historical and biological importance, however, the region is suffering heavy environmental impacts from limestone quarrying, sand extraction, domestic and industrial pollution, the burning of the natural vegetation, and other actions (Costa et al. 1998).

References

Ab'Sáber A (1977) Espaços ocupados pela expansão dos climas secos na América do Sul, por ocasião dos períodos glaciais Quaternários. Paleoclima 3:1–19

Alves CBM, Pompeu OS (2010) A fauna de peixes da Bacia do Rio das Velhas no final do século XX. In: Alves CBM, Pompeu PS (ds.) Peixes do Rio das Velhas: passado e presente. ARGVMENTVM, Belo Horizonte 3:167–189

Auler AS (1999) Karst evolution and paleoclimate of eastern Brazil. Thesis, University of Bristol

Auler AS, Wang X, Edwards RL, Cheng H, Cristalli PS, Smart PL, Richards DL (2004) Quaternary ecological and geomorphic changes associated with rainfall events in presently semi-arid northeastern Brazil. J Quat Sci 19:693–701

Auler AS, Piló LB, Smart PL, Wang X, Hoffmann D, Richards DA, Edwards RL, Neves WA, Cheng H (2006) U-series-dating and taphonomy of quaternary vertebrates from Brazilian caves. Palaeogeogr Palaeoclimatol Palaeoecol 240:508–522

Avila-Pires FD (1960) Roedores colecionados na região de Lagoa Santa, Minas Gerais, Brasil. Arquivos do Museu Nacional 50:25–46

Avila-Pires FD (1996) Observações gerais sobre a mastozoologia do Cerrado. Ann Acad Bras Ciências 38:331–340

Barnosky AD, Hadly EA, Bell CJ (2003) Mammalian response to global warming on varied temporal scales. J Mammal 84:354–358

Bauermann SG, Macedo RB, Behling H, Pillar V, Neves PCP (2008) Dinâmicas vegetacionais, climáticas e do fogo com base em palinologia e análise multivariada no Quaternário Tardio do sul do Brasil. Revista Brasileira de Paleontologia 11:87–96

Behling H (1995) A high-resolution Holocene pollen Record from Lago do Pires, SE Brazil: vegetation, climate and fire history. J Paleolimnol 14:253–268

Bergqvist LP, Abuhid VS, Del Giudice GML (2000) Mamíferos. In: Carvalho IS (ed) Paleontologia. Editora Interciência, Rio de Janeiro, pp 595–623

Birdlife International (2000) Threatened birds of the world. Lynx editions, Barcelona

Bonvicino CR, Bezerra AM (2003) Use of regurgitated pellets of barn owl (*Tyto alba*) for inventorying small mammals in the cerrado of central Brazil. Stud Neotrop Fauna E 38:37–41

Boroni NLM (2012) Roedores da região cárstica de Lagoa Santa, MG: implicações ambientais. Dissertation, Universidade Federal de Viçosa

Brain CK (1981) The hunters or the hunted? An introduction to African cave taphonomy. The University of Chicago Press, Chicago

Brien C (1996) Evaluation of conservation strategies for Neotropical parrots (Psittacidae): a literature survey—pet trade. http://mgfx.com/bird. Accessed 20 May 2016

Burton RF (1869) Explorations of the highlands of Brazil, with a full account of the gold and diamond mines. Tinsley Brothers, London

Campbel JA, Lamar WW (1989) The venomous reptiles of Latin America. University Press, Cornell

Campos VA, Oda FH, Juen L, Barth A, Dartora A (2013) Composição e riqueza de espécies de anfíbios anuros em três diferentes hábitats em um agrossistema no Cerrado do Brasil central. Biota Neotrop 13:1–9

Cartelle C (1999) Pleistocene mammals of the Cerrado ang Caatinga of Brazil. In: Eisenberg JF, Redford KH (eds) Mammals of the neotropics. The Central Neotropics, Chicago, pp 27–46

Cartelle C (2012) Das grutas à luz: os mamíferos pleistocênicos de Minas Gerais = from the caves to the light: the pleistocene mammals of Minas Gerais. Bicho do Mato Editora, Belo Horizonte

Cavalcanti LF, Lima MF, Meguerditchian I, Medeiros RCS, Lamblém HS (2011) Plano de Ação Nacional para conservação do Patrimônio Espeleológico nas áreas cársticas da Bacia do Rio São Francisco. In: Congresso Brasileiro de Espeleologia, Ponta Grossa

Christiansen MB, Pitter E (1997) Species loss in a forest community near Lagoa Santa in southeastern Brazil. Biol Cons 80:23–32

Colli GR, Bastos RP, Araujo AFB (2002) The character and dynamics of the Cerrado herpetofauna. In: Oliveira PS, Marquis RJ (eds) The Cerrados of Brazil: ecology and natural history of neotropical savanna. Columbia University Press, New York, pp 223–241

Conservation International (1999) Ações Prioritárias para a Conservação da Biodiversidade do Cerrado e Pantanal

COPAM Conselho de Política Ambiental (2010) Deliberação Normativa COPAM nº 147, de 30 de abril de 2010: Aprova a Lista de Espécies Ameaçadas de Extinção da Fauna do Estado de Minas Gerais. Diário do Executivo do Estado de Minas Gerais, 04/05/2010

Costa CMR, Herrmann G, Martins CS, Lins LV, Lamas IR (1998) Biodiversidade em Minas Gerais: um atlas para sua conservação. Fundação Biodiversitas, Belo Horizonte

Cunha OR, Nascimento FP (1978) Ofídios da Amazônia. X: as cobras da região leste do Pará. Publicações avulsas do Museu Emílio Goeldi, Pará 31:1–218

Davis SJM (1987) The archaeology of animals. Yale University Press, New Haven

De Oliveira P (1992) A palynological record of Late Quaternary vegetational and climate change in Southeastern Brazil. Thesis, Ohio University

Deus JAS, Rodrigues RS, Ferreira CCD (1997) Preservação da Área Carstica de Lagoa Santa/MG através da educação ambiental. Geonomos, Belo Horizonte 5:49–53

Dutra TL (2000) Paleoecologia. In: Carvalho IS (ed) Paleontologia. Editora Interciência, Rio de Janeiro, pp 145–154

Emmons LH, Vucetich MG (1998) The identity of Winge's Lasiuromys villosus and the description of a new genus of echimyid rodent (Rodentia: Echimyidae). Am Mus Novit 3223:1–12

Faunmap Working Group (1996) Spatial response of mammals to Late Quaternary environmental fluctuations. Science 272:1601–1606

Figueira JEC, Barbosa MM, Krettli LG (2004) No rastro de quem ainda resta: Educação Ambiental para alunos do ensino fundamental da APA Caste de Lagoa Santa, Minas Gerais. In: Encontro de Extensão da UFMG, Belo Horizonte

Fonseca GAB, Herrmann G, Leite YLR, Mittermeier RA, Rylands AB, Patton JL (1996) Lista anotada dos mamíferos do Brasil. Occas Pap Conserv Biol 3:1–34

Fundação Biodiversitas (1996) Plano de Gestão: Área de Proteção Ambiental Carste de Lagoa Santa - MG. Belo Horizonte, 41 pp

Giaretta AA, Menin M, Facure KG, Kokubum MNC, De Oliveira Filho JC (2008) Species richness, relative abundance, and habitat of reproduction of terrestrial frogs in the Triângulo Mineiro region, Cerrado biome, southeastern Brazil. Iheringia, Série Zoologia 98:181–188

Grelle CEV, Fonseca MT, Moura RT (1996) Diagnóstico da Mastofauna. APA Carste de Lagoa Santa. Fundação Biodiversitas. Unpublished report

Herrmann G, Kohler HC, Duarte JC, Carvalho PGS (1998) APA Carste de Lagoa Santa - Volume III: meio biótico. IBAMA/CPRM, Belo Horizonte

Hirsch A, Costa BMA (2005) Mamíferos: diversidade e representatividade. In: Goulart EMA (ed) Navegando o Rio das Velhas, das minas aos gerais. Instituto Guaicuy-SOS Rio das Velhas, Belo Horizonte, pp 605–629

Hubbe A, Haddad-Martim PM, Hubbe M, Mayer EL, Strauss A, Auler AS, Piló LB, Neves WA (2011) Identification and importance of critical depositional gaps in pitfall cave environments: the fossiliferous deposit of Cuvieri Cave, eastern Brazil. Palaeogeogr Palaeoclimatol Palaeoecol 312:77–78

Kipnis R (2002) Foraging societies of eastern central Brazil: an evolutionary ecological study of subsistence strategies during the terminal Pleistocene and early middle Holocene. Thesis, University of Michigan

Kopp K, Signorelli L, Bastos RP (2010) Distribuição temporal e diversidade de modos reprodutivos de anfíbios anuros no Parque Nacional das Emas e entorno, estado de Goiás, Brasil. Iheringia, Série Zoologia 100:192–200

Laming-Emperaire A, Prous A, Moraes AV, Beltrão MC (1975) Grottes et Abris de la Region de Lagoa Santa, Minas Gerais, Brésil: premier rapport de la Mission archéologique franco-brésilienne de Lagoa Santa. Cahiers d'archéologie d'Amérique du Sud 1. Paris: École pratique des hautes étude, 185 pp

Lanari CU (1909) Ossadas humanas fósseis encontradas numa caverna calcarea das vizinhanças de Mocambeiro. Annaes da Escola de Minas 11:15–35

Ledru M (1993) Late Quaternary environmental and climatic changes in Central Brazil. Quatern Res 39:90–98

Ledru M, Braga PIS, Soubiès F, Fournier M, Martin L, Suguio K, Turcq B (1996) The last 50.000 years in the Neotropics (Southern Brazil) evolution of vegetation and climate. Paleogeography, Paleoclimatology, Paleoecology 123:239–257

Liais E (1872) Climats, Geologie, Faune et Geographie Botanique du Brésil. Garnier Freres, Paris

Lima TF, Morais MS (2006) Contribuições para o desenvolvimento de plano de manejo em ambiente cavernícola - Gruta do Maquiné: Um estudo de caso. Geonomos 14:45–53

Lins L, Machado RB, Vasconcelos MF (1998) Avifauna, pp 32–36. In: Herrmann G, Kohler HC, Duarte JC, Carvalho PGS (eds) Estudo do meio biótico da APA Carste de Lagoa Santa., Ibama, Companhia de Pesquisa de Recursos Minerais, Belo Horizonte, 92 pp

Lund PW (1840) View of the fauna of Brazil previous to the last geological revolution. Mag Nat Hist 4, 1–8, 49–57, 105–112, 153–161, 207–213, 251–259, 307–317, 373–389

Lund PW (1950) Notícia sobre ossadas humanas fósseis achadas numa caverna do Brasil. 1844. In: Paula Couto C (ed) Memórias sobre a paleontologia brasileira. Instituto Nacional do Livro, Rio de Janeiro, pp 465–488

Machado ABM, Fonseca GAB, Machado RB, Aguiar LMS, Lins LV (1998) Livro vermelho das espécies ameaçadas de extinção da fauna de Minas Gerais. Fundação Biodiversitas, Belo Horizonte

Martins NLB, Pardinas UFJ, Lessa G (2018) Vanished Lagoa Santa: rodent species loss in the last two centuries. Mastozoologia neotropical/J Neotrop Mammal 25:17–33

Mattos GT, Andrade MA, Freitas MV (1994) Nova lista de aves do estado de Minas Gerais. Fundação Acangaú, Belo Horizonte, 21 pp

Mills JN, Ellis BA, Mckee KT, Maiztegui JI, Childs JE (1991) Habitat associations and relative densities of rodent populations in cultivated areas of central Argentina. J Mammal 72:470–479

Mittermeier RA, Myers N, Gil PR, Mittermeier CG (1999) Hotspots: earth's biologically richest and most endangered terrestrial ecoregions. Conservation International, CEMEX

Mones A (1991) Monografía de la Familia Hydrochoeridae (Mammalia, Rodentia). Sistemática-Paleontología-Bibliografía. Courier Forschungs-Institut Senckenberg 134:1–235

Morais ARD, Signorelli L, Gambale PG, Kopp K, Guimaraes LD, Vaz-Silva W, Ramos J, Bastos RP (2011) Anfíbios anuros associados a corpos d'água do sudoeste do estado de Goiás, Brasil. Biota Neotrop 11:355–363

Morais AR, Bastos RP, Vieira RRS, Signorelli L (2012) Herpetofauna da Floresta Nacional de Silvânia, um remanescente de Cerrado no Brasil Central. Neotrop Biol Conserv 7:114–121

Motta-Junior JC, Talamoni SA (1996) Biomassa de presas consumidas por *Tyto alba* (Strigiformes: Tytonidae) durante a estação reprodutiva no Distrito Federal. Ararajuba 4:38–41

Muller P (1979) Campo cerrado-forest or savanna? GeoJournal 3:5–25

Neves WA, Piló LB (2003) Solving Lund's dilemma: new AMS dates confirm that humans and megafauna coexisted at Lagoa Santa. Curr Res Pleistocene 20:57–60

Nogueira C, Valdujo P, França F (2005) Hábitat variation and lizard diversity in a Cerrado area of Central Brazil. Stud Neotrop Fauna Environ 40:105–112

Nogueira C, Colli GR, Martins M (2009) Local richness and distribution of the lizard fauna in natural hábitat mosaics of the Brazilian Cerrado. Austral Ecol 34:83–96

Nomura F, Maciel NM, Pereira EB, Bastos RP (2012) Diversidade de anuros (Amphibia) em áreas recuperadas de atividade mineradora no Cerrado e de plantio de Eucalyptus urophyla no Brasil Central. Biosci J 28:312–324

Oda FH, Bastos RP, Sá Lima MA (2009) Taxocenose de anfíbios anuros no Cerrado do Alto Tocantins, Niquelândia, estado de Goiás: diversidade, distribuição local e sazonalidade. Biota Neotrop 9:219–232

Pardiñas UF, Cirignoli S (2002) Bibliografía comentada sobre los análisis de egagrópilas de aves rapaces en Argentina. Ornitol Neotrop 13:31–59

Pardiñas UF, Moreira GJ, Garcia-Esponda CM, Santis LJ (2000) Deterioro ambiental y micromamiferos durante el holoceno en el nordeste de la estepe patagónica. Rev Chil Hist Nat 73:9–21

Passacantili MG (2008) Influência antrópica em ecótonos floresta e campo da Mata Atlântica do Brasil: análise de micropartículas carbonizadas em solos superficiais. Dissertation, Universidade de Guarulhos

Paula Couto C (1950) Memórias sobre a paleontologia brasileira. Instituto Nacional do Livro, Rio de Janeiro, pp 465–488

Paula Couto C (1953) Paleontologia Brasileira. Mamíferos. Instituto Nacional do Livro, Rio de Janeiro

Paula Couto C (1975) Mamíferos fósseis do Quaternário do sudeste brasileiro. Boletim Paranaense de Geociências 33:89–132

Perez CP (2009) Peleoecologia de mamíferos viventes como ferramente na caracterização do ambiente holocênico de Lagoa Santa. Dissertation, Universidade de São Paulo

Peters JÁ, Orejas-Miranda B (1970) Catalogue of Neotropical Squamata. Part I snakes. Bull Unites States Natl Mus 297:1–347

Pinto OMO (1952) Súmula histórica e sistemática da ornitologia de Minas Gerais. Arquivos de Zoologia 8:1–51

Polignano MV (2001) Uma viagem ao projeto Manuelzão e à bacia do Rio das Velhas: Manuelzão vai à escola. Projeto Manuelzão Belo Horizonte, 64 pp

Reitz EJ, Wing ES (2001) Zooarchaeology. Cambridge University Press, Cambridge

Ribeiro CM (2007) "O recado do Morro" e a geografia de Minas Gerais. Caderno de Geografia 17:121–140

Ricklefs RE (2001) A economia da natureza. Editora Guanabara Koogan, Rio de Janeiro

Rocha RG, Ferreira E, Leite YL, Fonseca C, Costa LP (2011) Small mammals in the diet of barn owls, *Tyto alba* (Aves: Strigiformes) along the mid-Araguaia River in central Brazil. Zoologia 28:709–716

Roda SA (2006) Dieta de *Tyto alba* na Estação Ecológica do Tapacurá, Pernambuco, Brasil. Rev Bras Ornitol 14:449–452

Rodrigues M, Gomes HB (2004) Range extension for the Greyheaded Tanager *Eucometis penicillata* in south-east Brazil. Bull-Br Ornithol Club 124:177–183

Rodrigues M, Goulart FF (2005) Aves Regionais: De Burton Aos Dias De Hoje. In: Goulart EMA (ed) Navegando o Rio das Velhas, das minas aos gerais. Instituto Guaicuy-SOS Rio das Velhas, Belo Horizonte, pp 588–603

Rodrigues M (2008) Noteworthy bird records at Lagoa Santa, southeastern Brazil. In: Revista Brasileira de Zoologia. Available via DIALOG. http://www.scielo.br/scielo.php?script=sci_arttext&pid=S0101-81752008000100020&lng=en&nrm=iso. Accessed 26 Apr 2016

Rodrigues M, Michelin VB (2005) Riqueza e diversidade de aves aquáticas de uma lagoa natural no sudeste do Brasil. Revista Brasileira de Zoologia 22:928–935

Rodrigues M, Carrara LA, Faria LP, Gomes HB (2005) Aves do Parque Nacional da Serra do Cipó: o Vale do Rio Cipó, Minas Gerais, Brasil. Revista Brasileira de Zoologia 22:326–338

Salgado-Labouriau ML, Casseti V, Ferraz-Vicentini KR, Martin L, Soubiès F, Suguio K, Turcq B (1997) Late Quaternary vegetacional and climatic changes in Cerrado and palm swamp from central Brazil. Palaeogeogr Palaeoclimatol Palaeoecol 128:215–226

Sazima I (1989) Comportamento alimentar da jararaca, Bothrops jararaca: encontros provocados na natureza. Ciência e Cultura 41:500–505

Scheibler DR, Christoff AU (2007) Habitat associations of small mammals in southern Brazil and use of regurgitated pellets of birds of prey for inventorying a local fauna. Braz J Biol 67:619–625

Schobbenhaus C, Campos DA, Queiroz ET, Winge M, Berbert-Born ML (2002) Sítios Geológicos e Paleontológicos do Brasil. DNPM/CPRM - Comissão Brasileira de Sítios Geológicos e Paleobiológicos (SIGEP). Brasília, 554 pp

Schoener TW (1968) Sizes of feeding territories among birds. Ecology 49:123–141

Servant M, Maley J, Turcq B, Absy ML, Brenac P, Fournier M, Ledru MP (1993) Tropical forest changes during the Late Quaternary in African and South American lowlands. Glob Planet Chang 7:25–40

Sick H (1965) A fauna do Cerrado. Arquivos de Zoologia do Estado de Sao Paulo 12:71–93

Sick H (1993) Birds in Brazil: a natural history. Princeton University Press, Princeton

Silva JMC (1995) Birds of the Cerrado Region, South America. Steenstrupia 21:69–92

Soubiès F, Suguiu K, Martin L, Leprun JC, Servant M, Turcq B, Fournier M, Delaune M, Sifeddine A (1991) The Quaternary lacustrine deposits of the Serra dos Carajás (state of Pará, Brazil)—ages and other preliminary results. Proc Glob Chang S Am Dur Quat 8:223–243

Souza DP, Asfora PH, Lira TC, Astúa D (2010) Small mammals in barn owl (*Tyto alba* – Aves, Strigiformes) pellets from Northeastern Brazil, with new records of Gracilinanus and Cryptonanus (Didelphimorphia, Didelphidae). Mamm Biol 75:370–374

Travassos LEP, Guimarães RL, Varela ID (2008) Áreas cársticas, cavernas e a Estrada Real. Pesquisas em Turismo e Paisagens Cársticas 1:107–120

Trolle M, Bissaro JR, Prado HM (2007) Mammal survey at a ranch of the Brazilian Cerrado. Biodivers Conserv 4:1205–1211

Uetanabaro M, Souza FL, Landgref Filho P, Beda AF, Brandão RA (2007) Anfíbios e répteis do Parque Nacional da Serra da Bodoquena, Mato Grosso do Sul, Brasil. Biota Neotrop 7:279–289

Valdujo PH, Silvano DL, Colli G, Martins M (2012) Anuran Species Composition and Distribution Patterns in Brazilian Cerrado, a Neotropical Hotspot. S Amn J Herpetol 7:63–78

Vanzolini PE (1986) Levantamento herpetológico da área do estado de Rondônia sob influência da BR 364. Programa Polonoroeste, Brasília - MTC/CNPq. 50 pp

Voss RS, Carleton MD (1993) A new genus for *Hesperomys molitor* Winge and *Holochilus magnus* Hershcovitz (Mammalia, Muridae) with an analysis of its phylogenetic relationships. Am Mus Novit 2085:1–9

Voss RS, Myers P (1991) *Pseudoryzomys simplex* (Rodentia: Muridae) and the significance of Lund's collections from the caves of Lagoa Santa, Brazil. Bull Am Mus Nat Hist. Am Mus Novit 206:414–432

Walter HV (1948) The pre-history of the Lagoa Santa region. Velloso & Cia. Ltda, Belo Horizonte

Walter HV (1958) Arqueologia da região de Lagoa Santa. Sedegra, Rio de Janeiro

Warming E (1908) Lagoa Santa: contribuição para a Geographia Phytobiologica. Imprensa Oficial do Estado de Minas Gerais, Belo Horizonte

Wikiaves (2008) WikiAves, a Enciclopédia das Aves do Brasil. http://www.wikiaves.com.br. Accessed 31 Aug 2016

Winge H (1888) Jorfundne og nulevende Gnawere (Rodentia) fra Lagoa Santa, Minas Gerais, Brasiliens: med udsigt over gnavares indbyrdes slaegtskab. E Museo Lundii 1:1–200

Yalden DW (1982) When did the mammal fauna of the British isles arrive? Mammal Rev 12:1–57

The Waterbirds and Dynamics of Lagoa Santa Karst Temporary Lakes

José Eugênio Côrtes Figueira, Paula Fernanda Albonette de Nóbrega, Tulio Dornas, Janaina Aparecida Batista Aguiar, Leonardo Lopes, and Maria Auxiliadora Drumond

Abstract

We provide a review of how waterbirds utilize the temporary lakes of the karst area of Lagoa Santa in southeastern Brazil. The area is punctuated by dozens of shallow depressions of different sizes and shapes, which are located at different altitudes, sometimes fed by creeks and connected in some situations to the endokarst by sinks. With the onset of the wet season, in years with typical rainfall levels, these depressions become lakes, most of which disappear during the dry season. Species richness, abundance, and foraging guilds of waterbirds vary according to lake area, depth, and available food types. Common species appear as soon as a thin film of water accumulates in the shallow depressions, while more specialized or migratory birds respond to water level, food availability, and migratory movements. Birds may use different lakes on different occasions, and they may abandon drying lakes, and even the entire region when lake conditions change with the advance of the dry season. We describe the behavior of the assemblage of birds associated with the flooding and receding waters of lakes of different sizes. We update the species list of nonpasserine waterbirds for this karst ecosystem, call attention to the growing threats to it and its sensitive lakes, and describe our initiative to protect it.

Keywords

Ecological succession • Foraging guilds • Karst depressions • Migratory species • Nonequilibrium assemblages • Phase shifts • Ramsar Site • Species abundances • Species composition • Species richness-area • Stopover • Threats

J. E. Côrtes Figueira (✉) · M. A. Drumond
Departamento de Genética, Ecologia e Evolução, Instituto de Ciências Biológicas, Universidade Federal de Minas Gerais, Belo Horizonte, MG 31270-901, Brazil
e-mail: cortesfigueira@gmail.com

M. A. Drumond
e-mail: alatus@ufmg.br

P. F. Albonette de Nóbrega
Departamento de Enfermagem, Faculdade de Ciências da Saúde, Universidade do Estadode Mato Grosso, Diamantino, MT 78400-000, Brazil
e-mail: pnbio@yahoo.com.br

T. Dornas
Bolsista PNPD/CAPES no Programa de Pós-graduação em Ciências do Ambiente, Universidade Federal do Tocantins, Palmas, TO 77001-090, Brazil
e-mail: tuliodornas@yahoo.com.br

J. A. B. Aguiar
Instituto Estadual de Florestas, Belo Horizonte, MG 31630-900, Brazil
e-mail: janaina.aguiar@meioambiente.mg.gov.br

L. Lopes
Laboratório de Biologia Animal, Instituto de Ciências Biológicas e da Saúde, Universidade Federal de Viçosa, Campus Florestal, Diamantino, MG 35690-000, Brazil
e-mail: leo.cerrado@gmail.com

1 Environmental Heterogeneity and Biodiversity in Lagoa Santa Karst

Many wetlands are temporary, passing through annual cycles of flood and drought related to rainfall and thus, oscillate between terrestrial and aquatic phases (Junk 2013). This oscillation favors the formation of heterogeneous mosaics of habitats (Amezaga et al. 2002; Blanco 1999; Gibbs 2000), which serve as refuges for a considerable part of the world's biodiversity, including many endangered species (Junk 2013; Junk et al. 2006). Annual and pluriannual cycles of often dramatic drought and flood occur with the temporary lakes found in the Karst Environmental Protection Area of Lagoa Santa ("Área de Proteção Ambiental Carste de Lagoa Santa", hereafter APA Carste) (Fig. 1). The APA Carste is located in one of the most important Brazilian examples of a karstic environment developed in Neoproterozoic carbonatic lithotypes of the Sete Lagoas Formation (Bambuí Group, in the São Francisco Basin, Minas Gerais State) (Berbert-Born 2002), and is situated in the transition between two important

A. S. Auler and P. Pessoa (eds.), *Lagoa Santa Karst: Brazil's Iconic Karst Region*, Cave and Karst Systems of the World, https://doi.org/10.1007/978-3-030-35940-9_5

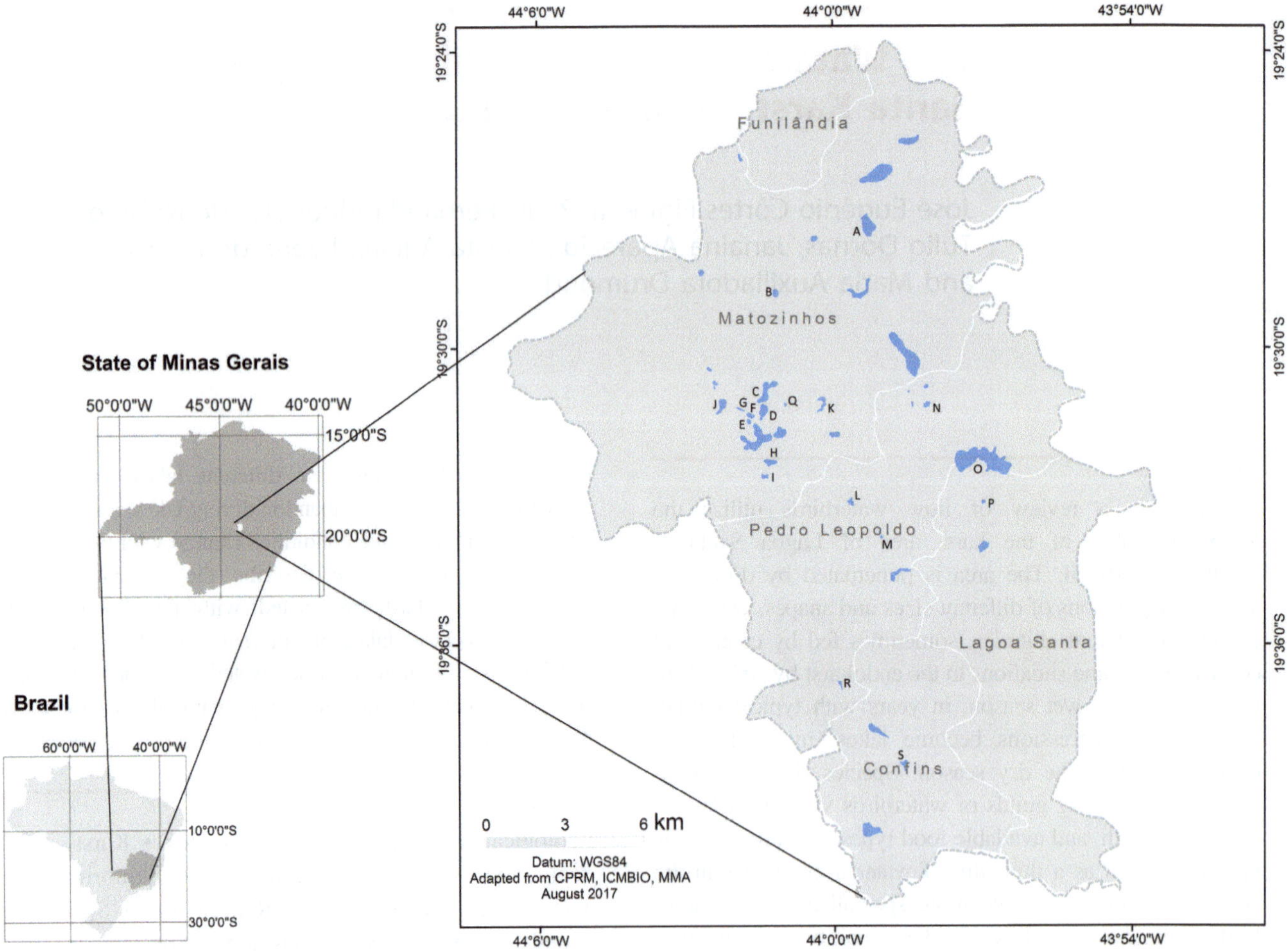

Fig. 1 Location of APA Carste, Minas Gerais, Brazil, and its temporary lakes. Letters indicate the lakes studied by Nóbrega (2015). A: Porcos; B: Curtume; C: da Fazenda Santo Antônio; D: Pedrinha; E: Santo Antônio I; F: Santo Antônio II; G: Pequena; H: Maria Angélica; I: Vargem de Pedra; J: da Fazenda do Marinheiro; K: Cerca Grande; L: Baú; M: Abrigo da Samambaia; N: Ribeira; O: Sumidouro; P: Sangradouro; Q: Caetano; R: Lapa Vermelha; S: Vargem Bonita

Neotropical biogeographic provinces (Morrone 2014) and hotspots of biodiversity (Myers et al. 2000), the Atlantic Forest and the Brazilian Cerrado.

Despite more than a century of heavy anthropic degradation (Viveiros and Martins 2002; Martins 2008), the APA Carste is characterized by high environmental heterogeneity, being comprised of a mosaic of gallery forests, remnants of savanna, and semi-deciduous and deciduous forests, with the latter being associated with large limestone outcropping massifs covered by rupicolous vegetation in their drier parts (IBAMA 1998a, see chapter "The Vegetation of Lagoa Santa Karst", this volume). In the last decade only 20% of the native vegetation of APA Carste remained with most of it having been replaced by large clearings and pastures of exotic grasses (Alt 2008). Even so, the remaining landscape units and fragments of native vegetation are very biodiverse. Surveys in the 1990s estimated that the APA Carste contains about 600 species of plants, 16 species of snakes, 42 species of mammals, and 216 species of birds, with 37 of the latter being waterbirds (IBAMA 1998a). More recent authors (Dornas 2004; Rodrigues and Michelin 2005; Nóbrega 2015, this study) recorded an additional 19 species of waterbirds in the region, bringing the total number of bird species to 235, which corresponds to about 30% of the avifauna of the state of Minas Gerais (Mattos et al. 1993) and 27% of the avifauna of the Cerrado (Silva 1995; Silva and Santos 2005). In addition, a rich fauna of invertebrates, including troglophiles and troglobites, has been identified in some of the more than 500 karst caves of the region (IBAMA 1998a, b, see chapter "Biospeleology of the Lagoa Santa Karst", this volume).

The entire APA Carste is an integral part of the Velhas River hydrographic basin, which in turn belongs to the larger basin of the São Francisco River. Although the APA Carste borders the Velhas River to the northeast and the Mata Creek to the south and west, the density of watercourses within its limits is relatively low and consists of only four main creeks (IBAMA 1998c, d).

In addition to the creeks, and a few flooding lagoons on the margins of the Velhas River, there are at least 40 floodable closed depressions of karstic origin within APA Carste, including dolines, uvalas, and a polje (IBAMA 1998c, d; Piló 2003, see chapter "Karst Landforms in the Lagoa Santa Area ", this volume). In years of regular or above-average rainfall, these karst depressions are filled with the accumulation of rainwater and/or groundwater arising from lower altimetric heights (Berbert-Born 2002; IBAMA 1998d; Kohler and Karfunkel 2002). The lakes formed are thus small and shallow, but vary greatly in area and shape, and are located at different altitudes (Berbert-Born 2002). These lakes distinguish the karst of Lagoa Santa because lakes are not common in typical tropical karst areas (Auler and Piló 2015).

Dozens of species of waterbirds were collected in the region by the naturalists Peter W. Lund and Johannes T. Reinhardt between 1834 and 1855 (Reinhardt 1870; Krabbe 2007), and many of these species continue to be recorded there today (IBAMA 1998a; Dornas 2004; Rodrigues and Michelin 2005; Nóbrega et al. 2015).

In this chapter, one of the largest lakes in APA Carste is used to exemplify how the assemblage of waterbirds changes over the course of the hydrological cycle. We will associate species richness, species composition, birds abundance, and foraging guild with the flood and drought phases of the lake. We will also compare some of these attributes of this assemblage with the attributes of bird assemblages found in other lakes of the APA Carste. In addition, we will update the list of waterbirds of the study region, identify threats to the lake system, and discuss ongoing conservation efforts.

2 The Cycle of Flood and Drought of the Karst Lakes

The climate of the region can be classified as Cwa of Köppen (Alvares et al. 2014), and is characterized by rainy summers, from October to March, and dry winters, with an average annual rainfall of 1,287 mm, although there can be very rainy (2,000 mm) or very dry (500 mm) years (IBAMA 1998c). The lakes of APA Carste undergo annual or pluri-annual flood and drought cycles determined by the rainfall regime and the level of the water table of the karst aquifers (IBAMA 1998c, d; Kohler and Karfunkel 2002; Sampaio 2010). After the end of the rainy season, the lakes may dry completely in a few weeks or months or they may remain with some water, depending on the size and depth of the karstic depression and its connections with the endokarst and surface creeks.

All the data presented below depicting temporal variation in the area of lakes of APA Carste were obtained by tracking their border coordinates with a Garmin GPSMAP 76CSx as they dried or filled, and processed with GPS TrackMaker version 13.9.

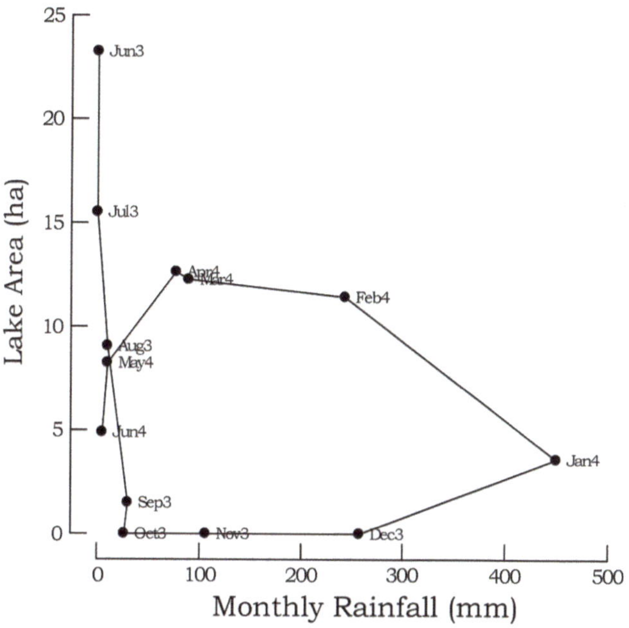

Fig. 2 The influence of monthly rainfall on the area of the Fazenda Santo Antônio Lake in APA Carste, Brazil, from June 2003 (Jun3) to June 2004 (Jun4) (after Dornas 2004)

The study of Fazenda Santo Antônio Lake (Dornas 2004), one of the largest lakes of APA Carste, shows the time lag between rainfall and variation in the area of the lake (Fig. 2).

After the summer rains of 2002–2003, the lake completely dried in just over three months and remained dry for another two months, in spite of the beginning of rains in October. The lake began to fill again in December. In 2003, the lake area reached its maximum in April, with an area half the size of January 2003, before it began to dry again. The longtime lag observed (around three months) suggests that the appearance of the lake is not due to the direct accumulation of rainwater in the karst depression but is likely dependent on the recharge of the karst aquifer.

Nóbrega (2015) compared the flood–drought cycle of several lakes of APA Carste (part of which is shown in Figs. 3 and 4), between the years of 2011 and 2013, starting in the rainy season and continuing through the dry season to the following rainy season. She showed that the temporary lake system of APA Carste mirrors an archipelago of habitats (flooded depressions that increase or decrease in an area with the arrival of rains or the advancement of the dry season), surrounded by a non-usable matrix of pastures, savanna vegetation, and forests.

The duration of the temporary lakes of APA Carste is directly proportional to their areas. The more extensive lakes maintain water for more time, but two of the lakes, Vargem Bonita and Santo Antônio, which had their sinkholes intentionally blocked with rocks, became more perennial. On the

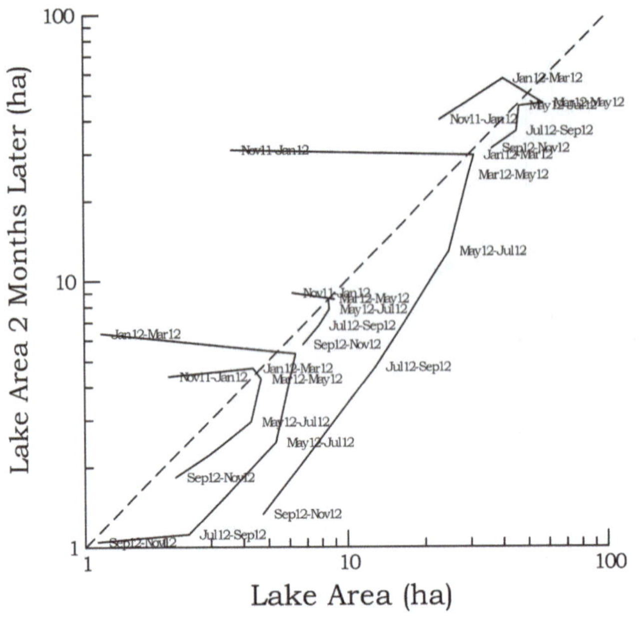

Fig. 3 Phase diagram for five temporary lakes of APA Carste (from right to left, S, C, H, K, and R, as in Fig. 1), from November 2011 (Nov11) to September 2012 (Sep12). The dashed line is a reference for no variation in lake area (or total stability) between samplings. Thus, the area of lake C was 2.6 ha in November 2011, but with the rains, its area increased to 30.1 ha, as measured in January 2012. From January 2012 to March its area varied little, but from March to July the lake area dropped considerably, first from 28.6 ha to 23.9 ha in May, indicating the beginning of the dry season, and them from 23.9 ha to 11.9 ha in July 2012. For the sake of clarity, some month/year labels were not displayed (after Nóbrega 2015)

other hand, small, shallow lakes with no connection to surface watercourses or the karst aquifer are short-lived and very unstable (Fig. 4) and, therefore, are practically devoid of aquatic macrophytes, mollusks, and fish. Carapaces of tiny planktonic crustaceans, probably ostracods (JEC Figueira pers. obs.), which exhibit dormant forms that allow them to survive temporary bodies of water (Radzikowski 2013), were found in one of the smallest lakes that had just dried.

Assuming that karst depressions at lower altitudes should be closer to groundwater, we tested the prediction that their flooded areas would be more stable. To do this, a linear regression model of the coefficient of variation (CV) of the lake area was regressed on initial area. The residuals of this regression were then correlated with mean altitudes of the karst depressions. Contrary to our prediction, the non-significant result ($P > 0.05$) suggests that the stability of the lakes is not influenced by the proximity of the water table or perhaps there is not a single water table. On the other hand, sediments carried from the surrounding hillsides may have filled the bottom of some karst depressions, generating time lags in their cycles (P Pessoa, pers. comm.).

3 The Transition Between Terrestrial and Aquatic Phases of Temporary Lakes

More than 100 years after the studies made by the Danish naturalist Johannes Eugenius B. Warming (1841–1924), the vegetation of the APA Carste floodable depressions remains poorly studied with only short descriptions being found in IBAMA (1998a). During the fieldwork of Nóbrega (2015), several lakes in different successional stages or the same lake in different seasons were visited, which allowed observations of vegetation differences and changes to be made.

Owing to different historical uses and hydrological dynamics, the lakes of APA Carste have varied vegetation cover around and within the floodable areas (IBAMA 1998a). They can possess floating macrophytes, such as *Salvinia*, *Pistia*, and *Eichhornia,* as well as macrophytes rooted in the bottom with emergent leaves, such as *Nymphaea* and *Nymphoides*, and rooted emergents, some with dense coverage, such as *Typha* and *Juncus*. They can also have plants associated with wet, soggy soils, such as *Ludwigia* and *Cyperus* (IBAMA 1998a).

As the water level drops, newly exposed soils are gradually covered by terrestrial herbaceous vegetation that advances like a "green wave". Legumes and other shrubby plants intolerant of soggy soils may grow and form very distinct bands in the higher and drier areas. Some herbs probably originate from seeds deposited in the soil during the previous dry season when their progenitors reproduced, suggesting that they are plants with a therophyte life cycle (JEC Figueira, pers. obs.). Warming (1908) states that 14% of the plants in the lakes are "annual plants", which may be interpreted as therophytes. Therophytes are common and exhibit seasonal bursts, for example, in the floodplains of the Brazilian Pantanal (Rebellato et al. 2012). Thus, karstic depressions become large fields, gradually dominated by species with different degrees of affinity for moisture, but are intolerant to soggy soils, and which grow rapidly and enter the reproductive phase quickly.

Some months later, in years of intense and regular rainfall, these fields flood and the terrestrial herbaceous vegetation gradually dies and disappears interrupting ecological succession, while the aquatic vegetation returns to dominate. *Polygonum* is one of the most abundant components of the vegetative coverage of the Sumidouro Lake during flooding but dies as the drought advances, surviving temporally in depressions or near the creek.

During phase transitions (from dry to flooding and back to dry), space is partitioned between aquatic and semiaquatic plants and terrestrial plants. Shrub plants that do not tolerate excessive moisture languish, but their dry branches serve as perches and nesting sites for birds.

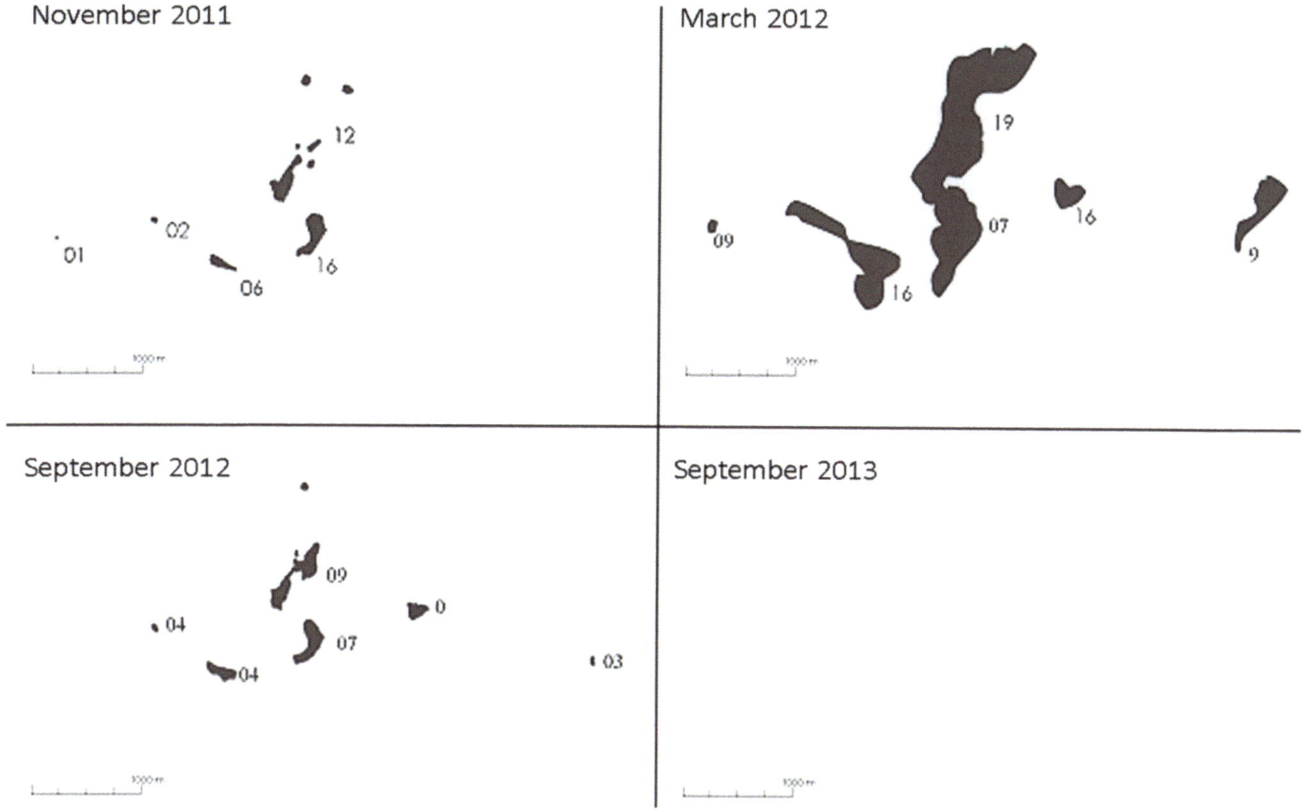

Fig. 4 Variation in the area of some lakes of the Mocambeiro region of APA Carste, between November 2011 and September 2013. Note the coalescence or separation of lakes and the unequal rate of reduction in lake areas that were originally similar, suggesting a strong influence of local topography and/or endokarst. Values indicate the number of species of waterbirds for each sampling period—see text (after Nóbrega 2015)

Naturalists of the 19th and early 20th centuries left few accounts of these lakes but Warming (1908) described the flood and drought of Sumidouro. He associated the appearance of field vegetation and the attraction of piscivorous birds with drought conditions [translated from Portuguese]: "*Most notable of these, however, is Quinta Lake* [= Sumidouro Lake], *which in the time of the waters becomes very large, but by the end of the drought, when water has been drained by subterranean channels in the calcareous rocks of Sumidouro, it diminishes until it is only a creek with small pools surrounded by green meadows where numerous cattle graze and where noisy flocks of Passa jacana* [= *Jacana jacana*] *and pink egrets, Platalea ajaja, feast on fish"*. In the 1930s, the naturalist Álvaro A. da Silveira (1867–1945) described the region of the Poções Cave [translated from Portuguese]: *"Surrounded by limestone, the basin of Bom Jardim has at its bottom a small lake fed by water that flows from a spring by the side of the road (…). The brackish water that comes out of it is not enough to fill the lake and make it overflow (…). On occasion of the rains, this overflow occurs and the water flows to the same creek that disappears in the sinkhole near the farm. Although this spring does not have any surface connection with any other watercourse, fish, such as traíras, bagres and lambaris appear in abundance when there are floods of Velhas River, which occurs three leagues distant."* (IBGE 1939). In the 1960s, the archeologist Aníbal Mattos (1886–1969) described what seems to have been troglobiont fish flushing in "Vargem do Mucambo" [translated from Portuguese]: *"…we have had the occasion to go through all its empty bed and watch the phenomenon of the outbreak of water. The amount of small fish that comes in this gush is fantastic. The inhabitants of the place pick them up in baskets, in great abundance (local people usually say that these small fish are blind). For years this lake remains full, until one day it mysteriously empties again*" (Mattos 1961).

These historical accounts describe the features of the karst landscape and facets of the unique behavior of karst lakes: upwellings and sinks, green fields covering the drying lake bed, the appearance of fish from the endokarst (still reported in some lakes by local residents) and the attraction of piscivorous birds.

From even further back in time are cave paintings on limestone rock walls estimated to be 3,800 to 8,000 years old (Baeta 2011), which portray, among other things, anthropomorphs, deer, lizards, armadillos, monkeys, and birds.

Among the bird paintings situated in the Cerca Grande Cliff, one resembles a cormorant (Neotropical Cormorant) with open wings, while others situated in the same limestone outcrop and in the shelter of Sumidouro Lake resemble herons (perhaps a Great Egret). Curiously, herons are more frequent in lakes that are drying, while cormorants are associated with flooded lakes. Thus, these two ancient paintings reference the cycles of droughts and floods that have been repeated since thousands of years ago, and probably helped to shape the lifestyle of the first inhabitants of the region since the late Pleistocene, when the climate became warmer and humid (see Araujo et al. 2005). These residents may have been hunter-gatherers and their prey may have included several species of medium and small-sized mammals (e.g., deer, anteaters, peccaries, monkeys, opossums, tapirs, and capybaras), birds (Greater Rhea, Red-legged Seriema, tinamous, guans, gallinules, and egrets), reptiles (terrestrial turtles and tegu lizards) and fish (Neves and Piló 2008; Silva 2013). It is noteworthy that capybaras and rails are still found in karst lakes, while tapirs (which also eat aquatic plants—Talamoni and Assis 2009), are known in the region only from the fossil record (Perez 2009). In addition, shells of the freshwater mussel *Diplodon* (family Hyriidae) are still found in the region and may have been used as cutting tools or ornaments by early inhabitants (Prous 1986/1990).

4 Waterbirds Associated with Karst Lakes

In order to evaluate and compare the richness, composition and abundance of waterbird species, Sumidouro, the largest lake of APA Carste, Grande Lake and Mares Lake were inventoried in October 1995 (IBAMA 1998a). Three other studies were designed to reveal the temporal dynamics of the avifauna in relation to the flood–drought cycles of several lakes in the region. Rodrigues and Michelin (2005) conducted monthly censuses of the waterbirds at Sumidouro Lake from June 1999 to December 2002. Dornas (2004) performed weekly censuses in Fazenda Santo Antônio Lake between June 2003 and September 2004. Nóbrega (2015) carried out censuses every two months for two years in 20 lakes that included some of the smallest and some of the largest.

The last three studies showed that the different phases of the lake cycle (filling, full, drying, and dry) create specific environmental conditions that have different plant formations and food resources associated with them. Because the size of a lake determines its duration, these same variables can be expected to vary with size as well. Consequently, different phases of the cycle, or different sizes of lakes, attract different species of waterbirds that find shelter, food, and/or nesting sites, or simply use them as a stopover during migration (Table 1). Thus, several attributes of the aquatic bird assemblage, such as species richness and abundance, and foraging guilds (groups of species that share common food types obtained in related ways), are expected to vary markedly throughout seasons and among lake sizes.

To investigate this, we fitted linear regression models relating species richness (S) to lake area (A), at the end of the dry season of 2011 (November), at the end of the rainy season of 2012 (March), and in between these two months (January 2012). The best fit was obtained with the semi-log species-area model: $S = z\ \log(A + 1) + k$ (Scheiner 2003). Thereafter, using the pooled data and a linear regression model with a dummy variable (u) indicating the months (November, January, and March), $S = z\ \log(A + 1) + k + vu\ \log(A + 1) + u + \varepsilon$, we compared the slopes and the intercepts of the previous models. Using a similar procedure, we tested for differences in abundance (log(N)) related to bird species richness (S), for the same aforementioned periods of time. We also performed linear regression with the semi-log species-area model for a single lake as it filled and dried during the study. Regression analyses were performed using Systat version 10 (Wilkinson 2000).

In addition, data of Dornas (2004) were reanalyzed to evaluate how species richness and abundance, and foraging

Table 1 Species of waterbirds recorded in the APA Carste, Minas Gerais, Brazil. Only nonpasserines are included in this list. Species for which we found evidence of breeding in the area are indicated by an asterisk. Species sequence and taxonomy follow Piacentini et al. (2015)

Taxa	English name	Historical records	Modern records
Anseriformes			
Anhimidae			
Anhima cornuta	Horned Screamer	2	
Anatidae			
*Dendrocygna viduata**	White-faced Whistling Duck		4, 5, 7, 9
*Dendrocygna autumnalis**	Black-bellied Whistling Duck		4, 5, 7, 9
Cairina moschata	Muscovy Duck	2	4, 7, 9

(continued)

Table 1 (continued)

Taxa	English name	Historical records	Modern records
Sarkidiornis sylvicola	Comb Duck		7
*Amazonetta brasiliensis**	Brazilian Teal	2, 3	4, 5, 7, 9
Netta erythrophthalma	Southern Pochard		5, 7, 9
Nomonyx dominicus	Masked Duck	2, 3	5, 7, 9
Podicipediformes			
Podicipedidae			
Tachybaptus dominicus	Least Grebe	1, 2, 3	4, 9
*Podilymbus podiceps**	Pied-billed Grebe	2, 3	4, 5, 7, 9
Ciconiiformes			
Ciconiidae			
Ciconia maguari	Maguari Stork	2, 3	5, 6
Jabiru mycteria	Jabiru	1, 2, 3	
Mycteria americana	Wood Stork		5, 6, 7, 9
Suliformes			
Phalacrocoracidae			
Nannopterum brasilianus	Neotropic Cormorant	2	4, 5, 7, 9
Anhingidae			
Anhinga anhinga	Anhinga	2, 3	4, 7, 9
Pelecaniformes			
Ardeidae			
Tigrisoma lineatum	Rufescent Tiger-Heron	1	9
Cochlearius cochlearius	Boat-billed Heron	2, 3	
*Nycticorax nycticorax**	Black-crowned Night Heron	2, 3	4, 5, 7, 9
*Butorides striata**	Striated Heron	2, 3	4, 5, 7, 9
Bubulcus ibis	Cattle Egret		4, 7, 9
Ardea cocoi	Cocoi Heron	1, 2, 3	4, 5, 7, 9
Ardea alba	Great Egret	1, 2, 3	4, 5, 7, 9
Syrigma sibilatrix	Whistling Heron		4, 5, 7, 9
Pilherodius pileatus	Capped Heron	2, 3	7
Egretta thula	Snowy Egret	2, 3	4, 5, 7, 9
Threskiornithidae			
Plegadis chihi	White-faced Ibis	2, 3	
Phimosus infuscatus	Bare-faced Ibis	2, 3	4, 5, 7, 9
Theristicus caudatus	Buff-necked Ibis		7, 9
Platalea ajaja	Roseate Spoonbill	2	4, 5, 6, 7, 9
Accipitriformes			
Pandionidae			
Pandion haliaetus	Osprey		10
Accipitridae			
Rostrhamus sociabilis	Snail Kite	2, 3	4, 7
Gruiformes			
Aramidae			
Aramus guarauna	Limpkin		4, 5, 7, 9
Rallidae			
*Aramides cajaneus**	Gray-necked Wood-Rail	1, 2, 3	4, 5, 9

(continued)

Table 1 (continued)

Taxa	English name	Historical records	Modern records
Aramides saracura	Slaty-breasted Wood-Rail	2, 3	4
Porzana flaviventer	Yellow-breasted Crake	2, 3	
Mustelirallus albicollis	Ash-throated Crake	2, 3	4
Pardirallus nigricans	Blackish Rail	2, 3	4
*Gallinula galeata**	Common Gallinule	2, 3	4, 5, 7, 9
*Porphyrio martinicus**	Purple Gallinule	2, 3	4, 7, 9
Porphyrio flavirostris	Azure Gallinule	2, 3	
Charadriiformes			
Charadriidae			
Vanellus cayanus	Pied Lapwing	1, 2, 3	7, 9
*Vanellus chilensis**	Southern Lapwing	2, 3	4, 7, 9
Pluvialis dominica		2, 3	7
Charadrius semipalmatus	Semipalmated Plover		8, 9
Charadrius collaris	Collared Plover	1, 2, 3	4
Recurvirostridae			
Himantopus mexicanus	Black-necked Stilt		10
Himantopus melanurus	White-backed Stilt	2, 3	4, 5, 7, 9
Scolopacidae			
Gallinago paraguaiae	South American Snipe	1, 2, 3	4, 5, 7, 9
Tringa solitaria	Solitary Sandpiper	2, 3	5, 9
Tringa melanoleuca	Greater Yellowlegs		7, 9
Tringa flavipes	Lesser Yellowlegs	2, 3	4, 5, 7, 9
Calidris melanotos	Pectoral Sandpiper	3	
Jacanidae			
*Jacana jacana**	Wattled Jacana	2, 3	4, 5, 7, 9
Sternidae			
Sternula superciliaris	Yellow-billed Tern	2, 3	
Gelochelidon nilotica	Gull-billed Tern		8, 9
Coraciiformes			
Alcedinidae			
Megaceryle torquata	Ringed Kingfisher	3	4, 5, 7, 9
Chloroceryle amazona	Amazon Kingfisher	3	4, 5, 7, 9
Chloroceryle americana	Green Kingfisher		4, 7, 9

Sources 1—Burmeister (1856); 2—Reinhardt (1870); 3—Krabbe (2007); 4—IBAMA (1998b); 5—Rodrigues and Michelin (2005); 6—Rodrigues (2008); 7—Dornas and Figueira (2012); 8—Nóbrega et al. (2015); 9—Nóbrega (2015); and 10—this study

Notes (1) Records of *Laterallus melanophaius* Rufous-sided Crake, *Gallinago undulata* Giant Snipe, and *Bartramia longicauda* Upland Sandpiper presented by Warming (1908) for Lagoa Santa and cited by Dornas and Figueira (2012) were not accepted here. These species were recorded by Lund and Reinhardt outside the limits of APA Carste, sometimes hundreds of kilometers away (Reinhardt 1870; Krabbe 2007); (2) The two American taxa included in the genus *Himantopus* have traditionally been considered as conspecific and included in the polytypic species *H. himantopus*, which is widely distributed throughout the world (Hayman et al. 1986). More modern taxonomic treatments of these two taxa have varied greatly among authors, some have chosen to split them into independent species, a treatment adopted by Piacentini et al. (2015). During our fieldwork, and in previous field campaigns to APA Carste, a number of individuals with the diagnostic characters of *H. melanurus* (see Hayman et al. 1986) were observed, and for which there are several records from municipalities neighboring Lagoa Santa. Some few individuals observed in APA Carste showed the typical characters of *H. mexicanus*, suggesting that the two species occur in sympatry in the study area, a fact already recorded for Belo Horizonte (see portal Wikiaves—www.wikiaves.com.br). However, since individuals with intermediate characters between the two species have already been recorded in south-central Minas Gerais (Pinto in Hellmayr and Conover 1948), and given the extensive individual and age variation exhibited by the species, we advise readers that the sympatric occurrence of the two species in the APA Carste has not yet been unequivocally confirmed, and that the inclusion of both of them in the list is tentative

* indicate bird species with some evidence of breeding in the temporary lakes

Table 2 Aquatic bird foraging guilds proposed in this study (bird nomenclature follows Piacentini et al. 2015)

Foraging guilds	Description	Genera
Active tactile waders —(ATW)	Feed by tactile groping while walking in shallow waters. Fish are the staple food, but diets are quite variable, including all sorts of invertebrates and vertebrates	*Mycteria* and *Platalea*
Active visual waders —(AVW)	Feed while wading in shallow waters or along the shoreline, gleaning their prey from the substrate. The two species of *Himantopus* also frequently forage by pecking and plunging. Insects are the staple food, but other invertebrates are frequently eaten	*Tringa* and *Himantopu–s*
Aquatic vegetation gleaners—(AVG)	Forage while walking over emergent or floating vegetation in marshy areas, gleaning insects and seeds, but also ingesting a wide range of plant and animal foods	*Porphyrio*, *Gallinula*, and *Jacana*
Dabbling ducks—(DAD)	Feed chiefly by swimming and dabbling in shallows close to the shore or bank. Seeds and other plant parts are their staple food, but insects, snails and other invertebrates are also eaten	*Amazonetta* and *Cairina*
Diving ducks—(DID)	Feed chiefly by swimming and up-ending or diving, especially when hidden among water plants. Seeds and other plant parts are the staple food	*Netta* and *Nomonyx*
Dry land foragers—(DLF)	Forages by probing in soft soil or in bunch grass tussocks, or by pecking at the surface of hard soil. These species are frequently found in dry grasslands, but also explore damp areas. Insects are the staple food, but all sorts of invertebrates and even small vertebrates are also ingested	*Theristicus*, *Syrigma*, and *Bubulcus*
Fresh marsh Mollusk-eater (FM)	Feed-in shallow waters, on thick mats of floating vegetation or on the shoreline, capturing prey during visual or tactile searching. Specialized mollusk eaters, ingesting snails and mussels	*Aramus* and *Rostrhamus*
Mud-probers—(MP)	Feed by vertical probing, eating insects, worms and other invertebrates	*Gallinago*
Plunge divers—(PD)	Sit-and-wait predators, search for food from water side perches, from where they plunge-dive into the water. Fish are the staple food, but insects and other invertebrates are also eaten	*Megaceryle* and *Chloroceryle*
Shoreline ground gleaners—(SGG)	Forage along the shoreline, actively walking and gleaning their prey from the ground. Insects are the staple food, but several groups of invertebrates are frequently eaten	*Vanellus* and *Pluvialis*
Small near-ground passerines—(SNP)	Small generalist insectivore birds that capture their prey on the ground or on shrubbery near water	*Fluvicola* and *Arundinicola*
Standing visual waders—(SVW)	Feed while wading in shallow waters or perched along the margins of water bodies. These are visual-oriented predators that strike prey while standing (waiting for potential prey to appear) or, more infrequently, walking slowly. Fish are the staple food, but insects and several other groups of invertebrates, as well as toads, lizards, snakes and small mammals, are eaten to some extent	*Ardea*, *Egretta*, *Butorides*, *Nycticorax*, *Pilherodius*
Underwater pursuit swimmers—(UPS)	Forages in shallows mainly by pursuit-diving from water's surface, using feet for propulsion. Fish are the staple food, but the grebe also ingests chiefly insects	*Nannopterum* and *Podilymbus*
Wading ducks—(WD)	Species in this guild are long-legged and feed equally well on land or on water. These species feed by various methods, dabbling in the shallows while wading, swimming and up-ending, and even by diving. Seeds and other plant parts are the staple food, but insects, snails, and small fish are eaten to some extent	*Dendrocygna* and *Sarkidiornis*

guilds, change from the wet season to the dry season and back to the wet season again. The classification of foraging guilds proposed here (Table 2) takes into account the main microhabitat, searching strategies, and diet types based on Hancock and Kushlan (1984), Hayman et al. (1986), Madge and Burn (1988), Fry et al. (1992), Hancock et al. (1992), Sick (1997), Taylor and van Perlo (1998), and Rodewald (2015). We noticed that substantial intra-guild variation does exist, as well as intraspecific variation in the behavior and diet of each species in relation to geography and season, among other factors.

Nonmetric multidimensional scaling (NMDS) was used to show the temporal changes in the aquatic bird assemblage. NMDS is an unconstrained ordination method that summarizes complex multivariate relationships among assemblages and graphically displays them within a small number of dimensions; it is known to perform very well when relationships are nonlinear (Clarke 1993; Kwak and Peterson 2007). NMDS was performed using PC-ORD (McCune and Mefford 2011). Prior to the analysis, bird abundances were log transformed and standardized for the calculation of Bray-Curtis distances. The number of dimensions was defined by the final stress values that were compared to randomized runs of the data set. The NMDS ordinated the 23 species according to their abundance in the middle and end of each month from June 2003 to September 2004. Instead of using species names, we labeled the NMDS plot with their foraging guilds (Table 2), to facilitate a functional interpretation. Spearman's rank correlations, performed with Systat, were used to test the following hypotheses: (i) the number of foraging guilds, and (ii) foraging guild composition vary with lake area as the number of species increases or decreases. For the second hypothesis (ii), we correlated the NMDS scores of axis II (the one which had the greatest correlation between ordination distances and distances in the original n-dimensional space) with the lake area.

4.1 Variation in Species Composition, Richness, and Abundance

When the lakes are not dry they attract gallinules, wood storks, cormorants, snail kites, ducks, sandpipers, stilts, jacanas, ibis, spoonbills, and kingfishers, among others. Several of these birds are migratory, but it is not known where they come from or where they go when the lakes dry. It is suspected that they migrate to other wetland areas of Minas Gerais, such as the lakes along the banks of the São Francisco River and/or its tributaries Bambuí and Paracatu (Faria et al. 2009), or even other Brazilian states, where there are seasonal concentrations of waterbirds, such as Peixe Lagoon (Rio Grande do Sul), Paraná River valley (Paraná), Araguaia River (Tocantins), and the Pantanal (Mato Grosso do Sul and Mato Grosso) (see Figueira et al. 2011).

The largest karst lakes have comparatively large littoral and limnetic zones and can reach several meters in depth (6.9 m was registered in Sumidouro in the decade of 1980). Their beds and banks can possess plants of different successional stages, a mixture of live and/or dead macrophytes and terrestrial plants, resulting from the retraction and advancement of the water.

Larger lakes also possess a great range of food resources, including fish, frogs, mollusks, other benthic invertebrates and the macrophytes themselves, with their leaves, seeds, and stalks, as well as the terrestrial and aquatic invertebrates that feed and shelter on them. Consequently, when different lakes are compared simultaneously, the larger the lake, the higher the species richness and the greater are their abundances. This same pattern is also observed for a single lake as it fills or dries (Figs. 5 and 6). Guadagnin et al. (2009) also found higher species richness and abundances of waterbirds in larger compared to smaller remnant wetlands, and Sebastián-González and Green (2013) described a similar pattern for waterbirds in a network of artificial temporary ponds.

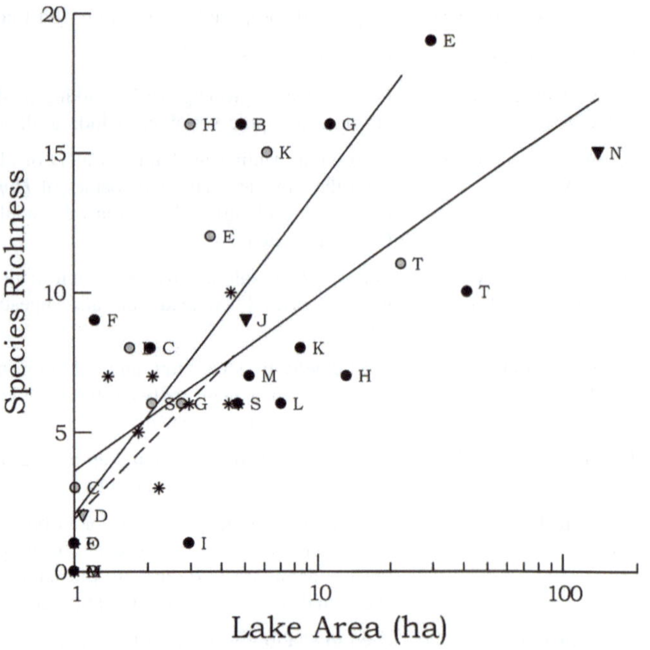

Fig. 5 Species-area relationships for temporary lakes at the end of the rainy season of 2011 (closed circles, lower slope line, S = 1.34 + 0.86 log(Area), p-value < 0.001, n = 16) and at the end of the dry season of 2012 (empty circles, steeper slope line, S = 1.04 + 1.50 log(Area), p-value < 0.001, n = 15), where S is the number of species of birds. Triangles indicate lakes that were not sampled on both occasions. Asterisks and dashed line indicate the behavior of a single lake (K) as it fills or dries. Letters indicate the lakes as in Fig. 1

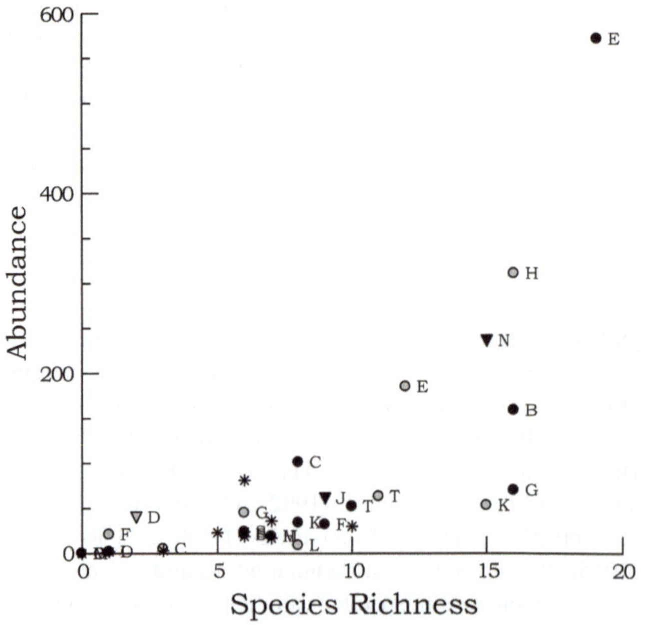

Fig. 6 The increase in waterbird abundance with species richness. Letters and symbols indicate the lakes as in Fig. 1

The species-area relationships for the end of the dry season of 2011, the end of the wet season of 2012 (Fig. 5), and in between these two time periods, do not differ (p-values = 0.165 for differences among intercepts and 0.605 for differences among slopes). No significant differences were also found among the intercepts and the slopes of the straight lines relating abundances (log(N)) to species richness (S), among the end of the rainy season of 2011, the middle of 2012 and the end of the dry season of 2012 (p-values = 0.066 for differences among intercepts and 0.104 for differences among slopes) (Fig. 6). The curvilinear relationship between abundance and species richness is a consequence of the arrival and departure of migratory birds and the short movements birds make among lakes in groups whose size can vary from a few to dozens of individuals. These results suggest a continuous adjustment of waterbird assemblages among filling or drying lakes, largely determined by lake size.

4.2 Types and Temporal Variations in Foraging Guilds

Snail kites (which belong to the FM guild) are attracted by lakes where Apple snails (mollusks of the genus *Pomacea*, family Ampulariidae) occur, but they also need trees or shrubs for perching. Cormorants (UPS) swim in flocks and dive in search of fish that are swallowed whole when they return to the surface. Solitary Kingfishers (PD) plunge into lakes to capture small fish. Dabbling (DAD) and diving ducks (DID) dip/submerge their heads, or even dive, in search of parts of plants, seeds, mollusks, and insects. Edge vegetation may be an important requirement for some species, such as the Purple Gallinule (AVG), which does not forage in open environments and needs to hide amidst vegetation. Aquatic macrophytes serve as shelter and their emergent leaves support birds like the Wattled Jacana (AVG), which walk on them in search of seeds, insects, mollusks, and spiders, aided by their long toes that distribute their weight. The depth of the lake can be more decisive than its area for wading birds: depths above 20–30 cm are already a barrier for some species, such as the Snowy Egret, the White-backed Stilt and even for some larger species. Temporary lakes also receive migratory birds fleeing the winter of the Northern Hemisphere, such as the Osprey (which make flybys to catch fish) and the Greater-yellow Legs, which winter in South America. In addition to these, errant migrants are sporadically observed, such as the Gull-billed Tern, a piscivore that was seen dipping swiftly to pluck prey from the water surface, and the Semipalmated Plover, which hunts invertebrates on the muddy banks (Nóbrega et al. 2015). Limpkins (FM) walk on the shallow shore of lakes and include in their diet, in addition to *Diplodon* mussels, Apple snails, whose shells are pierced for the extraction of the soft body.

As drought progresses, smaller lakes dry faster (Figs. 3 and 4) and are soon abandoned by birds who can concentrate in those lakes that still have water and are usually more perennial. This fact was well documented in the work of Sumidouro Lake by Rodrigues and Michelin (2005). In the dry season, there was a higher concentration of birds in terms of abundance, but not species richness. These birds probably came from smaller lakes that had already dried up. This suggests that several species of waterbirds of the karst lake system have a metapopulation dynamics determined by the size of the lakes. Sebastián-González et al. (2015) showed that Little grebes (*Tachybaptus ruficollis*, Podicipedidae) have metapopulation dynamics in irrigation ponds also related to pond sizes.

Drying lakes can be attractive places for foraging by birds such as the Wood Stork and the Roseate Spoonbill (ATW), as food becomes concentrated. Along the margins, which gradually become covered by creeping ground plants, birds that feed on jumping of flying insects, tadpoles, frogs, etc., can be found. This includes species such as the Buff-necked Ibis, Whistling Herons, and Cattle Egrets (all belonging to the DLF guild), which were also present during the flood when they foraged along the adjacent margins and pastures. Essentially terrestrial birds, such as Picazuro Pigeons (*Patagioenas picazuro*), Southern Caracaras (*Caracara plancus*), Turkey Vultures (*Cathartes aura*), and Black Vultures (*Coragyps atratus*), may also become more frequent in these dryer conditions. Picazuro Pigeons probably seek seeds in the newly exposed lakebed, while Caracaras and Vultures seek frogs, fish, bird eggs, and carcasses. At this time, sporadic insect clouds, possibly Chaoboridae, eventually emerge from newly exposed lakebeds.

With the first rains at the end of the dry season, herons and other predatory waterbirds explore the floodwaters, which are usually filled with toads and frogs that soon appear in huge densities and croak in a deafening chorus. On the other hand, filling lakes can become less attractive for the Wood Stork and the Roseate Spoonbill, because, in addition to the greater water depth, the little food that is available (mainly fish) becomes dispersed, which explains the absence of these birds. However, Cormorants and some ducks may achieve large numbers, mainly in deep lakes, since they search for fish, invertebrates, plants or seeds while diving underwater. Some of these predictable temporal changes in the bird assemblage were observed in the lake studied by Dornas (2004) (Fig. 7).

As the area of the lake studied by Dornas (2004, lake C) increased with the rains or decreased as rains ceased, the numbers of bird species ($r_s = 0.882$, p-value < 0.001, $n = 22$) and foraging guilds ($r_s = 0.807$, p-value < 0.001, $n = 22$) also increased or decreased, confirming our hypothesis.

Figure 8, which is a synthesis of Fig. 7, reveals the temporal trajectory of the bird assemblage discriminated into

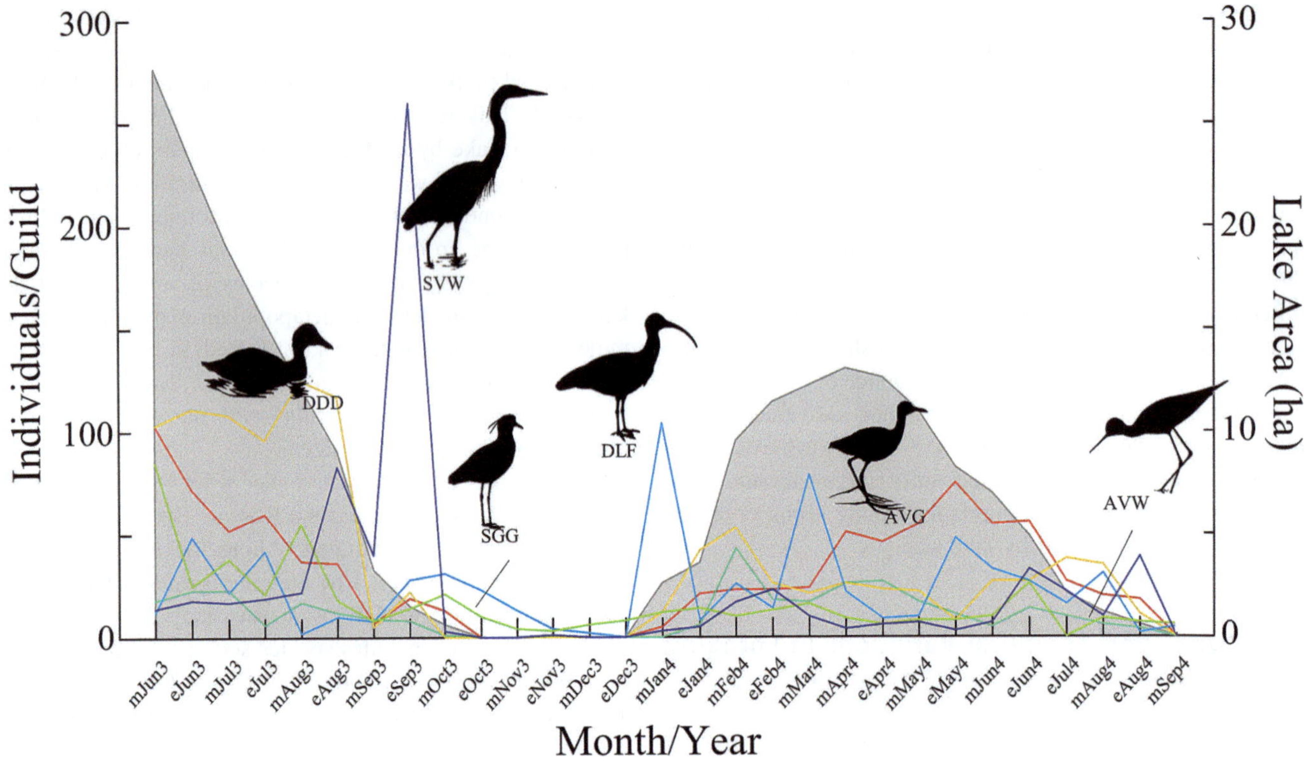

Fig. 7 Variation in water surface area (gray polygons) of the Fazenda Santo Antônio Lake (lake C) and the abundances of waterbirds belonging to different foraging guilds. DDD = dabbling and diving ducks, AVG = aquatic vegetation gleaners, SGG = shoreline ground gleaners, DLF = dry land forager, AVW = active visual waders, SVW = standing visual waders (after Dornas 2004—see Table 2 for guild descriptions). Birds silhouettes redrawn from Gwynne et al. (2010)

foraging guilds. In the upper quadrant are lakes with greater species richness and abundance of waterbirds belonging to the guilds WD, DD, AVG, AVW, UPS, and SVW.

Due to oscillations in water surface area and, consequently, of food resources, bird traffic between lakes and the arrival or departure of migratory birds, the species composition and the guild structure of the bird assemblage varies, but relatively little, while the area of its water surface remained larger than about 5% of its maximum area (Fig. 9). As the area reduces due to the lake drying, changes become more pronounced as the obligatory waterbirds abandon it, resulting in a simplified assemblage dominated by the guilds DLF and SGG in the lower quadrant (Fig. 8).

Guadagnin et al. (2005) and Figueira et al. (2011) described similar seasonal variations in bird assemblages, the first in the composition and abundance of waterbirds in a fragmented wetland, also related to their areas; the second, in the trophic structure of terrestrial and aquatic birds in a large landscape mosaic of the Brazilian Pantanal wetland.

Several theories have attempted to explain how biological communities are structured and which processes determine their biodiversity (Chesson and Case 1986). Karst lakes may better fit models where strong seasonal and stochastic fluctuations in environmental conditions lead to marked and unpredictable changes in resource levels through time, making population dynamics density-independent much of the time. Birds must opportunistically track those resources through time by adjusting their behavior (e.g. switching their preferred food items when only alternative food items are available or using a wide variety of food items) and distribution (e.g. moving between lake patches or transiting between lakes according to their area, depth, and successional stage; or escaping through migration when environmental conditions deteriorate and food resources become scarce). Thus, the seasonal variation observed in the waterbird assemblages of the karst lake system seems to corroborate the exploitation and escapism hypothesis proposed by Cumming et al. (2012), for the floodable Okavango River.

Birds can also adjust their reproductive effort (e.g. laying a smaller clutch size or reducing the number of breeding attempts or even giving up breeding for the season, according to lake size and duration). These factors favor the coexistence of noninteracting species, which promotes continuous changes in biodiversity, and therefore, waterbirds of the karst lakes may be regarded as belonging to nonequilibrium assemblages (Wiens 1984, 1992). Birds like the Whistling Heron and the Southern Lapwing are probably amongst the most adaptable and less dependent on water resources and, consequently, occur in the region throughout the year. On the other hand, the Wood Stork and the Roseate Spoonbill do not tolerate extreme variation in resource levels and commonly employ the strategy of escaping through migration.

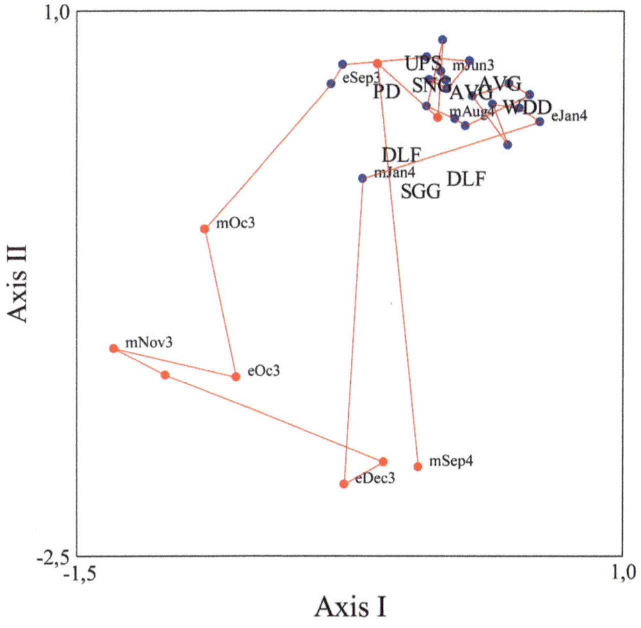

Fig. 8 Temporal trajectory of the aquatic bird assemblage of Fazenda Santo Antônio Lake (lake C), determined by nonmetric multidimensional scaling (NMDS) from June 2003 to September 2004 (letters m and e indicate middle and end of the months). Close points indicate that the assemblages possess similar composition and abundances of species, and the acronyms indicate the foraging guilds (Table 2) that characterize the different months. Blue colors indicate surface water areas greater than 5% of the maximum area measured at the beginning of the study (January 2003). When filled, the lake possesses birds mainly from the guilds WD, DD, AVG, AVW, UPS, and SVW, and in the drought it possesses the guilds DLF and SGG, which were also present during the flood, but in smaller abundance relative to the other guilds. For the sake of clarity some guild acronyms were not shown

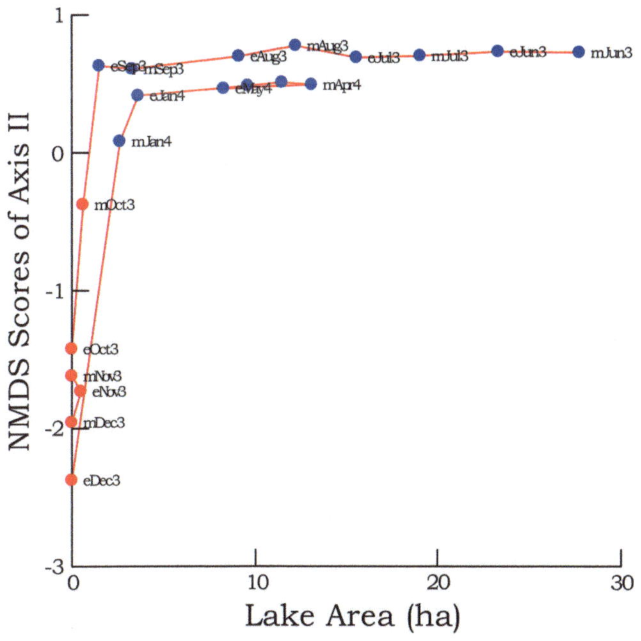

Fig. 9 Relationship between the scores of axis II of the NMDS and water surface area (rs = 0.688, p-value < 0.001, n = 26)

Plates 1 and 2 synthesize in images several aspects of the APA Carste temporary lakes, including phase shifts, waterbird assemblages and vegetation succession presented and discussed in this chapter.

5 Threats, Challenges and Local Conservation Efforts

The system of karst lakes at APA Carste is extremely fragile. The lakes, especially the smaller ones, are highly vulnerable to disturbances, mainly at their catchment basins, with urbanization as the most disruptive threat in the region (Nóbrega 2015). This threat is common to other karst systems as well (e.g. Hardt 2008; Milanovic 2002; Urich 2002), resulting in the suppression of native vegetation, contamination of surface or groundwater by effluent and solid waste, and changes in the flow and volume of rainwater drainage due to impermeability of paving or the filling of lakes for road construction. Urbanization also has the effect of suppressing areas already converted into agriculture and pasture that may have positive effects on the richness of waterbirds (Nóbrega 2015), since they provide alternative microhabitats for foraging (Dias and Burger 2005; Navedo et al. 2013).

Decreased availability of pastures in peripheral urban areas, where residents maintain rural habits, such as for free-range raising of livestock and horses, also has the consequence of increasing the pressure on the use of the basin or the immediate surroundings of lakes, especially in periods of decreased water or droughts. In addition to compacting the soil and reducing the capacity for water infiltration, the use of these areas for grazing and watering animals modify the natural cycles of change in the composition of the vegetation by the foraging itself or by the introduction of exotic species.

Despite all the direct and indirect negative consequences of urbanization, one can expect that this threat will increase in the region in the future since urban development and industrialization has been stimulated by current public policies that favor demographic growth (PDDI-RMBH, 2011a, b, c). Cement industries continue to destroy limestone outcrops with their forests and rupicolous vegetation; eucalyptus monocultures advance over the last remnants of the Cerrado. In addition, the expansion of the international airport in the city of Confins (one of the largest airports in Brazil), and the implementation of an industrial airport plans to destroy part of one of the last and largest native forest areas in the region at a time when the rains appear to become more unpredictable and scarce (ANA 2015), compromising the recharge of the karst aquifer.

This amount and variety of impacts are quite common in Environmental Protection Areas (APA), when compared to the other categories of the Brazilian System of Conservation

(a)
(b)
(c)
(d)
(e)
(f)
(g)
(h)
(i)
(j)

◀ **Plate 1** Sumidouro Lake at four different times of the year: **a** on March 2012, at the end of the rainy season; **b** in August 2013, at the peak of the dry season—note the creek that crosses the bare lakebed and disappears in a sinkhole (not shown); **c** in May 2014, at the beginning of the dry season, with the lakebed covered by aquatic plants, with a predominance of *Polygonum* and *Ludwigia* growing along the margins of the creek; **d** December 2016, at the beginning of the rainy season, after three years of low rains—the lakebed is now a heterogeneous field dominated by hygrophilous herbs and the remains of terrestrial, hygrophilous and aquatic plants which died during high and low water levels. *Polygonum* sp. now persists in wet patches while *Ludwigia* sp. has disappeared. **e** A mixed flock of Neotropic Cormorants (*Nannopterum brasilianus*) and Great Egrets (*Ardea alba*) at Sumidouro Lake during highwater; **f** Watermarks on a limestone fence in the lakebed of Sumidouro Lake at the peak of the 2013 dry season—note the advance of a "wave" of herbaceous terrestrial vegetation in the recently exposed lakebed behind the fence, and the bordering by Cerrado vegetation in the far background; **g** A Gull-billed Tern capturing a small fish during high water at Sumidouro Lake in October 2012; **h** Black-bellied Whistling duck (*Dendrocygna autumnalis*) with ducklings; **i** White-backed Stilt (*Himantopus melanurus*) feeding after a short period of rising water in Sumidouro Lake; **j** Roseate Spoonbills (*Platalea ajaja*) foraging in Sumidouro Lake at the peak of the 2013 dry season

Units ("Sistema Nacional de Unidades de Conservação"—SNUC). This is because, despite APA being *"an area, in general extent, with a certain degree of human occupation, endowed with abiotic, biotic, aesthetic or cultural attributes especially important for the quality of life and well-being of human populations"* and having *"basic objectives to protect biological diversity, discipline the process of occupation and ensure the sustainability of the use of natural resources"* [translated from Portuguese] (Brasil 2000), they are composed of public and/or private land, and the establishment of rules or regulations for private lands is restricted by constitutional limits. Therefore, as highlighted by the former manager and environmental analyst of APA Carste, Ricardo de Magalhães Barbalho (pers. com.), the private properties located within the APA Carste would have basic legal obligations to preserve the Areas of Permanent Preservation ("Áreas de Preservação Permanente"—APP), such as the margins of lakes and rivers and hilltops, and a Legal Reserve of 20% of the area of the property (see Brasil 2012). The property owner would have the right to use the rest of the area, even if the vegetation has to be suppressed, and since the APA Carste management plan does not contain any restrictions on the suppression of vegetation, only the general legislation is adopted. However, zoning of APA Carste (IBAMA 1998e, f) can contribute directly or indirectly to the conservation of vegetation, such as a Cultural Heritage Protection Zone ("Zona de Proteção do Patrimônio Cultural"), which restricts intensive animal husbandry, intensive agriculture, land parceling for urban purposes, and implementation of industries. In addition, more than 90% of the area of the APA is protected as a Special Protection Area ("Área de Proteção Especial"), designated for the protection of springs, and cultural, historical, landscape and archeological heritage (Minas Gerais 2009), which makes its native vegetation extremely protected, and which discourages the implementation of economic activities, if standards are met.

Despite all the impacts, APA Carste still holds a significant richness of waterbird species, nearly the same as that registered by Lund and Reinhardt more than 160 years ago, suggesting a resilient system in the face of disturbances. This richness is due in part to the flood–drought cycles, which determine constant changes in habitat and provide different types of resources for waterbirds, while the resilience relies in part on the ability of these birds to transit between the numerous lakes of the region and even to other wetlands of Brazil.

Seven species recorded in the 19th century (Reinhardt 1870) do not have modern records for the study area. These species are *Anhima cornuta*, *Cochlearius cochlearius*, *Plegadis chihi*, *Porzana flaviventer*, *Prophyrio flavirostris*, *Calidris melanotos*, and *Sternula superciliaris*. The lack of current records for these species in the study area does not necessarily represent a major conservation concern, since they all fall into at least one of the following categories (Sick 1997): (i) very rare and locally distributed species in Minas Gerais, even historically; (ii) species at the limit of their known geographic distribution, where they occupy marginal habitats; (iii) species considered vagrants by the authors of the historical records.

Waterbirds are, with some remarkable exceptions, animals of high vagility, broad geographic distributions and high tolerance to anthropic impacts, and thus are poor indicators of the condition of aquatic ecosystems (Stotz et al. 1996). Therefore, the concern of the conservationist should be directed towards the use of the land and the maintenance of water resources in the region. The consecutive years of drought observed during this decade demonstrated the fragility of the lake system of the study area, which has not been visited by some species even though they are still found in the region, such as *Jabiru mycteria,* or have received only small populations of more abundant species, such as *Platalea ajaja.*

The lake system of the karst should be viewed as an important inland stopover site for many of these birds, but particularly migratory species. Stopover sites are habitats that provide migrant birds a place to refuel and rest during their journeys. Inland stopover sites may not concentrate the large numbers of birds that wintering or breeding sites do, nevertheless, they are critical areas for the survival of these migrant birds (Winker et al. 1992). We know very little about the importance of the karst system as an inland stopover site. In addition, reproduction was observed in several species of waterbirds and in different lakes, generally the

(k)
(l)
(m)
(n)
(o)
(p)
(q)
(r)
(s)
(t)
(u)
(v)
(w)

◀ **Plate 2** **k** *Polygonum* sp. in cracked soil temporarily resists the advance of the dry season; **l** Cyperaceae and other terrestrial plants covering the lakebed where they grow and reproduce rapidly before the water rises; **m** Shrubs that are sensitive to flooding die and are used as perches by Chestnut-capped blackbirds (*Chrysomus ruficapillus*) and other birds; **n** An adult Snail-kite (*Rostrhamus sociabilis*) searching for Apple snails (*Pomaceae* sp.); **o** Shells of Apple snail shells at the base of a feeding-perch after the snails themselves were eaten by Snail kites in Cercado lake, Prudente de Morais; **p** Cyperaceae and other soft green plants covering the lakebed attract horses that are followed by Cattle egrets in the Santo Antônio Farm Lake, Mocambeiro; **q** A Mixed flock of Great Egrets (*Ardea alba*), Snowy Egrets (*Egretta thula*) and Wood Storks (*Mycteria americana*) foraging in the creek that crosses the lakebed of Sumidouro Lake, at the peak of the 2013 dry season; **r** Macrophytes of Sangradouro Lake: rooted emergent (*Cyperus*, *Juncus*), rooted in the bottom with emergent leaves (unidentified) and floating (*Eichhornia* sp.); **s** Limpkins (*Aramus guarauna*) searching for bivalves of the genus *Diplodon* in Sumidouro Lake at the peak of the 2013 dry season; **t** Fish, snails, frogs, insects, worms, bird eggs and carcasses, like the one shown here (probably a young Striated Heron—*Butorides striata*) in Maria Angelica Lake, are among the several animal food types that attract aquatic and terrestrial birds to the lakes; **u** A flock of White-faced Whistling Ducks (*Dendrocygna viduata*) leaving the Santo Antônio Farm Lake in the afternoon, and heading to a resting place; **v** and **w** Rock paintings of unidentified birds, possibly an egret and a cormorant, in the Cerca Grande State Park (both rock paintings reproduced by Maria Irene Neves and Sílvia Gaia Santana, from a panel at Peter Lund Museum)

bigger and more persistent in the landscape, highlighting their importance as reproductive sites.

APA Carste is a rich and unique, but highly threatened ecosystem. Because the temporary lakes are small, shallow, and well dispersed within its boundaries, they may be thought of as landscape "sentinels" of the weather, of climate changes, and of irresponsible (or responsible) local human actions. Large-scale conservation efforts are urgently needed because few areas of native vegetation and lakes (as well as their environments, springs, and other bodies of water) are within integral protection conservation units such as parks, natural monuments, and wildlife refuges which comprise only 6.4% of the area of APA Carste. The APA Carste includes two different categories of State Conservation Units: parks (Sumidouro and Cerca Grande) and natural monuments (Lapa Vermelha, Vargem de Pedra, Experiência da Jaguara, Santo Antônio, and Várzea da Lapa). Thus, incentives to implement existing conservation units and create other units in strategic areas for biodiversity conservation are urgently needed and the most important way of preserving species and guaranteeing the continuation of local ecological processes.

In an effort to help preserve this region, an educational project involving several public primary schools of the municipalities of the region, the Karst Wings' Web Project ("Projeto Rede Asas do Carste"—https://www.facebook.com/redeasasdocarste/, accessed 02 February 2018), was implemented in 2015. The monitoring of waterbirds in the temporary lakes by students [guided by monitors, their professors, a team of three subcommittees that fight for the preservation and restoration of the main rivers of the region (Carste and Ribeirão da Mata river basin subcommittees), the Rio das Velhas river basin committee and professors from the Federal University of Minas Gerais], is the background for several educational activities and valuation of the karst ecosystem.

In addition, it has been proposed to the Brazilian Environmental Ministry ("Ministério do Meio Ambiente"—MMA), that part of the APA Carste be designated as a wetland of international importance by the Ramsar Convention, since at least four criteria of the convention are satisfied (Figueira et al. 2014): (i) it contains a representative, rare, or unique example of a natural or near-natural wetland type found within the appropriate biogeographic region; (ii) it supports vulnerable, endangered, or critically endangered species or threatened ecological communities; (iii) it supports populations of plant and/or animal species important for maintaining the biological diversity of a particular biogeographic region; and (iv) it supports plant and/or animal species at a critical stage in their life cycles, or provides refuge during adverse conditions (specific criteria based on waterbirds). The application was approved in May 2015 by the National Council of Wetlands of Brazil "Conselho Nacional das Zonas Úmidas do Brasil" (CNZU) and was officially designated Lund Warming Ramsar Site by the Ramsar International Committee on June 05 2017 (https://rsis.ramsar.org/ris/2306, accessed February 26 2018).

Acknowledgements We thank the Instituto Estadual de Florestas for financial (20101010402410) and logistic support; the Sumidouro State Park for logistic support; Idea Wild for donation of equipment; the Instituto do Patrimônio Histórico e Artístico Nacional for access to the archeological sites of Lapa Vermelha and Cerca Grande; and the Fundação de Amparo à Pesquisa do Estado de Minas Gerais for funding. We thank Conselho Nacional de Desenvolvimento Científico e Tecnológico (CNPq) for a scholarship PIBIC for TD between 2003/2004, for a doctoral scholarship for PFAN, and a research fellowship for LEL (#305401/20149). Vanessa Monteiro and Ariadna de Souza Assis provided help in the field. We also thank Marilia Cunha Lignon, consultant hired by MMA to draft the first version of the Lund Warming Ramsar Site proposal; Ricardo Barbalho, who provided information on current environmental legislation in the APA; Ana Maria Lima from Instituto Estadual de Florestas (IEF), who modified the map of the APA Carste showing the studied lakes; Marcos Rodrigues, who provided useful suggestions and data about some bird species; Erika Suzanna Bányai, who provided reproductions of the cave paintings; José Hein who allowed the access to the lake of his farm; and Esther Sebastián-González, for her careful review and valuable suggestions.

References

Alt L (2008) Efetividade sócio-ambiental da APA Carste de Lagoa Santa - MG: uma avaliação a partir de suas ferramentas de planejamento e gestão. Dissertation, Universidade Federal de Minas Gerais

Alvares CA, Stape JL, Sentelhas PC, Gonçalves JLM, Sparovek G (2014) Köppen's climate classification map for Brazil. Meteorol Z 22:711–728

Amezaga JM, Santamaria L, Green AJ (2002) Biotic wetland connectivity—supporting a new approach for wetland policy. Acta Oecol 23:213–222

ANA (2015) Conjuntura dos Recursos Hídricos: Informe. Agência Nacional de Águas, Brasília

Araujo AGM, Neves WA, Piló LB, Atui JPV (2005) Holocene dryness and human occupation in Brazil during the "Archaic Gap". Quat Res 64:298–307

Auler A, Piló LB (2015) Lagoa Santa Karst: cradle of Brazilian cave studies. In: Vieira BC, Salgado AAR, Santos LJC (eds) Landscapes and landforms of Brazil. Springer, Dordrecht, pp 183–190

Baeta AMM (2011) Os grafismos rupestres e suas unidades estilísticas no Carste de Lagoa Santa e Serra do Cipó - MG. Thesis, Universidade de São Paulo

Berbert-Born M (2002) Carste de Lagoa Santa, MG - berço da paleontologia e da espeleologia brasileira. In: Schobbenhaus C, Campos DA, Queiroz ET, Winge M, Berbert-Born MLC (eds) Sítios Geológicos e Paleontológicos do Brasil, vol 1. DNPM/CPRM - Comissão Brasileira de Sítios Geológicos e Paleobiológicos (SIGEP), Brasília, pp 415–430

Blanco D (1999) Los humedales como habitat de aves acuaticas. In: Malvárez AI (ed) Topicos sobre Humedales Subtropicales y Templados de Sudamerica. Universidad de Buenos Aires, Montevideo, pp 219–228

Brasil (2000) Lei nº 9.985 Regulamenta o art. 225, § 1º, incisos I, II, III, e VII da Constituição Federal, institui o Sistema Nacional de Unidades de Conservação da Natureza e dá outras providências. Diário Oficial [da República Federativa do Brasil], vol 138 (19-07-2000) Seção I. Brasília

Brasil (2012) Lei nº 12.651. Dispõe sobre a proteção da vegetação nativa; altera as Leis nos 6.938, de 31 de agosto de 1981, 9.393, de 19 de dezembro de 1996, e 11.428, de 22 de dezembro de 2006; revoga as Leis nos 4.771, de 15 de setembro de 1965, e 7.754, de 14 de abril de 1989, e a Medida Provisória no 2.166-67, de 24 de agosto de 2001; e dá outras providências. Diário Oficial [da República Federativa do Brasil], vol 102 (28/05/2012) Seção I. Brasília

Burmeister H (1856) Systematische uebersicht der thiere Brasiliens, welche während einer Reise durch die provizen Von Rio de Janeiro und minas Geraës gesammelt ober beobachtet wurden, Vögel (Aves), vols 2, 3. Georg Reimer, Berlin, Germany

Chesson PL, Case TJ (1986) Overview: nonequilibrium community theories: chance, variability, history, and coexistence. In: Diamond J, Case TJ (eds) Community ecology. Harper & Row, New York, pp 227–239

Clarke KR (1993) Non-parametric multivariate analyses of changes in community structure. Aust J Ecol 18:117–143

Cumming GS, Paxton M, King J, Beuster H (2012) Foraging guild membership explains variation in waterbird responses to the hydrological regime of an arid-region flood-pulse river in Namibia. Freshwater Biol 57:1202–1213

Dias RA, Burger MI (2005) A assembléia de aves de áreas úmidas em dois sistemas de cultivo de arroz irrigado no extremo sul do Brasil. Rev Bras Ornitol 13:63–80

Dornas T (2004) Sazonalidade, riqueza e abundância de espécies de aves aquáticas associadas a uma lagoa temporária da APA Carste de Lagoa Santa, MG. Graduation Report, Universidade Federal de Minas Gerais

Dornas T, Figueira JEC (2012) Aves aquáticas da Área de Proteção Ambiental Carste de Lagoa Santa, Minas Gerais, Brasil: uma síntese comparativa entre os estudos de Peter Wilhelm Lund e Johannes Theodor Reinhardt em meados do século XIX e estudos contemporâneos. Cotinga 34:5–14

Faria LCP, Carrara LA, Amaral FQ, Vasconcelos MF, Diniz MG, Encarnação CD, Hoffmann D, Gomes HB, Lopes LE, Rodrigues M (2009) The birds of Fazenda Brejão: a conservation priority area of Cerrado in northwestern Minas Gerais, Brazil. Biota Neotrop 9:223–240

Figueira JEC, Mourão FA, Coelho AS (2011) Habitat heterogeneity and climatic seasonality structure the avifauna trophic guilds in the Brazilian Pantanal wetland. Can J Zoolog 89:1206–1213

Figueira JEC, Nóbrega PFA, Aguiar JAB, Lima AMS, Dornas T, Drumond MA, Lignon MC (2014) New Ramsar Site Proposal - Peter Lund Karst, Minas Gerais State, Brazil. Ministério do Meio Ambiente, Brasília

Fry CH, Fry K, Harris A (1992) Kingfishers, bee-eaters and rollers. Christopher Helm, London

Guadagnin DL, Maltchik L, Fonseca CR (2009) Species–area relationship of neotropical waterbird assemblages in remnant wetlands: looking at the mechanisms. Diversity Distrib 15:319–327

Guadagnin DL, Peter ÂS, Perello LFC, Maltchik L (2005) Spatial and temporal patterns of waterbird assemblages in fragmented wetlands of Southern Brazil. Waterbirds 28:261–272

Gibbs JP (2000) Wetland loss and biodiversity conservation. Conserv Biol 14:314–317

Gwynne JA, Ridgely RS, Society WC, Tudor G (2010) Wildlife conservation Society Birds of Brazil: The Pantanal and Cerrado of Central Brazil. Comstock Pub. Associates, New York

Hancock JA, Kushlan JA (1984) The Herons handbook. Harper & Row, New York

Hancock JA, Kushlan JA, Kahl MP (1992) Storks, ibises and spoonbills of the world. Academic Press, London

Hardt R (2008) Sistema cárstico e impactos antrópicos: considerações sobre o manejo. Paper presented at the 1º Simpósio de Pós-Graduação em Geografia do Estado de São Paulo, Universidade Estadual Paulista, Rio Claro, 17–19 November 2008

Hayman P, Marchant J, Prater T (1986) Shorebirds: an identification guide to the Waders of the world. Houghton Mifflin Company, Boston

Hellmayr CE, Conover B (1948) Catalogue of birds of the Americas and the adjacent islands in Field Museum of Natural History, part I, number 3, Jacanidae, Rostratulidae, Haematopodidae, Charadriidae, Scolopacidae, Recurvirostridae, Phalaropodidae, Burhinidae, Thinocoridae, Chionididae, Stercorariidae, Laridae, Rynchopidae, Alcidae. Field Mus Nat Hist Pub Zool Series 13:1–383

IBAMA (1998a) Meio Biótico, vol II. IBAMA/CPRM, Belo Horizonte

IBAMA (1998b) Patrimônio Espeleológico, Histórico e Cultural, vol III. IBAMA/CPRM, Belo Horizonte

IBAMA (1998c) Meio Físico – Estudos hidrológicos e qualidade das águas superficiais, vol I. IBAMA/CPRM, Belo Horizonte

IBAMA (1998d) Meio Físico - Síntese da Geologia, Recursos Minerais e Geomorfologia, vol I. IBAMA/CPRM, Belo Horizonte

IBAMA (1998e) Zoneamento Ambiental. IBAMA/CPRM, Belo Horizonte

IBAMA (1998f) Gestão Ambiental. IBAMA/Fund, BIODIVERSITAS/CPRM, Belo Horizonte

IBGE (1939) As Grutas em Minas Gerais. Departamento Geral de Estatística, Divisão de Estatística Fisiográfica e da Viação, Estado de Minas Gerais. Oficinas Gráficas da Estatística, Belo Horizonte

Junk WJ (2013) Current state of knowledge regarding South America wetlands and their future under global climate change. Aquat Sci 75:113–131

Junk WJ, Brown M, Campbell IC, Finlayson M, Gopal B, Ramberg L, Warner BG (2006) The comparative biodiversity of seven globally important wetlands: a synthesis. Aquat Sci 68:400–414

Kohler HC, Karfunkel J (2002) The quaternary morphogenesis of the Lagoa Santa tropical karst, Minas Gerais State, SE Brazil. Theor Appl Karst 15:93–99

Krabbe N (2007) Birds collected by P.W. Lund and J.T. Reinhardt in south-eastern Brazil between 1825 and 1855, with notes on P.W. Lund's travels in Rio de Janeiro. Rev Bras Ornitol 15:331–357

Kwak TJ, Peterson JT (2007) Community indices, parameters, and comparisons. In: Guy CS, Brown ML (eds) Analysis and interpretation of freshwater fisheries data. American Fisheries Society, Bethesda, pp 677–763

Madge S, Burn H (1988) Waterfowl: an identification guide to the ducks, geese and swans of the world. Houghton Mifflin Company, Boston

Martins ML (2008) História Ambiental e recursos hídricos nos domínios do carste. Cadernos Manuelzão. Edição Especial Lagoas Cársticas, Belo Horizonte

Mattos A (1961) O Homem das Cavernas de Minas Gerais. Editora Itatiaia, Belo Horizonte

Mattos GT, Andrade MA, Freitas MV (1993) Nova Lista de Aves do Estado de Minas Gerais. Fundação Acangaú, Belo Horizonte

McCune B, Mefford MJ (2011) PC-ORD. Multivariate analysis of ecological data. Version 6. MjM Software, Gleneden Beach, Oregon

Milanovic P (2002) The environmental impacts of human activities and engineering constructions in karst regions. Episodes 25:13–21

Minas Gerais (2009) Lei Estadual nº 18043. Modifica o decreto nº 20.597, de 4 de Junho de 1980, que define Área de Proteção Especial, situada nos municípios de Lagoa Santa, Pedro Leopoldo e Matozinhos, para fins do disposto no art. 13 da lei federal nº 6.766, de 19 de dezembro de 1979. Diário Oficial do Estado de Minas Gerais (DOEMG), vol (23/01/2009). Belo Horizonte

Morrone JJ (2014) Biogeographical regionalisation of the Neotropical region. Zootaxa 3782:1–110

Myers N, Mittermeier RA, Mittermeier CG, Fonseca GAB, Kent J (2000) Biodiversity hotspots for conservation priorities. Nat 403:853–858

Navedo JG, Arranz D, Herrera AG, Salmón P, Juanes JA, Masero JA (2013) Agroecosystems and conservation of migratory waterbirds: importance of coastal pastures and factors influencing their use by wintering shorebirds. Biodivers Conserv 22:1895–1907

Neves WA, Piló LB (2008) O Povo de Luzia - Em Busca dos Primeiros Americanos. Globo, São Paulo

Nóbrega PFA (2015) Aves aquáticas da Área de Proteção Ambiental Carste de Lagoa Santa: ecologia e conservação. Universidade Federal de Minas Gerais, Thesis

Nóbrega PFA, Aguiar JAB, Figueira JEC (2015) First records of *Charadrius semipalmatus*, Bonaparte 1825 (Charadriidae) and *Gelochelidon nilotica* Gmelin 1789 (Sternidae) in the State of Minas Gerais, Brazil. Braz J Biol 75:451–454

PDDI-RMBH (2011a) Plano diretor de desenvolvimento integrado da região metropolitana de Belo Horizonte - propostas de políticas setoriais, projetos e investimentos prioritários, vol 1. Governo de Minas Gerais, Belo Horizonte

PDDI-RMBH (2011b) Plano diretor de desenvolvimento integrado da região metropolitana de Belo Horizonte - propostas de políticas setoriais, projetos e investimentos prioritários, vol 2. Governo de Minas Gerais, Belo Horizonte

PDDI-RMBH (2011c) Plano diretor de desenvolvimento integrado da região metropolitana de Belo Horizonte - propostas de políticas setoriais, projetos e investimentos prioritários, vol 5. Governo de Minas Gerais, Belo Horizonte

Piacentini VQ, Aleixo A, Agne CE, Mauricio GN, Pacheco JF, Bravo GA, Brito GRR, Naka LN, Olmos F, Posso S, Silveira LF, Betini GS, Carrano E, Franz I, Lees AC, Lima LM, Pioli D, Schunck F, Amaral FR, Bencke GA, Cohn-Haft M, Figueiredo LFA, Straube FC, Cesari E (2015) Annotated checklist of the birds of Brazil by the Brazilian Ornithological Records Committee/Lista comentada das aves do Brasil pelo Comitê Brasileiro de Registros Ornitológicos. Rev Bras Ornit 23:91–298

Piló LB (2003) Mapa arqueológico da APA Carste de Lagoa Santa. Universidade de São Paulo, São Paulo

Perez CP (2009) Paleoecologia de mamíferos viventes como ferramenta na caracterização do ambiente holocênico de Lagoa Santa, MG. Dissertation. Universidade de São Paulo

Prous A (1986/1990) Os moluscos e a arqueologia brasileira. Arq Mus Hist Nat UFMG 11:241–298

Radzikowski J (2013) Resistance of dormant stages of planktonic invertebrates to adverse environmental conditions. J Plankton Res 35:707–723

Rebellato L, Nunes da Cunha C, Figueira JEC (2012) Respostas da comunidade herbácea ao pulso de inundação no Pantanal de Poconé, Mato Grosso. Oecol Austr 16:797–818

Reinhardt J (1870) Bidrag til Kundskab om Fuglefaunaen i Brasiliens Campos. Videnskabelige Meddelelser fra den Naturhistoriske Forening i Kjöbenhavn

Rodewald PE (2015) The birds of North America. https://birdsna.org. Accessed 15 May 2016

Rodrigues M (2008) Noteworthy bird records at Lagoa Santa, southeastern Brazil. Rev Bras Zool 25:150–153

Rodrigues M, Michelin VB (2005) Riqueza e diversidade de aves aquáticas de uma lagoa natural no sudeste do Brasil. Rev Bras Zool 22:928–935

Sampaio JLD (2010) Inventário digital da APA (Área de Proteção Ambiental) Carste de Lagoa Santa e algumas implicações. Thesis, Pontifícia Universidade Católica de Minas Gerais

Scheiner SM (2003) Six types of species-area curves. Glob Ecol Biogeogr 2:441–447

Sebastián-González S, Alexander KL, Sánchez-Zapata JA, Botella F (2015) Key factors controlling the metapopulation dynamics of the Little Grebe Tachybaptus ruficollis in an artificial habitat network of irrigation ponds. Ardeola 62:151–161

Sebastián-González E, Green AJ (2013) Habitat use by waterbirds in relation to pond size, water depth, and isolation: lessons from a restoration in Southern Spain. Rest Ecol 22:311–318

Sick H (1997) Ornitologia Brasileira. Editora Nova Fronteira, Rio de Janeiro

Silva JMC (1995) Birds of the Cerrado region, South America. Steenstrupia 21:69–92

Silva JMC, Santos MPD (2005) A importância relativa dos processos biogeográficos na formação da avifauna do Cerrado e de outros biomas brasileiros. In: Scariot A, Souza-Silva JC, Felfili JM (eds) Cerrado: Ecologia, Biodiversidade e Conservação. Ministério do Meio Ambiente, Brasília, pp 220–233

Silva L (2013) O antropogênico e o geogênico na préhistória de Lagoa Santa: estudo geoarqueológico da Lapa do Niáctor. Dissertation, Universidade Federal de Minas Gerais

Stotz DF, Fitzpatrick JW, Parker TA III, Moskovits DK (1996) Neotropical birds: ecology and conservation. University of Chicago Press, Chicago

Talamoni SA, Assis MAC (2009) Feeding habit of the Brazilian tapir, *Tapirus terrestris* (Perissodactyla: Tapiridae) in a vegetation transition zone in south-eastern Brazil. Zool 26:251–254

Taylor B, van Perlo B (1998) Rails: a guide to the rails, crakes, Gallinules and coots of the world. Pica Press, Crowborough, Sussex

Urich PB (2002) Land use in karst terrain: review of impacts of primary activities on temperate karst ecosystems. Sci Conserv 198:1–60

Viveiros E, Martins ML (2002) As paisagens, os recursos naturais e o modo de vida na região de Fidalgo e Sumidouro (1860–1930). Cronos: Rev Hist 5:57–76

Warming E (1908) Lagoa Santa: Contribuição para a Geographia Phytobiologica. Imprensa Oficial do Estado de Minas Gerais, Belo Horizonte

Wiens JA (1984) On understanding a non-equilibrium world: myth and reality in community patterns and processes. In: Strong DR Jr, Simberloff D, Abele LG, Thistle AB (eds) Ecological communities: conceptual issues and the evidence. Princeton University Press, Princeton, pp 439–457

Wiens JA (1992) The ecology of bird communities. Cambridge University Press, Cambridge

Wilkinson L (2000) SYSTAT 10: statistics II. SPSS Inc, Chicago, IL

Winker K, Warner DW, Weisbord AR (1992) Daily mass gains among woodland migrants at an inland stopover site. Auk 109:853–862

Soils of the Lagoa Santa Karst

Luís B. Piló and Selma Simões de Castro

Abstract

Red Latosols are the predominant soil type in the Lagoa Santa Karst High Plains and they can be up to 10 m deep on the high and middle slopes. Representative profiles of the soils have been described morphologically and micro-morphological, granulometrical and chemical analyses have been conducted. Two horizon groups were repeatedly identified in the surveyed toposequences: an upper red horizon and a lower yellow one. In both cases, the granulometric analysis revealed the predominance of a highly clay-like texture. Chemical analyses showed that the soils are acid with a low permeable base totals (S) and low cationic exchange capacity (CEC), indicating soils pedogenetically well evolved. That is corroborated by the Ki relations indicating the presence of kaolinite and therefore a process of mono-sialitization (secondary silicate formation) typical of tropical climates. The source material of both the red and the yellow horizons was correlated to meta-siltites and meta-argillites of the Serra de Santa Helena Formation deposited over limestones of the Bambuí Group. It can be concluded that the karst modeling of the Lagoa Santa Karst High Plains evolved due to a set of inter-relations between the epikarst, the surface karst and the underground karst with various pieces of evidence reinforcing the idea of a connected system in which the soil cover plays a major role.

Keywords

Lagoa Santa Karst • Soils • Pedogenesis • Epikarst

L. B. Piló (✉)
Instituto do Carste e Grupo Bambuí de Pesquisas Espeleológicas, Rua Conselheiro Mata 35/301, Belo Horizonte, Minas Gerais 31010-510, Brazil
e-mail: lbpilo7@gmail.com

S. S. de Castro
Department of Soil Science, Superior School of Agriculture Luiz de Queiroz, University of São Paulo, São Paulo, Brazil
e-mail: selma.castro@uol.com.br

1 Introduction

Well-developed soil is usually composed of the super-positioning of two groups of horizons each one characterized by very distinct mineralogical and structural properties. The upper one nearer to the surface and known as the *solum* is made up of horizons A and B. They are the result of innumerable mineral and structural alterations brought about by various aspects that control their formation, including geomorphological processes and biological activity. An accumulation of humified organic material is frequently found in horizon A. The deeper set also known as the C horizon, horizon of alteration or saprolite, is the result of the progressive alteration of the subjacent rock or sediment of which it conserves at least some of the primary minerals that are still susceptible to alteration and sometimes even maintain the structure of the rock itself (Delvigne 1983).

As regards their formation, the soils that develop on carbonate rock formations differ from soils formed on other types of rock by the notable dissolution of $CaCO_3$, its removal by leaching processes and sometimes, its re-precipitation when it may even form new horizons rich in calcium (Pedro 1976, 1987). In that regard, Lamouroux (1972) stressed that the weathering of limestone rocks occurs through a process of pellicular alteration in which the disaggregation and dissolution of the rocks are conditioned by mineralogy, texture, structure, porosity, and climate. That pellicular alteration process leads to the accumulation of an insoluble residue of the carbonate rock which is normally only found in small quantities in the pure limestones (<10%). White (1988) states that the insoluble portion of limestone rocks mainly consists of SiO_2, Al_2O_3, and Fe_2O_3 and that silica may be present in the form of chert, authigenic, or detritic grains of sand or as a component of other silicates; and that among the silicate minerals most commonly found in carbonate rocks are kaolinite and illite. Aluminum and iron may be present that are adsorbed to clay

A. S. Auler and P. Pessoa (eds.), *Lagoa Santa Karst: Brazil's Iconic Karst Region*, Cave and Karst Systems of the World, https://doi.org/10.1007/978-3-030-35940-9_6

minerals or in the form of hydrated oxides (gibbsite, goethite).

According to Pedro (1976), in the evolution of surface karst landscapes in wet climate conditions, there is a relative accumulation of clays essentially inherited from the limestone rock which liberates them during its alteration. Duchaufour (1968) explains that in that situation of the dependence of the residual product, the limestone-derived soils are very old and consequently they are almost always polycyclic. In his studies of soils overlying limestone formations in Switzerland, Germany, Bulgaria, Pakistan and the former country of Yugoslavia, Ciric (1967) demonstrated that direct relationship of dependence on the type of soils formed on the mode of alteration of the subjacent limestone. The dolomitized limestones or marls (with from 30 to 40% of residues) are associated with mineral soils formerly denominated rendzine but nowadays classified among the chernozemic soils in the Brazilian soil classification system (EMBRAPA 2013) and in the mollisols group in US Soil Taxonomy. Such soils result from the greater intensity of formation and their profiles vary from moderately deep to deep. However, the mineral part of soils that develop over pure limestone rocks (with $\geq 90\%$ of $CaCO_3$) originates from the chemical decomposition of the rock itself in the form of an insoluble residue notably in organogenic and non-carbonate soils with slow formation processes and not very deep profiles.

Nevertheless, in many cases, it may be that there is no genetic relation at all between the *solum stricto sensu* and the subjacent altered material or even with the limestone rock found at the base of the profile. In such cases the material that originated the soils may have been allochthonous, that is to say, the hard rock at the base of the profile was covered by other geological materials of varied origins from which the soil developed (White 1988; Bouyer 2004). Outstanding among such materials are alluvial mud (Olson et al. 1980), volcanic ash (Comer 1974), aeolian sediments (Yaalon and Ganor 1973; Muhs et al. 2007) or metapelite rock settled on top of the limestone (Lund 1837; Boulet et al. 1992). In that regard, the residual (autochthonous) or detritic (allochthonous) origin which gave rise to the soils over limestones rocks have been the object of a long debate that began with the discussion of the genesis of the *Terra Rossa* of the Mediterranean region (MacLeod 1980) and which went on in several countries such as Australia (Foster et al. 2004), Barbados (Ahmad and Jones 1969), Bahamas and Florida (Muhs et al. 1990), China (Feng et al. 2009), Brazil (Lynch 2009) and others. A mixed origin for such soils, that is to say, soils originated by materials that had a co-participation of both autochthonous and allochthonous materials, is another concept that several authors have defended for some time now (Yaalon 1997).

Brazil is a country with a tropical wet (equatorial) and sub-tropical (savannah) climate and it has an extensive area of carbonate rock outcrops estimated by Auler et al. (2001) to cover about 190,000 km^2, most of it covered by soils. In a far-reaching biographic survey, Shinzato (1998) reported great diversity among the soils directly influenced by the alteration of limestones rocks, among them cambisols, vertisols, rendzines, and lithosols. According to the Brazilian soil classification system, the SiBCS (EMBRAPA 2013) the last two are now called *chernossols* (chernozems) and litholic neosols. Podzols were also reported (*argissolos* in the SiBCS) and latosols which, according to that author, could have been influenced by material of allochthonous origin. However, in addition to the fact that there have been few studies of soils overlying limestone rock, such studies have also failed to contribute much towards a better understanding of the relations between the soils and the karst, especially in regard to morphogenesis versus pedogenesis relations.

Earlier studies of soils in the Lagoa Santa Karst High Plain (Boulet et al. 1992; Piló 1998, 2004; Shinzato 1998, and others) reported a clear predominance of latosols. Latosols are generally correlated with oxisols (American soil taxonomy) and ferralsols (WRB system). In the tropical regions of Brazil, latosols are considered to be geochemically and pedogenetically well evolved with deep, homogeneous profiles, whatever the substrate, and they predominate in more elevated geomorphic surfaces that are smoothed and geomorphologically stable, being mainly located on the highest thirds or tops of slopes (EMBRAPA 2013). The occurrence of cambisols and podzols (*argissolos*) is more limited but they have also been registered for the Lagoa Santa Karst (IBAMA-CPRM 1998; Shinzato 1998).

The aim of this chapter is to present and characterize the soils of the Lagoa Santa Karst High Plains as being an outstanding component of the regional karst as a whole and clearly influencing the genesis and evolution of the surface karst, the epikarst and the underground karst.

2 Study Area

The Lagoa Santa Karst is located approximately 40 km north of Belo Horizonte, the capital of the state of Minas Gerais, in Brazil's southeast macro-region, and it encompasses parts of the municipalities of São José da Lapa, Lagoa Santa, Pedro Leopoldo, Matozinhos, Prudente de Morais, Funilândia and Sete Lagoas. The classic Lagoa Santa Karst High Plain corresponds to the eastern and southeastern sectors of the Lagoa Santa Karst and its landscape is marked by the presence of doline fields and limestone outcrops (Fig. 1). The predominant soils in the area are deep, homogenous latosols, geochemically and morphogenetically

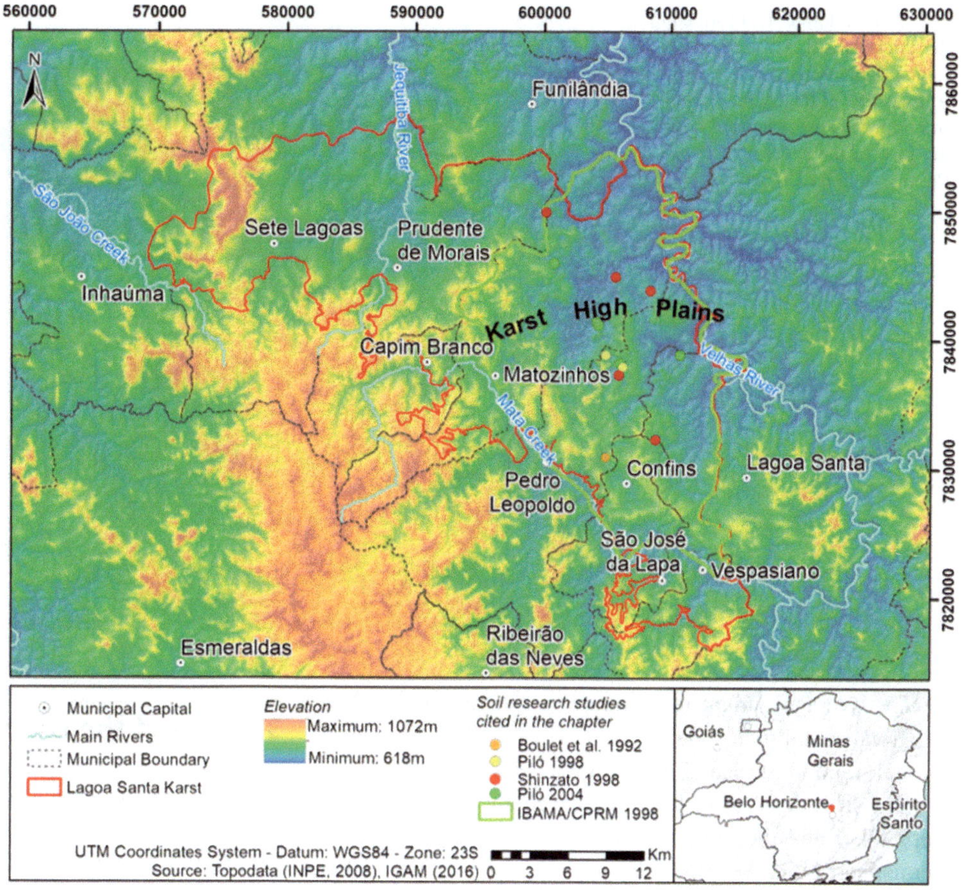

Fig. 1 Location of the Lagoa Santa Karst and the areas of soil research referred to in this chapter

well evolved but little is known about their material of origin and they are very different from the red Mediterranean soils that usually typify karst soils. That is why the attention of this chapter has been devoted to the soils of the Lagoa Santa Karst High Plains.

The regional geology is comprised of two formations of the Neoproterozoic Bambuí Group: the Serra de Santa Helena and the Sete Lagoas formations (Branco and Costa 1961; Dardenne 1978). On the top, there is the Serra da Santa Helena Formation made up of a metapelite sequence consisting of meta-siltites and meta-argillites (Tuller et al. 1992). The Sete Lagoas Formation is divided into two members: the Lagoa Santa Member and the Pedro Leopoldo Member. The Lagoa Santa Member consists of a carbonate sequence predominantly represented by calcarenites. The calcarenites are dark gray limestones with an abundance of calcite. These rocks are usually quite pure with over 95% $CaCO_3$ content (Table 1) and a considerable propensity for karst formation. The Pedro Leopoldo Member is formed by a carbonate sequence of calcisiltites and subordinately, by calcarenites and loams. The calcisiltites are impure limestones, sometimes containing silica (Table 1), with overall gray color with darker intercalations (Tuller et al. 1992). At the base of the carbonate sequence, there are igneous and metamorphic rocks (Auler 1994).

Table 1 Average percentage contents of $CaCO_3$, $MgCO_3$, SiO_2, Al_2O_3, and Fe_2O_3 in carbonate rocks of the Bambuí Group in the Lagoa Santa Karst High Plains (Piló 1998)

Parameters	Members	
	Limestone—Lagoa Santa Member (32 samples)	Limestone—Pedro Leopoldo Member (6 samples)
$CaCO_3$	96.34	67.08
$MgCO_3$	1.37	3.03
SiO_2	0.34	23.0
Al_2O_3	0.25	2.95
Fe_2O_3	0.18	1.75
Total	98.48	97.81

Fig. 2 **a** A doline field formed in the thick latosol cover of the Lagoa Santa Karst High Plains; **b** limestone cliff in which the Sumidouro Cave is located beside the lake of the same name. Photos: Ataliba Coelho

The average annual temperature in the region is 23 °C and relative humidity varies from 60 to 77%. The average annual rainfall is 1,355 mm with a seasonal distribution pattern. The dry season lasts for five months, from May to September and the rainy reason occurs during the southern hemisphere summer, especially in the months of December, January, and February (Nimer and Brandão 1989). That 3-month period registers the highest water surplus and surface, sub-surface, and subterranean runoff activity.

The karst landscape of Lagoa Santa lies in a plateau domain on the interfluvial block between Mata Creek and Velhas River at altitudes varying from 650 to 900 m (Kohler 1989). It displays a well-developed tropical karst with certain specific surface features, thick soil cover and the presence of doline fields (Fig. 2a). On one side of the dolines, limestone rock may come to the surface in the form of limestone cliffs (Auler and Piló 2015). The base of such formations normally coincides with the doline bottom which may or may not be occupied by lakes (Fig. 2b) and in many cases entrances of horizontal caves occur. Outstanding among them are caves with a network or anastomotic pattern showing clear evidence of paragenetic evolution (Coutard et al. 1978; Auler 1994, see Chapter "Caves and Speleogenesis in the Lagoa Santa Karst", this volume).

The caves can be up to 2 km long and inside there are considerable deposits of red and yellow clastic material alternating with stalagmitic crusts that mark the various phases of (clastic and chemical) sedimentation. Important sedimentary facies often bear the remains of extinct Pleistocene fauna species (Lund 1845; Cartelle 2012; Auler et al. 2006).

3 Soils of the Lagoa Santa Karst High Plains

The soils of the Lagoa Santa Karst High Plains were analyzed on the basis of four toposequences denominated Baú I, Baú II, Baú IV (Piló 1998) and Lapa Vermelha—LV (Boulet et al. 1992), in which 17 trenches and 75 boreholes were analyzed. The toposequences were examined to identify the role of surface processes in the lateral differentiation of the soils on the slopes and to observe the relative importance of geochemical and pedological processes in the evolution of the relief, in addition to evaluating whether the pedological covers are in dynamic equilibrium in the current landscape (Boulet et al. 1993; Ruellan and Dosso 1993; Queiroz Neto 2000) and their relations with the limestone substrate. Shinzato's (1998) classical analysis of seven profiles was taken into consideration of as well as the semi-detailed mapping of the Lagoa Santa Karst Environmental Protected Area in which 18 trenches and 40 boreholes were analyzed (IBAMA-CPRM 1998). The locations of the study areas are displayed in Fig. 1.

In the toposequences, the dominant presence was of thick layers of soils up to 10 m deep and consisting mainly of latosols with varying colors, textures, and structures. Two horizons were outstanding and repeated in the toposequences: an upper red horizon and a lower yellow one (Boulet et al. 1992; Piló 1998). The same super-position was observed in several cuts made for the highways that cross the Lagoa Santa Karst High Plains (Fig. 3).

On the tops and high slopes that super-positioning is even more apparent because they are not subject to the influence

Fig. 3 Highway cut exposing the two main sets of pedological horizons of the Lagoa Santa Karst High Plains: the red upper horizon and the yellow lower one

of doline which fosters soil transformations through hydromorphy and nodulation. The red horizons (2.5 YR) can be up to 7 m in depth on the top of the Baú II toposequence, before they make vertical contact with the underlying yellow horizons. In the case of the LV toposequence, the red latosol in the upper third of the slope goes down to a depth of 8 m getting shallower on the downside of the slope (Boulet et al. 1992). In the Baú IV toposequence, the red horizons are not so thick and they make contact with the yellow horizons at a depth of less than 2.5 m (Fig. 4).

It is worth noting in Fig. 4 that the doline truncates the soil but the soil carries on to the downhill side keeping up the same pattern among the horizons in relation to the current topography which indicates that the doline formed after the latosol was developed. Also, beneath the doline there is a notable depression of the limestone substrate. The illustration highlights two interesting features; first, the soil horizons in the doline are concordant with one another and with the topography of the doline, and second, that they are discordant with the topography of the substrate which suggests there may have been a collapse, a common occurrence in the region, and subsequently a progressive sedimentation.

The toposequences Baú I, II, and IV revealed a predominance of very homogeneous latosols with a very clayey texture (over 60% clay content) structured in blocks near to the surface and gradually turning to the fine granulate texture (micro-aggregate) in the red horizons (B1). The yellow horizons are resistant to the auger, structured in blocks and with a strong tendency to be prismatic, suggesting the presence of expansive clays. In the red horizons porosity is intergranular and in the yellow ones, it is fissural and apparently less developed.

Boulet et al. (1992) identified two nodular horizons at a depth of more than 7 m and interpreted them as two stable positions of a former water table prior to the latosol-forming process on top of the yellow horizons in which the nodular horizons were found (Fig. 5). A nodular horizon was also recorded for the Baú IV toposequence, which was continuous and located at the level of contact between the red soil and the yellow soil (Piló 1998), possibly indicating the top of the temporary, perched, epikarst aquifer.

In the lower slope of the toposequence LV, Boulet et al. (1992) reported a thinning of the latosol cover and the rising of the yellow horizon nearer to the surface. That, however, does not correspond to a toposequence differentiation but instead, the present-day topography has been truncated by erosion which has removed the red latosol from the vicinity of the doline (Fig. 5). In the toposequences Baú I, II and IV, the horizons in the lower part of the slope are arranged parallel to the surface and there are no clear signs of erosion or any important differentiation among the brown, red and yellow horizons.

Figure 5 differs from Fig. 4 in various aspects. First, the doline is further down the slope than in the earlier figure and the latosol cover in Fig. 5 is interrupted in the middle third of the slope and disappears which suggests truncation resulting from erosion. The volumes of compacted clay are embedded in a micro-aggregate matrix like those of the latosols uphill which suggests the occurrence of a morpho-pedological adjustment of the cover by reorganization of the material in

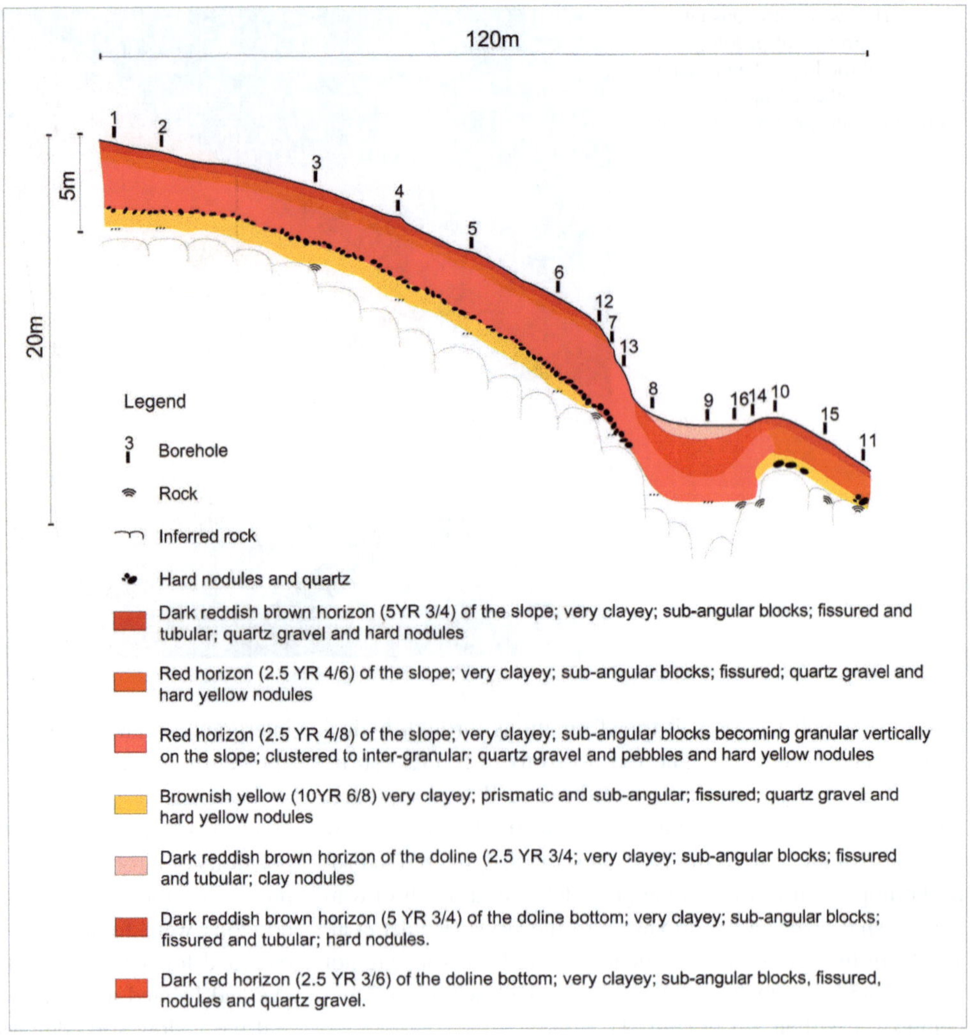

Fig. 4 Baú IV Toposequence showing the red and yellow horizons parallel to one another and to the current topographical surface (Piló 1998)

the light of the erosive truncation it suffered further down where the presence of variegated material on the edge of the doline suggests the material underwent hydromorphy.

The horizons in the bottom of the doline are more complex, more numerous, sometimes concentrated and may present nodules associated with the seasonal presence of the water table. Colors vary from reddish brown, to yellowish brown, to stained and variegated and textures are marked by increased participation of silt and sand. In all the doline bottoms, organic soils (organosols) buried under sediments were found.

Reconstruction using geometry approximation for the horizons surveyed and present in the toposequences showed that they are laid out more or less parallel to the present-day topographical surface meaning that they are concordant among one another and with the surface. According to Boulet et al. (1993), that kind of soil cover with no discordance among horizons that are parallel to the topography of the surface shape of the relief are formations in equilibrium with the prevailing pedoclimatic conditions, that is to say, they have evolved over time in climatically and tectonically stable conditions so that the transformation sequence from the bedrock to the soil surface is maintained. That means their organization reveals continuity in the alteration and pedogenetic processes insofar as they show no evidence of lateral transformation fronts which would result from more significant climate changes or from the downcutting of the base level associated with them or from tectonic events. On the contrary, when there is any evidence of notable differentiation as in the case of the LV toposequence, it is probably due to the influence of erosive truncation phenomena.

The Baú I and II toposequences are examples of a complete vertical transformation system still in equilibrium with the current karst model. The Lapa Vermelha toposequence would seem to show evidence of a small transformation due to the presence of the doline where there is erosive

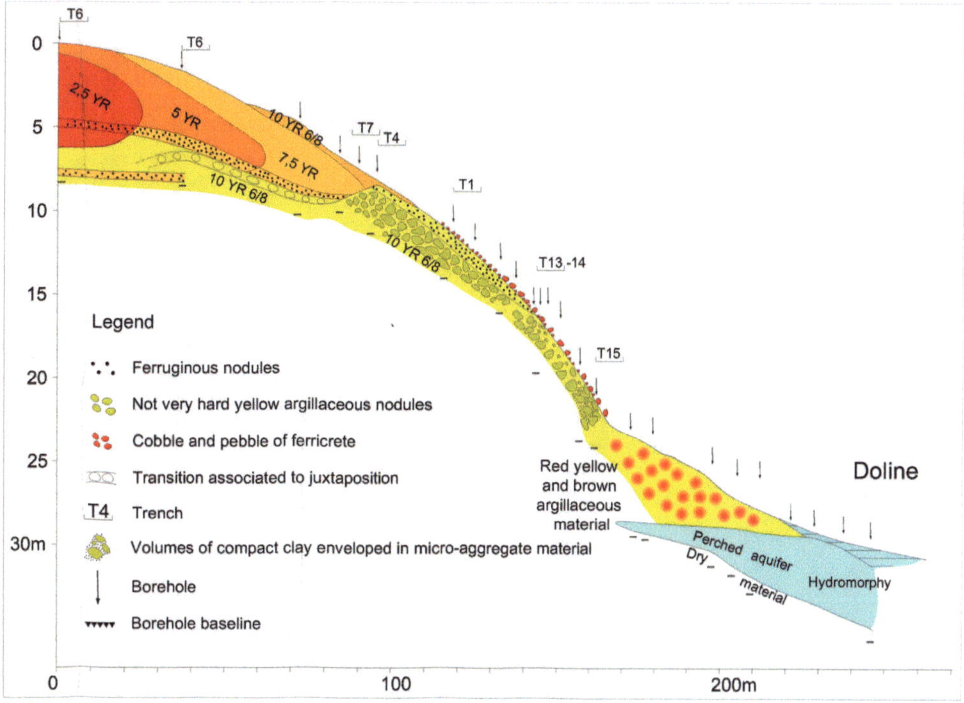

Fig. 5 Toposequence of the Lapa Vermelha. Adapted from Boulet et al. (1992)

truncation halfway down the slope probably due to changes in the underground karst.

4 Micromorphology of the Soils

Observation of the micromorphology of the Bau I, II, and IV toposequences revealed a distribution of porphyric types for all the soils that were analyzed. The superficial red horizons (2.5 YR 4/4 e 4/8) found in the TR-3 and TR-6 profiles are either granular or crumb microstructures with a highly developed pedality (Fig. 6, Table 2). Porosity is predominantly intergranular leading to a high overall porosity (Piló 1998). In the layer from 0.50 to 1.00 m of the TR-6 profile horizon, the presence was observed of iron-clay material lining the pores which is usually interpreted as a sign of illuviation. However, those features only occur locally and do not mean that the horizon as a whole is an illuvial horizon (Fig. 6c).

The microstructure of the yellow horizon of TR-6 is of the angular blocks type with a moderately developed pedality (Fig. 6b). Porosity is predominantly fissural but considerably less developed than that found in the red soils of the surface. In the yellow horizon close to the contact with the underlying rock the micromorphology can be almost solid.

The upper set of red horizons is marked by the predominance of a microstructure that is enaulic (micro-aggregate) and piled, as is typical for red latosols with strong inter-aggregate porosity. Micro-aggregated is the term usually used to denominate the granular structure that typifies latosols but on the microscopic scale it can be spheroid, ovoid or polyhedral and it may stem from various different origins (Buol and Eswaran 1978; Stoops 1983; Stoops and Buol 1985) so accordingly, they have been attributed specific names. The microstructures of the lower set of the lower yellow horizons are predominantly fissured to solid porphyric with fissure porosity (Fig. 6d). According to Ruellan and Dosso (1993), angular microstructures are formed by flocculation and/or concretization followed by splitting. They may have the appearance of blocks, cubes or prisms favored by the presence of 10–30% of argillaceous material.

According to Ruellan and Dosso (1993), the mechanisms that originate the aggregation of pedological material are: (i) flocculation of the constituents, especially clays, associated to the presence of organic material and bivalent cations (Ca^{++}, Mg^{++}) or trivalent ones (Al^{+++}) in the adsorption complex; (ii) cementation of the constituents by organic material, iron, calcium, silica, or biological activity; (iii) fissuring of the flocculated or cemented domains, above all associated to the expansion and contraction of clays in desiccation/humectation conditions that particularly affect the expansive clays.

Thus the red horizons would be of geochemical origin (flocculation and cementation), representing a more evolved stage of pedogenesis, while the yellow ones formed by fissuring represent a less evolved stage. In addition, those

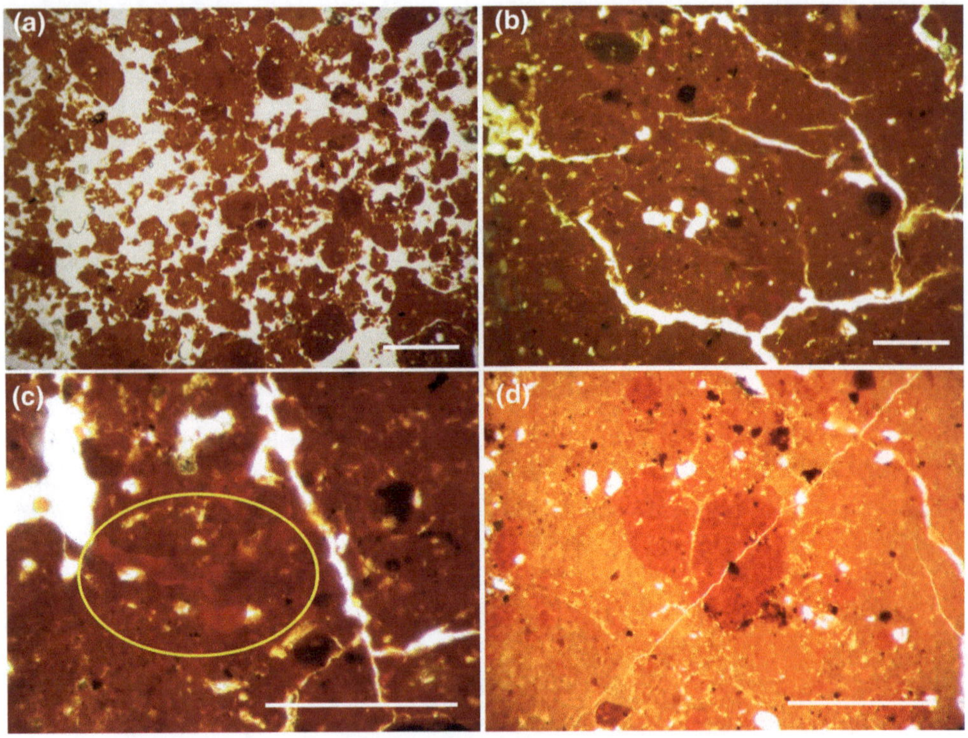

Fig. 6 Microphotographs of soil horizons of the Lagoa Santa Karst High Plains: **a** granular (enaulic) and fragmentary (porphyric) microstructure in horizons **B** and **C**, respectively (TR-3) showing the presence of small polyhedral fragments filling the spaces in a matrix of much larger fragments; **b** porphyric microstructure fissured in sub-angular blocks in the horizon (TR-06); **c** scattered illuvial ferri-argillaceous coating entirely filling a pore in horizon **B** (TR-06); **d** solid porphyric microstructure with poorly developed pedality from the deep part of the yellow horizon. Scale = 500 μm

Table 2 Main microstructure characteristics of the A and B horizons of the red latosol of profile TR-3

Horizons	Depth in cm	Color	Microstructure (pedes)	Pedality	Porosity
A	0–60	Reddish brown (5 YR 4/4)	Fissural, blocks and closed enaulic	Moderately developed	Fissures and channels (25%)
B1	60–100	Reddish brown and Red (2.5 YR 4/4, 4/6)	Dense enaulic and fragmentary porphyric	Highly developed	Closed intergranular/Fissures and channels (35%)
B2	100–180	Red (2.5 YR 4/8)	Small enaulic and fragmentary porphyric	Highly developed	Small intergranular (40%)

characteristics, especially in the case of inter-aggregate porosity, indicate an internal drainage that is faster and more efficient in the red latosols and slower in the yellow ones; a situation which may even lead to the formation of a perched epikarst aquifer draining downhill and that could lead to the formation of nodules. On the other hand, they do not indicate any lithological discontinuity so the possibility cannot be ruled out of there being genetic affiliation among them whereby the yellow ones would correspond to the saprolite material of origin, evolving to become the red soil.

5 Granulometric Composition

Granulometric analysis of the soil samples of the studied toposequences revealed a predominance of very argillaceous soils with a participation of clays of up to as much as 95% (heavy clay). A greater participation of silt fractions only occurs in the soils around the rims and in the bottoms of the dolines and can be as high as 23% at the base of profile TR-9 in which total sand content is 32%, partly constituted by quartz, clay, iron and manganese nodules; these last three constituting the pseudo-sand. No differences in the textures were found between the red set and the yellow set (Piló 1998).

Likewise, Shinzato (1998) reported little variation among the profiles of the Lagoa Santa Karst. The clay contents in the B horizon are quite homogenous and the soils all belong to a single texture class: highly argillaceous (Fig. 7). In profiles P5, P6, P13, and P14, it can be seen that there is a gradual increase in clay content with increased depth but not sufficient to determine any significant textural difference and the form of the curve is similar to that observed in the latosols. In all of the analyzed profiles, the sand fraction

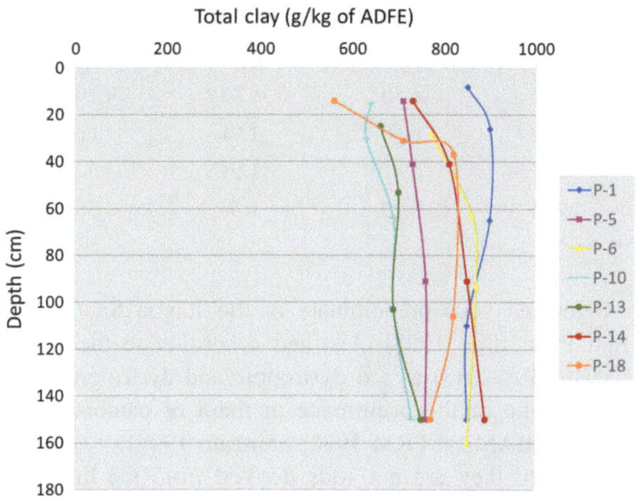

Fig. 7 Clay content variations by depths in the profile analyzed in the Lagoa Santa Karst High Plains (Shinzato 1998)

decreases gradually with an increase in depth. In most of the soils, coarse sand prevails over fine sand.

In the soil mapping undertaken in the Lagoa Santa Karst Environmental Protected Area by IBAMA-CPRM (1998) all the soils identified—latosols, cambisols, podzsols, and gleysol also showed very clayey textures.

6 Soil Mineralogy

Analysis of the resistant materials is based on the principle that minerals like zircon, tourmaline, garnet, anatase, rutile quartz, albite, and microcline do not suffer any expressive alteration during soil formation and are therefore considered to be resistant (Moniz 1972). The occurrence of these minerals and their distributions in the profile have allowed interpretations as to the affiliation of the soils under analysis. That aspect was analyzed in two soil horizons of the Lagoa Santa Karst High Plains: the red horizon and the yellow horizon and the following minerals were identified:

- Red horizon—milky quartz, hyaline quartz, black tourmaline, green tourmaline, limonitized pyrite, garnet, and rutile;
- Yellow horizon—milky quartz, hyaline quartz, black tourmaline, limonitized pyrite, garnet, and rutile.

It can be seen that the red and yellow horizons possess the same assembly of resistant minerals, which only corroborates the idea that those soils developed on top of the same material of origin. It should be stressed that it is a mineralogical assembly that cannot possibly be a residual product of the limestone (Piló 1998) but it is compatible with the meta-siltites and meta-argillites overlying the Sete Lagoas Formation as being the material of origin of the latosols. The same conclusion was reached at by Lund in the mid-nineteenth century (Piló 2002) and by Boulet et al. (1992) and Shinzato (1998).

7 Soil Chemistry

Chemical analysis showed that the soils are predominantly acid with pH values ranging from 4.3 to 5.7 in the high and mid slopes. Acidity is reduced at greater depths and laterally in the downslope direction which is a typical pattern for tropical soils. The soils in the doline bottoms are less acid (Piló 1998). Shinzato (1998) in turn, reported pH values for regional soils in the range of 4.7–6.9. In the studies undertaken by the IBAMA-CPRM (1998), pH was found to vary from 4.8 to 6.3 (Table 3).

Values for soil organic matter (SOM) are higher in the superficial levels and decrease at greater depths as was to be expected. However, the highest values are those of the organic horizons in the doline bottoms where they can be up to 7.6% (Piló 1998). The sum of the permutable bases (S) is very low in the soils and so is the cation exchange capacity (CEC) (Table 3). The values of the latter are less than 15 meq/100 g, indicating the predominant presence of kaolinite. Due to the higher CEC values of the organic

Table 3 Chemical analysis for a typical latosol profile of the Lagoa Santa Karst High Plains according to the IBAMA-CPRM (1998) soil mapping study

Profile	Horizon	pH in H_2O	C (%)	Ca^{++}	Mg^{++}	S	Al^{++++}	H^+	T	V (%)	Sat. Al (%)	CEC of clay cmol (+) kg^{-1} of clay
				cmol (+) kg^{-1} of ADFE								
P1 Latosol	A	6.3	1.89	4.74	0.1	4.95	0	2.71	7.66	65	0	–
	AB	5	1.16	0.71	0.02	0.77	0.6	4.01	5.38	14	44	0.2
	BA	5.2	0.96	0.26	0.02	0.31	0.6	8.33	9.24	3	66	5.5
	Bw1	4.8	0.76	0.1	0.01	0.13	0.45	3.55	4.13	3	78	0.8
	Bw2	5.1	0.6	0.16	0.01	0.19	0.25	3.17	3.61	5	57	1.1

Table 4 SiO_2, Al_2O_3, and Fe_2O_3 percentages and Ki and Kr indexes in five latosol horizons in the Lagoa Santa Karst High Plains (Piló 1998)

B Horizons	SiO_2	Al_2O_3	Fe_2O_3	Ki	Kr
45-100 TR-6	34.64	30.91	5.64	1.91	1.71
100-240 TR-6	32.70	30.60	6.70	1.82	1.59
240-300 TR-6	29.80	26.52	6.40	1.91	1.65
40-90 TR-2	35.01	31.11	5.41	1.91	1.72
90-160 TR-2	32.18	29.76	6.84	1.84	1.58

material, the horizons with higher SOM content have somewhat higher S values.

According to Shinzato (1998), CEC values for the clay are lower than 24 cmol (+) kg^{-1} of clay, classifying them as low-activity clays. Furthermore, since the threshold distinguishing latosols from cambisols is 13 cmol (+) kg^{-1} of clay, the soils analyzed are within the range normally accepted for latosols and in concordance with the classification of the Brazilian system SiBCS (EMBRAPA 2013). Similar figures were obtained by the studies of CETEC (1994) and IBAMA-CPRM (1998), as shown in Table 3. In regard to base and aluminum saturation percentages (V% and M%), according to Piló (1998), regional soils can be alic, eutrophic or dystrophic with a predominance of this last.

The relations between silica and alumina in the soils showed Ki values (SiO_2/Al_2O_3) lower than 2 confirming the presence of minerals of the kaolinite group in both yellow and red horizons (Table 4).

Shinzato's (1998) results showed that the mineral fraction of the soils in the region that were analyzed is predominantly made up of argillaceous minerals of the kaolinite group given that the values he obtained for the molecular Ki ratio were close to 2.

In the TR-6 profile (Baú IV top sequence) there were three main horizons namely, a dark red-yellow B horizon (45 cm to 1.0 m), a bright red B horizon (1.0–2.0 m), and a strongly yellowish-brown horizon B (2.4–3.0 m). The oxides percentages and the molecular Ki and Kr ratios are presented in Table 4.

Two red B horizons (2.5 YR) in the TR-2 profile of the Baú II top sequence were analyzed, situated at depths of 40–90 cm and 90–1.65 m. SiO_2 values were 35.01 and 32.18%, respectively. Those figures are very close to the figures obtained for the TR-6 profile. Al_2O_3 values were 31.11 and 29.76%, respectively; again very similar to the figure for the TR-6 profile. Fe_2O_3 figures were 5.41 and 6.84%. Ki values were 1.91 and 1.84 and those for Kr were 1.72 and 1.58, respectively. The Kr values are indicative of the presence of iron oxides/hydroxides.

Those percentages confirm the existence of a mono-sialitization/ferralization environment. An analysis of the overall set of results shows that highly weathered and pedogenized soils predominate in the Lagoa Santa Karst High Plain (Piló 1998) classified according to the SiBCS (EMBRAPA 2013) as red dystrophic and dystroferric latosols in spite of the occurrence in them of cambisols and acrisols (IBAMA-CPRM 1998; Shinzato 1998).

In short, they are not soils derived from the limestone rocks and therefore the red color of their upper horizons does not mean that they are the equivalent of the Mediterranean *Terra Rossa* soils as Sans (1973) and Auler (1988) proposed for the region, but instead, they are derived from the oxic-ferralitization that is typical of sub-humid tropical regions.

8 Age of the Organic Horizons and the Charcoal Present in the Soil

According to Lowe and Walter (1997), sediments that accumulate in lakes and ponds are either of allochthonous or autochthonous origin. The former are partially derived from organic production of the lake ecosystem and partially from rainwater runoff from the slopes of the lakes' drainage basins in which case they include both organic and inorganic materials.

Lakes rich in mineral nutrients have a high rate of organic material production and create eutrophic environments. Deposits formed in such contexts are dark-colored and rich in organic material that is usually of autochthonous origin. In lakes where such nutrients are less present, organic production is low (oligotrophic) and the predominant sediments are allochthonous.

In the case of the Lagoa Santa Karst, the bottoms of the dolines are the main sites of deposition in the dynamics of the karst surface relief as confirmed by the organic horizons to be found buried at the bottom of the dolines.

For sediment investigation purposes, four temporary lakes were selected, namely, Baú, Santo, Sumidouro, and Cerca Grande. The investigation of the sedimentary records in the bottoms of the karst lakes made it possible to obtain seven radiocarbon datings (Table 5). It must be emphasized that the dates obtained represent the average for all the components present in the respective sample after the latter

Table 5 Datings for organic material and charcoals buried in the soil cover of the Lagoa Santa Karst

Sample	Beta number	Site	Material	Depth (cm)	Method	Conventional radiocarbon age
TR-04	103615	Baú	Peat	168	Standard	580 ± 60 yr BP
LAGSANTOB	178557	Santo	Peat	35	AMS	430 ± 70 yr BP
LAGSANTOH	178558	Santo	Peat	145	AMS	810 ± 50 yr BP
LS-115	162531	Sumidouro	Peat	115	AMS	1,570 ± 70 yr BP
LS-275	162532	Sumidouro	Peat	275	AMS	2,630 ± 40 yr BP
CGSONO.10-OS	159240	Cerca Grande	Peat	50	AMS	2,800 ± 40 yr BP
CGS0 N.11	159241	Cerca Grande	Peat	210	AMS	9,680 ± 230 yr BP
CGS9NO.08	159238	Cerca Grande	Charcoal	125	Standard	2,320 ± 60 yr BP
CGS10NO.09	159239	Cerca Grande	Charcoal	210	Standard	8,690 ± 140 yr BP

had undergone treatment to remove carbonates and roots. According to the information made available by Beta Analytic Radiocarbon Dating (http://www.radiocarbon.com) ages obtained for organic materials pretreated in that way always correspond to the minimum ages.

Two facies were identified in the Cerca Grande Lake (UTM 604423 E – 7841173 S) that were typical of the organic production of a lake ecosystem with datings of 9,680 ± 230 years BP (depth 210 cm) and 2,800 ± 40 years BP (depth 50 cm). Two organic facies were also identified at Sumidouro Lake (UTM 610899 E – 7838879 S) with datings of 2,630 ± 40 years BP (depth 275 cm) and 1,570 ± 70 years BP (depth 115 cm). In the perched lake at Baú (UTM 606160 E 7838200 S) the radiocarbon dating obtained was 580 ± 60 years BP (depth 180 cm). At the Santo perched lake (UTM 600882 E 7845965 N) another two facies were identified with datings of 810 ± 50 years BP (depth 145 cm) and 430 ± 70 years BP (depth 35 cm). Unlike the aspect of the facies in the Cerca Grande, Sumidouro and Baú lakes, the organic levels at the Santo Lake were distinguished by the frequent presence of charcoal showing an allochthonous origin for the organic contribution to the facies records, probably stemming from burning of the vegetation.

During the borings drilled to identify soils around the Cerca Grande Lake (UTM 604295 E 7841465 N) two horizons rich in charcoal were found, one at a depth of 210 cm dated 8,690 ± 140 years BP and the other at a depth of 125 cm dated 2,320 ± 60 years BP (Table 5).

The ages showed that at the beginning of the Holocene there were two lakes occupying dolines. It was found that the bottoms of the dolines are important sites of deposition in the Lagoa Santa Karst and the rate of sediment accumulation varies from 0.02 to 0.29 cm/year. It must be pointed out that those depositing processes are not continuous over time or space and the rates mentioned here are merely referential. It was also found that the soils may have a colluvional contribution as indicated by the charcoals buried at a depth of 2 m at Cerca Grande.

It should be noted that with the occurrence of dolines at the beginning of the Holocene the landscape would have evolved slowly without going through any great erosive processes but certainly, there would have been laminar erosion that would have supplied the material found in the dolines. The latosols would have been developing in a way quite compatible with the tropical climate for at least 10,000 yr BP.

9 Soil and Epikarst Drainage

The epikarst is the upper portion of the carbonate rocks immediately under the cover of unconsolidated material (alterites, soils, sediments, etc.) and is marked by the presence of a network of fissures (cracks) opened up by karst processes (Williams 1985; Pessoa 2005; Ford and Williams 2007). The epikarst corresponds to the vertical extension of the soils (Bakalowicz 2012).

The macroscopic and microscopic analyses of the horizons investigated by Piló (1998) and Boulet et al. (1992) make it possible to propose a hypothetical model for the drainage dynamics of the epikarst zones of the upper and middle slopes of the Karst High Plains: predominantly fast and vertical in the red horizons (rapid hydrolysis) and a slower lateral circulation in the yellowish horizons (slow hydrolysis) in contact with the epikarst. In the case of the lower slopes and the bottoms of the dolines occupied by temporary lakes, both drainage and pedological organizations are usually more complex and there may be nodular levels due to the fluctuations in the epikarst aquifer or levels with ferricrete formations, or truncation of soil horizons caused by erosion, and buried organic horizons as registered by Boulet et al. (1992) and Piló (1998).

The first thing to be pointed out is the micro-aggregated structures of the red horizons, which are typically associated with free vertical drainage. In that context, there is a well-developed (inter-aggregate) structural porosity showing

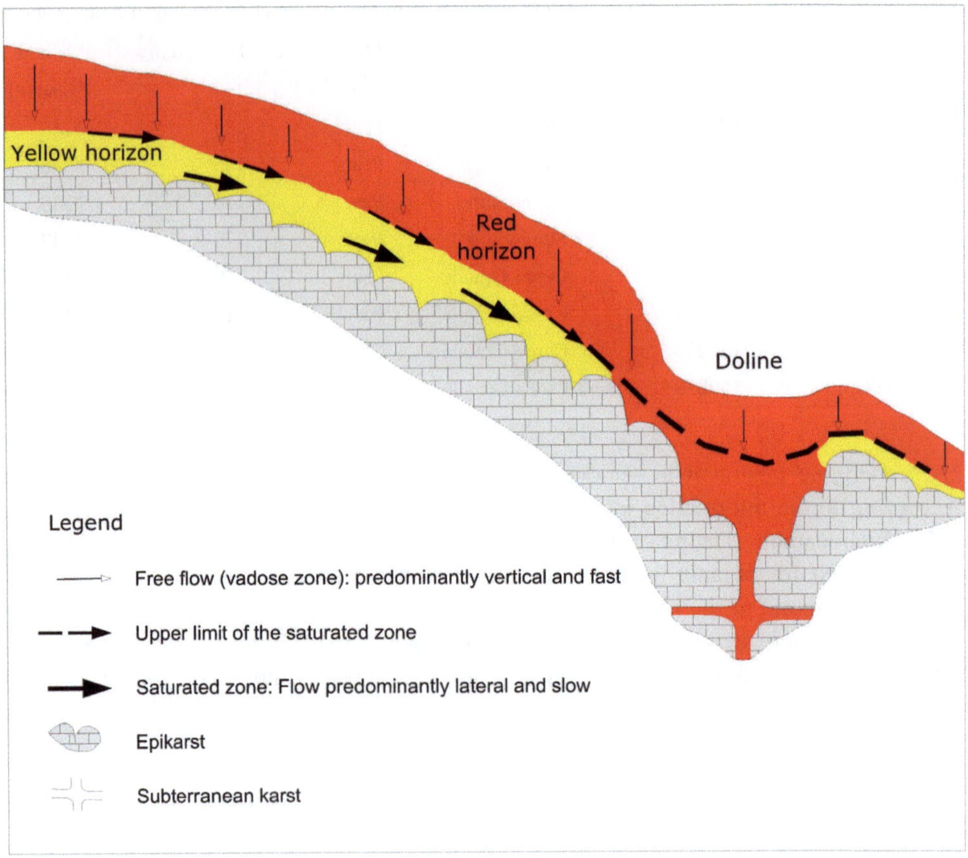

Fig. 8 Simplified model of water circulation in the soils and the epikarst in the upper and middle slopes of the karst High Plains

the possibility of the occurrence of a free internal drainage. The red horizons are also considered to be a favorable environment for the formation of hematite given the more rapid hydrolysis and the effect of ferri-hydrate dehydration processes, all of which tend to support that interpretation. The yellow horizons, on the other hand, have a much more closed microstructure between the sub-angular blocks resulting from the installation of a very fine fissural porosity. As seen above, the micro-morphological analyses clearly indicate that structural differentiation.

In that regard, after the infiltration on the slopes, the red horizons would be responsible for a flow vector that was predominantly vertical, initially associated with capillarity and then to gravity. As infiltration increases, so a fast (vertical) gravitational flow would be established in the red horizons going down to near the level of contact with the yellow horizons which would represent the top of the saturated zone. In those lower horizons, there would be a reduction in the velocity of flow as the level neared the point of contact with the epikarst and that would lead to the formation of a lateral base flow in the downslope direction.

High recharge rates in the rainy season or even resulting from other heavy rainfall events would generate an excess of the recharge water stored in the epikarst and especially in the porosities of the yellow horizons. Once the base of the soil in contact with the underlying rock becomes saturated, especially in view of the poor capacity for drainage into the underground karst, it is possible that a perched epikarst aquifer might form which would usually be of a temporary nature. In that case a slow, predominantly lateral, drainage process would occur over the surface of the rock and be eventually exported through preferential drainage routes usually facilitated by vertical and horizontal cracks that allow for the transmission of water from the epikarst to the underground karst. In the deeper part of the epikarst, the saturated zone may actually enter into contact with the karst aquifer as in the case registered by Pessoa (2005). The flows inferred above are illustrated in Fig. 8.

Ford and Williams (2007) stated that most of the chemical energy involved in dissolving the carbonated rocks is spent near to the surface because of the proximity of the main source of CO_2 production, which is the soil. The same authors report that 80% of the chemical denudation in the watershed takes place within the first 10 m of carbonate rock and the efficiency of the dissolution gradually diminishes as the distance from the surface (and the source of CO_2) increases. Water dynamics in the epikarst represent a mechanism that is essential for the development of the surface karst and the underground karst as well (Bakalowicz 2012). In the Lagoa Santa Karst, large volumes of red and yellow

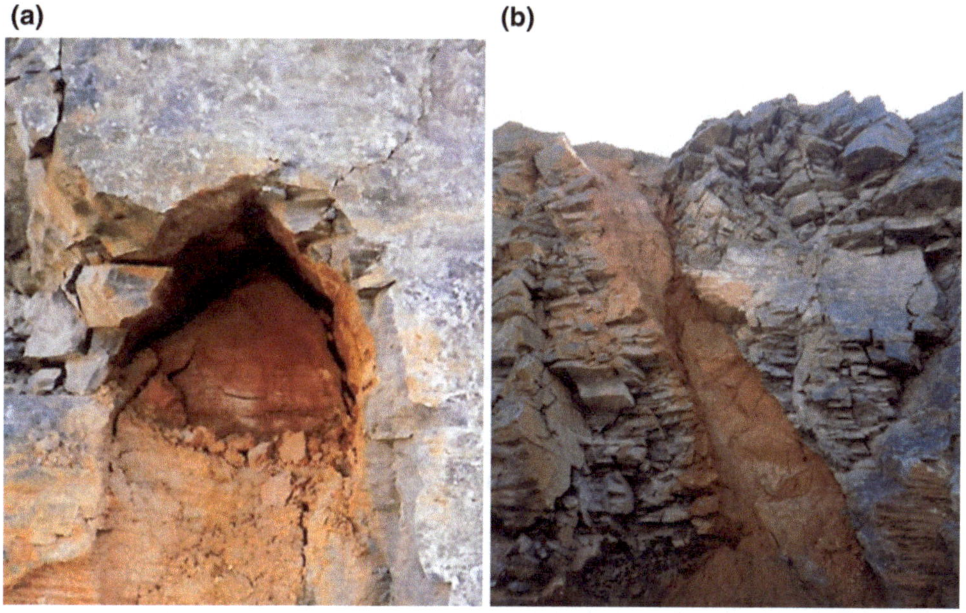

Fig. 9 Sites of mining/quarrying activity exposing red and yellow soils introduced into the underground karst

soils have been introduced into the underground karst, via the epikarst (Fig. 9) directly influencing the paragenesis processes which are frequent in the region's caves and also playing a role in the dominant anastomotic (Palmer 1991) pattern in them.

Because of its soil cover, the observation of the epikarst is difficult. However, the areas where the cover has been removed by limestone quarrying serve as good points for analyzing the epikarst morphology. It is very apparent that the yellow soils predominate in the zones of contact with the epikarst while the red soils are more superficial. However, it is possible at times for the red soils to be in contact with the epikarst.

The larger shapes are measurable in meters and represented by a sequence of residual shapes of the pinnacle type. In the top areas, they form a set that is less individualized and more labyrinth-like. The further down the slope one goes, the more individualized those shapes become. Given that description, the karst is a typical example of a covered pinnacle-type karst (Fig. 10a). The rounded tops of the pinnacles follow the slope of the surface.

Micro-shapes groove the covered epikarst and define four different typologies for the karrens of soil covered karst. Karrens in which the joints are horizontal are the most frequently found showing plan-parallel grooves measurable in

Fig. 10 Epikarst exposed by limestone quarrying showing the main forms elaborated in the sub-surface: pinnacles (**a**) and karrens with horizontal joints (**b**). Figure 10b shows how the karrens are more developed in contact with the red soils

millimeters or centimeters and following the well-marked foliation of the rock (Fig. 10b). Alveolar karrens are made up of small horizontal cavities with sizes ranging from a few millimeters to a few centimeters. Such pockets left by dissolution are usually set out in an East–West direction which is the orientation of the mineral lineation and/or stretch developed in the foliation plane of the rock. It was also found that they usually begin in the horizontal joints. Kamenitzas were also found especially on the tops of the pinnacles. They are closed basins in the rock a few centimeters wide and high.

In the light of the profiles of the soils in contact with the karst's lateral walls, it was possible to show that the karren are more highly developed in the surfaces in contact with the red soils than in those in contact with the yellow soils (Fig. 10b). That is due to less intense hydrochemical dynamics in the contact with the yellow soils and that is corroborated by the microstructure and the porosity micro-morphologically observed in the thin section.

The slow drainage rate in the yellow horizons in contact with the limestone rock is what would be responsible for the elaboration of such a well-developed epikarst relief in the Lagoa Santa Karst High Plain especially in view of the time of residence of infiltration water rich in CO_2.

Those morphologies are indicative of a more dynamic zone of the local karst. In that zone of contact between soil and epikarst, with its great lateral extension, the water circulation is more effective for karst formation than in the superficial zones and is actually the main mechanism lowering the karst relief as well as being the main driver of superficial and subterranean karst dynamics.

10 Conclusions

The soil horizons studied in the toposequences were relatively parallel to the topographic surface with no apparent current (lateral) transformation fronts and therefore represent a system of vertical transformation of the substrate into the soil. Such features have been interpreted as representing a state of equilibrium of the morphogenesis and pedogenesis with the currently stable climatic and tectonic conditions.

Analyses of the macro and micro-morphology of those main toposequence horizons clearly indicate the differentiation of structures between the red horizons and the yellow ones and it can be deduced that the yellow horizons are at a less evolved stage of pedogenesis than the red ones. Furthermore they make it possible to deduce an internal drainage model for the soils in the high and middle slopes: fast vertical drainage in the red horizons (rapid hydrolysis) in the more superficial positions and lateral drainage along the zone of contact between soil and epikarst in the base of the yellow horizons (slow hydrolysis). On the other hand, the analyses also make it reasonable to propose the hypothesis that the red soils originate from the yellow ones and the presence of relics of yellow material in the mass of red latosol tends to corroborate that.

The epikarst relief was found to be well developed and formed by covered pinnacles thus testifying to a karst formation process more expressive than in the surface and underground karst. The current general geometry of the upper and middle slopes of the Karst High Plains reflects the development of the epikarst.

The red and yellow horizons derive from the same material of origin which is the source of the continuous pedological cover on top of the limestone rocks. That material, in turn, corresponds to the alteration and pedogenesis of meta-siltites and meta-argillites of the Serra de Santa Helena Formation which, in the stratigraphic sequence, are situated on top of the limestone rocks of the Sete Lagoas Formation corroborating the reports of various other authors. The altered metapelites evolved into red mono-sialithic/ferralithic (oxidic) latosols by means of a vertical transformation system in the climatic conditions that prevailed in the Quaternary period, especially the Holocene epoch. The dolines are therefore more recent than the latosols that they have truncated, controlled by the underground karst but re-capped by the sediments coming from the lateral removal of the latosols. Thus, the contribution of the limestone is insignificant in the pedogenetic evolution of the region considering that the limestones consist of more than 95% of $CaCO_3$, whereas the clay content of the soils and their mineralogical assemblages are incongruent with those characteristics, especially in such a short time as the Holocene.

Given the expressive introduction of volumes of soil via the epikarst into the underground network, there is an increase in the clastic sedimentary dynamics in the conduits, influencing the morphology patterns of the caves and favoring paragenesis. The materials that fill the cave passages also suggest that their two main sources are the two latosol horizons, which shows that the extinct megafauna remains were deposited along with those materials especially through the bottoms of the dolines. The deeper closed depressions represent the points of greatest hydric flows to the underground karst (accelerated dissolution) with an important introduction of cover material into the underground karst and the configuration of the surface concavity of the slopes (doline formation) and the outcropping of limestone cliffs at the same time as the morpho-pedogenesis of the interfluvium covered by the latosols.

The studies have revealed that the karst modeling of the Lagoa Santa Karst High Plains evolved in the light of a set of inter-relations involving the epikarst, the superficial karst and the underground karst thereby reinforcing the idea of a coupled system with special importance attributed to the soil cover. At the same time, that indicates incompatibility with the idea that the limestone rock is the source material but nevertheless, it does condition an evolution of the surface and subterranean landscapes favored by the contact between the soil and the limestone substrate.

References

Ahmad N, Jones RL (1969) Genesis, chemical properties and mineralogy of limestone-derived soils, Barbados, West Indies. Trop Agric 46:1–15

Auler A (1988) Caracterização geral do carste de Lagoa Santa. In: Congresso de Espeleologia da América Latina e Caribe. Roteiro de Excursão. Belo Horizonte, 16 pp

Auler A (1994) Hydrogeological and hydrochemical characterization of The Matozinhos - Pedro Leopoldo Karst, Brazil. Dissertation, Western Kentucky University

Auler A, Rubbioli EL, Brandi R (2001) As grandes cavernas do Brasil. Rona Editora, Belo Horizonte, p 230

Auler A, Piló LB, Smart PL, Wang X, Hoffmann D, Richards DA, Edwards RL, Neves WA, Cheng H (2006) U-series dating and taphonomy of Quaternary vertebrates from Brazilian caves. Palaeogeogr Palaeoclimatol Palaeoecol 240:508–522

Auler AS and Piló LB (2015) Lagoa Santa Karst: Cradle of Brazilian Cave Studies. In: Vieira BC, Salgado AAR, Santos LJC (eds) Landscapes and landforms of Brazil, pp 183–190

Bakalowicz M (2012) Epikarst. In: White WB, Culver D (eds) Encyclopedia of caves, pp 284–288

Boulet R, Kohler HC, Filizzola EF, Malta IM (1992) Estudo da cobertura pedológica de uma vertente adjacente à uvala do Conjunto Cárstico da Lapa Vermelha de Lagoa Santa, MG. Congresso da Associação Brasileira de Estudos do Quaternário. Imprensa Universitária, Belo Horizonte, pp 59–64

Boulet R, Lucas Y, Fritsch E, Paquet H (1993) Géochimie des paysages: le rôle des couvertures pédologiques. In: Coll. "Sédimentologie et Géochimie de la Surface" à la Mémoire de Georges Millot. Académie des Sciences et du Cadas, pp 55–76

Bouyer DG (2004) Soils on carbonate karst. In: Gunn J (ed) Encyclopedia of caves and Karst science. Fitzroy Dearborn, New York/London, pp 1400–1403

Branco JJR, Costa MT (1961) Roteiro da excursão Belo Horizonte – Brasília. UFMG - Instituto de Pesquisas Radioativas, Belo Horizonte

Buol SW, Eswaran H (1978) The micromorphology of oxisoils. In: Proceedings Vth international workshop meeting on soil micromorphology, M. Delgado Editora, Granada, vol 1, no 1, pp 325–347

Cartelle C (2012) Das grutas à luz. Os mamíferos pleistocênicos de Minas Gerais, Bicho do Mato Editora, p 236p

CETEC—Centro Tecnológico de Minas Gerais (1994) Caracterização pedológica da região Sete Lagoas-Lagoa Santa. Série Cartas Temáticas, Belo Horizonte, 59 pp

Ciric M (1967) Characteristics of soil formation on limestone and principles of limestone classification. Soviet Soil Sci, pp 57–64

Comer JB (1974) Genesis of Jamaican bauxite. Econ Geol 69:1251–1264

Coutard JP, Kohler HC, Journaux A (1978) Comentário do Mapa do Carst. Região de Pedro Leopoldo-Lagoa Santa, MG, Brasil. Laboratoire D'analyse et de Cartographie des Formations Superficielles de l'université de Caen, et le Centre de Géomorphologie du CNRS à Caen - France. Escala, 1:50.000

Dardenne A (1978) Síntese sobre a estratigrafia do Grupo Bambuí no Brasil Central. Congr Bras Geol Recife 30:597–610

Delvigne J (1983) Alteração. Instituto de Geociências da USP, São Paulo. 150 pp

Duchaufour P (1968) Évolution des sols conditionnée par le calcaire. In: L'évolution des Sols Essai sur la dynamique des profils. Masson et Cie, Éditeurs, Paris, vol 5, pp 50–57

EMBRAPA Centro Nacional de Pesquisa de Solos (2013) Sistema Brasileiro de Classificação de Solos. EMBRAPA-SPI Rio de Janeiro, 306 pp

Feng JL, Cui ZJ, Zhu LP (2009) Origin of terra rossa over dolomite on the Yunnan-Guizhou Plateau, China. Geochem J 43:151–166

Ford DC, Williams P (2007) Karst hydrogeology and geomorphology. Unwin Hyman, London

Foster J, Chittleborough DJ, Barovich K (2004) Genesis of a Terra Rossa soil over marble and the influence of a neighbouring texture contrast soil at Delamere, South Australia. Supersoil, University of Sydney, Australia

IBAMA-CPRM—Instituto Brasileiro do Meio Ambiente e dos Recursos Naturais Renováveis - Companhia de Pesquisas de Recursos Minerais (1994) Informações Básicas para a gestão territorial: região de Sete Lagoas-Lagoa Santa. Caracterização Pedológica, 57 pp

Kohler HC (1989) Geomorfologia cárstica na região de Lagoa Santa-MG. PhD Thesis, Universidade de São Paulo

Lamouroux M (1972) Étude de sols formés sur roches carbonatées. Pédogenèse fersiallitique au Liban. Thèse Sc., Strasbourg et Mém. ORSTOM

Lynch LS (2009) Gênese e geoquímica de solos em ambiente cárstico no cerrado da região de Planaltina de Goiás. Universidade Federal de Viçosa, Tese de Doutorado

Lowe JJ, Walker MJC (1997) Reconstructing quaternary environments. Pearson Education Limited, England, 446 pp

Lund PW (1837) Primeira Memória sobre a fauna das cavernas. In: Paulo Couto C (ed) Memórias sobre a Paleontologia Brasileira. Instituto Nacional do Livro. Rio de Janeiro, vol 3, pp 107–130

Lund PW (1845) Notice sur des ossements humains fossils, trouvés dans une caverne du Brésil. Mémoires de la Societé Royale des Antiquaires du Nord 1849:49–77

MacLeod DA (1980) The origin of the Red Mediterranean soils in Epirus, Greece. J Soil Sci 31:125–136

Moniz AC (1972) Mineralogia dos solos. In: Moniz AC (ed) Elementos de Pedologia. EDUSP, São Paulo, pp 391–408

Muhs DR, Bush CA, Stewart KC (1990) Geochemical evidence of Saharan dust parent material for soils developed on quaternary limestone of Caribbean and Western Atlantic islands. Quaternary Res 33:157–177

Muhs DR, Budahn J, Prospero JM, Carey SN (2007) Geochemical evidence for African dust inputs to soils of western Atlantic islands: Barbados, the Bahamas and Florida. J Geophys Res 112:F02009

Nimer E, Brandão AMPM (1989) Balanço hídrico e clima da região dos cerrados (ed) FIBGE Rio de Janeiro, 166 pp

Olson CG, Ruhe RV Mausbach MJ (1980) The terra rossa limestone contact phenomena in karst, southern Indiana. Soil Sci Soc Am J 44:1075–1079

Palmer AN (1991) Origin and morphology of limestones caves. Geol Soc Am Bull 103:1–21

Pedro G (1976) Sols argileux et argiles; elements généraux en vue d'une introduction à leur étude. Science du Sol 2:69–84

Pedro G (1987) Géochimie, mineralogie er organisation des sols. Aspects coordonnés des problèmes pédogénétiques. Cahier ORSTOM, Série Pédologie, 23(3):169–186

Pessoa PFP (2005) Hidrogeologia dos aquíferos cársticos cobertos de Lagoa Santa. Universidade Federal de Minas Gerais, MG. Tese de Doutoramento

Piló LB (1998) Morfologia cárstica e materiais constituintes: dinâmica e evolução da depressão poligonal Macacos-Baú. Universidade de São Paulo, Tese de Doutoramento

Piló LB (2002) P. W. Lund e a geomorfologia cárstica de Lagoa Santa. O Carste 17(1):12–17
Piló LB (2004) Registros sedimentares no fundo das dolinas ocupadas por lagoas temporárias. Unpublished report FAPESP/LEEH, 55 pp
Queiroz Neto JP (2000) Geomorfologia e Pedologia. Rev Bras de Geomorfologia 1:58–67
Ruellan A, Dosso M (1993) Regards sur le sol. Foucher, Paris. 192 pp
Sans LMA (1973) Classificação e mineralogia de solos da Região de Sete Lagoas - MG. Dissertation, Escola Superior de Agricultura "Luis de Queiroz", USP
Shinzato E (1998) O Carste da Área de Proteção Ambiental de Lagoa Santa (MG) e sua influência na formação dos solos. Universidade Estadual do Norte Fluminense, Thesis
Stoops G (1983) Micromorphology of oxic horizon. In: Bullock, Murphy CP (eds) Work meeting on soil micromorphology. Academic Publis., Rothamsted, London, vol 2, pp 419–440
Stoops G, Buol SW (1985) Micromorphology of oxisols. In: Douglas LA, Thompson ML (eds) Soil micromorphology and soil classification. Madison, pp 105–119
Tuller MP, Ribeiro JH, Danderfer A (1992) Mapeamento geológico da área do Projeto VIDA. CPRM, Belo Horizonte
White WB (1988) Geomorphology and hydrology of Karst Terrains. Oxford University Press, 464 pp
Williams PW (1985) Subcutaneous hydrology and the development of dolines and cockpit Karst. Z Geomorph N.F 29:63–482
Yaalon DH (1997) Soils in the Mediterranean region: what makes them different? Catena 28:157–169
Yaalon DH, Ganor E (1973) The influence of dust on soils during the quaternary. Soil Sci 116:146–155

Karst Landforms in the Lagoa Santa Area

Augusto S. Auler

Abstract

The karst landscape in the Lagoa Santa area is characterized by a full suite of typical karst landforms, with some peculiar features commonly not found in other karst areas. The karst is developed in Proterozoic carbonates of the Sete Lagoas Formation. The pure upper Lagoa Santa Member, which covers 44% of the area, displays the majority of the karst features, while the impure lower Pedro Leopoldo Member outcrops at the borders and exhibits subdued karst landforms. The Lagoa Santa Karst is mostly a covered karst, with only 11.5% of the total area presenting karstified outcrops, which are more frequent in the dissected areas toward the east. Water recharge is predominantly autogenic, as the region displays limited noncarbonate catchment areas, with drainages directed toward the Velhas River, the main base level, and Mata Creek. Seven geomorphological domains can be recognized in the area, with karst landforms being more developed in the karst plateaus. Due to its covered nature, many karst landforms, such as karren, towers, and corridors, initially developed under the soil and were gradually exposed at the surface. Karst landforms thus exhibit a strong inheritance of subsoil processes. The most common karren type is the joint karren, which developed along subhorizontal bedding planes and joints and formed initially under the soil, with a later evolution at the surface. Typical karst landforms occur, such as various karren types, dolines, dry valleys, swallets and springs. In addition to these karst forms, the area is characterized by peculiar limestone cliffs, which represent sharp vertical breaks in an otherwise undulating soil-covered landscape. These cliffs tend to present swallets or dolines at their bases, causing progressive lowering at the base and increasing cliff exposure. Large depressions mostly occur close to the base levels. The unusual occurrence of lakes is also noteworthy. Three types of lakes can be recognized: (i) Sinkhole lakes formed due to the plugging of swallets by impermeable soil-derived material, (ii) epikarst lakes in the lower portions of the area and (iii) lakes over metapelite. The Lagoa Santa Karst has experienced a long evolutionary history dating back to the period when the upper carbonate sequences were exposed, probably well within the Tertiary, assuming regional uplift rates. The karst shows strong control both in the tectonic framework and in the direction of the hydraulic gradient toward the NE. Depressions, groundwater flow, and other karst features show well-marked alignment, especially when the joint direction, bedding dip, and hydraulic gradient are concordant.

Keywords

Lagoa Santa • Karst landforms • Limestone cliffs • Covered karst • Karst lakes

1 Introduction

The caves of the Lagoa Santa Karst were the magnet that initially attracted saltpeter miners in the eighteenth century and then Danish scientist Peter W. Lund to the area, starting a long-term interest in this scientifically important karst region. The cave paleontological findings of Lund drew the attention of the international community to this then-remote area; many scientists during the nineteenth century followed in his footsteps and used Lund's findings in Lagoa Santa as a reference for further studies (see chapter "History of Research in the Lagoa Santa Karst", this volume). Later, urban development in the surrounding area and the inauguration of the state of the Minas Gerais capital, Belo Horizonte, in 1897 facilitated access to the area, making it a convenient research site for many scientists of diverse disciplines.

A. S. Auler (✉)
Instituto do Carste, Rua Barcelona 240/302, Belo Horizonte, MG 30360-260, Brazil
e-mail: aauler@gmail.com

A. S. Auler and P. Pessoa (eds.), *Lagoa Santa Karst: Brazil's Iconic Karst Region*, Cave and Karst Systems of the World, https://doi.org/10.1007/978-3-030-35940-9_7

The karst landforms in the Lagoa Santa area were initially studied as a background for more comprehensive cave studies. Reports by Lund (1837a, b, 1841a, b, 1845) and his guest Johannes Reinhardt (1867) commented on the dissolutional landforms. Among the several European scientists who visited Lagoa Santa in the nineteenth century, Liais (1872) provided brief comments on the karst features of the Lagoa Santa Karst. In the early twentieth century, much of the information on the karst landforms were produced by archaeological and paleontological research expeditions by members of the Academy of Sciences of Minas Gerais, including Arnaldo Cathoud, Harold Walter and Aníbal Mattos—see the review in Costa (2017) and the National Museum of Rio de Janeiro (Jorge Padberg-Drenkpohl), in addition to contributions by local engineer Cássio Lanari (1909). Geomorphologists Elzio Dolabela (1958) and Getúlio Barbosa (1961) also contributed brief comments on the karst landforms in the area, and archaeological missions by American (Hurt 1960) and French (Laming-Emperaire et al. 1974) teams described karst features in the context of their excavations.

Research focused on the karst landscape did not start until the late nineteenth century, through a detailed study of karren by Jean Tricart (1956) and the production of a karst map by Heinz Kohler, Jean Pierre Coutard and André Journaux (Kohler et al. 1978), which contained a comprehensive summary of the karst landforms (Coutard et al. 1978). This same group provided important insights into the genesis of limestone cliffs (Journaux 1977) and into general karst evolution (Kohler 1989; Kohler and Karfunkel 2002). Denes Balasz (1984) commented on the soil stripping of many of the karst outcrops. A more quantitative approach has been provided since the 1990s through research focused on hydrology (Auler 1994; Pessoa 1996, 2005), morphometry and pedology (Piló 1998), karren landforms (Knez et al. 2012), geochronology and karst evolution (Auler 1999) and geotechnical hazards (Galvão et al. 2015). A general overview of the karst landforms in the Lagoa Santa area is presented by Berbert-Born (2002) and Auler and Piló (2015).

The Lagoa Santa Karst (Fig. 1) is developed in carbonate and metapelite rocks of the Bambuí Group. The karst landforms occur in both the Pedro Leopoldo and the Lagoa Santa members of the Sete Lagoas Formation. The former comprises impure limestones (insoluble content between 7 and 24%) and is less prone to display developed karst landforms and caves (Peñaranda 2016). The latter, however, comprises very pure limestone, with an insoluble content between 1 and 10% (Peñaranda 2016), and thus presents the majority of karstified outcrops in the area. The thickness of the carbonates in the area reaches approximately 150 m as observed in well logs presented by Pessoa (2005). However, close to the limit with outcrops of the crystalline basement, especially in its western portion near Mata Creek, the thickness does not exceed a few tens of meters. Limited noncarbonate terrains occur in the area, mostly in elevated areas at the northern boundary, such as the Santa Helena Ridge, providing surface runoff that sinks upon reaching the carbonate outcrops. The limited amount of recharge coming from the noncarbonate areas characterizes the Lagoa Santa Karst as mostly an autogenic karst. The base level of the Lagoa Santa Karst is represented mainly by the Velhas River, one of the major tributaries of the São Francisco River. The Velhas River marks much of the eastern boundary of the karst. In the central area close to the municipalities of Pedro Leopoldo and Matozinhos, Auler (1994) calculated that approximately 88% of the discharge issuing from springs flows toward the Velhas River; the remaining 12% flows toward the smaller Mata Creek, a tributary of the Velhas River that delimits the western boundary of the karst zone. Surface streams are relatively rare in the area. Small creeks, notably Samambaia, Palmeiras/Mocambo, Gordura and Jaguara, occur, many of which display underground sections and karst springs. The tectonic imprint in the Lagoa Santa Karst is marked by joints with directions of N10-20E, N30-40E, and N80-90W. This tectonic control, inherited from the Brasiliano orogeny at c. 600 Myr, is responsible for the orientation of many karst features, notably the limestone cliffs.

Although the area is named after the town of Lagoa Santa, located at its easternmost limit, much of the karst area extends over other municipalities, especially Matozinhos, Pedro Leopoldo, Confins, São José da Lapa, Prudente de Morais, Funilândia and Sete Lagoas. However, the "classic" core zone of the Lagoa Santa Karst can be considered as the area belonging to the municipalities of Matozinhos, Pedro Leopoldo and Lagoa Santa (Fig. 1). This is the area where most of the geomorphological, archaeological and paleontological studies have been performed. Much of this central area has now been protected through a number of conservation units at the state and federal levels. However, environmental pressure due to population increase, groundwater pollution and mining is growing in the area and in its surroundings (Auler 2016). The Lagoa Santa Karst is both the best-studied and most threatened karst area in Brazil (Auler 2016). This chapter will present an updated review of the karst landforms in the Lagoa Santa area.

2 Karst Morphology

The Lagoa Santa Karst comprises a subdued karst landscape, with altitudes decreasing gradually from west to east toward the main base level, the Velhas River (Fig. 1), which marks the eastern limit of the karst. To the west, the boundary is usually considered to be Mata Creek, a tributary of the Velhas River. To the northwest, however, the limit of the karst is

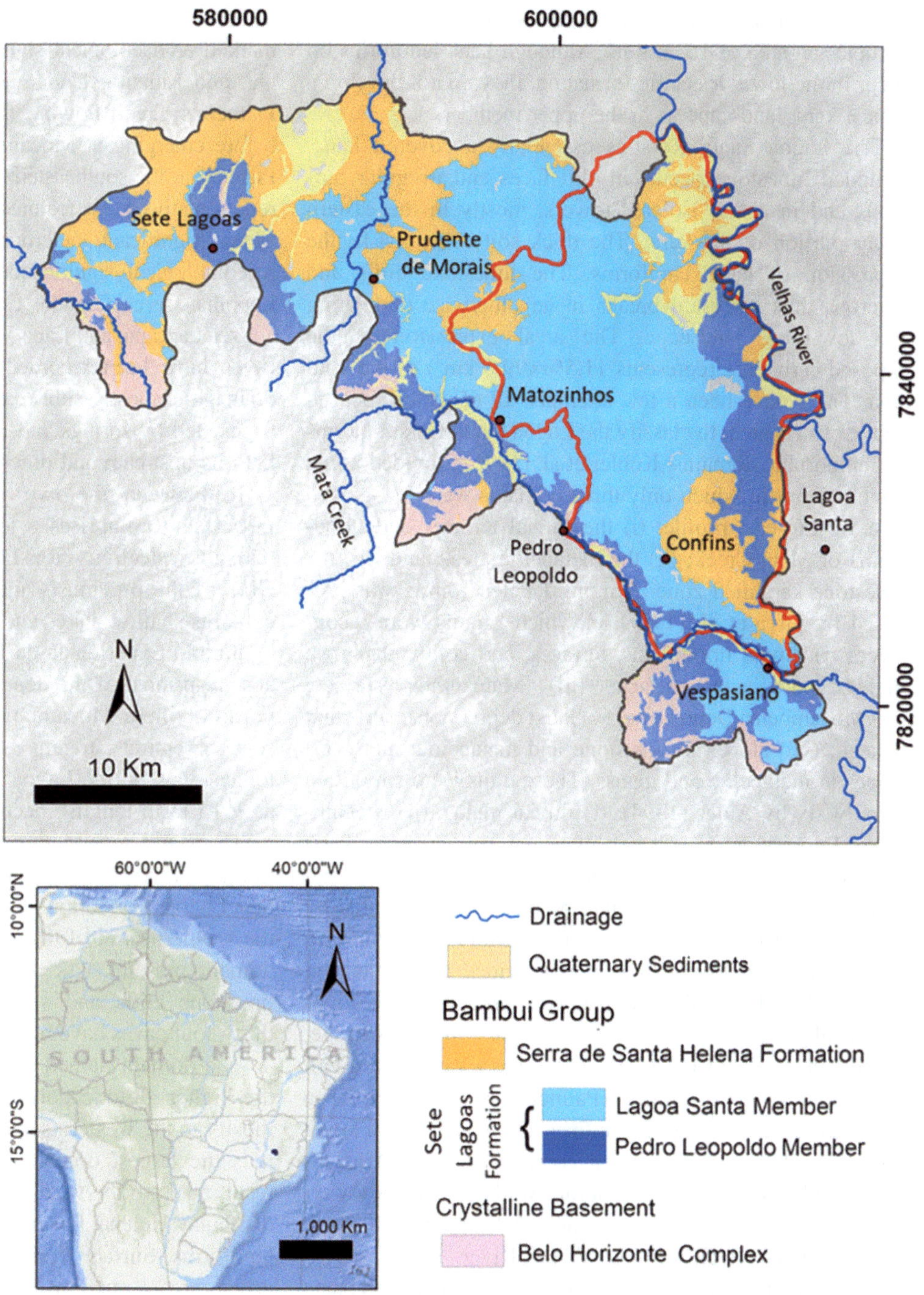

Fig. 1 Location of the Lagoa Santa Karst, showing the main cities, geology, hydrography, and the central area, which roughly coincides with the environmental protection area (APA Carste de Lagoa Santa) marked in red

taken as the zone where the limestone becomes less frequent and granitic rocks predominate. The northern limit of the karst is represented by the contact with the metapelites of the Serra de Santa Helena Formation. The southern limit of the Lagoa Santa Karst is represented by the contact between the limestone and the crystalline (granite, gneiss) basement. The area is not entirely covered by limestone. Major metapelite zones occur mostly at the Santa Helena Ridge (the type locality of the Serra de Santa Helena Formation), in an elevated zone in the western central area, to the north and east of the town of Prudente de Morais and between the towns of Lagoa Santa and Confins (Fig. 1). Metapelite zones always represent elevated areas, reaching altitudes of 1074 m asl, and are thus the headwaters of small streams. The proportion of noncarbonate areas within the limits of the Lagoa Santa Karst represents approximately 39% (335.5 km^2), compared to 533.3 km^2 for the carbonates. Karst groundwater circulation does occur under the metapelite areas, as indicated by some springs such as the Poço Verde in the southeastern portion of the area. Most of the well-developed karst occurs in the purer upper member, which has a much broader areal occurrence (44%–379.7 km^2). The lower Pedro Leopoldo Member is nearly always covered by soil, with very limited outcrops, and occupies an area of 153.6 km^2. Although karst landforms do occur in the lower less pure formation, they do not develop a typical karst landscape as in the upper member.

The Lagoa Santa Karst is essentially a covered karst. Denuded limestone occurs in cliff faces and irregular outcrops and in a few isolated towers, mostly in the eastern lower portion of the karst. The thick soil cover masks the expression of karst landforms. The limestone cliffs and outcrops thus represent breaks in an otherwise undulating and low-lying landscape. The areal expression of the exposed carbonate represents 11.5% (99.6 km^2) of the total area. There have been a few attempts, all of which focused on the central area, to classify the Lagoa Santa Karst in terms of geomorphic domains. Kohler et al. (1976) provided a first basic attempt in which only three domains were recognized in a west-to-east profile: (i) the fluvial terraces and floodplains of Mata Creek; (ii) the metapelite areas, and (iii) the limestone karstified zone. A more detailed zoning was proposed by Silva et al. (1987), in which 8 units were recognized: (i) fluvial floodplains, terraces, and colluvial ramps; (ii) Lagoa Santa surface; (iii) Matozinhos surface; (iv) Mocambeiro Depression; (v) karst depressions; (vi) karst plateau; (vii) slopes in limestone and metapelite; and (viii) slopes in metapelite and gneiss. These units were simplified to only six by Auler (1994): (i) fluvial plains; (ii) Mocambeiro Depression; (iii) karst plateaus; (iv) covered karst surfaces; (v) metapelite surfaces; and (vi) igneous and metamorphic domain. A distinct proposal was presented by Kohler (1989) and later adapted by Dutra et al. (1998). A simplified version of these was provided by Kohler and Karfunkel (2002) and comprised four domains: (i) a topographically higher area up to 850 m asl at Ferradores Ridge, comprising covered karst and presenting a fluviokarst with blind valleys and dolines (Palmeiras-Mocambo); (ii) a semicircular belt comprising dolines and uvalas; (iii) a lower elevation plateau containing a complex system of asymmetric dolines aligned along abrupt walls and (iv) a lowermost compartment at approximately 650 m asl with many poljes and ponors close to the Velhas River.

A new proposal covering the entire area of the Lagoa Santa Karst is presented in this review and is based on field observations and digital cartography analysis using ArcGIS 10.3 and Google Earth Pro. The criteria used to delineate the domains were hypsometry, slope, hydrology, lithology, structural framework, and residual karst features. Areas with similar characteristics were clustered into the same domain (Fig. 2). Seven domains have been recognized. Karst plateaus occur as two separate domains: the Palmeiras Karst Plateau and the Samambaia Karst Plateau. The plateaus display a typical karst landscape and feature the more developed karst landforms, the largest outcrops and the larger number of caves. The Palmeiras Plateau occupies 145.3 km^2 with 669 recorded caves. This area contains some of the most representative outcrops, many with marked archaeological significance, such as Poções, Escrivânia, and Limeira. Underground drainage predominates, with Palmeiras Creek flowing through a series of karst windows before emerging at Mocambo Spring. The Samambaia Karst Plateau, in the southeastern portion of the area, shows equally representative karst features, with a total extent of 93.5 km^2. Many significant outcrops, with both archaeological and paleontological significance, occur in this domain, such as Lapinha, Cerca Grande, Caianga, Baú, Vargem do Macaco, Várzea da Lapa, and Lapa Vermelha. Four hundred eighty-four caves have been recorded in the Samambaia Plateau. The drainage is mostly subaerial and is controlled by Samambaia Creek, fed by springs and displaying underground sections in both its upstream and downstream portions.

In between the two karst plateaus, the Mocambeiro Depression comprises a low-lying area (below 700 m asl) with a low declivity; it is limited in the N and S by limestone cliffs of the previously mentioned karst plateaus. The area contains shallow lakes and residual hills such as Vargem da Pedra and Experiência da Jaguara. Active drainage is nearly absent in most of the depression, except in the northeastern portion, where Mocambo Creek and Jaguara Spring represent perennial streams. Forty-five caves, including the longest cave in the Lagoa Santa Karst (Santo Antonio Cave at 2615 m in length), occur in this domain.

The Mata Creek Marginal Karst Plateau corresponds to areas developed over the less pure Pedro Leopoldo Member of the Sete Lagoas Formation. It corresponds mostly to areas close to the southern limit of the area, extending for 167.5 km^2. This domain presents less developed karst landforms, although limestone cliffs and sparse outcrops occur. One hundred thirty-four caves have been identified in this domain.

The domain named "Covered Karst" comprises much of the higher altitude portion of the area, which corresponds mostly to the western sector. With a total area of 273.7 km^2, it is the largest domain, and the presence of soil, mostly resulting from the weathering of the metapelites of the Serra de Santa Helena Formation (Piló 1998), masks the karst landforms. Surface drainages occur as well as lakes. Some urban areas, such as Sete Lagoas and Matozinhos, are located in this domain.

The two other domains represent mostly non-karst landscapes. The Santa Helena Fragmented Plateau occurs in the NW portion of the area, marked by NE–SW aligned ridges related to the metapelites of the Serra de Santa Helena Formation. Some restricted dolines occur in this domain, which covers a total area of 248.4 km^2. Fluvial plains have limited occurrence and develop along the major drainages in the area, the Velhas River and Mata Creek. These are low-declivity areas in which sedimentation processes predominate. Characteristic landforms in the geomorphic domains are shown in Fig. 3.

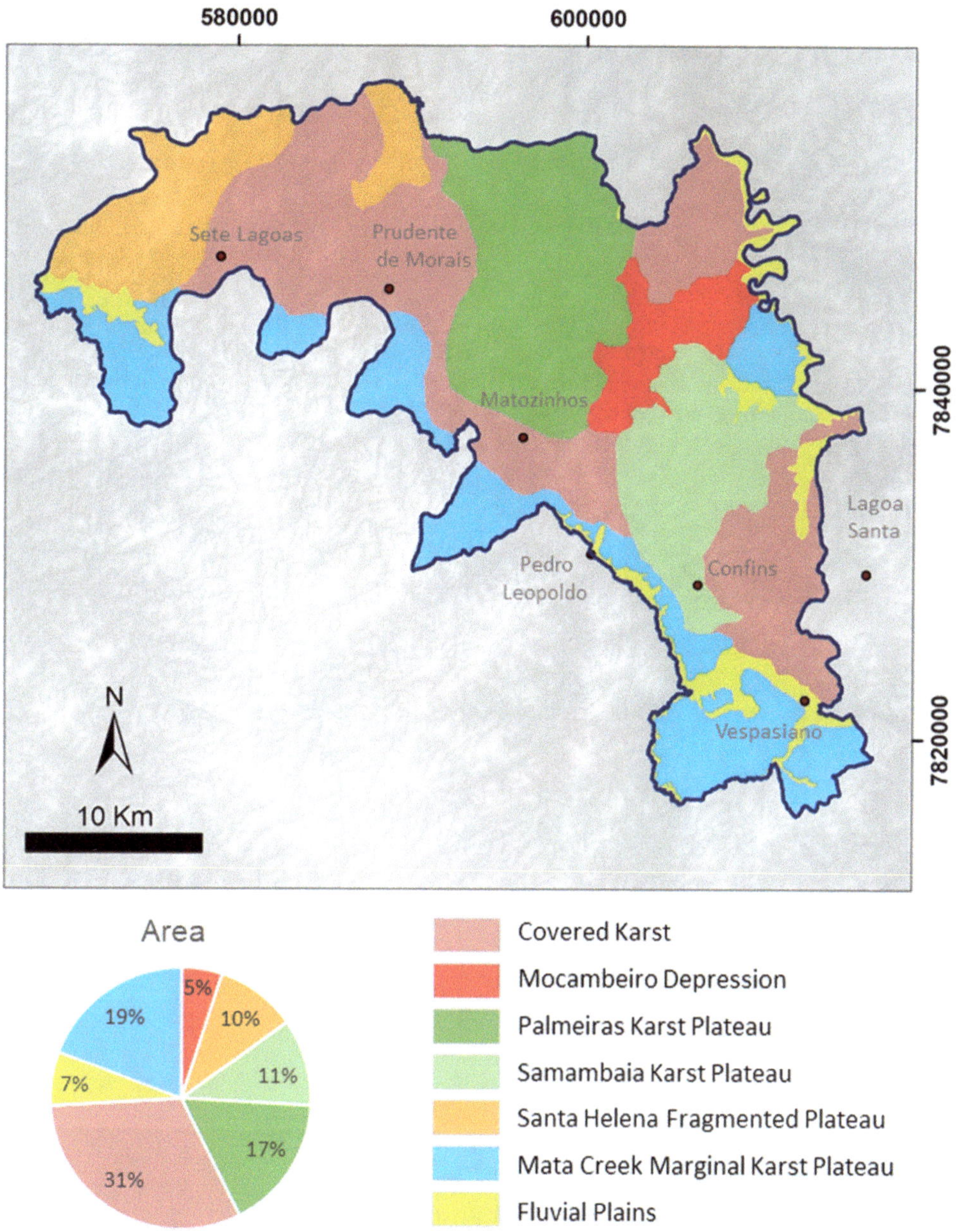

Fig. 2 Geomorphic domains of the Lagoa Santa Karst

3 Karst Landforms

Due to the covered nature of the Lagoa Santa Karst, classic karst landforms are not as evident here as they are in other karst areas (Fig. 4) because they are confined to selected areas of exposed limestone. Some landforms start under the soil and are modified following exposure at the surface. Considering that the karst is gradually stripped of soil, there is a relationship between karst development and the age of exposure. The transition from more rounded forms (subsoil features) in recently denuded outcrops to sharp jagged outcrops can be observed in the area.

3.1 Subsoil Features and Inheritance

Observations in many quarries in the area indicate that some of the karst landforms, of various dimensions, have their origins in the epikarst under the soil cover (Fig. 5). In fact, the general morphology of much of the karst landscape, including the towers and corridors and some karren types, are acquired under the soil and are gradually modified (at various degrees) after exposure at the surface. The importance of soil-covered dissolution processes has been commented on by Tricart (1956), Journaux (1977) and Auler (1994) and described in more detail by Piló (1998) and Knez et al. (2012). Landforms that have their origins and earlier

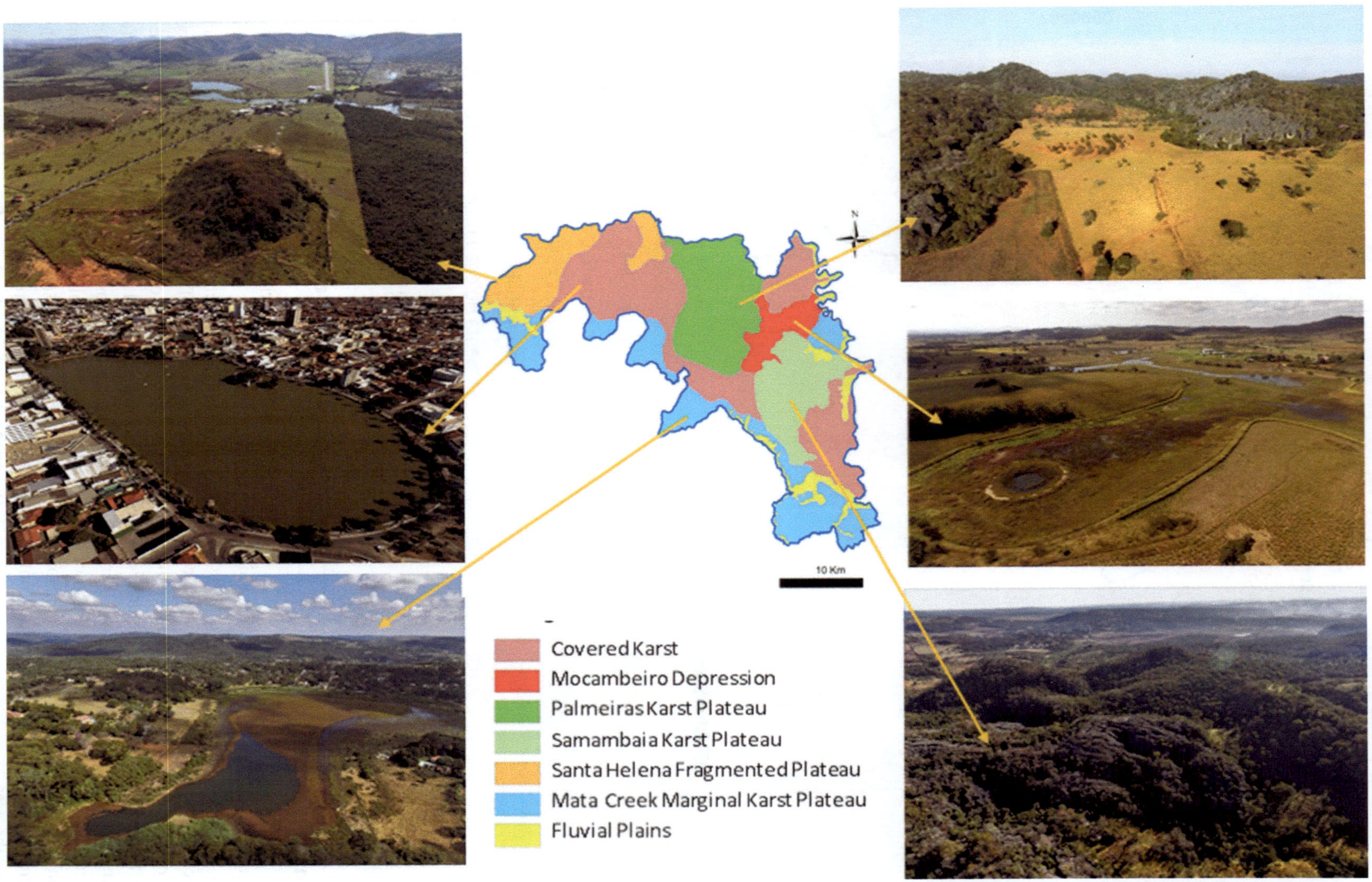

Fig. 3 Some characteristic features of the Lagoa Santa Karst geomorphic domains. Clockwise from upper right: outcrops amidst pasture in the Palmeiras Karst Plateau; epikarst lakes in the Mocambeiro Depression; outcrops in the Samambaia Karst Plateau; fluvial features in the Mata Creek Marginal Karst Plateau; Paulino Lake in the urban area of Sete Lagoas, Covered Karst; Polje-like depression in the Santa Helena Fragmented Plateau. In all domains, severe anthropogenic modifications have taken place, especially deforestation for conversion into pasture for cattle ranching and urbanization

Fig. 4 Covered karst in the Samambaia Karst Plateau, showing a few isolated outcrops (arrows) in the process of being exposed

Fig. 5 Quarry face showing the development of subsoil pinnacles

development in soil-covered conditions will be described in this section.

3.1.1 Joint Karren

This karren type was first described by Tricart (1956) and comprises elongated, centimeter-long dissolution pockets along subhorizontal planes, which were also described as schichtenkarren by Kohler (1989). This is perhaps the most frequent type of karren in the area and is due to the exploitation of well-marked subhorizontal horizons represented by foliation and bedding planes. It is clearly observed in limestone recently exposed due to soil stripping during quarrying. Under the soil, it tends to be more evenly distributed and less jagged (Fig. 6a). The soil cover homogenizes the distribution of acidic water, promoting a more uniform development of the karren. Piló (1998) noted that

Fig. 6 **a** Subsoil joint karren following recent soil removal due to quarrying operations. **b** Ubiquitous presence of joint karren in an outcrop

the upper reddish soil horizon, due to more efficient hydrochemical action, favors better development of joint karren when compared to the lower yellowish soils. Following surface exposure, selected horizons will be favored, and uneven subaerial dissolution will occur, resulting in deeper and more pronounced karren. This karren type is extremely common and occurs in nearly all vertical or subvertical limestone outcrops (Fig. 6b).

3.1.2 Alveolar Karren

According to Piló (1998), these alveoli develop initially under the soil and later evolve under subaerial conditions. First described by Tricart (1956), like joint karren they also tend to develop over subhorizontal partition planes but are oval shaped, with the lower surface inclined toward the exterior (Fig. 7). They may resemble cave scallops, but they are deeper and less regular and do not occur in groups or in areas subject to water flow. Alveolar karren over vertical joints also occur, although less frequently (Piló 1998). As with the joint karren, the transition from soil cover to the surface results in sharper and more pronounced features. Alveolar karren are less frequent than joint karren but may have a common or parallel development. The coalescence of closely spaced alveolar karren can result in forms similar to joint karren (Fig. 7b). It is possible that both the alveolar karren and the joint karren cannot develop solely at the surface, requiring an initial enlargement phase under the soil. Alveolar karren are very common in the Lagoa Santa Karst and occur in most vertical to subvertical limestone outcrops.

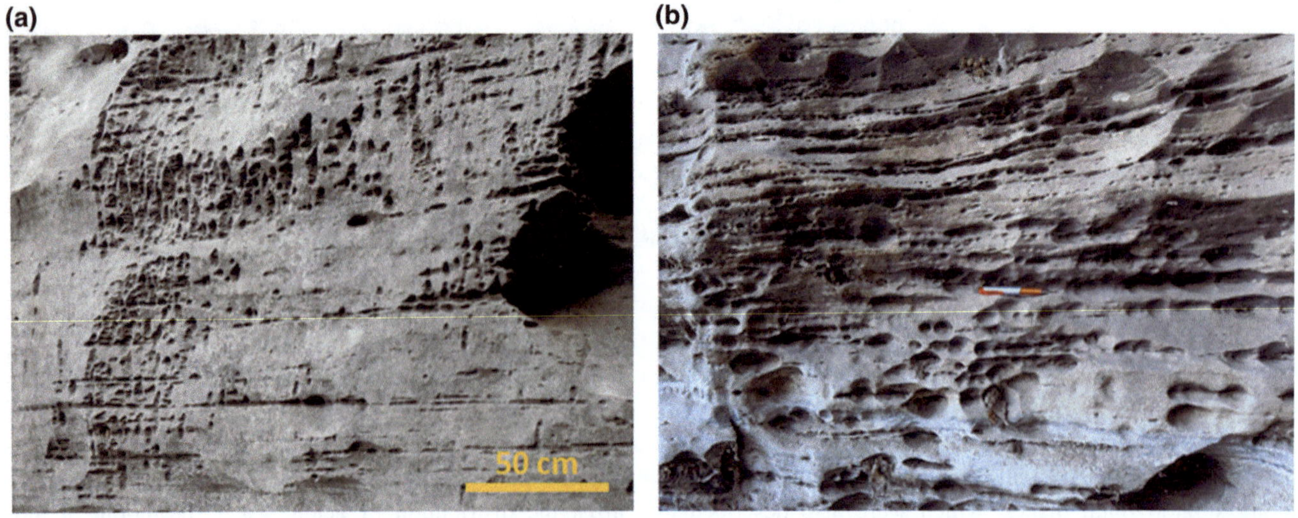

Fig. 7 Alveolar karren. Note that the coalescence of alveolar karren can result in joint karren

3.1.3 Subsoil Channels, Pits and Cups

In a detailed work at the Lapinha outcrop, Knez et al. (2012) described a series of karren subtypes associated to some extent with the soil cover (Fig. 8). Most features appear to be generated not under deep soil cover in the epikarst but rather in situations where there is thin soil or debris cover at the surface. Channels (rundkarren) form either entirely under the soil or when only their bottoms are covered by soil while the surrounding rock is exposed. They have an inverted omega shape and can have multiple levels (Knez et al. 2012). These authors comment that they may originate entirely within the rock mass and become exposed by the later removal of the upper rock floor. Subsoil pits are described by Knez et al. (2012) as being one to five centimeters in diameter, with larger cups. Their genesis is attributed to the percolation of water through the soil and the dissolution of vertical joints; their cross sections are circular or elliptical along the joints (Knez et al. 2012). Several other subtypes of karren, some subaerial, are shown in Fig. 8.

Fig. 8 Sketch showing several types of karren observed at Lapinha outcrop. (1) subsoils channel, (2) subsoil cup, (3) network of tubes along contact, (4) subsoil cup or kamenitza, (5) rain flutes, (6) notches along contacts, (7) vertical jaggedness, (8) half-funnel shaped notch at the end of channels, (9) channel from the interlaminar tube, (10) ceiling pocket, (11) pits formed by water trickles. From Knez et al. (2012)

3.1.4 Subsoil Pinnacles, Towers and Corridors

These forms undoubtedly originate under soil cover. Soil stripping at quarries shows that the limestone surface is very uneven, and pinnacle-like projections, deep grykes, and tower-like features are common (Fig. 9), as initially reported by Tricart (1956). The general resemblance between these subsoil forms and the ones observed at the surface strongly suggest that they represent stages of an evolutionary continuum. The knobs described by Kohler (1989) and Piló (1998) represent the partial exposure of the topmost portion of buried towers or pinnacles (Coutard et al. 1978). In between these pinnacles, the dissolution along joints creates deep grykes that may interconnect, forming a highly irregular array of corridors. These corridors between towers are preserved after exposure, resulting in an extremely irregular network in the interior portions of the outcrops. The quarries at São José da Lapa and Lagoa de Santo Antonio provide good examples of a highly irregular epikarst surface.

3.2 Subaerial Microforms

3.2.1 Rain Pits

This ubiquitous form was named "cellular karren" by Tricart (1956) and occurs in horizontal or weakly dipping surfaces. Normally, they are no less than a few centimeters in diameter

Fig. 9 **a** Pinnacles and towers evolving under the soil in a quarry face. **b** Pinnacles recently uncovered due to mine soil stripping

and less than one centimeter deep. However, coalescence or focused dripping can generate larger forms described by Knez et al. (2012) as “pits” (Fig. 8).

3.2.2 Condensation Pockets

These are small features that occur in ceilings at the entrance zones of caves. They comprise millimeter-wide and deep vertical holes that are unevenly spaced (Fig. 10). Because they grow upwards, their genesis is attributed to the dissolution of the limestone by hanging droplets created by condensation due to the climatic contrast between the cave atmosphere and the surface. Good examples occur at the northern entrance of Baú Cave.

3.2.3 Kamenitza

A kamenitza, or solution pan, is another landform that may bear the influence of soil processes but that does not originate deep under soil cover in the epikarst zone. Kamenitzas are a few centimeters in length and depth (Piló 1998). Their morphology is similar to that of other areas, comprising a flat floor, sometimes with water or a vegetal mat, and sharp undercut edges, sometimes with an outflow channel. They are not very frequent forms in the Lagoa Santa Karst and are restricted to upper portions of denuded limestone outcrops.

3.2.4 Rillenkarren

Although they represent a frequent type of karren elsewhere, rillenkarren are relatively uncommon in the Lagoa Santa Karst due to the covered nature of the area (Fig. 11). In recently exposed outcrops, there may not have been enough time for the proper development of rillenkarren. Therefore, well-developed rillenkarren and associated spitzkarren are rare. Knez et al. (2012) observed these forms at the Lapinha outcrop, and Tricart (1956) suggested that the vertical flutes may be overwritten by the faster developing joint karren.

3.2.5 Subhorizontal Tubes

Cylindrical subhorizontal tubes tend to be very common in the Lagoa Santa Karst. They may represent former protoconduits bisected by outcrop fragmentation. Although some horizons may be favored, in some outcrops, they occur at various levels (Fig. 12). Their diameter ranges from a few centimeters to 10 cm, although larger, human-scale phreatic tubes occur. The tubes are controlled mostly by subhorizontal planes (Dutra et al. 1998), but vertical joints also play a role. The large number of such tubes and the fact that they are not always associated with an anastomotic channel may bring into question their role in the initiation of large caves. Because their length and orientation are difficult to measure, it is difficult to ascertain whether they represent fragmented portions of larger former groundwater routes or more local forms restricted to single outcrops. Since they are isolated high above the water table and do not show evidence of water flow, they are interpreted as being relict phreatic features. Knez et al. (2012) described another type of tube, this one draining surface water from the outcrops. Water flows through these channels, leaving wall channels where they intersect the surface.

3.2.6 Nondirectional Scallops

Several limestone cliffs display widespread dissolution cusps similar to scallops. Lund (1841b) was the first to notice these forms. These scallops are usually up to 5 cm wide and do not indicate a flow direction. Due to the common occurrence of lakes associated with these cliffs (Auler 1995), they are interpreted as being formed by lake action, created by episodes of water rising and lowering. Good examples are identified in the cliffs at Lapa da Vargem, Escrivânia and Lapa do Santo (Fig. 13).

Fig. 10 Condensation pockets in overhanging surfaces at cave entrances

Fig. 11 Karstified outcrop showing rillenkarren on the Palmeiras Karst Plateau

Fig. 12 Subhorizontal tubes evolving into caves

3.2.7 Other Karren Forms and the Role of Vegetation

Several other karren forms occur in the Lagoa Santa Karst, although many of them appear to be variants of the ones already described in this work. Both Tricart (1956) and Knez et al. (2012) commented on vertical cylinders, "curtain-like" subvertical grooves or wall pockets that cross the outcrops vertically. When they are several meters deep, they may represent grykes expanded in a soil-covered situation. Knez et al. (2012) also described ceiling pockets in overhanging walls created by dripping water from fissures (Fig. 8).

Roots are frequently associated with limestone outcrops in the Lagoa Santa Karst (Fig. 14), and in some situations, they take advantage of existing karren, developing through

Fig. 13 Nondirectional scallops at Lapa da Vargem cliff

Fig. 14 Occurrence of roots along the Limeira outcrop displaying several types of karren

cylindrical tubes. The role of roots in generating karren features is, however, unproven. Tricart (1956) assigned the genesis of curtain-like vertical grooves to the dissolution action of roots, while Knez et al. (2012) suggested that roots may accelerate the expansion of some types of karren, mostly tubes. Due to the short-lived nature of these trees, the ancient surface exposure of the limestone and the fact that many of these features actually start deep in the subsurface, the action of roots probably plays only a minor role and does not significantly alter the morphology of these karren forms.

3.3 Subaerial Macroforms

3.3.1 Rinnenkarren

Rinnenkarren is not a common type of karren in the Lagoa Santa Karst. The area lacks extensive inclined barren limestone surfaces exposed for enough time to allow the generation of this karren type. When exposed in cliffs, the vertical or overhanging nature of the outcrop precludes the generation of runnels. A few examples have been observed at the Macacos Depression in the Palmeiras Karst Plateau.

3.3.2 Small Pavements and Knobs

Knobs and small pavements are transitional forms between epikarst subsoil towers and larger outcrops. They represent the small, upper, exposed portion of a much wider limestone mass still under soil (Fig. 15). These landforms usually

Fig. 15 Small pavements and knobs outcropping in an otherwise covered karst

Fig. 16 Examples of towers in the Lagoa Santa Karst. **a** Vargem da Pedra. **b** Pedra da Orelha

occur in the upper or middle portion of slopes (Piló 1998) and can range from a few centimeters in length and height to up to a few meters high. Because they represent the "tip" of a larger subsurface mass, they tend to become wider as they become exposed. Knobs occur when the vertical dimension is significant, while pavements are essentially bedding plane-oriented small outcrops. Joint karren are ubiquitous in the knobs, and the dissolutional widening of the subhorizontal planes can lead to the separation of limestone slabs.

3.3.3 Towers

Limestone towers in the Lagoa Santa Karst are not very frequent, although excellent examples exist, such as those at Vargem da Pedra and Pedra da Orelha (Fig. 16). Towers are

Fig. 17 A conical limestone hill developed in the flat polje-like depression north of Sete Lagoas

residual remnants of once much larger outcrops that became fragmented and isolated. They can reach up to 20 m in height. Joint karren tend to be abundant. Some towers are located close to the parent massif, while others, such as the Caetano outcrop, are utterly isolated in the landscape and could be described as hums (Coutard et al. 1978). In some situations, residual limestone towers are associated with a depression, which may contain a lake. The individualization and final decay of the towers may be associated with the dissolutional action of lakes or the development of expanded joints between the tower and the massifs. As mentioned before, the initiation of the towers may be associated with pinnacles in the epikarst, although the pinnacles tend to be wider toward the base.

3.3.4 Conical Hills

Conical hills are not widespread in the Lagoa Santa Karst. They comprise residual hills, usually soil covered, with a diameter of up to 800 m; they are developed mostly over the less pure Pedro Leopoldo Member. Good examples occur at the large depression to the north of Sete Lagoas (Fig. 17).

3.3.5 Large Outcrops

Although mostly a covered karst, large outcrops occur, especially in the karst plateaus (Fig. 18), away from the zones predominantly covered by soil. In the higher catchment areas to the west, the karst plateaus display abundant karst landforms, although they tend to be essentially soil and vegetation covered. The dimension of these outcrops varies considerably, as they tend to amalgamate with neighboring massifs, forming elongated outcrops. The top portion is invariably concave regardless of whether soil covered or denuded; this portion presents deciduous arboreal vegetation and may display a cliff on one of its sides (Fig. 19). Dolines, sometimes filled with lakes, can occur at the base. As soil stripping progresses from the sides toward the top, it is common to find partially barren outcrops, such as the outcrop to the west of the village of Mocambeiro and the outcrop of Baú. Entirely barren outcrops occur at Experiência da Jaguara and in the Gordura area. The outcrop of Lapinha, studied by Balasz (1984) and Knez et al. (2012), is mostly devoid of soil. In an area in which the original vegetation has been removed, giving way to pasture land, the outcrops resemble vegetated "islands" amidst a grassy landscape. The evolution of the limestone outcrops involves not only dissolutional fragmentation but also mechanical processes. The detachment of large portions of outcrops has been observed in recent times, and cones of large blocks occur at the base.

3.3.6 Cliffs

Imposing limestone cliffs are the most characteristic landform of the Lagoa Santa Karst (Reinhardt 1867; Coutard et al. 1978). These cliffs comprise vertical to slightly overhanging faces that are remarkably rectilinear and may reach heights of up to 60 m. The overhanging nature of the cliffs and the occurrence of caves make them particularly favorable for hosting archaeological sites such as Santo or Cerca Grande. Their surfaces are usually covered with a whitish

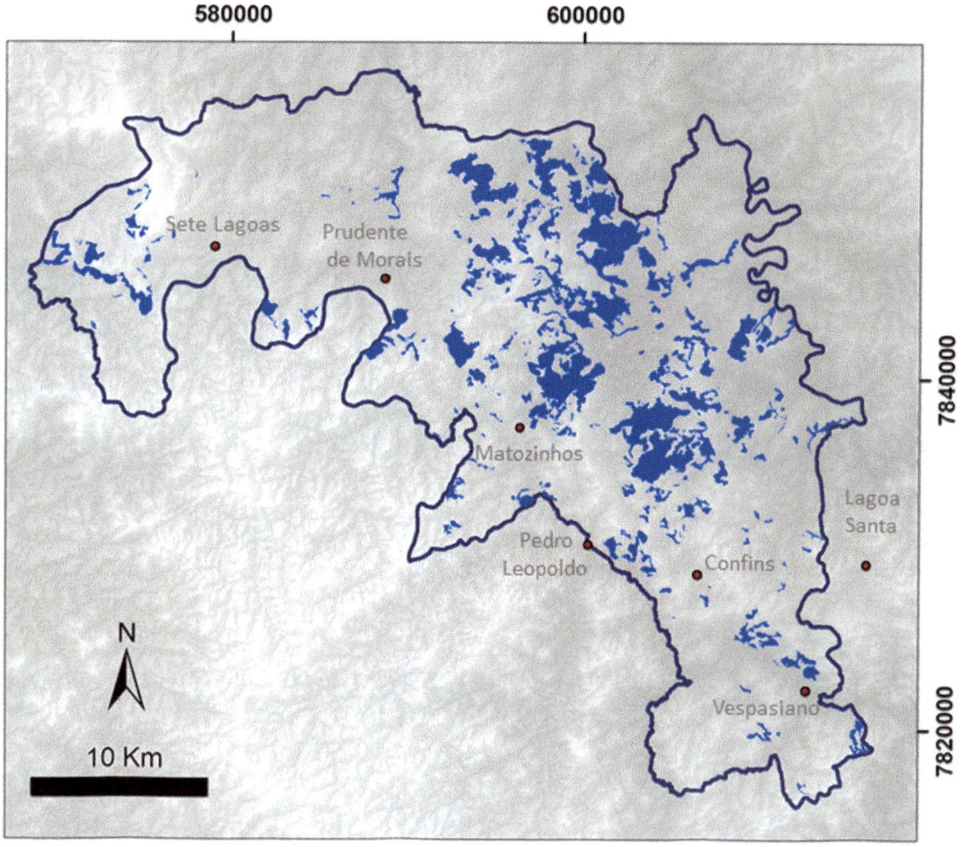

Fig. 18 Limestone outcrops (in blue) in the Lagoa Santa Karst

Fig. 19 Examples of outcrops. **a** Cerca Grande. **b** Outcrop near Sangradouro

(calcite) to reddish (iron) tone that provides a peculiar color that helps them stand out in the landscape. Large stalactites adorn their faces. A survey of 216 cliffs shows that nearly all are quite regular, showing linear faces. The morphometry of the cliffs (Table 1) indicates that 88% developed in the Lagoa Santa Member and few are associated with caves. The average length is 98 m. The dominant orientation is NNW–SSE (340°), approximately perpendicular to the hydraulic gradient of the area. Their bases are frequently associated with asymmetric and sometimes water-filled dolines and cones of fallen blocks (Reinhardt 1867). They normally represent the exposed portions of partially covered concave hills. They seem to be best developed in the central area and in the areas surrounding the towns of Matozinhos and Pedro

Table 1 Morphometry of limestone cliffs

Cliffs		Associated with cave	Pedro Leopoldo member	Lagoa Santa member
Linear	187	63	21	166
Irregular	29	8	4	25
Total	216	71	25	191
	100%	33%	12%	88%

Fig. 20 Some representative examples of limestone cliffs. **a** Macacos-Baú. **b** Cerca Grande. **c** Lapa do Santo. **d** Sumidouro

Leopoldo, and they are associated with the karst plateaus. Some of the best-known examples occur at Lapa Vermelha, Cerca Grande, Caetano, Poções, Escrivânia, and Campinho (Fig. 20).

The differential denudation that created the sharp contrast between the base and the vertical cliff face has been attributed by Tricart (1956) to the weathering of a metamorphic complex (the crystalline basement) in lateral contact with the limestone by faults. Dolabela (1958) presented a hypothesis (based on information provided by geologist Djalma Guimarães) in which a more impure and shaly limestone is in lateral gradational contact with the limestone. Being more susceptible to weathering, it would be denuded more quickly, creating a cliff in the purer limestone. The mechanisms proposed by Tricart (1956) and Dolabela (1958) are not supported by more recent geological studies.

Journaux (1977) proposed a model (Fig. 21) in which sinking points tend to occur at major joints, creating a slightly depressed area bordered by a small cliff. The limestone cliff, no longer having a significant soil cover that can provide acidic soil and vegetation-derived CO_2, becomes less susceptible to dissolution ("immunized", according to Journaux 1977). The progressive deepening of the sinking point may leave older cave passages perched at the cliff face. The overhanging nature of the cliffs is thus due not only to the joint orientation but also to the continuous dissolutional action at its base (Journaux 1977). Unlike other karst landforms in the area, there is no evidence of the early subsoil control of the cliffs; they appear to be generated at the surface. Another type of cliff has its genesis associated with fluvial valleys and is thus elongated and curvilinear along the meandering valleys; examples include the cliffs on the left side of Mata Creek and to the north of the industrial district of Vespasiano, and the extensive cliff associated with Jaguara Creek a few kilometers before its mouth at the Velhas River (close to Jaguara Velha Farm).

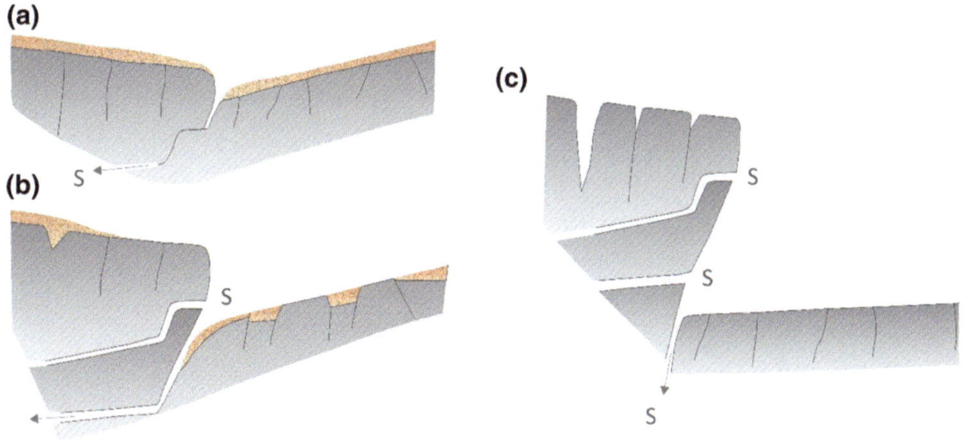

Fig. 21 Schematic sequential mode of development of limestone cliffs in the Lagoa Santa Karst according to Journaux (1977). The progressive deepening of the swallet (*S*) leaves older sinking points perched at the cliff face

3.4 Karst Depressions and Lakes

The Lagoa Santa Karst is known for its subdued undulating landscape punctuated with lakes, in which the limestone outcrops appear as isolated elevations (Lund 1841a; Lanari 1909). Karst depressions of various dimensions occur throughout the area. Small-scale depressions (dolines) tend to be more common in the karst plateaus. Dolines also occur in the lower areas to the east, although they are shallower and wider.

3.4.1 Dolines

Dolines are a conspicuous feature of the Lagoa Santa Karst. Coutard et al. (1978) and Kohler (1989) concentrated their studies in the central part of the Lagoa Santa Karst and noted that many of the dolines are asymmetric, presenting an outcrop on one side. Using data from these studies, and after careful analyses of Google Earth images, 876 larger dolines were identified within the area; they are unevenly distributed (Fig. 22), with a general density of 1 large doline per km^2. Piló (1998) presented a morphometric study of dolines in his 5.13 km^2 study area in Macacos-Baú, located in one of the karst plateaus. In this area, 97 dolines were identified at a 1:10,000 scale, with a density of 18.0 per km^2, an average depth of 6.4 m and a mean area of 5460 m^2 (Piló 1998). Approximately 27% of the dolines are circular, and the remaining dolines are elongated, according to the structures oriented N–S and N60-80W (Piló 1998). In the Macacos-Baú area, dolines limited on one side by limestone outcrops comprise 48% of the sample set. The coalescence of dolines, forming uvalas, occurs frequently, and smaller dolines inside a basin of larger ones are also a recurring feature. It is especially common to have elongated large dolines along limestone cliffs; these dolines are controlled by the cliff orientation, as at Lapa Vermelha, where the dolines were named uvalas by Kohler (1989). In these situations, however, it is not clear if there was a coalescence of separate dolines (a requirement for having a uvala) or if the entire depression evolved gradually as a single landform. As mentioned previously, the evolution of limestone cliffs is, according to Journaux (1977), linked to the development of a depression (swallet) at its base, suggesting why cliffs are so common on the sides of dolines.

At lower elevations, dolines are also quite frequent but tend to become shallower and are commonly filled with lakes. Kilometer-long depressions that are only a few meters deep occur along the Mocambeiro Depression. Solution and subsidence dolines are certainly the most frequent genetic types (Fig. 23a), but collapse dolines have been identified at a few sites, mostly associated with urban areas. In the area surrounding the town of Sete Lagoas, a well-studied collapse doline was interpreted as being caused by the natural collapse of a 7 m-high cave located 58 m below the surface (Silva 1988) (Fig. 23b). It is, however, unlikely that this collapse, as well as 17 other subsidence/collapse occurrences identified in the area since 1988 (Galvão et al. 2015), has no relationship with the heavy groundwater drawdown in the area.

3.4.2 Karst Lakes

The common occurrence of lakes in the Lagoa Santa Karst is due to (i) the existence of a thick, impermeable pedological cover that protects the bottom of depressions, favoring the accumulation of water; (ii) a low hydraulic gradient, with few elevations, leading to proximity with the water table; (iii) the existence of areas where the limestone is covered by metapelite rocks, favoring water accumulation in depressions and (iv) seasonal climate, in which the meteoric recharge in the wet months considerably exceeds the infiltration capacity. Lakes were outlined using Google Earth's historical and present images. The outcome was then combined with previous information collected by Kohler (1989), resulting in a total of 199 lakes in the area (Fig. 24).

Auler (1994, 1995) distinguished three main types of lakes in the Lagoa Santa Karst: water table lakes, sinkhole lakes, and lakes over metapelite. A newly updated classification is proposed in this work, modifying the previous

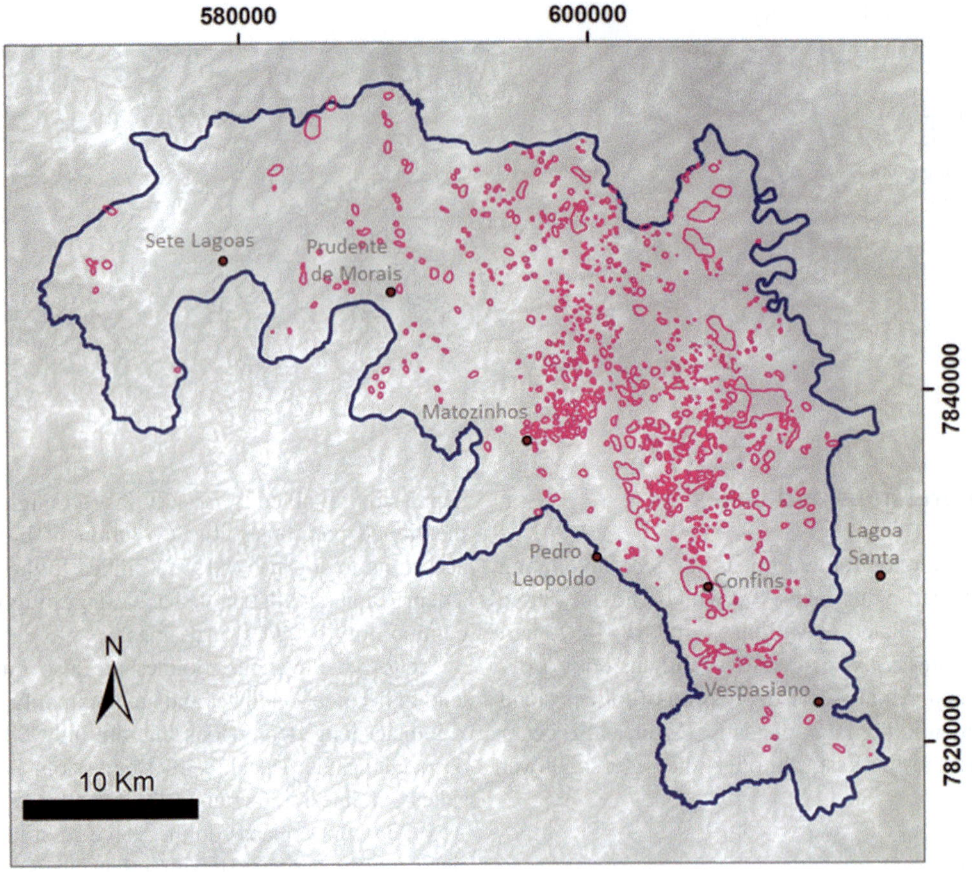

Fig. 22 Distribution of dolines in the Lagoa Santa Karst

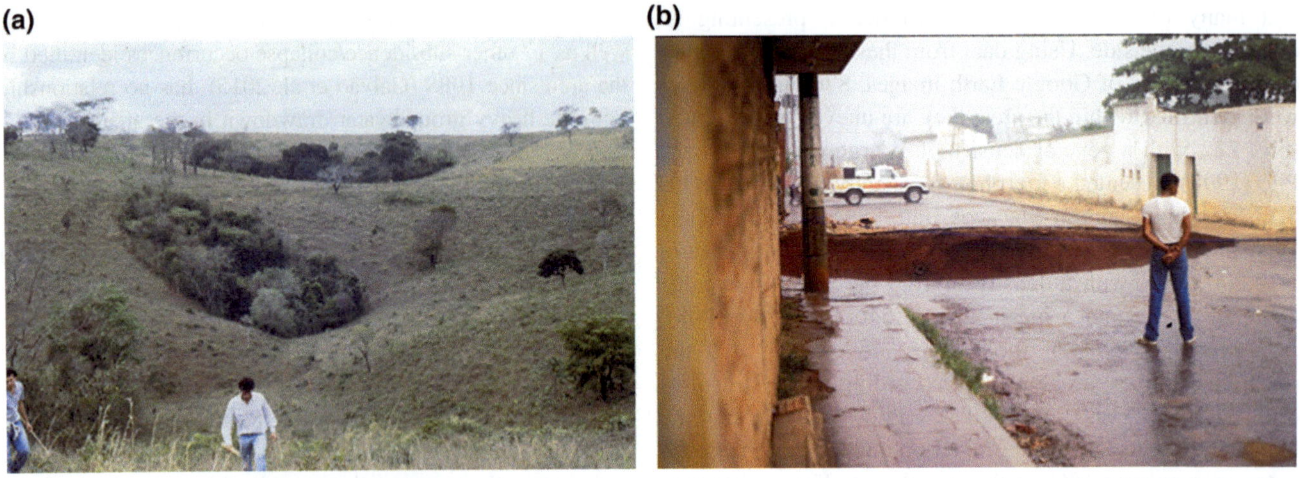

Fig. 23 **a** Solution dolines near Lapa Vermelha, Samambaia Karst Plateau. **b** Collapse doline in the urban area of Sete Lagoas

designation of "water table" lakes. In this new version, the three lake types recognized in the Lagoa Santa Karst are epikarst lakes, sinkhole lakes, and lakes over metapelite. Epikarst lakes were previously described as "water table lakes" by Auler (1994, 1995). However, the discovery of vadose river caves below some of these lakes in the Mocambeiro Depression demonstrates that they are not related to the water table but to a perched epikarst water level. These lakes may experience variations in level due to both meteoric inputs as well as the rise in the water table/the flooding of cave passages. These lakes are thus related mostly to the epikarst but may also receive contributions

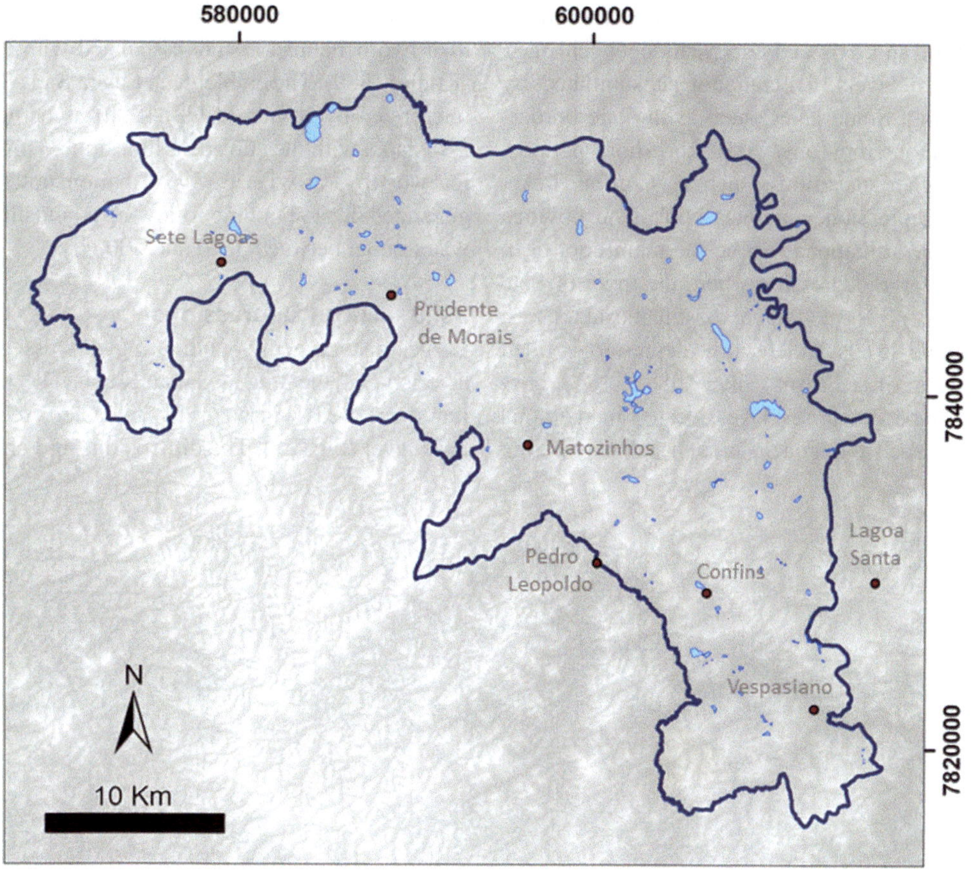

Fig. 24 Distribution of significant lakes in the Lagoa Santa Karst

Fig. 25 Examples of the three types of lakes in the Lagoa Santa Karst. **a** Epikarst lakes at Mocambeiro Depression. **b** A sinkhole lake (mostly dry) associated with the entrance of Vermelha Cave. **c** Lagoa dos Olhos and in the foreground, Lagoa Santa, lakes located over metapelites

from the water table rises. Good examples of this type of lake are found in the Mocambeiro Depression (Fig. 25a).

The second type is commonly located in the karst plateaus and comprises sinkhole lakes with floors capped by breakdown or impermeable clay (Auler 1995). Sinkhole lakes are usually located at the bottom of asymmetrical dolines and are limited on one side by a cliff (Fig. 25b). Good examples are the lakes at Lapa Vermelha, Lagoa dos Mares, and Lagoa de Santo Antonio. The level of these lakes varies considerably and does not appear to closely follow rainfall intensity. Ancient watermarks on the cliffs attest that in the past, they have experienced levels well above the present recorded maximum. Coutard et al. (1978) argued for a lake periodicity of approximately 15 years between maximum and minimum levels, a claim not substantiated by consistent monitoring. Some lakes, such as the one bordering the Escrivânia outcrop, only exist for short periods. A few lakes at higher elevations, such as Confins Lake, occupy a shallow depression without a cliff or an obvious swallet. Impermeable soil appears to be the main control in these lakes. At the large depression where the town of Sete Lagoas (Seven Lakes) is located, there exist namesake lakes, which present stable levels and occupy depressions filled with more than 20 m of sediment (Silva 1988).

The third lake type comprises lakes located in marginal metapelitic surfaces, such as Lagoa Santa itself and neighboring Lagoa dos Olhos (Fig. 25c). These lakes have stable levels, and their hydrochemistry is not associated with carbonate aquifers (Kohler 1989). The emblematic Lagoa Santa Lake was initially interpreted by Kohler (1978, 1989) as being due to collapse associated with the subjacent limestone. However, palynological work by Parizzi et al. (1998) demonstrated that the lake was generated due to ponding following a landslide that blocked its water outlet between 3100 and 5000 kyr.

3.4.3 Large Depressions

Large, flat, or gently undulating depressed areas are characteristic of the Lagoa Santa Karst. The expected trend is that these areas will occur in the eastern sector, closer to the base level (Fig. 26). The dimensions and characteristics vary

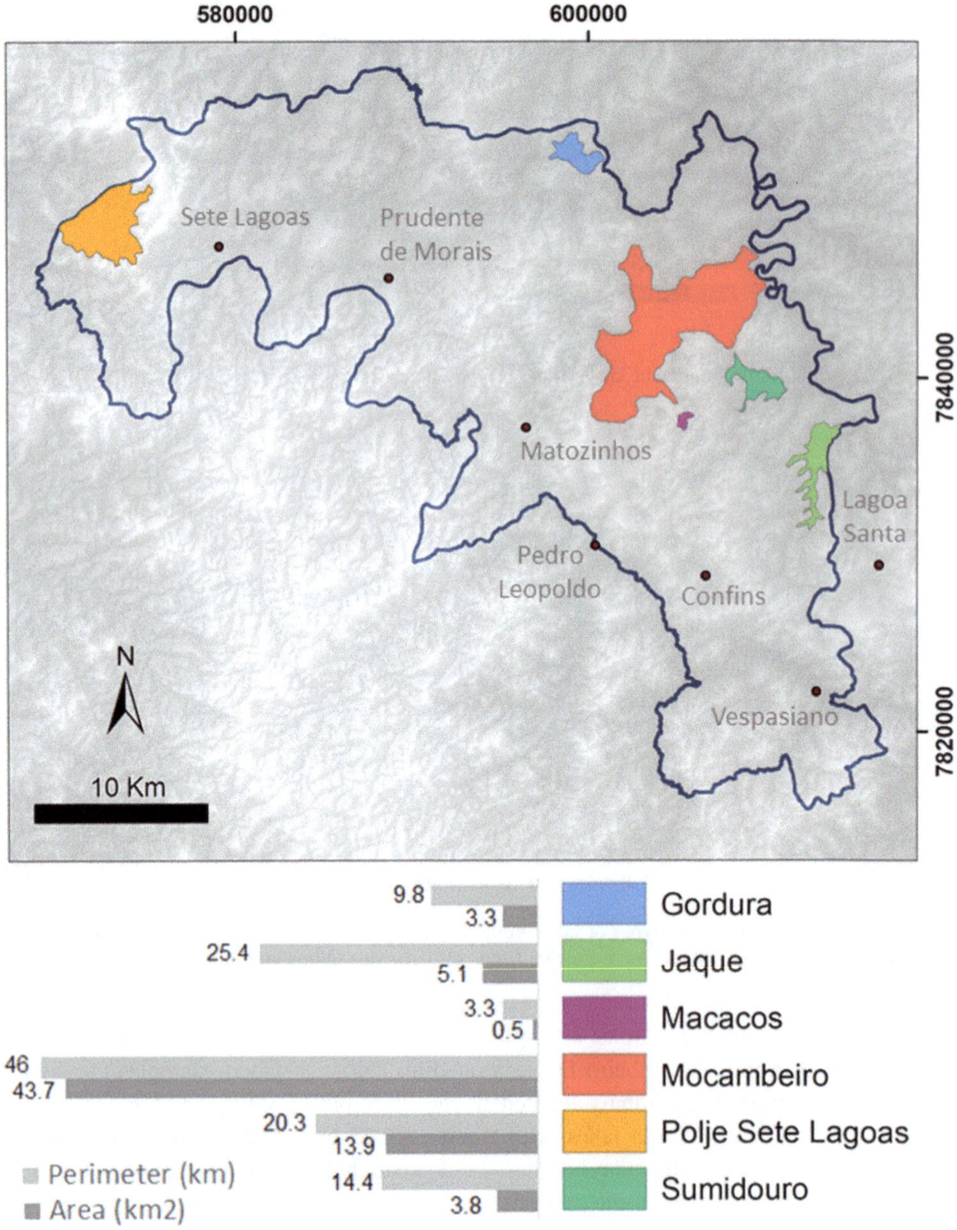

Fig. 26 Location of large depressions in the Lagoa Santa Karst

Table 2 Morphometry of some large depressions in the Lagoa Santa Karst

Name	Area (km^2)	Perimeter (km)
Mocambeiro	43.7	46.0
Sumidouro	3.8	14.4
Gordura	3.3	9.8
Jacque	5.1	25.4
Macacos	0.5	3.3
"*Polje*" Sete Lagoas	13.9	20.3

considerably (Table 2), and it is possible that they are controlled by more than one genetic mechanism. The best-known depression, Mocambeiro Depression, has been described by most recent researchers since Coutard et al. (1978). Although a depressed area, Mocambeiro Depression is not closed at its eastern limit and contains numerous limestone outcrops and cliffs. Kohler (1989) classified it as a "semi-polje". This depression shows continuity with low-lying areas surrounded by outcrops to the southwest, such as the one that leads to Morro Redondo Cave and the one directed toward Poções. Therefore, Mocambeiro Depression is interpreted here as a mature depression generated by dissolution controlled by tectonic structures combined with doline coalescence. Smaller depressions in the karst plateaus, such as Macacos, are interpreted as the result of doline coalescence with borders marked by cliffs and control of outcrops by tectonic structures.

The Sumidouro Lake depression is fed by the fluvial valley of Samambaia Creek and contains an ephemeral lake (Fig. 27a). It is bound to the east by a residual outcrop containing the well-known Sumidouro Cave. It has been interpreted by Kohler (1989) and Piló (1998) as a polje. Because it lacks many of the classic features of poljes, Auler (1994) defined it as a swallet largely modified by fluviokarst processes. In the northern portion of the Lagoa Santa Karst, dimensionally notable flat depressions share more similarities to the poljes in the classic Dinaric Karst. The large circular depression immediately to the northwest of Sete Lagoas contains hums and is surrounded by higher slopes of both limestone and metapelite (Fig. 27b). The slightly sloping low area where the town of Sete Lagoas is located is also noteworthy, as it was interpreted by Galvão et al. (2016) as a graben.

3.5 Fluvial Features

Remnants of the Serra de Santa Helena metapelites overlying the carbonates allow for some volumetrically restricted allogenic flow toward the carbonate. These streams generate classic fluvial landforms as they enter the carbonate. In the recharge zone in the lower elevation areas, springs give rise to streams that meander toward the Velhas River. The lower course of Samambaia, Jaguara, Gordura, and Mocambo creeks display classic fluvial features.

3.5.1 Dry Valleys, Blind Valleys, and Karst Windows

The most representative dry valley is the E–W elongated valley immediately to the east of the town of Matozinhos. This valley ends in a cliff that contains a dry cave hanging a few meters above the floor. Another characteristic dry valley occurs to the NW of Poções, meandering along for 1.2 km until a cave (Freática Cave) is reached (Fig. 28).

Fig. 27 Some of the major depressions of the Lagoa Santa Karst. **a** Sumidouro Lake depression. **b** "Polje" north of Sete Lagoas

Fig. 28 Dry valley NW of the Poções area

Blind valleys are not common features and usually have small dimensions. One of the most significant is the one that marks the first swallet of Palmeiras Creek. This creek sinks and reappears several times along the karst plateau, generating blind valleys and short karst windows until it reaches lakes known as Poções. Another major blind valley marks the swallet at Cascata I cave, draining a creek that originates from the metapelites. Small blind valleys are also associated with the swallet of Buraco do Medo cave. An interesting karst window occurs in the Cocho d'Água area (east of Lagoa de Santo Antonio), in which a short drainage rises and sinks, reappearing at Engenho spring. Another short section of a horizontal stream occurs at Bebida Farm and was first recorded by Liais (1872). The stream emerges from a spring and sinks a few meters beyond it.

3.5.2 Swallets

Swallets tend to occur in the karst plateaus, draining streams that sometimes originate from the metapelite. As already mentioned, the swallet of Cascata I Cave is among the most representative as well as the swallet associated with the downstream end of Samambaia Creek at Sumidouro Lake. The series of swallets and springs associated with Palmeiras Creek provide several significant examples, such as the one at Poções Cave (Fig. 29) and at the lakes also known as Poções.

Fig. 29 Swallet of Palmeiras Creek at Poções

3.5.3 Springs

Auler (1994) studied 12 springs in the central area of the Lagoa Santa Karst. Pessoa (2005) also monitored some springs at the southern limit of the karst. The major springs are associated with Gordura, Mocambo, Jaguara, and Samambaia creeks, although several other springs exist in the area. The majority of the springs drain toward the Velhas River. All springs of the Lagoa Santa Karst issue from flooded passages. Further details on karst springs are provided by Auler (1994), Pessoa (2005) and chapter "Karst Hydrogeology of the Lagoa Santa Area", this volume.

4 Evolution of the Lagoa Santa Karst

The Lagoa Santa Karst has undoubtedly undergone a long evolutionary history. In fact, the karst presently occurs in the basal formation of the Bambuí Group, implying that hundreds of meters of metapelites and carbonates of the upper formations have been previously removed. In most of the areas, the thickness of the carbonate does not surpass a few hundred meters. Near Pedro Leopoldo, the highly irregular relief of the crystalline basement can be observed lying just a few meters below the limestone. Because the noncarbonate sequences overlying the limestone have been nearly entirely removed, there is limited parent material subjected to pedogenic processes, and thus soil generation is restricted (Auler et al. 2009). The amount of soil removal largely surpasses soil generation, and the carbonate is thus gradually being exposed. The Lagoa Santa Karst is in a final evolutive stage, a "terminal karst" in the sense that as the base level drops and denudation progresses, it will eventually reach the nonpermeable basal gneiss and be entirely removed from the landscape.

Soil cover, hydraulic gradient, and tectonic structures together control the development of the karst. The large numbers of karst landforms that owe their initial genetic stages to epikarst processes attest to the crucial role played by soil cover. The lower impure Pedro Leopoldo Member does not display well-developed epikarst forms, as observed in quarry faces to the west and in the vicinity of Sete Lagoas; this demonstrates that soil cover, although undoubtedly very important for landform generation in the core area, may in some situations not be the sole or dominant mechanism.

Tricart (1956) and Balasz (1984) proposed that the Lagoa Santa Karst is presently being exhumed, and Balazs (1984) claimed the occurrence of several phases of burial and exhumation. However, as correctly noted by Kohler (1989), the karst has never been previously exposed; its present cover is due to the alteration of the upper metapelite formations with some transport due to colluviation (Coutard et al. 1978).

The general shapes of many landforms, including some karren types and towers/outcrops, owe their origins to epikarst processes under soil cover. Presently exposed landforms have, thus, a strong degree of inheritance. Due to the homogeneous nature of carbonate soil, the aggressiveness is evenly spread over the bedrock. The expression of tectonic structures is subdued in those subsoil features but becomes more developed at the surface, with the generation of linear features as cliffs. However, as suggested by Piló (1998), the lower areas in between outcrops may be initially generated under the soil through the influence of groundwater routes.

Some of the major features of the Lagoa Santa Karst, especially in the central areas of Matozinhos and Pedro Leopoldo, are oriented SW–NE along structures. This orientation is observed for large landforms such as the Mocambeiro Depression as well as smaller depressed areas such as the covered low-lying karst corridors in Poções and Morro Redondo or some of the cliffs at Cerca Grande and Caianga. This orientation agrees with the general hydraulic gradient of the area toward the Velhas River base level. Regionally, the existence of favorable joints and bedding planes aligned according to the hydraulic gradient creates suitable conditions for concentrated flow and thus increased dissolution. The initial process may start at smaller scales under the soil, but once established, the process will tend to be strengthened because the depressed areas will be kept mostly under the soil and thus should have higher dissolution rates than the more exposed area, which will receive dissolution only through meteoric processes. Collapse may further contribute to the process, as suggested by Kohler (1989), although scant evidence of major collapse processes is found in the Lagoa Santa Karst. The existence of large cave systems under the Mocambeiro Depression and active groundwater routes in the depressed area south of Funilândia shows that active groundwater circulation occurs along these depressions, demonstrating that conduit dissolution is underway in the subsurface. The local karst drainages of Gordura, Palmeiras/Mocambo, and Samambaia and their tributaries may act as local base levels controlling the direction of landforms in their surroundings. The asynchronous development of these local drainages may have led to changes in the base level over time (Auler 1998), which may have also affected the general configuration of some of the landforms. It should be noted that some landforms, although also controlled by the direction of tectonic structures and the hydraulic gradient, will be oriented perpendicular to the main hydraulic gradient direction. This is the case with limestone cliffs because their genesis, as interpreted by Journaux (1977), is related to the formation of a sinking point that will be directed along the hydraulic gradient but perpendicular to the cliff, as shown in Fig. 21. In the marginal areas to the south or north, the distance to the major base level and the existence of other minor base levels (Mata Creek) may provide distinct favorable directions for karst development.

Fig. 30 In this schematic view of the Lagoa Santa Karst, the subdued landscape slopes toward the Velhas River base level to the east. A few outcrops occur in an otherwise soil-covered landscape. Cliffs with water-filled sinkholes can be observed, as well as towers, dolines, springs, swallets, and lakes. Pinnacles can be observed at the contact between the soil and the carbonate. Sparse dissolutional features occur in the lower impure Pedro Leopoldo Member, while a full suite of karst landforms can be observed in the upper Lagoa Santa Member. Drawing by Bruna Meyer

In the north, the major role played by horst and graben structures (Galvão et al. 2016) may affect the general morphology of the karst. A general schematic diagram of the main karst features in the Lagoa Santa area is presented in Fig. 30.

Kohler (1989) believed that the wide valley of Mata Creek had been occupied by the much larger Velhas River in the past. In Kohler's (1989) hypothesis, the massive Velhas River would enter the karst in the Matozinhos region; flow underground through the central area; be responsible for the enlargement of caves, deposition of rounded clasts, and formation of terraces; and finally emerge as a spring in the drainage basin of the Paraopeba River to the north. The collapse of the underground channels would play a role in the general configuration of the depressed areas. For such a large river (mean discharge over 50 m^3/s) to flow through the karst would certainly result in major imprints not observed in the area. The rounded clasts found in the caves can be explained by the present streams, and there is no evidence of large fluvial terraces. Kohler's (1989) hypothesis assumes that in the past, there was no Velhas River to the east and thus no eastern base level to control the karst. The general trend of the development of the karst and the chronology of the cave sediments (Auler et al. 2009) suggest that the Velhas River has acted as the major base level in its present condition for a considerable time.

The timescales for the development of the Lagoa Santa Karst are limited to local estimates, and there is still no long-term comprehensive chronological study of the area. However, ^{230}Th dates in cave calcite reinforce the considerable antiquity of the karst. King (1956), based on correlations with ancient planation surfaces, assigned a Pliocene age to the older cave sediments. Auler (1994), through groundwater chemistry of carbonate solutes in springs, found denudation rates that vary considerably between the drainage basins: 10 mm/kyr for Samambaia, 79 mm/kyr for Palmeiras Mocambo and 140 mm/kyr for Moinho Velho, a spring tributary of the Samambaia basin. This author notes that anthropogenic effects could have affected some of these rates, although they are in accordance with denudation rates from other tropical areas. A series of ^{230}Th dates have been produced for the Lagoa Santa Karst (Piló 1998; Auler 1999; Auler et al. 2006, 2009; Hubbe et al. 2011; Haddad-Martim et al. 2017). Correlations of a few of these initial dates with the associated doline floor have led to the establishment of rates of doline deepening or (if extrapolated to a higher residual surface) to a more general age for the initial breaching of the metapelite surfaces and the start of karstification at 1.9 Myr (Piló 1998). However, as noted by Auler (1999), there are considerable uncertainties with these assumptions, including the fact that calcite precipitation may not have been level with the doline floors. The large variation in the ^{230}Th ages presented in other studies would lead to widely different rates of doline lowering and thus corroborate that this approach must be considered with caution. Unpublished data by Xianfeng Wang (personal communication) from Ballet Cave shows that all ages for the calcite over the sediment, >650 kyr, are beyond the limit of the ^{230}Th method, attesting to the large intersite variations (controlled mostly by paleoclimate conditions and not by geomorphic doline processes) and the considerable antiquity of cave sedimentation. Assuming that speleogenesis started long before the vadose zone processes associated with clastic sedimentation and calcite deposition (Auler et al. 2009), ages well within the Tertiary must be assumed for the initial karstification in the area.

Acknowledgements I have, over the past four decades, had the privilege of working in the Lagoa Santa Karst with a large number of people, from caving friends (especially from Grupo Bambuí de Pesquisas Espeleológicas), university colleagues, local environmentalists, staff from various mining companies, and many of the local farmers. Without their support, this review would not have been possible. Bárbara Zambelli helped immensely, especially in maneuvering the drone, helping organize figures, producing GIS and statistical analysis, and collaborating with the inventory of lakes, dolines, and depressions, in addition to providing overall help with the manuscript. The staff at Carste Ciência e Meio Ambiente, especially Tatiana Souza, Fernanda Pizani, Rafael Cruz, Érika Ribeiro, Felipe Loureiro, and Carolina Reis, provided crucial help during the long gestation of this manuscript. I acknowledge the helpful review by Tatiana Souza.

References

Auler AS (1994) Hydrogeological and hydrochemical characterization of the Matozinhos-Pedro Leopoldo Karst. Thesis, Western Kentucky University

Auler AS (1995) Lakes as a speleogenetic agent in the karst of Lagoa Santa, Brazil. Cave Karst Sci 21:105–110

Auler AS (1998) Base-level changes inferred from cave paleoflow analysis in the Lagoa Santa Karst, Brazil. J Caves Karst Stud 60: 58–62

Auler AS (1999) Karst evolution and palaeoclimate in eastern Brazil. Thesis, University of Bristol

Auler AS (2016) Cave protection as a karst conservation tool in the environmentally sensitive Lagoa Santa Karst, southeastern Brazil. Acta Carsologica 45:131–145

Auler AS, Piló LB (2015) Lagoa Santa Karst: cradle of Brazilian cave studies. In: Vieira BC et al (eds) Landscapes and landforms of Brazil. Springer, Dordrecht

Auler AS, Piló LB, Smart PL, Wang X, Hoffmann D, Richards DA, Edwards RL, Neves WA, Cheng H (2006) U-series dating and taphonomy of quaternary vertebrates from Brazilian caves. Palaeogeogr Palaeoclimatol Palaeoecol 240:508–522

Auler AS, Smart PL, Wang X, Piló LB, Edwards RL, Cheng H (2009) Cyclic sedimentation in Brazilian caves: mechanisms and palaeoenvironmental significance. Geomorphology 106:142–153

Balasz D (1984) Exhumált Trópusi Oskarszt Lapinha Vidéken (Minas Gerais, Brazília). Karst és Barlang 2:87–92

Barbosa GV (1961) Notícia sobre o karst na mata de Pains. Boletim Mineiro de Geografia 2:3–21

Berbert-Born MLC (2002) Carste de Lagoa Santa. Berço da paleontologia e da espeleologia brasileira. In: Schobbenhaus C, Campos DA, Queiroz ET, Winge M, Berbert-Born M (eds) Sítios Geológicos e Paleontológicos do Brasil. Brasília, CPRM, pp 415–430

Costa FWS (2017) The minas Gerais academy of sciences: Lund's inheritors. In: Da-Gloria et al (eds) Archaeological and paleontological research in Lagoa Santa. Springer, Cham

Coutard JP et al (1978) Map of Lagoa Santa Karst, description. Caen

Dolabela E (1958) Estudo das regiões cársicas. UFMG, Belo Horizonte

Dutra GM, Horta LS, Berbert-Born ML (1998) Levantamento espeleológico. In: APA Carste de Lagoa Santa. Patrimônio espeleológico, histórico e cultural, CPRM, vol 3, 71 pp.

Galvão P, Halihan T, Hirata R (2015) Evaluating karst geotechnical risk in the urbanized area of Sete Lagoas, Minas Gerais, Brazil. Hydrogeol J 23:1499–1513

Galvão P, Hirata R, Cordeiro A, Barbati D, Peñaranda J (2016) Geologic conceptual model of the municipality of Sete Lagoas (MG, Brazil) and the surroundings. An Acad Bras Ciências 88:35–53

Haddad-Martim PM, Hubbe A, Giannini PCF, Auler AS, Piló LB, Hubbe M, Mayer E, Wang X, Cheng H, Edwards RL (2017) Quaternary depositional facies in cave entrances and their relation to landscape evolution: the example of Cuvieri Cave, eastern Brazil. CATENA 157:372–387

Hubbe A, Haddad-Martim PM, Hubbe M, Mayer EL, Strauss A, Auler AS, Piló LB, Neves WA (2011) Identification and importance of critical depositional gaps in pitfall cave environments: the fossiliferous deposit of Cuvieri Cave, eastern Brazil. Palaeogeogr Palaeoclimatol Palaeoecol 312:66–78

Hurt WR (1960) The cultural complexes from the Lagoa Santa region, Brazil. Am Anthropol 62:569–585

Journaux A (1977) Geomorphologie et préhistoire. Méthodologie pour une cartographie de l'environement des sites préhistoriques; l'exemple de Lagoa Santa (Minas Gerais, Brésil). Norois 95:319–335

King LC (1956) A geomorfologia do Brasil oriental. Revista Brasileira de Geografia 18:147–265

Knez M, Slabe T, Travassos LEP (2012) Karren on laminar calcarenitic rock of Lagoa Santa (Minas Gerais, Brazil). Acta Carsologica 40:357–367

Kohler HC (1978) A evolução morfogenética da Lagoa Santa. In: Proceedings XXX Congresso Brasileiro de Geologia, Recife, pp 147–153

Kohler HC (1989) Geomorfologia cárstica da região de Lagoa Santa, MG. Dissertation, Universidade de São Paulo

Kohler HC, Karfunkel J (2002) The Quaternary morphogenesis of the Lagoa Santa tropical karst, Minas Gerais state, SE Brazil. Theor Appl Karstology 15:93–99

Kohler HC, Queiroz Neto JP, Coltrinari L, Ferreira RPD, Stanowski ISM, Cançado AM, Gomes D, Macieira FL, Nascimento NR (1976) Diferentes níveis de seixos nas "formações superficiais" da região de Lagoa Santa/MG. Anais 29 Congresso Brasileiro de Geologia, SBG, Ouro Preto, pp 343–347

Kohler HC, Cançado AM, Gomes D, Macieira FL, Nascimento N (1978) Carte du Karst. Laboratoire d'Analyse et Cartographie des Formations Superficielles de l'Université de Caen et le Centre de Geomorphologie, CNRS, Caen

Laming-Emperaire A, Prous A, Moraes AV, Beltrão MC (1974) Grottes et abris de la région de Lagoa Santa, Minas Gerais, Brésil. Cahiers d'Archéologie d'Amerique du Sud 1. EPHE, Paris

Lanari CU (1909) Ossadas humanas fósseis encontradas numa caverna calcária das vizinhanças do Mocambo. Annaes da Escola de Minas 11:15–35

Liais E (1872) Climats, geologie, faune et geographie botanique du Brésil. Garnier Freres, Paris

Lund PW (1837a) Om huler i kalksteen i det indre af Brasilien, der tildeels indeholde fossile knokler. Første Afhandling. Det Kongelige danske Videnskabernes Selskabs Naturvidenskabelige og Mathematiske Afhandlinger 6:209–248

Lund PW (1837b) Om huler i kalksteen i det indre af Brasilien, der tildeels indeholde fossile knokler. Anden Afhandling. Det Kongelige danske Videnskabernes Selskabs Naturvidenskabelige og Mathematiske afhandlinger 6:309–330

Lund PW (1841a) Blik paa Brasiliens Dyreverden før sidste Jordomvæltning. Første Afhandling. Det Kongelige danske Videnskabernes Selskabs Skrifter 8:29–60

Lund PW (1841b) Blik paa Brasiliens Dyreverden før sidste Jordomvæltning. Tredje Afhandling: Fortsættelse af Plattedyrene. Det Kongelige danske Videnskabernes Selskabs Skrifter 8:217–272

Lund PW (1845) Notice sur des ossements humains fossiles, trouvés dans une caverne du Brésil. Mémoires de la Societé Royale des Antiquaires du Nord 49–77

Parizzi MG, Salgado-Labouriau ML, Kohler HC (1998) Genesis and environmental history of Lagoa Santa, southeastern Brazil. Holocene 8:311–321

Peñaranda JR (2016) Condicionamento estrutural e litológico da porosidade cárstica da Formação Sete Lagoas, município de Sete Lagoas (MG). Dissertation, Universidade de São Paulo

Pessoa PFP (1996) Caracterização hidrogeológica da região cárstica de Sete Lagoas, MG: Potencialidades e riscos. Dissertation, Universidade de São Paulo

Pessoa PFP (2005) Hidrogeologia dos aquíferos cársticos cobertos de Lagoa Santa, MG. Thesis, Universidade Federal de Minas Gerais

Piló LB (1998) Morfologia cárstica e materiais constituintes: Dinâmica e evolução da depressão poligonal Macacos-Baú – Carste de Lagoa Santa, MG. Thesis, Universidade de São Paulo

Reinhardt JT (1867) De brasilianske Knoglehuler og de i dem forekommende Dyrelevninger. Tidsskrift for populaere Fremstillingen af Naturvidenskaben 4(193–207):291–353

Silva AB (1988) Abatimento de solo na cidade de Sete Lagoas, Minas Gerais, Brasil. Revista Água Subterrânea 12:57–66

Silva AB, Moreira CVR, César FM, Auler AS (1987) Estudo da dinâmica dos recursos hídricos da região cárstica dos municípios de Lagoa Santa. Pedro Leopoldo e Matozinhos, CETEC, Belo Horizonte

Tricart J (1956) O karst das vizinhanças setentrionais de Belo Horizonte. Revista Brasileira de Geografia 18:451–469

Karst Hydrogeology of the Lagoa Santa Area

Paulo Pessoa

Abstract

The Lagoa Santa karst region can be distinctly characterized from other karst environments due to its geologic framework, stratigraphy, and variable carbonate rock composition. The region is characterized by partially covered karst with a thick soil mantle and significant limestone outcrops. The lithological profile of the Lagoa Santa karst aquifer system exhibits abrupt changes in hydraulic conductivity, which necessarily results in a high degree of heterogeneity and anisotropy. Such parameters are intrinsic and powerful controllers of the velocities and directions of flow due to their close association with compositional variations among different strata. Variability in the distribution of pure and impure limestones in the Sete Lagoas Formation with depth results in a high degree of variability in the groundwater flow. Additionally, the tectonic fabric in the bedrock allows widespread interconnections along strata boundaries, although some hydrogeologic compartments might be separated from others. Moreover, infiltrating water percolating through intricate vadose pathways produces widespread recharge and connects various portions of the karst aquifer throughout the entire system, including deep epikarst zones. The karst hydrodynamics involve the flow of large volumes of groundwater from west to east. In this karst environment, the occurrence of springs, lakes, caves, and swallets is common. However, no systematic monitoring of the karst aquifer hydrodynamics has been performed to better understand the groundwater behavior and improve policies regulating the use of this groundwater. Throughout the entire area, approximately 600 pumping wells are the main source of groundwater. These groundwater wells serve 100% of the population's needs and represent an average exploitation of 1.5 m^3/s. Hydrogeological analyses and interpretations demonstrate that research in this region should focus on understanding the behavior of karst springs, which must be considered in integrated management schemes and assessments of the real state of the groundwater potential.

Keywords

Mantled karst • Hydrodynamics • Epikarst • Karst aquifer • Lagoa Santa karst

1 Introduction

The purpose of this chapter is to update the hydrogeological characterization of one of the best-known karst areas in Brazil (Auler and Piló 2015) and to document the state-of-the-art of available hydrogeologic data for the Lagoa Santa Karst aquifer system (LSKS). This overview includes data on precipitation, water levels, water quality, recharge, discharge, specific capacity, well yields, tracer test results, and a geological summary of the main attributes related to groundwater circulation and storage.

The history of human occupation in this region is intrinsically related to the environment, and archeological, paleontological, speleological, and geomorphological attributes represent the main values of the site, which is currently experiencing tremendous pressure related to urban expansion, increasing the risk of deleterious consequences for the karst aquifer system.

Important hydrogeologic studies on this region mostly include academic dissertations and private companies' technical reports for environmental licenses and impact assessments, and these sources represent the actual state-of-the-art information on the hydrogeology in the Lagoa Santa area. However, very recently, academic efforts have improved the knowledge of this area through a wide range of studies, including hydrogeological assessments of the karst aquifers. These assessments include a focus on the effective water level and a hydrochemical monitoring

P. Pessoa (✉)
Hidrovia Hidrogeologia e Meio Ambiente, Rua Albita 131/301, Belo Horizonte, MG 30310-160, Brazil
e-mail: ppessoa@hidrovia.com.br

A. S. Auler and P. Pessoa (eds.), *Lagoa Santa Karst: Brazil's Iconic Karst Region*, Cave and Karst Systems of the World, https://doi.org/10.1007/978-3-030-35940-9_8

program involving piezometers and discharge meters installed to monitor karst springs (Velasquez et al. 2018).

The Lagoa Santa Karst aquifer, which needs to be analyzed from the perspective of environmental legislation or basic technical guidelines for hydrogeological assessment, exhibits different groundwater behavior at different scales due to its intrinsic heterogeneity, anisotropy, and extreme variations in hydraulic conductivity, which pose restrictions on the scale of data collection (Ford and Williams 2007; Fleury 2009).

This heterogeneity often necessitates gathering measurements and specific geological information to identify the lithostratigraphic characteristics of the zones beneath the surface to improve groundwater circulation analysis. These lithostratigraphic data often come from boreholes, pumping well drilling data, aquifer tests, hydraulic head measurements, tracer tests, discharge and pumping rates, location data, and other sources (White 1998; Ford and Williams 2007).

At the surface, outcrops of various types of limestone are scarce in a regional sense but display impressive dissolution features, and these dispersed and sometimes concentrated dissolution features constitute the intriguing LSKS. Due to the naturally high porosity rate enhanced by the dissolution process, karst environments are commonly associated with an abundance of water, as evidenced by both the high degree of groundwater use and the presence of approximately 200 different karst lakes (see chapter "Caves and Speleogenesis in the Lagoa Santa Karst", this volume) and dozens of different karst springs in the region (Beato and Pessoa 1991). Indeed, the presence of water is intrinsically associated with the historic occupation, as evidenced by the wealth of human skeletal remains and artifacts found in the Lagoa Santa karst terrains (see chapter "The Archaeological Record of Lagoa Santa (East-Central Brazil): From the Late Pleistocene to Historical Times", this volume).

In this chapter, a review of the latest studies published on the hydrogeology of the Lagoa Santa Karst is presented and discussed.

1.1 Site Characterization

The Lagoa Santa area includes Neoproterozoic carbonates of the Sete Lagoas Formation, Bambuí Group, especially the pure limestone of the Lagoa Santa Member. Limestone cliffs, karst lakes, swallets, springs, and resurgences are notable karst features in this region and are associated with a widespread thick soil mantle, which covers large portions of the karst and originates from the weathering of metapelites.

The Lagoa Santa area, located 30 km north of Belo Horizonte city, the capital of Minas Gerais state, occupies the southeastern portion of the Bambuí carbonate sedimentary basin. The nine municipalities in the area are in non-karst areas (30%) and covered karst terrain (70%), and the groundwater dynamics are influenced by the lithostratigraphic and structural framework (Pessoa 2005). The study area is in the geological context of the São Francisco Basin (Pinto and Martins-Neto 2001; Reis et al. 2017), especially the carbonate strata of the Sete Lagoas Formation, which is part of the Bambuí Group.

The Lagoa Santa area is in the catchment of the São Francisco River. Locally, the study area includes a part of the Velhas River and Mata Creek basins, as well as a small portion in the extreme west of the municipality Sete Lagoas, where drainage flows to the Paraopeba River.

The occurrence and distribution of the LSKS coincides with the local geology (chapter "The Geology of Lagoa Santa Karst", this volume). Stratigraphically, the LSKS is composed mostly of pure limestones overlain by metapelites of the Bambuí Group. This package overlies the crystalline basal gneiss-migmatite complex, which crops out in the western and southern portions of the area (CPRM 2003; Galvão et al. 2016; Peñaranda 2016).

The climate of the Lagoa Santa area is characterized by very well marked seasons with warm summers and mild winters, and the area experiences a mean annual temperature of approximately 21.4 °C and mean annual precipitation ranging from approximately 1,290 mm, as noted by Patrus (1998). This climate type is tropical wet with dry winters and rainy summers, and approximately 90% of the total precipitation falls between October and March. The dry season occurs from April to September and does not exceed more than 12 days of rainfall during these 6 months. Hydrologic records for the area indicate that recurring patterns of droughts are followed by periods of heavy precipitation. The prolonged drought of 2011 through 2015 was followed by heavy precipitation in 2016. Daily precipitation data are available for the stations operated by the National Institute of Meteorology (INMET) and National Agency of Waters (ANA).

The main drainage network in the recharge catchment area of the LSKS is shown in Fig. 1. Considering the proximity to the metropolitan region of Belo Horizonte city, this figure also shows the main roads where traffic is intense and the locations of the urban districts and peripheral industrial sites.

Considering that rainfall has a strong influence on the recharge process and groundwater storage, data gathered from November 1941 to July 2017 are used to reflect the average precipitation parameters over these 75 years, as shown in Fig. 2. The highest frequency of rainy days occurs in December and January and the lowest frequency occurs in July.

The average temperature observed at the Lagoa Santa station is approximately 21.4 °C, with the highest temperatures varying between 23.2 and 29.3 °C and the lowest temperatures varying between 12.5 and 18.1 °C.

Recently, in the years 2012 to 2015 the region faced a water crisis induced by a dramatic nationwide decrease in

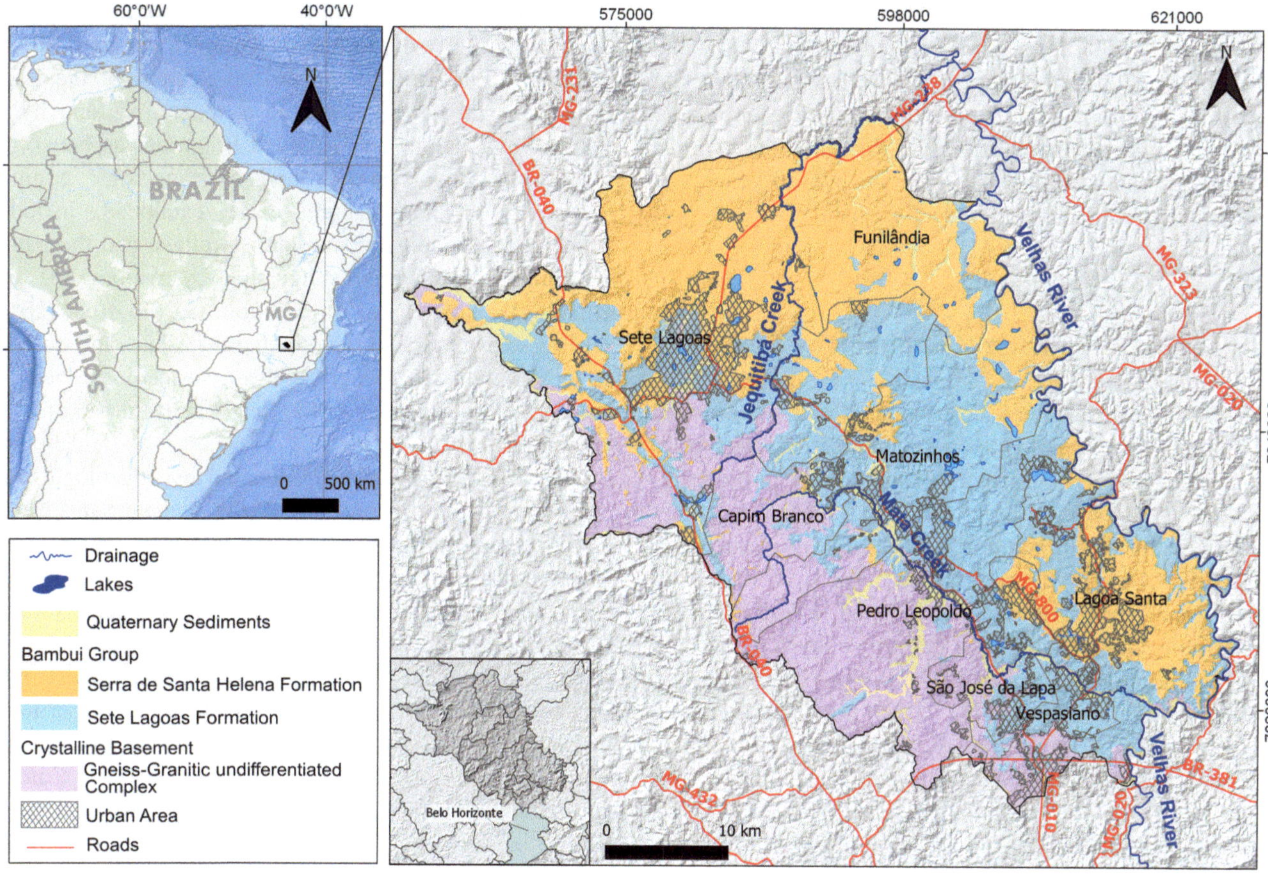

Fig. 1 Lagoa Santa Karst area along with municipal boundaries, urban zones, main drainages and lakes, geological attributes around the metropolitan region of Belo Horizonte city and the main roads crossing the karst area

Fig. 2 Historical annual values of rainfall from six regional stations around Lagoa Santa Karst, operated by ANA—National Agency of Waters (*Source* http://www.snirh.gov.br/hidroweb/publico/medicoes_historicas_abas.jsf)

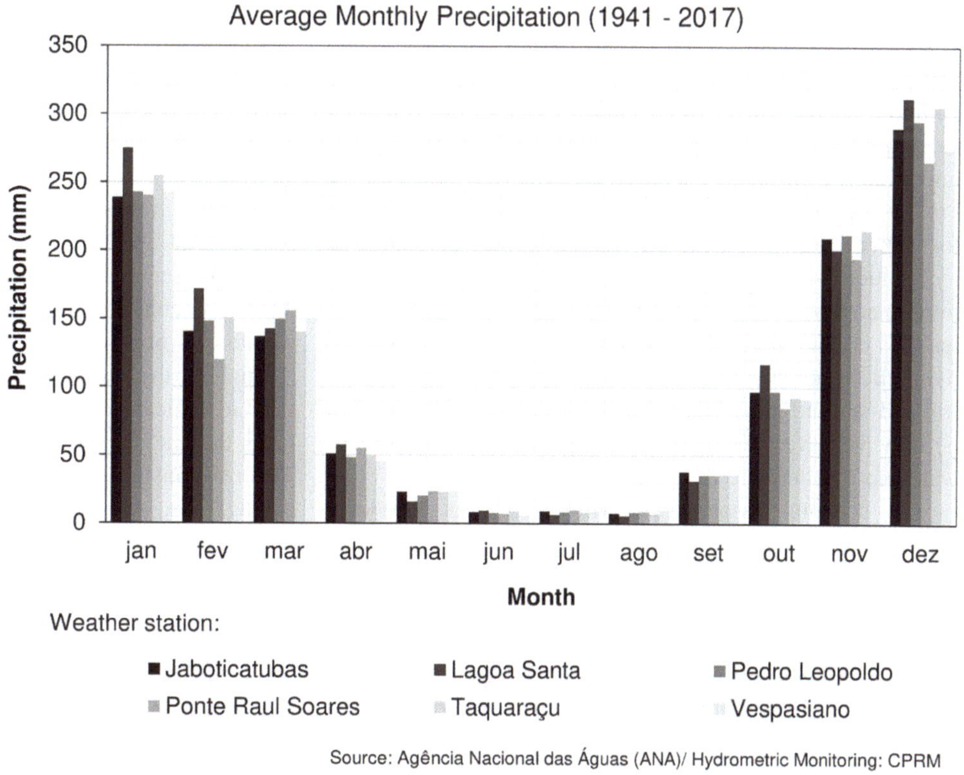

rainfall. At the studied weather stations, annual average value of only 550 mm was recorded in 2014, below the historical annual average. This year was also marked by the highest average maximum temperature in the region for the period analyzed, reaching 30 °C. For the minimum average temperature, the lowest value over the last 50 years was approximately 15°C at the Sete Lagoas station, according to Vieira (2015).

1.2 Lagoa Santa Karst Hydrogeology Review

In the recent decades, there has been an effort to understand the Bambuí Group from the perspectives of facies, sedimentary environments, sequence stratigraphy, chemostratigraphy, geochronology, isotope geology, paleontology, and tectonic evolution. In addition to the studies organized by CPRM/IBAMA (1998) and CPRM (2003, 2010), more in-depth technical analyses of the Sete Lagoas Formation and the Bambuí Group have been performed by Vieira et al. (2007), Martins and Lemos (2007), Babinski et al. (2007), Alkmin and Martins-Neto (2001, 2012), Kuchenbecker et al. (2013), Warren et al. (2014), Paula-Santos et al. (2015), Bittencourt et al. (2015), Kuchenbecker et al. (2015), Reis (2016), Reis et al. (2017) and Sanchez and Fairchild (2017).

Investigations concerned with developing a hydrogeological conceptual model have been performed by obtaining specific data available from borehole hydraulic head measurements, tracer tests, structural geology, and imagery interpretations (Pessoa and Mourão 1998; CPRM/IBAMA 1998; Pessoa 1996, 2005; Hidrovia 2016, 2017).

Considering that hydrogeologic studies are rare in the LSKS, the first systematic inventory of groundwater pumping wells and springs performed by the Brazilian Geological Survey (CPRM) in the early 1990s represents an important contribution. This inventory was supported by the Projeto VIDA (CPRM 1991), which aimed to catalog data for better industrial planning and environmental attribute recognition, and this project was the first to identify important aspects of groundwater uses by the eight municipalities around the LSKS. At that time, the first analyses of hydraulic heads were provided for more than 50 wells distributed around the Lagoa Santa Karst area. These data permitted the very first attempt to characterize the groundwater flow patterns to discharge zones.

The economic aspects and vulnerability of the karst region delineated by CPRM/IBAMA (1998) were the basis for creating a protected zone named APA Carste Lagoa Santa, whose main attribute is related to the maintenance and preservation of the karst environment (see chapter "Protection of the Environmental and Cultural Heritage in the Lagoa Santa Karst: History and Challenges", this volume). Pessoa (1996) performed an investigation of the potential and vulnerability of the Sete Lagoas city region, whose population was entirely dependent on groundwater until 2018.

Auler (1994) studied the central portion of the Lagoa Santa Karst area using tracer tests techniques and hydrochemical evaluations and revealed important aspects of the groundwater flow behavior and hydrogeologic compartments. Many karst features were cataloged, including springs, karst windows, caves and swallets, and the aquifer contribution areas were distinguished by mapping the main groundwater flow routes. In addition, Galvão et al. (2015a, 2017) used isotopic data, hydrochemical data, and borehole log observations to recognize the hydrogeologic behavior and land subsidence risk in the Sete Lagoas urban area and surroundings.

Pessoa (2005) presented a detailed hydrogeologic study of the southernmost portion of the Lagoa Santa region involving the evaluation of 154 drilling logs, among which 67 boreholes penetrated the full thickness of the carbonate strata. Data on the elevation of the top of the crystalline basement, the groundwater level, the flow direction, and several geochemical parameters were considered. This work provided substantial information on the southern portion of the mantled LSKS and revealed the existence of interstitial carbonate facies originating from tectonic stress. Comparisons among several borehole logs and geophysical investigations in the central and southern portions of the area indicate large stratigraphic differences.

The study by Pessoa and Auler (2015) highlights the importance of the influence of interstratal karst development in the southern portion of the study area, because these zones can control groundwater flow.

Furthermore, recent studies have focused on vulnerability zone mapping (Tayer 2016), which involves considering the geologic framework of the Bambuí Group and the crystalline complex (Andrade and Amorim 2018), the water budget of karst aquifers based on discharge measurements and dye tracing techniques (Teodoro 2018) and groundwater chemistry (Vieira 2018).

The presence of water is intrinsically associated with the historic occupation of the study area, as evidenced by the wealth of human skeletal remains and artifacts found throughout the Lagoa Santa karst terrain (see chapter "The Archaeological Record of Lagoa Santa (East-Central Brazil): From the Late Pleistocene to HistoricalTimes", this volume). This relationship between humans and water has existed for a long time, and many cities still rely entirely on groundwater. This extreme dependence on the karst aquifers is highlighted by the pumping well exploitation, which supplies 100% of the total water demand of approximately 570 thousand people (IBGE 2016; COPASA 2019). Table 1 shows the main parameters of the pumping wells exploiting the carbonate aquifer. In Sete Lagoas municipality, surface water from the Velhas River supplies part of the population

Table 1 Hydraulic characteristics of the pumping wells in the Lagoa Santa Karst system

Statistical variables	Depth (m)	Water level (WL)	Pumping level (PL)	Specific capacity (m^3/h/m)	Pumping rate (m^3/h)
Average	80 (612)	13.7 (381)	33.6 (348)	14.5 (299)	44.4 (362)
Maximum	396	102	135	420	327
Minimum	10	0	1.7	0.004	0.4
Median	80	9	30	2	20

Observations: (299) Number of data available from a total of 612 pumping wells in the Lagoa Santa area, which is composed of 9 municipalities (Sete Lagoas, Funilândia, Prudente de Morais, Pedro Leopoldo, Capim Branco, Matozinhos, Confins, Lagoa Santa and Vespasiano) *Source* CPRM (2018)

and will soon represent the second most important source of water.

These observations reflect the importance of analyzing groundwater dynamics and storage capacity. Furthermore, some areas of the LSKS have experienced increases in groundwater exploitation via pumping wells. This increase in groundwater exploitation in recent decades is now playing an important role in the development of new hydrodynamic behavior in the northern portion of the study area. As mentioned by Galvão (2016), in the urban area of Sete Lagoas city, there is a cone of depression caused by high well pumping rates, which are estimated to be 75,000 m^3/d.

In terms of groundwater quality, although karst systems are especially fragile and particularly vulnerable to surficial influences, the natural conditions of mantled karst systems can prevent or retard groundwater contamination, despite their high potential to develop very well-connected hydraulic networks. Tayer (2016) conducted intrinsic groundwater vulnerability analyzes of the karst environment in the LSKS, and due to the extensive soil coverage, only specific portions were mapped as highly vulnerable or very highly vulnerable zones.

The collection of all these data for reanalysis and interpretation made it possible to complement hydraulic information in a preliminary attempt to describe groundwater flow divides, routes and inferred internal limits of the hydrogeologic framework of LSKS, as discussed in the next several sections.

2 Lagoa Santa Karst Aquifer System

2.1 Karst System Framework—Regional Characterization

Any hydrogeologic system can be understood on the basis of its bedrock structure because water will always move along the easiest route. Through geologic time, many types of secondary porosity can form by means of hardening, weathering, and fracturing, and this porosity development can favor limestone dissolution beneath surface relief (Ford and Williams 2007). This lithostratigraphic framework, in conjunction with the soil coverage, is key to understanding groundwater behavior, hydraulic characteristics, storage compartments, and groundwater flow preferences.

Briefly, the Bambuí sedimentary basin featured marine carbonate deposition and the formation of a large platform on the São Francisco Craton in the upper Proterozoic (550 Myr). This location featured a particular tectonic style characterized by a group of structural features associated with a stepped fault pattern in the basement (Ribeiro et al. 2003; Tuller et al. 2010; Andrade and Amorim 2018). In fact, prior to carbonate deposition, in the southeasternmost portion of the São Francisco Craton, the crystalline rocks were affected by tectonic stresses. These Archean events were responsible for the development of the widespread regional fabric (including fractures and faults) that is readily observed in aerial photography and field observations of outcrops (see chapter "The Geology of Lagoa Santa Karst", this volume).

The limits of the LSKS have been well characterized in the southern, western, and eastern portions of the study area; however, to the north, complexities still remain due to hydrogeological uncertainties. Furthermore, additional issues of crucial importance to delimiting the LSKS remain unresolved, including the depth of the LSKS and the nature of the boundaries defining its inner limits. In addition, why does the stratigraphy differ between the central LSKS and the margins of the LSKS?

From a regional scale, the Velhas River valley, situated in the eastern portion of the LSKS, is oriented NNW-SSE. The orientations of the Mata Creek and Jequitibá Creek valleys in the southwestern and northwestern portions of the study area, respectively, conform to the regional arcuate structure of the faulted basement blocks (Pessoa 2005), with the down-dropped side toward the interior of the basin and the uplifted side toward the exterior of the basin (Fig. 3). The LSKS can be assumed to be a structural basin in which carbonates were deposited under distinct tectonic influences and intensities (Magalhães 1988; Ribeiro et al. 2003).

Three different kinds of rocks—crystalline rocks, limestones, and metapelites—represent the main stratigraphic units and play the most important roles in the current karst hydrodynamic behavior and flow routes. Thus, the structural

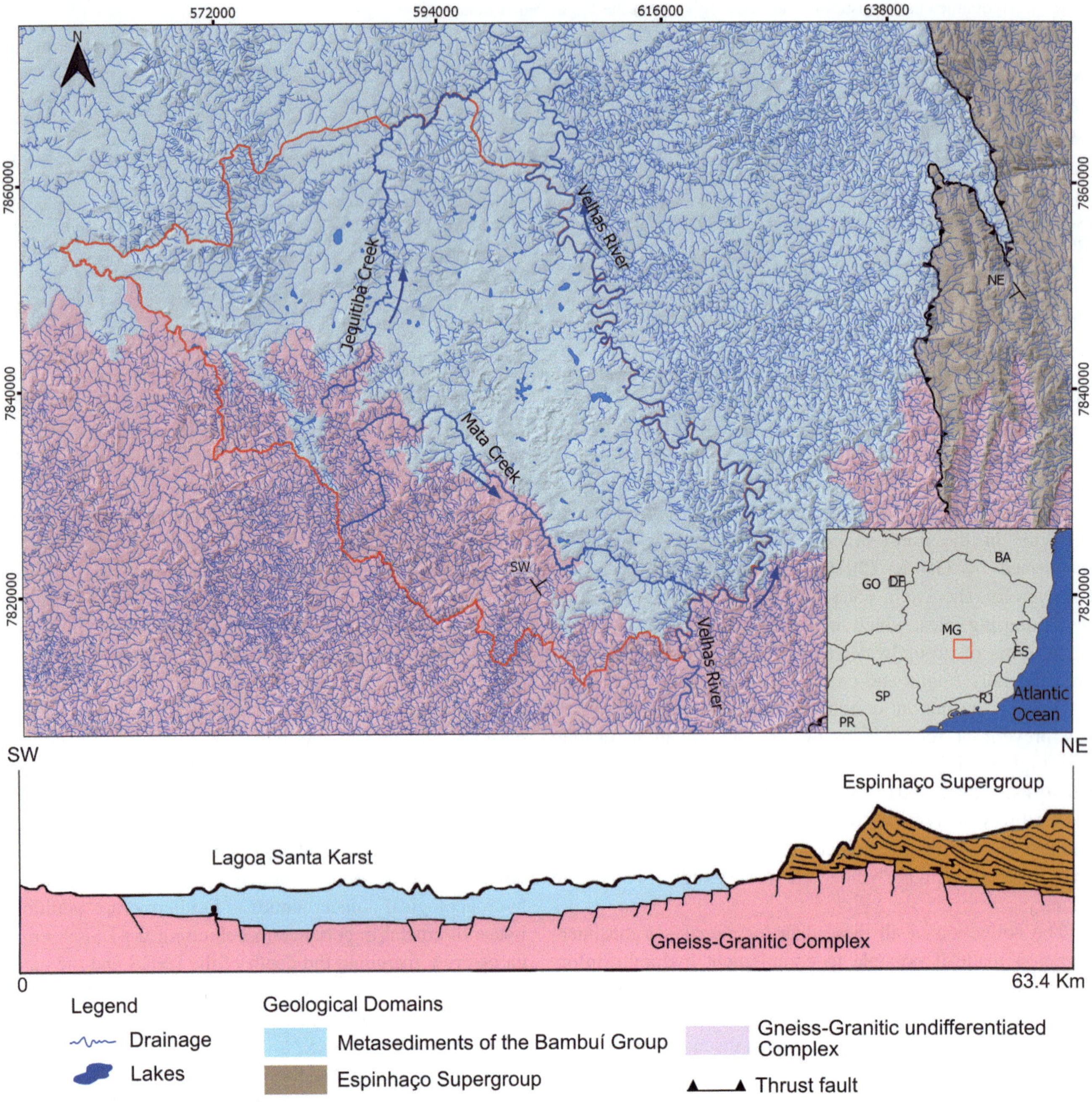

Fig. 3 Geologic framework of the Lagoa Santa Karst system (modified from Pessoa 2005). The boundary of the LSKS is shown in red, and the contour encompassing the Bambuí carbonate sedimentary basin and elevation of the top of the crystalline basement is specified. Created by a tectonic-metamorphic event, Espinhaço Ridge in the easternmost part of the framework is also shown

attributes and the initial geologic configuration are closely related, as discussed in chapter "The Geology of Lagoa Santa Karst", this volume.

As shown in Fig. 3, the external limits of the LSKS correspond to the Mata Creek valley in the west and south and to the Velhas River in the east, which functions as the regional base level. The northern portion of the area is characterized by the presence of Jequitibá Creek; the valley associated with this creek remains poorly studied in terms of its behavior, but it is assumed to be the local base level (i.e., where karst water converges) and to be aligned with the regional structure.

To understand the configuration of the study area, it is important to understand the hydraulic and physical conditions and to evaluate how they can be interpreted. Pessoa (2005) showed the differences in the entire carbonate deposit on the southeasternmost border of the Bambuí basin, using data from boreholes that reached the crystalline basement at

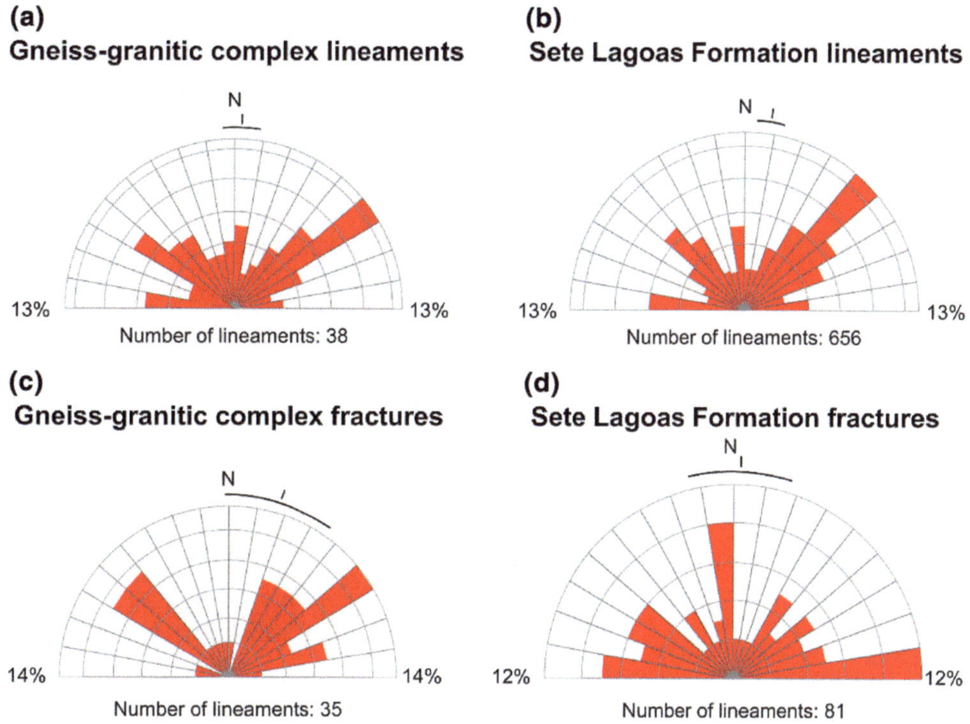

Fig. 4 Rose diagrams of the lineaments and fractures in the gneiss-granitic complex (**a**) and the Sete Lagoas Formation, Bambuí Group (**b**). The diagrams show the frequency of occurrence of the different directions of karst-related features in the massifs and **b** the occurrence of the fracturing directions. *Source* Andrade and Amorim (2018)

depths greater than 250 m. The elevation of the top of the basement decreases eastward, indicating that the carbonate thickness increases and that the contact between the carbonate and the crystalline complex increases in depth.

In addition, the investigations of Galvão et al. (2016) in the northernmost part of the LSKS, surrounding the Sete Lagoas city limits, showed evidence of irregularities in the basal crystalline rocks, where normal faulting structures placed different lithologies at the same elevations. These structures greatly influence the behavior of groundwater flow and are closely related to karst development through the formation of preferential paths, which affect the progression of dissolution, speleogenesis, and hydrogeologic reservoir boundaries and potentials.

Recent geologic studies have indicated an NW-oriented fabric based on measurements of the main fractures in the crystalline rocks and on aligned features in the satellite images (Ribeiro et al. 2016). The first phase of geotectonic deformation was characterized by normal faults, which were generated by topographic differences (horst and grabens). Thus, structures in the basement in part controlled the development of surface karstic features. In support of the existence of a structural relationship between the crystalline basement rocks and the Bambuí Group carbonate rocks, Andrade and Amorim (2018) observed strong associations among patterns of structures measured in the field, with major similarities in azimuths (N40-60E) in both types of rocks. Figure 4 summarizes the average directions of fractures in the Sete Lagoas Formation, and the results can be considered the most important trends for groundwater circulation.

Bedding planes are the other important geologic attribute; their intersections with fractures play a very important role in karst development by creating the major form of groundwater flow connection. The majority of the groundwater discharge occurs through very well-distributed outlets, such as those in the Velhas River and Mata Creek valleys. The basement composed of undifferentiated crystalline rocks possesses a very old and complex history of deformation processes. Among these processes, extensional faulting developed special discontinuities associated with NNW-SSE and ENE-WSW high-angle displacements. Structures with these orientations then influenced the accommodation that allowed Bambuí carbonate deposition.

These fracture planes in the crystalline basement may have been reactivated in the Tertiary period (Sawasato 1995), thereby imprinting their orientations on the massif carbonate platform. Consequently, the karst framework conduits form one of the main flow systems at present.

2.2 Hydrostratigraphic Characterization

The Lagoa Santa karst region is bounded on the eastern, lowest elevation side by the Velhas River, which represents the regional base level. Crystalline basement outcrops along the western and southern margins of the carbonate basin form a well-defined boundary of the sedimentary basin. To

Fig. 5 **a** Limestone massif exposed by mining activity in the southern portion of the LSKS. **b** Bare limestone cliffs with a thick soil mantle in the central LSKS

the north, outcrops of carbonate sediments end due to the presence of siltstones. The increase in the thickness of these rocks alters the relief texture and drives the karst deeper underground, where it is poorly constrained due to a lack of well-distributed geologic data from these greater depths.

Spanning approximately 1,600 km^2, the Lagoa Santa Karst aquifer comprises predominantly carbonate rocks that are locally covered by metapelites or a soil mantle in the extreme northern and southern portions of the area. However, in places, mantled karst is also adjacent to bare limestone hills, which occur throughout most of the central area (Fig. 5). In such situations, the two types of terrain are probably separated by a normal fault, which can contribute to the formation of cliff walls in limestone outcrops, as suggested by Ribeiro et al. (2016) and Andrade and Amorim (2018).

Some of the terrain underlain by carbonate might be associated with a down-dropped block related to faulting, and such areas are frequently covered by a thick soil mantle.

The study area involves carbonate and pelite rocks deposited over a gneiss-migmatite basement. The stratigraphy comprises four major sequences. The topmost Serra de Santa Helena Formation contains pelite rocks represented by siltites, mudstone with some sandstone, carbonate lenses of argillaceous limestone and fine-grained calcarenites. These rocks exhibit well-developed weathering and restricted groundwater circulation. This domain is classified as an aquitard.

The Lagoa Santa Member of the Sete Lagoas Formation contains pure micritic limestone with intercalations of siltite, breccia, stromatolites, and mylonitic zones. Subhorizontal laminations displaying calcite and quartz lenses are frequent. This member represents an aquifer with well-developed conduit flow.

The Pedro Leopoldo Member of the Sete Lagoas Formation contains calcisiltites, argillaceous limestone with sparitic to micritic textures, stromatolites with some intercalations of fine-grained calcarenites, and mylonites. Shear zones contain graphite and quartz veins. There is an occurrence of a basal marble. Overall, this sequence shows limited groundwater flow, as indicated by a fractured aquifer with karstified horizons.

The crystalline complex contains migmatites and granitoids and is classified as an aquiclude. These rocks comprise the crystalline basement complex that does not crop out within the area and were only described based on the cores of the numerous borehole logs that reached the basement. These rocks can be described as crystalline rocks with fine to medium crystals, the sporadic presence of feldspar porphyries, a quartz matrix, and local gneissic banding.

Figure 6 shows the lithologic types described above, using borehole samples collected at different depths from boreholes in which 57 piezometers were installed.

Table 2 presents the analytical results of Pessoa (2005) for core samples of carbonate rocks extracted from boreholes in the southern portion of the Lagoa Santa Karst. These samples are representative of the characteristics of the carbonate package.

According to Pessoa (2005), the variation in the mineralogy of the carbonate rocks present in the Sete Lagoas Formation can be mapped from the perspective of the composition. An in-depth understanding of the different types of carbonate rock makes it easier to understand the active karstification processes, the intensity of which is directly proportional to the calcium content and inversely proportional to the silica content. Therefore, the Lagoa Santa Member, with its average calcium carbonate ($CaCO_3$) content of more than 93%, exhibits the greatest potential for dissolution. In the Pedro Leopoldo Member, in which the silica content varies widely, the carbonate schists (with higher silica concentrations) tend to dissolve less readily than the calcium phyllites (with lower silica concentrations).

In addition, sharp compositional variations exist at the contacts between the Lagoa Santa and Pedro Leopoldo members within the Sete Lagoas Formation. To illustrate this situation, Fig. 7 presents two graphs that show the variations in composition, on the basis of the concentrations of calcium, magnesium, silica, aluminum, and iron oxide in hundreds of borehole core samples.

The observed variations in the concentrations of the major elements $CaCO_3$, $MgCO_3$, CaO, SiO_2, MgO, Fe_2O_3, and Al_2O_3, in conjunction with descriptions of the core samples, has revealed that a relationship exists between the rock composition and the landscape morphology, as discussed by Pessoa (2005). Considering that rainfall is evenly distributed throughout the study area, the relevant factors in the evolution of relief are related to the properties and the arrangement of the rocks, which together (Piló 1998) determine the fundamental and preferential paths in the evolution of the landscape. In addition, Rauch and White (1970) have shown through petrographic and chemical analyses that the ratio of the volume of voids is directly associated with the concentration of calcium carbonate.

As noted by Pessoa (2005), in the southern part of the Lagoa Santa area, the existence of a clearly defined lithostratigraphic column has enabled the characterization of this region in terms of the hydrogeologic heterogeneities and anisotropies caused mostly by lithologic variations. Such differences can lead to frequent variation in the hydraulic conductivity and groundwater flow patterns, especially the flow velocity and direction. Figure 8 shows two geologic cross sections showing the lithostratigraphic variations present in the southern part of the LSKS.

Analysis of the stratigraphic framework and distribution reveals a convergence of lithologic factors that contribute to remarkable differences in the geological contact patterns and structural behavior. These features play an important role in the hydrogeological karst system and groundwater routes. The main hydrodynamic attributes and properties of these units are known and well recognized from the available literature (Pessoa 1996, 2005; CPRM/IBAMA 1998; Ribeiro et al. 2003).

Thus, according to the stratigraphic data from borehole logs, pumping wells and field observations carried out by CPRM/IBAMA (1998), five main geological sketches were drawn to illustrate the nature of the lithologic types in the LSKS and surrounding non-karst areas, and these sketches are shown in Fig. 9. Each of them presents the lithologies encountered in different profiles and highlights the differences in groundwater intake, circulation patterns, and discharge. Based on the analyses in previous studies, in the LSKS, the majority of the surficial features are composed of or associated with the following types of materials:

- Carbonate rock outcrops (20%),
- Siltite and metasiltite outcrops (35%),
- Product of metasiltite rock alteration in the form of thick clayey-silt soil cover (40%), and

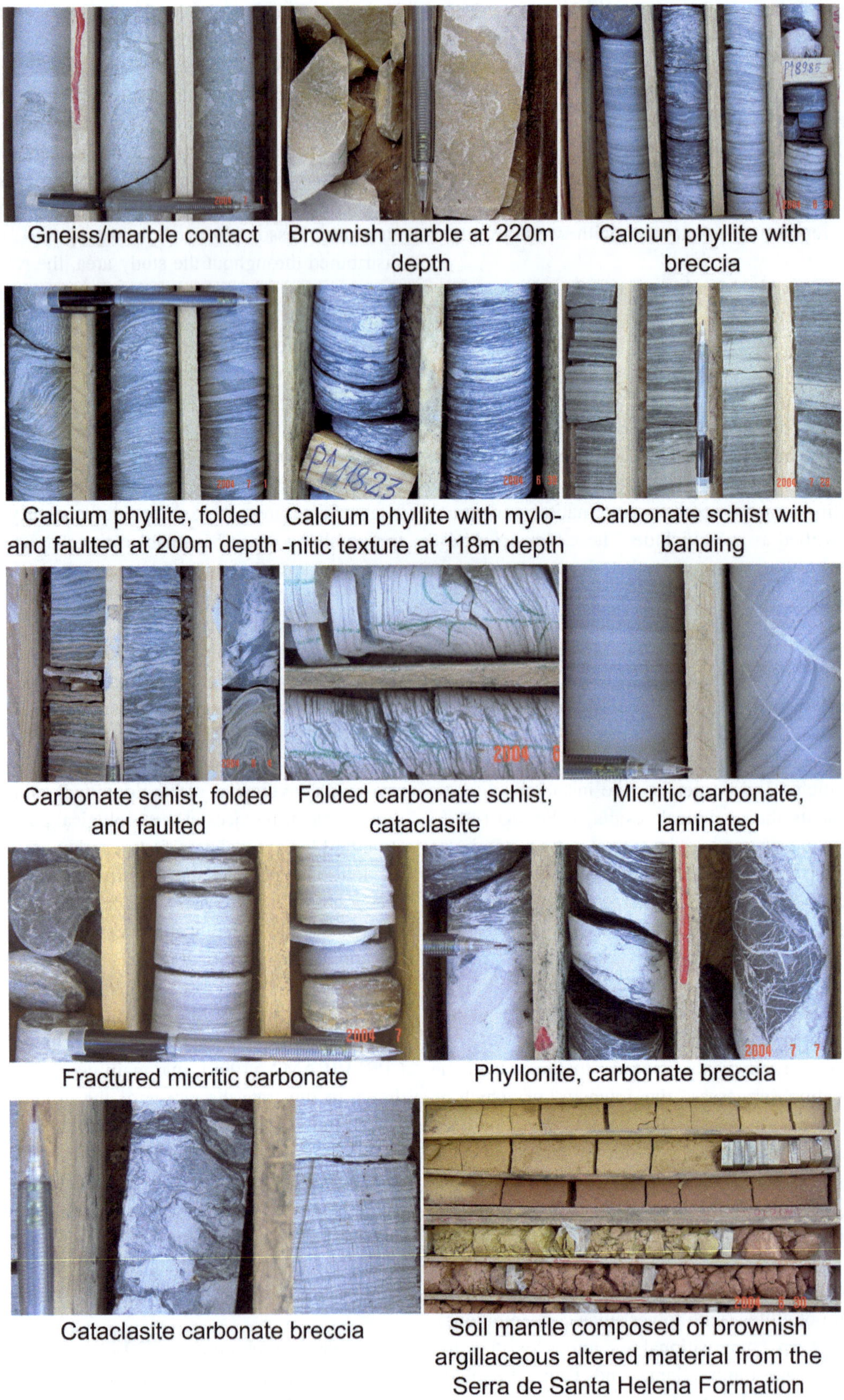

Fig. 6 Photographs of drilling samples from different depths and locations, all of them related to the studies carried out by Pessoa (2005). These samples are presented for the characterization of the covered karst region in the southern portion of the Lagoa Santa Karst system. The phyllites and schists are associated with the Pedro Leopoldo Member, and the micritic carbonate is associated with the Lagoa Santa Member (Sete Lagoas Formation)

Table 2 Compositional variations in the rocks associated with the Lagoa Santa Karst system

Local rock description	Bambuí group			Geochemical compositions (%)	
				Average	General
Argillaceous soils, red yellowish, ferruginous. Locally thick zones of gravel, with angular quartz pebbles	Serra de Santa Helena Formation (metasiltites, altered or not) (79 samples)			$CaO < 10$ $SiO_2 > 50$	$32 < SiO_2 < 90$ $0.08 < CaO < 8.7$ $0.0 < MgO < 2.1$ $2.4 < Al_2O_3 < 36.3$ $2.3 < Fe_2O_3 < 17.2$
Calcitic limestone or micritic carbonate	Sete Lagoas formation	Lagoa Santa member	Undifferentiated (869 samples)	$CaO > 90$ $SiO_2 < 10$	$0.38 < SiO_2 < 68$ $5.3 < CaO < 58$ $0.0 < MgO < 18.3$ $0.0 < Al_2O_3 < 16.1$ $0.0 < Fe_2O_3 < 6.8$
Calcium phyllites and carbonate schists		Pedro Leopoldo member			
Carbonate phyllites consisting of intense quartz and calcite veins with a calcitic matrix					

Source Pessoa (2005)

- Product of crystalline rock alteration in the form of silty arenaceous soil cover (5%).

In terms of recharge processes, Fig. 9 illustrates the variations among the lithological profiles in terms of the rainfall infiltration rates and percolation of groundwater into the karst aquifer system. Among the 5 sketches, type (a) represents a common pattern encountered all over the area, in which a granular soil mantle directly overlies carbonate rocks and is derived mostly from the metapelites of the Serra de Santa Helena Formation. Type (b) represents the situation in which the soil mantle overlies partially altered to massive metapelites. Type (c) is a tentative explanation for how a deep soil mantle can coexist beside massive cliff walls and suggests that faulting has affected the structure of the carbonate units. In this case, the soil could also originate from the impure carbonates present in the Pedro Leopoldo Member (carbonate schists or phyllites) through common weathering processes. Types (d) and (e) refer to environments in which both an altered metapelite mantle and granular soil originating from altered gneiss-migmatite directly overlie a massive crystalline basement.

2.3 Hydrogeologic Functioning of the Lagoa Santa Karst Aquifer

To describe the LSKS and its groundwater dynamics, it is important to be aware of two important factors: the interactions between topography and lithology variations and the spatial distribution of structures imprinted by the sedimentary, tectonic, and metamorphic fabrics developed over geologic time within the carbonate rocks. Additionally, the compositional and geomorphic aspects of the materials overlying, surrounding, and underlying carbonate-related relief are also very important.

Values for hydraulic conductivity and recharge potentials can be obtained from hydrogeological studies carried out by Pessoa (1996, 2005) and Galvão et al. (2015b). Here, typical values are taken for different recognized materials based on field measurements in similar areas during aquifer test evaluations.

Considering the main factors affecting groundwater dynamics, it is important to describe how the main hydrostratigraphic units behave in the LSKS. Thus, the crystalline basement, Sete Lagoas karst aquifer, Santa Helena aquitard, and soil cover aquifer are characterized individually. Table 3 shows a synthesis of the main values for several hydraulic parameters.

2.3.1 Crystalline Basement—Regional Aquifuge and Local Aquifer

The hydraulic conductivity in a karst aquifer system is known to be exponentially greater than that of a fractured crystalline basement (Freeze and Cherry 1979; Domenico and Schwartz 1998). For a hydrodynamic approach to a regional perspective, it is common to assume that crystalline rocks act as a type of aquifuge due to their absence of primary porosity and poorly interconnected fissures and fractures.

As the volume of groundwater circulation in a medium with limited storage and transmissivity is much smaller than that in karst aquifers, the crystalline lithology can transmit or store only comparatively small amounts of water because the massive granite and metamorphic rock contain few open fractures (Custodio and Llamas 2001). Secondary porosity is not well understood, but it is believed that it is either negligible or extends to only shallow depths. Hence, based on

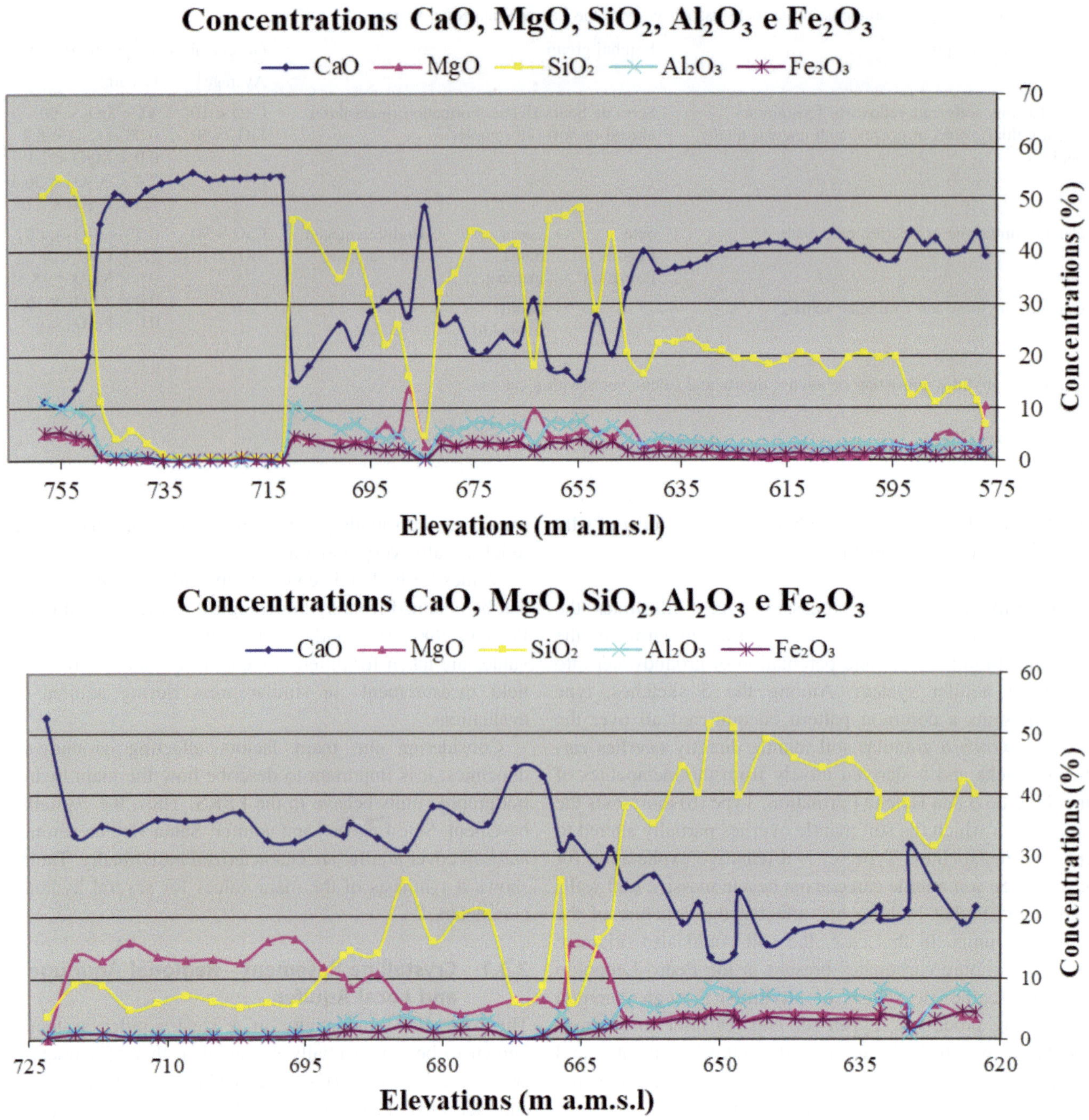

Fig. 7 Compositional variations among the carbonate rocks from core samples extracted from boreholes in the southern portion of the Lagoa Santa Karst system. Data collected and studied by Pessoa (2005)

the more permeable rocks that overlie the crystalline basement rocks, the elevation of the top of the crystalline rocks can be considered the lower limit of groundwater flow in regional terms.

This type of poor aquifer can be effectively described by comparison of pumping wells drilled into the crystalline basement and wells drilled into the carbonate rocks, as presented in Table 3. The average discharge values for these two types of wells are very different, with the values for the crystalline basement wells being at least ten times lower than those of the carbonate wells. As an estimate of the groundwater flow capacity, the hydraulic conductivity of the crystalline basement exhibits a narrow range of low values, approximately 1.0×10^{-7} to 1.0×10^{-5} m/s. According to a few pumping tests in groundwater wells drilled in this aquifer, discharges are only capable of supplying domestic

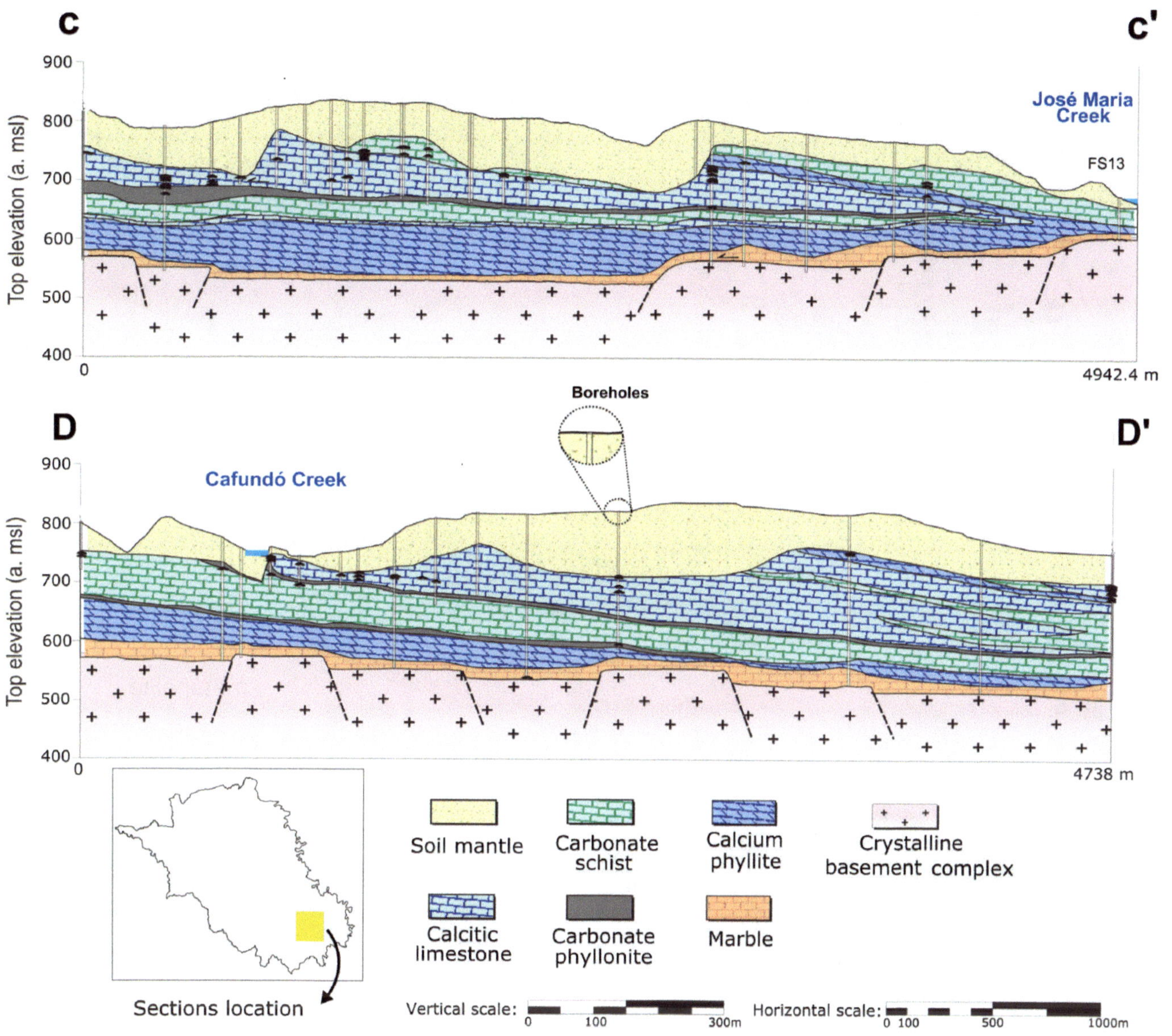

Fig. 8 Geologic cross sections constructed by Pessoa (2005), based on borehole logs

usage, with average discharges between 0.2 and 3.0 L/s (Company of the water supply of the Sete Lagoas city, personal communication).

Frequently, the crystalline rocks are covered by their own weathered granular materials, which can function locally as zones of improved recharge. This phenomenon occurs because the permeability and second porosity of the underlying hard crystalline rock are lower than those of the overlying weathered material. As the top of the crystalline basement reaches relatively high elevations along the western border of the LSKS, some recharge occurs from the weathered layer to the karst aquifer system, even if this process implies an extension of the hydrographic basin limits and the import of groundwater to the karst aquifer. This possibility is illustrated in the hydrogeologic section presented, where limestone strata are projected beyond the LSKS beneath elevation divides at the surface.

2.3.2 Sete Lagoas Aquifer

The configurations presented in Fig. 9 permit extrapolation and include a part of the crystalline hydrogeologic unit as an integrated marginal system associated with the carbonate environment, at least in some areas along the western border of the LSKS. It is interesting to observe that situations such as this one occur frequently in the carbonate environment, but indeed, this geologic framework must be considered critical to specific environmental approaches (see chapter "Environmental Problems in the Lagoa Santa Karst", this volume).

In the study region, high surface elevations occur in the western part of the LSKS, where the crystalline basement

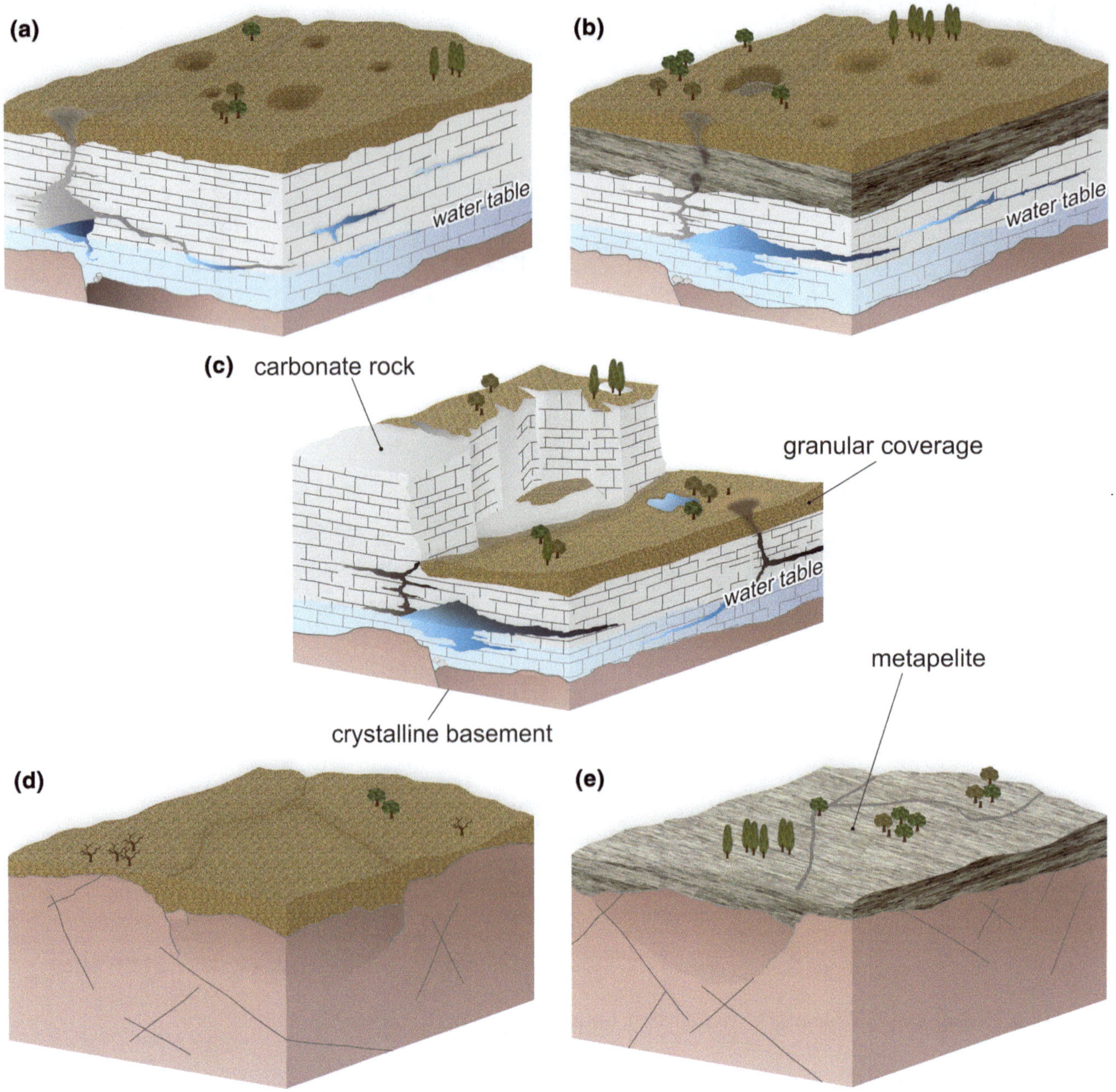

Fig. 9 Schematic lithologic profiles representing the Lagoa Santa Karst region (not to scale): **a** thick soil cover overlying the karst aquifer system; **b** thin soil cover overlying the Santa Helena aquitard, which in turn overlies the karst aquifer system; **c** semicovered karst aquifer system—a thick soil mantle coexists beside a limestone massif; **d** and **e** non-karst terrains—crystalline basement overlain by soil cover or Serra de Santa Helena metasediments, respectively

occurs at shallow depths or crops out. However, with distance downstream along Mata Creek, the surface elevations decrease, and outcrops occur along the southeasternmost margin of the LSKS where Mata Creek meets the Velhas River.

Considering that the Sete Lagoas Formation is associated with horizontal sedimentary deposits, which formed throughout the Bambuí sedimentary basin, it is easy to understand why so many places are regionally similar in terms of relief and topography. However, strong correlations between the thickness of the carbonate rock and the elevation of the top of the crystalline basement exist due to normal faulting.

Only part of the catchment zone has been studied (Auler 1994; Pessoa 1996; CPRM/IBAMA 1998; Pessoa 2005; Teodoro 2018). In some areas, the recharge process feeds and maintains perennial seasonal discharge at specific springs (Mocambo, Engenho, Samambaia). However, the catchments and hydrodynamic behavior of most important

Table 3 Hydraulic parameters gathered from pumping tests and field tests

	Hydraulic parameters	Maximum	Minimum	Average
Sete Lagoas aquifer	Transmissivity from pumping tests (m^2/d)	3590	90	913
	Hydraulic conductivity from pumping tests (m/s)	2.20E−04	1.85E−05	2.02E−04
Santa Helena aquitard	Transmissivity from pumping tests (m^2/d)	10	9.0	9.5
	Hydraulic conductivity from pumping tests (m/s)	3.00E−06	2.50E−06	2.75E−06
Gneiss-migmatite complex	Transmissivity from pumping tests (m^2/d)	38	1.5	20
	Hydraulic conductivity from pumping tests[b] (m/s)	89E−06	3.20E−08	4.61E−07
Soil mantle	Hydraulic conductivity from permeability field tests[a] (m/s)	2.90E−05	3.50E−08	6.07E−06

Source Pessoa (1996), 4 pumping test interpretations; Galvão et al. (2015b), 32 pumping test interpretations; [a]Castro (2011), 11 field test interpretations for shallow boreholes. [b]Freeze and Cherry (1979)

springs (Gordura, Jaguara, Poço Verde, Lagoa Bonita, Poço Azul, etc.) are not yet known.

Efforts must be made to acquire such data for an overall regional groundwater budget. For instance, the discharge has been monitored by some effective programs, such as the one carried out by Veláquez et al. (2018). For the main known springs in the LSKS, discharges vary from approximately 15 L/s to 6,000 L/s. Table 4 presents a list of the main outflow and inflow spots and their properties concerning the location and main discharge estimates.

Regarding outflow and circulation behavior, recharge contributes fully or in part to the outflow from springs. If recharge contributes only in part to a spring, the remaining water can contribute to processes deeper in the karst aquifer and can result in outflow in different locations in an interdependent system. Thus, some springs can be clearly characterized as full flow, overflow or underflow springs (Worthington 1991).

The Lagoa Santa Karst is composed of limestone with varying thickness and a spatially variable cover of weathered or nonsoluble pelitic rocks. Such circumstances must be considered in order to understand the flow pathways that potentially work as recharge conduits. However, such information can only be obtained through the investigation of the subsurface materials via geological drilling surveys or less conclusively through lithologic log analysis of pumping wells.

One of the most common features of the LSKS is the hundreds of karst lakes, which represent one of the most typical patterns of dissolution in this karst region. Understanding these features is important because the lakes can be either directly connected to the karst aquifer system or not directly connected with the karst aquifer. While many lakes are in mantled karst areas, others are the surface expression of the dissolution-induced doline formation process, which sometimes occurs beside limestone cliffs and is a typical speleogenesis processes (Auler 1995, chapter "Caves and Speleogenesis in the Lagoa Santa Karst" this volume).

The behavior of the limestone in the Sete Lagoas Formation is dominated by the presence of bedding planes in the pure carbonate rock (Lagoa Santa Member) and in the argillaceous carbonate rock (Pedro Leopoldo Member) and results in some differences in the dissolution potential. The combination of bedding planes and aligned vertical fractures makes this unit the main host of the pathways controlling groundwater flow.

Groundwater circulation develops along fractures and sedimentary bedding planes, which can be greatly enlarged in some specific directions, creating high values of hydraulic conductivity (Table 3), as can be observed in many caves and fractures opened by the dissolution of carbonate by water. As most bedding planes have very gentle dips, the intersections between structural features, such as faults and fractures, play a very important role in the flow behavior and dissolution process. In some outcrops in the LSKS, evidence of intense speleogenesis and associated groundwater flow following either bedding planes or vertical fractures can be clearly detected, as shown in Fig. 10.

The lithologic differences between the Lagoa Santa and Pedro Leopoldo members, especially the low silica content in the former, exert an important controlling force on the development of preferential groundwater flow pathways, as described before. Therefore, the majority of the dissolution features are present in the pure limestone rocks of the Lagoa Santa Member due to its high content of calcium carbonate.

Pessoa (2005) presented a very detailed characterization of lithostratigraphic borehole logs, describing groundwater entrances in the wells associated with preferential dissolution zones occurring between 680 m above mean sea level (a.m.s.l.) and 720 m a.m.s.l. The development of the karst groundwater flow system was also influenced by the geochemical differences at the contacts between calcium carbonates in the Lagoa Santa Member and carbonate schists in the Pedro Leopoldo Member.

The hydraulic conductivity values along these geologic contacts are shown in Table 3. In this case, the values for the

Table 4 Characteristics and locations of karst-related features exhibiting inflow or outflow

Feature name	Feature	UTM (SIRGAS 2000, Fuse 23)		Discharge (L/s)	
		East (X)	North (Y)	Teodoro (2018)	Auler (1994)
Lagoa Bom Jardim	Swallet	597825	7838540	192	–
Lapa Vermelha	Cave	605330	7831181	15	–
Samambaia	Spring	606395	7833972	113	150
Gordura	Spring	601795	7850545	696	–
Bebida	Swallet	597495	7849934	199	–
Goiabeiras	Swallet	600216	7852045	283	–
Gruta da Lagoa	Cave	601816	7850694	1.749	–
Mocambo	Spring	601724	7842699	822	725
Jaguara	Spring	594649	7844367	963	700
Palmeiras	Swallet	597338	7840233	10.7	–
Mocambo	Spring	601724	7842699	731	–
Cocho D'água	Karst window	604886	7834995	85	–
Água Fria	Cave	594220	7834716	0.089	–
Engenho	Spring	606428	7834617	121	100
Zé Irene	Swallet	599485	7843665	128	–
Palestina	Spring	608459	7838048	–	200
Gameleira	Spring	608750	7837924	–	50
Poço Azul	Spring	612926	7839343	–	200
Lagoa da Mata	Spring	613217	7839218	–	50
Moinho Velho	Spring	609372	7825285	–	150
Olho d'Água	Spring	616604	7822257	–	75
Cauê	Spring	602993	7830211	–	75

Lagoa Santa Member are higher than those for the Pedro Leopoldo Member. These lithotypes represent excellent aquifers that can provide pumping wells with large amounts of groundwater.

2.3.3 Santa Helena Aquitard

Along the northern and southern margins of the LSKS, the metasiltstone of the Serra de Santa Helena Formation crops out at the top of higher elevation relief. This unit represents an aquitard cover overlying the Sete Lagoas aquifer, and its thickness varies. Composed mainly by low-grade metamorphic argillaceous sediments, it acts as an impervious unit and cannot be considered a recharge zone. This unit can locally isolate the karst aquifer systems below it, thereby influencing the recharge process. Due to its highly clayey composition and consequently low permeability, the Santa Helena aquitard has not been considered for the installation of pumping wells because the depth to reach the karst aquifer results in unacceptably high drilling costs.

Throughout geologic history, in response to climate changes, gradual uplift, decrease in groundwater level and changes in the hydraulic gradient, differential erosion has taken place and has removed the siltite strata in many places within the LSKS. Thus, relatively low-elevation areas commonly exhibit soils that originate from the Serra de Santa Helena Formation via the weathering of the metasiltstones (Piló 1998).

2.3.4 Granular Undifferentiated Cover Aquifer

This aquifer unit is composed of weathered rock and mainly overlies limestone massifs. This porous medium forms extensive and thick soil layers that vary in composition from clayey to fine silty. The majority of the material is derived from the weathering of metasiltite in the Serra de Santa Helena Formation. However, soils originating from the Pedro Leopoldo Member at the base of the Sete Lagoas Formation are present in part of the region and are composed of argillaceous material overlying limestone rocks (Shinzato

Fig. 10 **a** Linear intersection formed by the junction of a bedding plane and the foliation. **b** Conduits developed in the wall oriented according to the alignment of intersecting features. **c** Development of elliptical conduits along bedding planes. Development of conduits (**a**, b and **d**) and a cave (**c**) along fracture planes. **e** Conduit partially filled by a breccia composed of fossiliferous carbonate

1998) because insoluble portions of the impure carbonates are completely different from the entirely soluble pure carbonates of the Lagoa Santa Member (Piló 1998).

Despite its broad regional distribution, soil is a very important agent that participates distinctly in the rainfall infiltration process and in the effective recharge distribution. Because of their generally high primary porosity and low permeability, the extensive contact zones between rock and soil allow both water percolation and enhancement of rock dissolution, resulting in intense karstification.

Additionally, the frequent presence of soil in caves and the paragenetic model of the evolution of the Lagoa Santa karst corroborate the intense participation of weathered cover materials in the development and evolution of karst terrain and in the associated hydrogeologic framework (see chapters "Soils of the Lagoa Santa Karst" and "Caves and Speleogenesis in the Lagoa Santa Karst", this volume).

The distribution of the lithostratigraphic units mapped at the surface is presented for the entire LSKS (Fig. 11), and the main aquifer units and catchment drainage basins are based on Auler (1994) and Teodoro (2018). The groundwater well distribution, major springs, and swallets throughout the LSKS are also shown. Each one of the catchment drainage areas is discussed in detail in the next section.

To summarize the main attributes of the entire karst system, Table 5 describes the lithologic and hydrogeological characteristics of each formation based on fieldwork performed by CPRM/IBAMA (1998).

2.4 Drainage Catchments and Groundwater Routes

The drainage basins located in the physiographic domains of this karst region have been studied to better understand the boundaries of the groundwater basin because the preferred flow paths associated with sinks and springs are not yet fully understood. Based on the results of a dye tracer study by Auler (1994), fieldwork in the karst region of Lagoa Santa and Matozinhos (Hidrovia 2016), and the research conducted by Teodoro (2018), the hydrogeological boundaries of some karst catchments in the LSKS have been identified. These proposed hydrogeological catchment domains are discussed below, and a detailed map showing the main results obtained from the tracer tests is presented in Fig. 12.

2.4.1 Domain of the Paraopeba Tributaries Drainage

The Paraopeba tributaries drainage domain is in the extreme northwest of the Lagoa Santa Karst area. Most of the drainages forming the basin of São João Creek flow over the crystalline basement, although a few small streams flow over areas underlain by the Serra de Santa Helena and Sete Lagoas formations. Pessoa (1996) documented the characteristics of the hydrogeological circulation in this area and concluded that the pattern is related to the Paraopeba River basin.

2.4.2 Domain of the Jequitibá Creek Drainage

The Jequitibá Creek catchment covers the northwestern end of the Lagoa Santa Karst area. The headwaters are located south of the Sete Lagoas municipality, and Jequitibá Creek has three main tributaries before the confluence with the Velhas River. The subbasins of these tributaries are partially within the study area. Most of these drainages are underlain by the Serra de Santa Helena Formation; however, in the headwaters, the limestones of the Sete Lagoas Formation appear, especially around the city of Sete Lagoas.

Galvão (2015a, b) proposed a system of horsts and grabens controlled by normal faults; in this system, the crystalline basement and overlying sediments are sectioned into several compartments with distinct depths, ranging from tens to hundreds meters in a few kilometers. The Sete Lagoas and Serra de Santa Helena formations are a part of this system and exhibit different thicknesses, thickening to the northeast. The sediment layers also decrease in elevation to the northeast. Most of the urban area of Sete Lagoas city is on a graben filled with carbonate rocks and unconsolidated sediments, whereas Santa Helena mountain is composed of the crystalline basement, which has been elevated to the surface. This structure has relevance for groundwater circulation and karstification processes because it influences the flow of groundwater.

In the urban area built on the Sete Lagoas Formation, there is evidence of dolines and karst collapses, which denote the existence of caves and active karstification processes.

2.4.3 Domain of the Gordura Creek Drainage

The Gordura Spring is a typical resurgence of groundwater at the end of the saturated karst system, and this spring flows continuously along a valley floor called Flor or Gordura Creek. The spring arises in the carbonates of the Lagoa Santa Member, flows through metapelites of the Serra de Santa Helena Formation, and crosses carbonates of the Pedro Leopoldo Member towards the Velhas River.

As noted in Table 5, this karst spring with high discharge values was studied by Teodoro (2018), who identified the origin of a portion of the contribution to the discharge.

This domain presents a complex system of groundwater flow and exhibits almost no surficial runoff. Instead, it features springs, resurgences, caves, and swallets. The main karst landforms in this catchment are the Goiabeiras Resurgence and Bebida Swallet. In the central portion of this

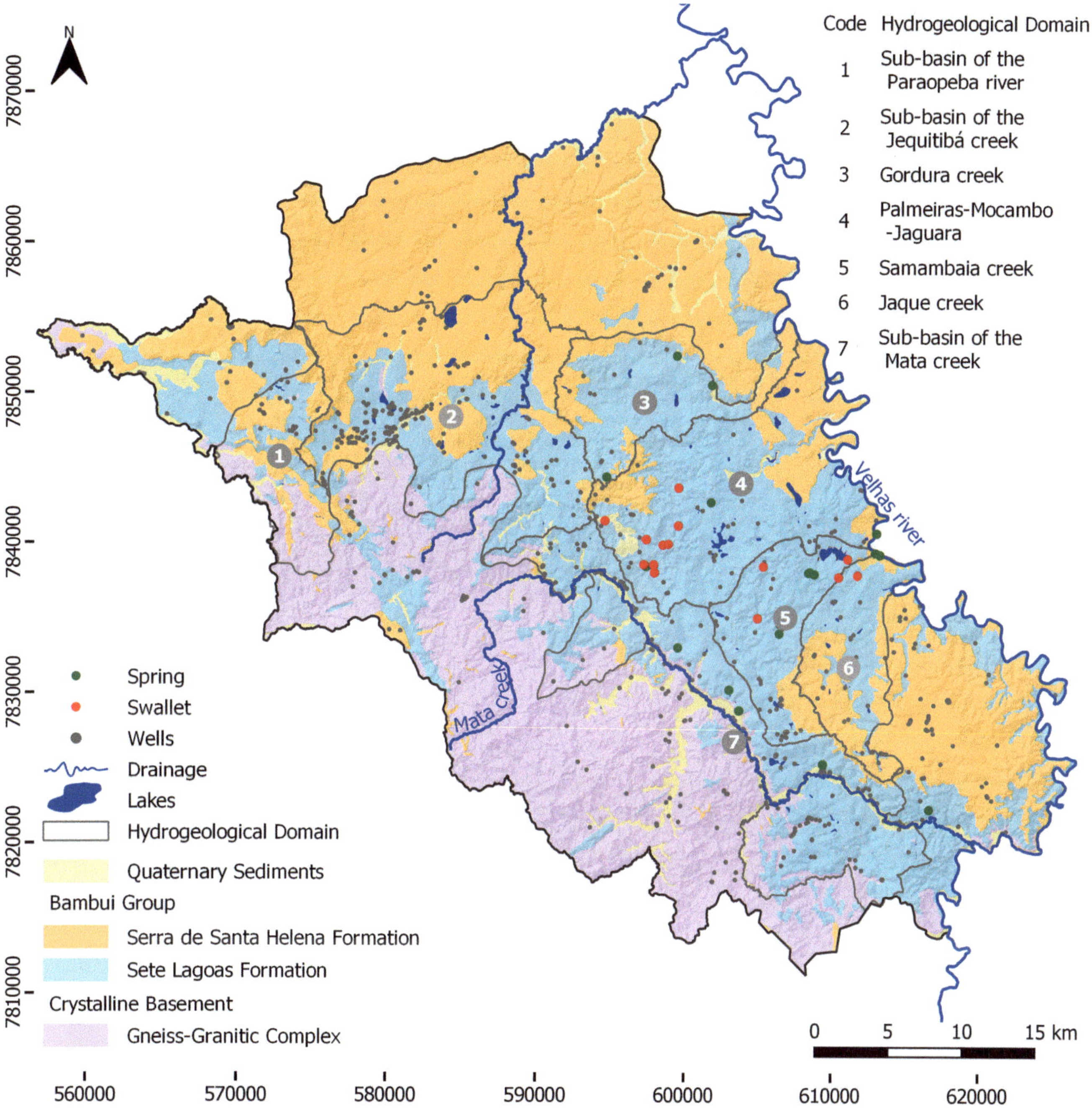

Fig. 11 Simplified hydrogeological map with the distribution of the aquifer units, groundwater wells, major surface features (such as karst springs, lakes or swallets), and the drainage network

domain, the water captured in the upstream karst structures overflows, resulting in the Gordura Resurgence, which is the largest of this type of feature in the region (Pessoa and Mourão 1998).

The elevation difference between the assumed regional base level (the Velhas River) and the top of the aquifer (600 m a.m.s.l.) in the western portion reflects a very low hydraulic gradient, resulting in a low-energy regime.

2.4.4 Domain of the Palmeiras-Mocambo-Jaguara Creeks Drainage

In the central portion of the LSKS, Palmeiras, Mocambo and Jaguara creeks join to form one highly integrated fluviokarst system that includes blind valleys, swallets, and caves in the vicinity of the Depression of Mocambeiro. This major fluviokarst system is composed of three minor subsystems, namely, the headwaters of the Bom Jardim Lake tributaries,

Table 5 Summary of the main hydrogeological characteristics of the Lagoa Santa Karst aquifer system

Geological formation	Aquifer category	Rock types	Description	Thickness (m)	Water quality
Undivided soil cover	Porous aquifer	Soil mantle	The granular material grouped under the generic name of unconsolidated cover is predominantly located in the unsaturated zone and has been analyzed only as an overlying porous material, where a small volume of the recharge is processed. The hydrogeological behavior of this unit is not well known, but the unit is believed to have very low primary permeability due to its high clay content and negligible secondary permeability due to the absence of extensive discontinuities	Extremely variable, from few meters till some dozens	No samples available
Serra de Santa Helena formation	Porous aquifer	Metapelites, siltites, phyllites	This unit is characterized mainly as an aquitard with very low permeability related to its high content of clayey material but with significant storage capacity. It has a poorly developed fissure porosity, which should reflect low values of hydraulic conductivity and small vertical recharge capacity between the soil mantle and the karst aquifer. Bedding plane (S_0) orientation preferential to N-S/10E, and schistosity (Sn) varies from N-S/25E	Based on 139 borehole drill logs: Average: 44 Maximum: 104 Minimum: 0–4.5	Electrical conductivity (EC) 28 samples 9.4 < EC < 346 Mean value EC = 120 3.1 < pH < 7.1 Mean value pH = 5.5
	Poor Aquifer or Aquitard Granular/fissure				Electrical conductivity (EC) 5 samples 94 < EC < 336 Mean value EC = 242 5.5 < pH < 7.6 Mean value pH = 6.7
Sete Lagoas formation Lagoa Santa member	Epikarst Conduit/dissolution		The epikarst contains extensive discontinuities formed by large irregularities due to the dissolution processes of the carbonates in contact with the base of the soil mantle and the top of the limestones, functioning as a well-developed zone and preferred groundwater flow paths at different depths, presenting high ranges of hydraulic conductivity	Varies from very thin (<5 m) to very thick, with a median of 40 m	81 samples Electrical conductivity (EC) 9 < EC < 529 Mean value EC = 226
	Fissure/karst aquifer	Pure limestone (> 90% $CaCO_3$)	The hydrogeological behavior of these units are mostly based on the presence of the vertical fractures intersecting bedding planes and connecting different units of pure and impure limestones, carbonate schists and marbles, and these fractures correspond to the elements controlling groundwater flows. The circulation of groundwater occurs through this secondary porosity, such as planes of fractures and fissures, which may show small to moderate openings or well-developed features greatly enlarged by intense dissolution. The hydraulic conductivity values are variable, resulting in high degrees of anisotropy and heterogeneity in the aquifer. Bedding plane (S_0) orientation preferential to N-S/10E. Schistosity (Sn) varies from N-S/45E.	35 a 135 m Average = 75 m	
Sete Lagoas formation Pedro Leopoldo member		Impure limestone (<66% $CaCO_3$)			
		Marble			

(continued)

Table 5 (continued)

Geological formation	Aquifer category	Rock types	Description	Thickness (m)	Water quality
Crystalline basement	Local aquifer Regional aquifuge	Gneiss-migmatite	The crystalline basement can behave as a very local aquifer, but due to its regional extent and its nonexistent primary porosity, it must be considered an aquiclude. Its secondary porosity is not well known but is believed to be insignificant. Thus, in relation to the much more permeable units located above, the crystalline basement can be considered the lower limit of the groundwater flow	Depth to the top of the basement Average = 209 m Maximum = 315 m Minimum = 57 m	–

Source CPRM/IBAMA (1998), Pessoa (2005)

Palmeiras Creek, and the Zé Irene karst window. Auler (1994) investigated the groundwater flow in this region and described detailed flow paths according to results obtained from tracer tests.

The headwaters of Palmeiras Creek are on the Santa Helena aquitard. This flow is an allogenic contribution dissecting the rocks of the Lagoa Santa Member and eventually sinks downstream into the Palmeiras Swallet.

Bom Jardim Lake is a collection of other minor tributaries from the Matozinhos peripheric urban area. Similar to Palmeiras Creek, overflow comes out sequentially into Ballet Cave, Poções Cave, and Poções Spring and then flows directly to the Mocambo Spring. All the groundwater coming out of the Mocambo Spring flows superficially through small channels until it joins the water flowing out of Jaguara Spring, and this combined flow migrates downstream towards the Velhas River.

The studies carried out by Teodoro (2018) showed a groundwater connection between the Palmeiras-Mocambo system and the Mocambeiro Depression outlet area. Additionally, the results show a connection between the Palmeiras-Mocambo and Jaguara karst catchment since the tracer injected into the Palmeiras Swallet was detected in Poções Cave and Mocambo and Jaguara springs. In addition to these springs, the tracer was also detected in one well located at Santo Antônio de Cerca Grande farm. Furthermore, the tracer injected into the Bom Jardim swallet was detected at the same points described above. These results indicate flow dispersion and splitting into a deeper circulation pattern towards the Jaguara Spring and to the Mocambeiro Depression (Fig. 12).

The subsystems associated with Bom Jardim Lake and Palmeiras Creek tend to flow to the NE, whereas those associated with Zé Irene tend to flow toward the SE; thus, both flow toward the Jaguara spring. The entire system can be interpreted as an overflow/ full flow (underflow) system, according to Worthington (1991), in which the springs associated with the subsystems located in the upstream portion are overflow springs and the Jaguara spring is a full flow spring. Based on the distinct pathways described and the karst aquifer framework, schematic hydrogeologic cross sections are shown in Fig. 13, and these cross sections focus on the groundwater connection potentials related to lithostratigraphic variations in the crystalline basement and carbonate.

2.4.5 Domain of the Samambaia Creek Drainage

The hydrogeological basin of Samambaia Creek encompasses the region of Confins, Lapa Vermelha Cave, and Sumidouro Lake and is an area of abundant karst features, and the geologic characteristics of this basin are dominated by the carbonates of the Lagoa Santa Member. Samambaia Creek corresponds to the local base level in the northeastern portion of the karst plateau. There are four major springs along Samambaia Creek. The spring farthest upstream is the Samambaia Spring (Fig. 14a), and in the downstream direction, there are Engenho (Fig. 8.14b), Gameleira and Palestina springs; the latter two are close to Sumidouro Lake, into which the fluviokarstic Samambaia Creek empties.

Auler (1994) hypothesized that the water accumulating in the lake located on the edge of the Lapa Vermelha Cave massif sinks into the cave and is connected to the Samambaia Spring. This hypothesis was confirmed by a recent study conducted by Velásquez et al. (2018). Additionally, the results of this study revealed that the discharge entering the Lapa Vermelha Cave is not enough to maintain the Samambaia spring; thus, during part of the year, this spring must be fed by contributions from other recharge areas.

The water circulation in Samambaia Creek is characterized by the intense interaction between groundwater and surface waters along flow paths that connect different fluviokarst systems.

Sumidouro Lake forms seasonally via the accumulation of water from upstream areas. The lake is very shallow and drains via a small channel to a swallet located in the

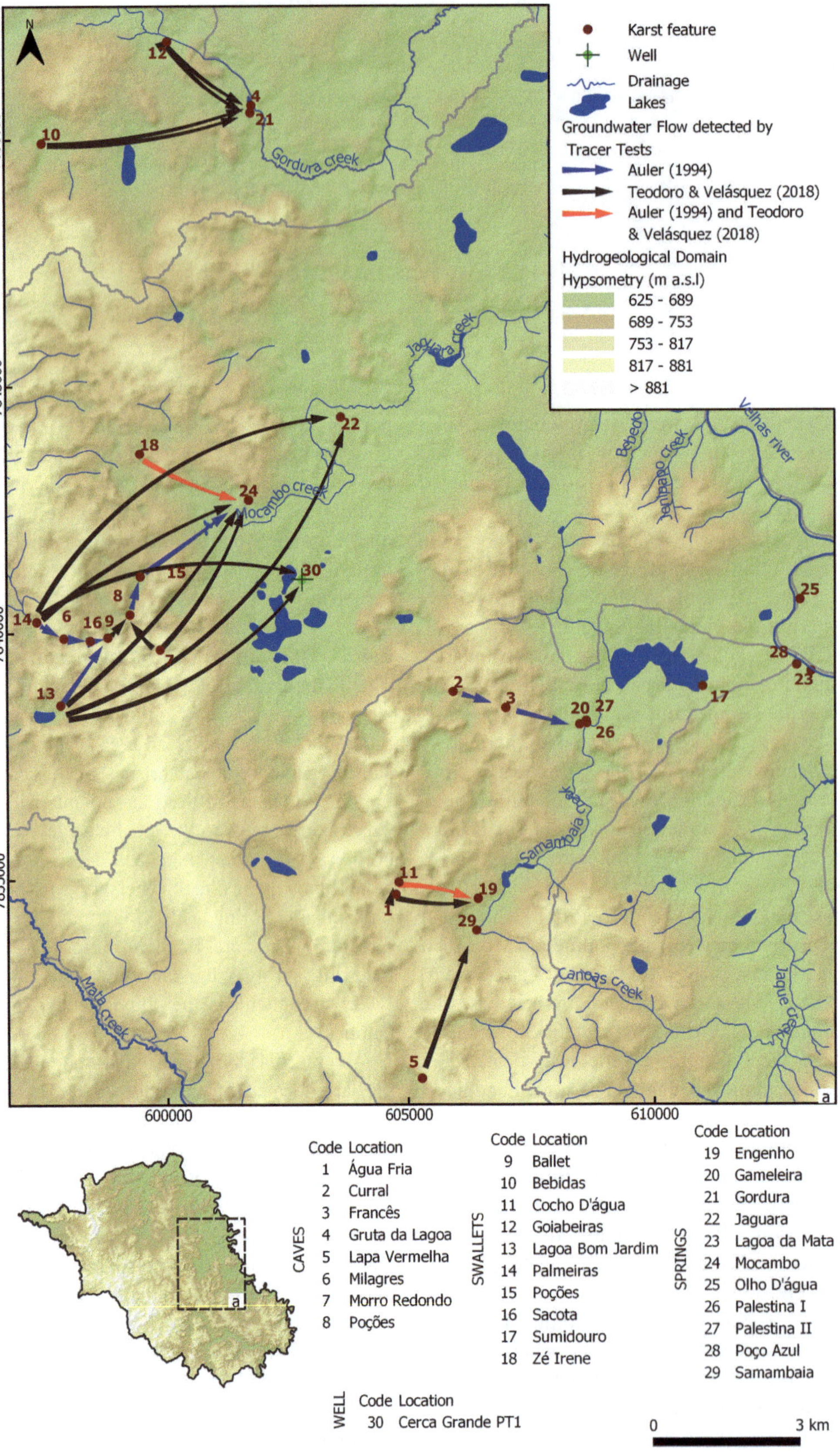

Fig. 12 Hydrogeological catchment domains based on the results of tracer tests

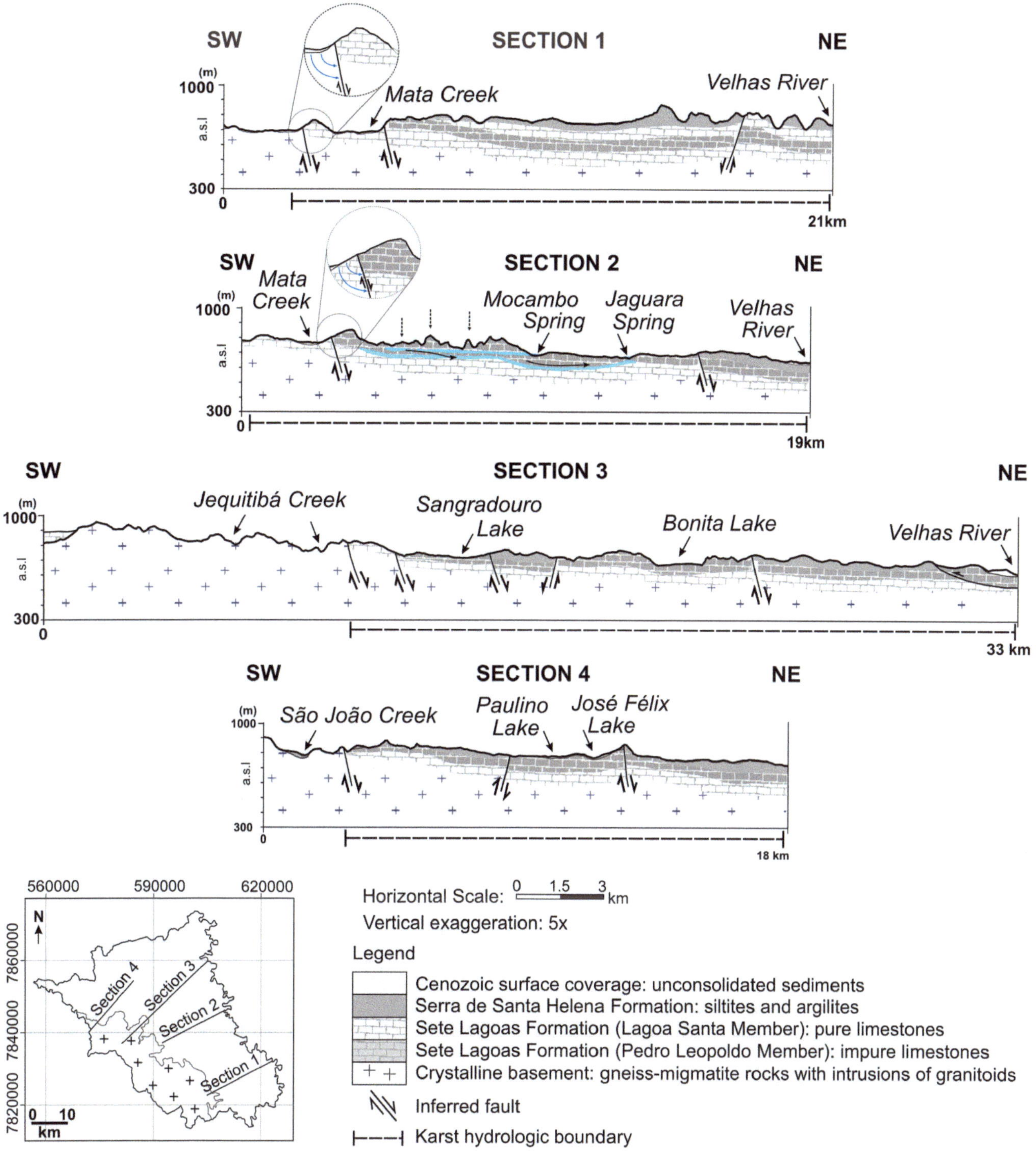

Fig. 13 Hydrogeologic cross sections exemplifying the different forms of groundwater flow connections, considering the capacity of the crystalline system to contribute to the karst aquifer and the differences among distinct hydrographic basins (São João, Jequitibá and Mata creeks), due to the lithostratigraphic, morphologic, and surface elevation conditions

southwestern portion of the lake (Fig. 15). The route of these waters flowing out of the Sumidouro Lake is not well known, and tracer tests performed by Auler (1994) and Teodoro (2018) did not yield strong positive results. However, the lake is thought to be connected to Poço Azul Lake, located on the banks of the Velhas River.

Fig. 14 Main springs contributing to Samambaia Creek: **a** Samambaia spring and **b** Engenho spring

2.4.6 Domain of the Jaque Creek Drainage

This basin is located in the northwest of Lagoa Santa city, in the south-central portion of the LSKS. The headwaters are the Confins region on the aquitard formed by the Santa Helena Formation, and the main tributaries are Fidalgo and Capão de Santana creeks. Water collects in the valley of Jaque Creek underlain by the Sete Lagoas Formation, across which it flows until the confluence with the Velhas River. The karst features in this subbasin are swallets and lakes, which occur in the domain of the Sete Lagoas Formation, including both the Pedro Leopoldo and Lagoa Santa members.

The hydrogeological limits of the basin are considered to be the same as those of the hydrographic basin. A proposed hypothesis that has yet to be tested suggests that the contribution of the recharge into the Sete Lagoas aquifer around the Confins region in the southwest portion of the basin flows to the Jaque Creek valley at depth. However, there is no documentation of this flow path yet, and the only support for this model is the favorable orientation of carbonate bedding planes and some discontinuities that are coincident with the trend of regional flow towards the Velhas River. Another possibility that should be considered is that the water loss from the Samambaia basin forms recharge directly

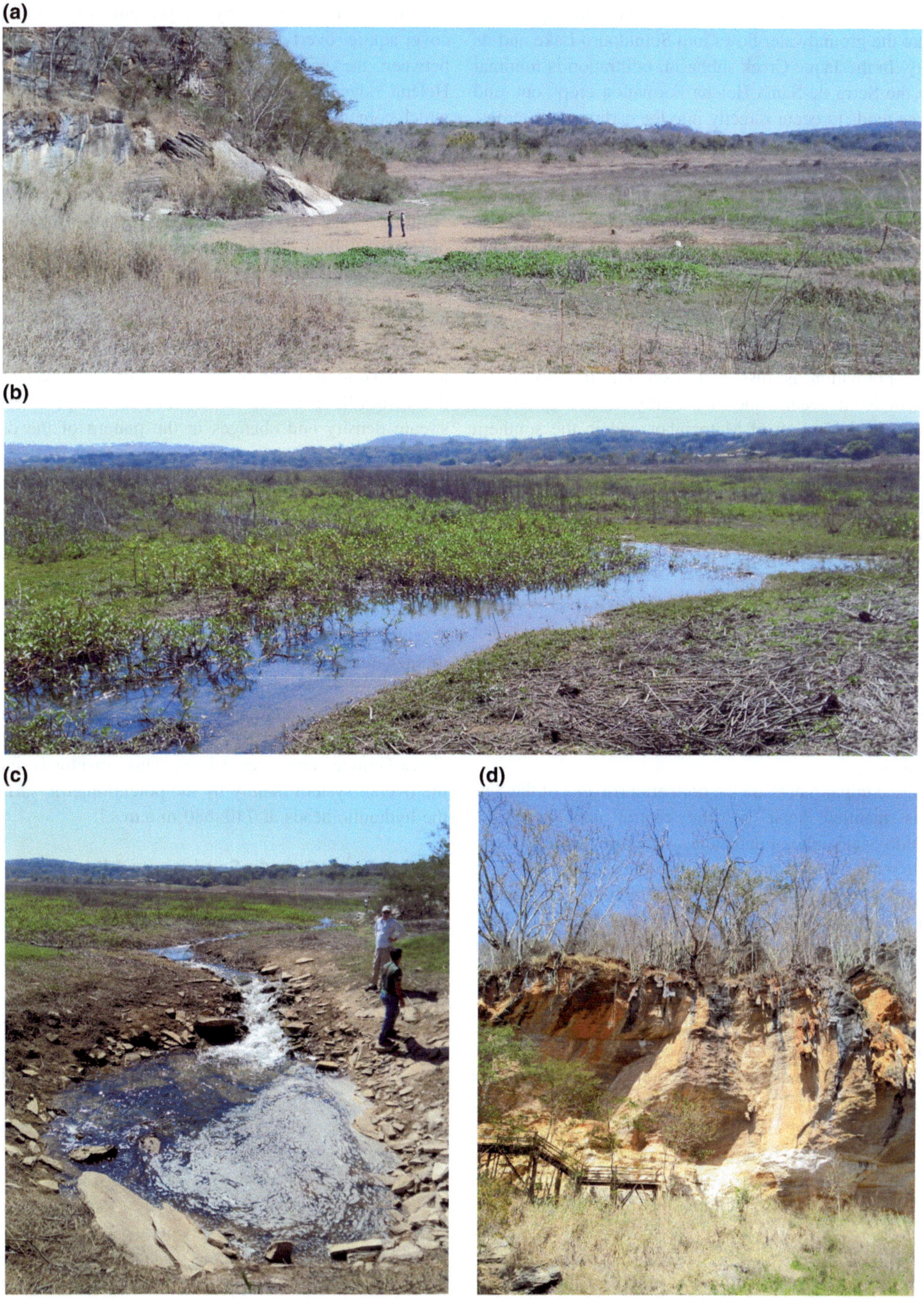

Fig. 15 Sumidouro lake. **a** Photograph of the dry lakebed. **b** Channel formed along the lakebed. **c** The swallet into which the water in Sumidouro Lake drains. **d** Carbonate cliff walls beside the lake

through the carbonate outcrops northwest of the Jaque basin and that the groundwater flows into Sumidouro Lake and its vicinity. In the Jaque Creek subbasin, infiltration is minimal where the Serra de Santa Helena Formation crops out, and recharge tends to occur directly into the carbonate outcrops.

2.4.7 Domain of the Mata Creek Drainage

Mata Creek corresponds to the western boundary of the karst area in several areas. In contrast, the western boundary is marked by high surface elevations. In this domain, the crystalline basement, the Sete Lagoas Formation, and alluvial deposits occur. To the north, the karst area includes several parallel drainages that are oriented N-S to NE-SW and that flow into Mata Creek. Along the valleys, the Sete Lagoas Formation is dissected, exposing the crystalline basement downstream, which is partially covered by alluvial deposits. In the region of Matozinhos and in the southern portion of Prudente de Morais, the Pedro Leopoldo Member and the crystalline basement rocks appear. Exposures of the Lagoa Santa Member and Serra de Santa Helena Formation are limited.

2.5 Hydrogeologic Conceptual Model

The analysis of the different hydrostratigraphic sequences described in the study area reveals that three major compartments with similar hydrogeologic behavior can be defined due to their stratigraphic patterns, as shown in Fig. 16.

These compartments are the northernmost area of the Sete Lagoas mantled karst (i), the central area of Pedro Leopoldo-Matozinhos-Funilândia, hereafter called Cerca Grande semicovered karst (ii) and the southern area of Lagoa Santa mantled karst (iii). Due to the similarities in terms of depth of the crystalline rocks and thickness of the carbonates, both the northern and southern portions of the Lagoa Santa Karst region are still covered by metasedimentary rocks belonging to the Serra de Santa Helena Formation and exhibit higher elevations. In contrast, the central semicovered karst area has both a thick soil cover and bare limestone cliffs. The mantled karst areas have a high potential for the formation of confined flows, as the limestone rocks dip and become thicker to the east and northeast of the mantled area, while the central area behaves as a very free aquifer system. At least, the crystalline basement has special participation in the scenery, as it coexists in a conspicuous hydrogeological position.

2.5.1 Sete Lagoas-Funilândia Mantled Karst Compartment

The Sete Lagoas-Funilândia mantled karst compartment has predominantly two main types of hydrostratigraphic profiles. The first is characterized by the presence of a granular soil cover aquifer overlying the limestones in all regions situated between the eastern low-elevation border of the Santa Helena ridge and the western margin of Jequitibá Creek, which coincides with a graben. The second one occurs along the eastern margin of Jequitibá Creek, where the relief is mainly represented by the Santa Helena aquitard overlying the limestones at greater depths. The eastern margin of Jequitibá Creek is thought to feature a shallow upper surface of the crystalline basement, which creates an intermediate divide between the basin and the Velhas River valley.

Therefore, the hydrogeological model proposed for the Sete Lagoas-Funilândia mantled karst area represents the northern edge of the LSKS. Thus, the margins of this basin are defined to the north by the valleys of Jequitibá Creek and the Velhas River and to the south by an abrupt difference in stream density and changes in the pattern of the drainage system.

This region exhibits the effects of E-W-oriented normal faults, which resulted in uplifted sides of blocks and higher relief to the north and down-dropped sides of blocks to the south. Furthermore, the degree of erosion and degradation differs between the sides. This region appears to form a preferential pathway with high hydraulic conductivity, potentially linking the central urban area of Sete Lagoas municipality to the eastern portion of LSKS. This zone might cross the Prudente de Morais municipality limits and Gordura spring catchment area, potentially reaching the Velhas River valley. This corridor marks the entrance to the Cerca Grande semicovered karst. This corridor is related to the overall system mainly by the potentiometric surfaces of the hydraulic heads at 740–680 m a.m.s.l.

2.5.2 Cerca Grande Semicovered Karst

The Cerca Grande semicovered karst is the most important compartment in the entire system. Distinct from the other two compartments, this compartment features abundant springs, dolines, and lakes, through which large volumes of groundwater exit and enter the karst system. Such a wealth of karst features reveals differences in the nature and hydrodynamic behavior between this system and the other two compartments. In the Cerca Grande semicovered karst, the exposure of limestone bedrock has enhanced the dissolution process and has led to karst landscape development and a high potential for aquifer effective recharge.

Compared to the northern and southern areas, the central area of the Cerca Grande semicovered karst has a different pattern of flow circulation. This unique flow pattern may be related to the thicker carbonate strata due to the much deeper surface of the faulted crystalline basement. The thicker carbonate strata produce a more homogeneous thickness and composition of the carbonate rocks than is present in the southern and northern portions, which exhibit greater

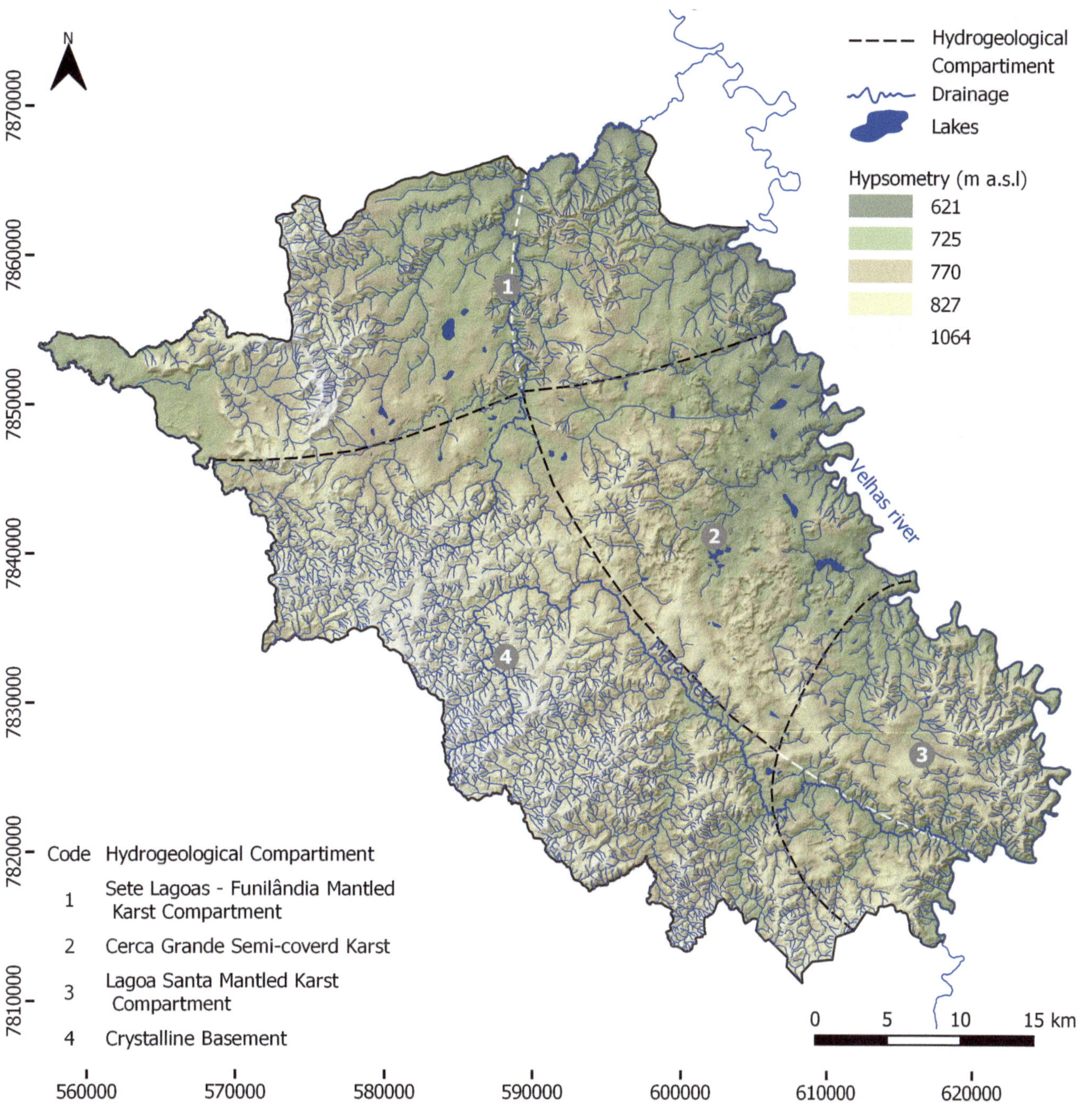

Fig. 16 Lagoa Santa Karst system showing four different compartments. (1) Sete Lagoas mantled karst, (2) Cerca Grande semicovered karst, (3) Lagoa Santa mantled karst, (4) Crystalline basement (*Source* Adapted from Pessoa (1996), CPRM/IBAMA (1998), Pessoa (2005))

influences of the stepped crystalline basement, more intense tectonic–metamorphic interferences and variable carbonate compositions, as documented by Pessoa (2005). The tectonic effects led to the alignment of some drainages, especially Mata Creek, Jequitibá Creek, and the Velhas River, whose valley orientations are related to the occurrence of basement regional faulting (Pessoa 2005; Galvão et al. 2016).

In this central area, some discontinuities can be observed, as shown in Fig. 16, which reveals the proper boundaries of the domain, as well as smaller internal differences. The behavior evidenced by drainage alignment in the boundaries of this domain may have limited the impact of the stepped pattern of the faulted crystalline basement. Dissolution could produce sharp anisotropies throughout the area.

2.5.3 Lagoa Santa Mantled Karst Compartment

The Lagoa Santa mantled karst compartment is one of the most studied portions of the LSKS and, to the north of Sete

Lagoas, conforms to the patterns associated with the stepped crystalline basement and high topography produced by the Serra de Santa Helena Formation. Pessoa (2005) presented interpolated data from borehole logs demonstrating that the crystalline basement decreases in elevation towards the east and exhibits a stepped pattern at depth.

Other data collected from boreholes showed the presence of small isolated grabens existing in this domain. This finding supports the configuration proposed by Galvão et al. (2015a) on the northern border of the LSKS. Small grabens were also described by Karfunkel et al. (2006) in the Vespasiano area, just inside this compartment, corroborating the observations in the north and south.

2.5.4 Crystalline Basement

This compartment deserves to be evaluated due to its hydrogeological position along this western portion of the entire aquifer system, considering its high-top elevation occurrences and potential to interfere in the groundwater flow that moves eastwards. Due to the differences of hydraulic heads, fissure, and porous matrix interconnections and carbonate layers that even appear downwards, it becomes imperative more attention on that portion, mainly the stripe coincident to the border limits and its geological contact which one shows obliterated due to the Mata Creek valley.

2.6 Groundwater Quality

Water sampling campaigns in wells and springs were carried out by CPRM/IBAMA (1998) with the goal of measuring groundwater quality in several spots in the LSKS. In total, 82 seasonal samples were collected from 37 pumping wells and springs.

First, these water samples were classified based on the major ions, and the results showed that the groundwater presented a very homogeneous chemical composition. In these samples, calcium and bicarbonate prevail, with the following ionic relations: $Ca^{++} \gg Mg^{++} > Na^{+}$ and $HCO_3^{-} \gg Cl- > SO_4$. The groundwater in the karst aquifers presented good physical–chemical quality, with local areas of lower quality, and few elements exceeded the potability standards.

The existence of element concentrations that did exceed the environmental standards may be related to old wells with degraded cement seals and corroded pipes, which may allow the presence of excessively high element concentrations detected seasonally in some analyses. In the case of springs, there are several potential sources of interference. Contamination generally arises from anthropogenic activities, including waste disposal and effluents released on the surface. Such activities can increase the chance of contamination of the aquifer and alter the chemical quality of the groundwater. The water quality results gathered are summarized in Table 6.

The groundwater has pH values between 6.5 and 8.1, with a mean of 7.3. The temperature ranges from 21 to 25 °C, with a mean of 22.7 °C, and the electrical conductivity ranges from 9 to 412 µS/cm, with a mean value of approximately 303 µS/cm.

Within the Sete Lagoas city limits, the proximity to the crystalline basement leads to the presence of water with higher concentrations of magnesium, sodium, and potassium. The detection of nitrate in some samples collected from wells (Galvão et al. 2017), especially in the central region of the urban area, clearly demonstrates the pollution of the groundwater and the effects of the dense human population on the groundwater quality.

Table 6 Results of the groundwater quality analyses performed on samples from the Lagoa Santa Karst system

Aquifer	Statistics	pH	HCO^{3-}	Ca^{2+}	Cl-	Mg^{2+}	K^{+}	Na^{+}	$SO4^{2-}$	Total water hardness	Alkalinity	EC
			mg/L							mg $CaCO_3$/L	mg $CaCO_3$/L	µS/cm
Karst	Average	7.00	142.1	44.0	2.7	3.4	0.8	3.3	4.0	106.3	116.4	269.1
	Amount	254.00	193.0	193.0	193.0	193.0	193.0	193.0	193.0	137.0	137.0	257.0
	Max.	8.40	454.9	114.5	21.9	38.0	3.4	29.0	27.0	245.1	441.1	680.0
	Min.	5.29	9.8	3.2	0.0	0.1	0.1	0.1	0.0	0.0	8.0	20.9
Basement	Average	6.75	79.3	11.8	1.1	4.2	1.1	8.4	5.5	30.9	76.1	142.8
	Amount	21.00	12.0	12.0	12.0	12.0	12.0	12.0	12.0	5.0	5.0	21.0
	Max.	8.50	209.2	43.0	4.5	8.2	2.4	22.9	30.4	57.4	94.1	355.0
	Min.	5.10	34.4	2.4	0.0	1.3	0.4	2.8	0.0	3.9	56.1	60.0

Source Pessoa (1996, 2005) and Galvão (2015)

3 Final Remarks

The LSKS can be characterized based on its geologic framework. The carbonate rock mostly covered by a thick soil mantle, and the scarce limestone outcrops, and the stepped crystalline basement have led to considerable heterogeneity in the aquifers. Thus, differences in the lithological profiles are associated with abrupt changes in hydraulic conductivity. These conditions, including the compositional variations among different strata, act as an intrinsic and powerful controller of the velocities and directions of flow. Such differences are especially distinct between the northern and southern carbonate basin compartments.

Variations in the distribution of pure and impure limestones in the Sete Lagoas Formation with depth can result in variations in the flow behavior of the groundwater. Similarly, the presence of a tectonic fabric in the bedrock allows widespread interconnection along strata, although some hydrogeologic compartments are separated from others. Moreover, infiltrating water in the vadose zone produces widespread recharge and connects various portions of the karst aquifer throughout the entire system.

Understanding of groundwater systems and their hydrodynamics has focused on the way in which flows move in space and time. The quantity of circulating groundwater is related to the ease with which the surface medium absorbs rainfall. Additionally, the stratigraphic framework of the area imposes boundaries, which can result in discharge zones, outlet spots or restrictions on the groundwater movement itself. In the Bambuí carbonate basin, heterogeneity and anisotropy are high due to the sedimentary variations that exist throughout the carbonate profile.

Despite the horizontally continuous pattern of the depositional sequence in the Bambuí carbonate basin, the compositional differences among the strata and the tectonic fabric play important roles in groundwater behavior.

The numerous dry open channels in limestone outcrops represent a high degree of porosity. In contrast, the soil cover broadly distributed throughout the LSKS interacts with these outcrops, enhances dissolution beneath the surface, and contributes to active flow behavior. Such contact zones are one of the most important to the lateral propagation of groundwater to the aquifer. This zone is known as the epiphreatic zone, which represents an intermittently saturated upper portion of the profile that is inundated seasonally by rapidly infiltrating water (Ford and Williams 2007). In the wet period, infiltrating water percolates vertically downwards through the unsaturated (vadose) zone between the ground surface and the karst water table, filling some fissures and joints.

It is important to highlight that effective recharge volumes are necessarily much greater along massive outcrops, where fracturing representing a high degree of secondary porosity facilitates water rainfall percolation; instead, effective recharge can also occur through areas covered by clayey-silt soil (see Piló 1998; Shinzato 1998).

The analysis of the karst hydrodynamics indicates that large volumes of groundwater move from west to east and that the majority of this water infiltrates into limestone cliffs through intricate and sometimes deep epikarst flow paths. The water infiltrating into the karst from the preferential recharge zones fills up rock joints and fracture spaces until the impounding effect of the rock underlying the karst raises the water level to the elevation of the local base level. From that point, the water stored in the karst flows out of the karst system, forming karst springs (Veress 2016).

In this context, the Cerca Grande semicovered karst area represents one of the most important recharge zones of the entire LSKS, mostly due to a large number of limestone outcrops, which allow water infiltration from rainfall. This process markedly enhances the effective recharge process because this compartment presents the most density of limestone outcrops in LSKS.

Seasonally, the karst water table reaches its highest point during periods of high precipitation and is lower during dry periods. This fluctuation in the water table height characterizes the epiphreatic zone, and the phreatic zone is defined as the zone below the lowest water table height. The phreatic zone extends down below the karst spring level and might characterize, at deeper levels lower than the regional base level, a karst water zone where the hydraulic gradient is very low (Veress 2016). This zone can be observed in some piezometers in the Mocambeiro Depression, where the water level is a few meters above the regional base level.

In this scenario, it is very important to understand that the central area of the Cerca Grande semicovered karst presents a higher capacity for water circulation and recharge since rainwater has a greater ability to infiltrate into the epiphreatic zone. Connections can occur between adjacent hydrogeologic basins situated far from the infiltration sites, which means that most of the carbonate aquifer, if not all, is integrated into one large karst aquifer system.

Such a heterogeneous environment deserves urgent attention in terms of both storage capacity and exploitation. The stratigraphic characteristics and nature of the geologic variability in the LSKS contribute both advantages and disadvantages in terms of exploitation.

These characteristics make it imperative to understand the particular behavior of the LSKS. However, spatial and seasonal information on the karst aquifer hydrodynamics are lacking, and such basic information must be provided for water use regulation and the management of karst land occupation. Hydrogeological analyses and interpretations demonstrate that this region needs a specific focus on karst spring behavior, which must be considered for integrated management and understanding of the real state of groundwater potentials.

Acknowledgements The advances in the understanding of the connections in the karst system presented by means of intensive groundwater exploitation demonstrate the usefulness of the efforts to catalog specific hydrogeologic data in the LSKS conducted by the Brazilian Geological Survey (CPRM) almost 30 years ago. Therefore, the author emphasizes his satisfaction with CPRM colleagues, especially Fernando Antonio de Oliveira (the VIDA Project's founder), and collaborative colleagues from the Hidrovia company, Giancarlo Ramos, Tamires Nogueira, Leonardo Faria, Larissa Martins, and Rogerio Lucas, who are appreciated for the help with figure drafting and composition. Thanks are also extended to COPASA-MG, for help with groundwater use information and Geraldo (Gagá) due to his ability in driving the drone.

References

Alkmim FF, Martins-Neto M (2001) Bacia intracratônica do São Francisco: Arcabouço estrutural e cenários evolutivos. In: Pinto CP, Martins-Neto MA (eds) Bacia do São Francisco: Geologia e Recursos Naturais. Sociedade Brasileira de Geologia, Belo Horizonte

Alkmim FF, Martins-Neto M (2012) Proterozoic first-order sedimentary sequences of the São Francisco craton, eastern Brazil. Mar Pet Geol 33(1):127–139

Andrade IB, Amorim JB, (2018) Mapeamento Geológico com Ênfase na Relação Hidráulica entre o Complexo Granítico-Gnáissico e o Grupo Bambuí. Final Graduation Report, Universidade Federal de Minas Gerais

Auler A (1994) Hydrogeological and Hydrochemical characterization of the Matozinhos - Pedro Leopoldo Karst, Brazil. MSc thesis, Western Kentucky University

Auler AS (1995) Lakes as a speleogenetic agent in the karst of Lagoa Santa,Brazil. Cave Karst Sci 21(3):105–110

Auler AS, Piló LB (2015) Lagoa Santa Karst: Cradle of Brazilian cave studies. In: Vieira BC et al (eds) Landscapes and landforms of Brazil. Springer, Dordrecht

Babinski M, Vieira LC, Trindade RIF (2007) Direct dating of the Sete Lagoas cap carbonate (Bambuí Group, Brazil) and implications for the Neoproterozoic glacial events. Terra Nova 19:401–406

Beato AC, Pessoa PFP (1991) Cadastro dos pontos d'água. Belo Horizonte: CPRM [Relatório interno]

Bittencourt J, Kuchenbecker M, Vasconcelos A, Meyer KEB (2015) O registro fóssil das coberturas sedimentares do cráton do São Francisco em Minas Gerais. Geonomos 23(2):39–62

Castro E (2011) Estudo Hidrogeoquímico Ambiental em Região Cárstica no Município de Sete Lagoas—Minas Gerais: Investigação da Contaminação de Águas Superficiais e Subterrâneas por Nitrato e Atrazina na Área do Centro Nacional de Pesquisa de milho e Sorgo da EMBRAPA. Universidade Federal de Minas Gerais, Thesis

Copasa MG (2019) Informações dos Sistemas de Abastecimento da COPASA MG. Unpublished report, Entorno da Região Cárstica de Lagoa Santa

CPRM—Companhia de Pesquisa de Recursos Minerais/IBAMA—Instituto Brasileiro do Meio Ambiente e dos Recursos Naturais Renováveis (1998) Área de Proteção Ambiental Carste de Lagoa Santa—Meio Físico. CPRM/IBAMA, Belo Horizonte

CPRM—Companhia de Pesquisa de Recursos Minerais (2003) Mapeamento geológico da região de Sete lagoas, Pedro Leopoldo, Matozinhos, Lagoa Santa, Vespasiano, Capim Branco, Prudente de Morais, Confins e Funilândia, Minas Gerais—Escala 1:50.000. Ribeiro JH, Tuller MP, Filho A, Padilha AV, Córdoba VC (eds) Projeto Vida—Viabilidade Industrial e Defesa Ambiental. CPRM, Belo Horizonte

CPRM—Companhia de Pesquisa de Recursos Minerais (2010) Projeto Sete Lagoas-Abaeté, Estado de Minas Gerais: texto explicativo. Manoel Pedro Tuller, José Heleno Ribeiro, Nicola Signorelli, Wilson Luis Féboli, Júlio Murilo Martino Pinho (Orgs), CPRM-BH. Belo Horizonte

CPRM—Serviço Geológico do Brasil (2018) Sistema de Informação de Águas Subterrâneas—SIAGAS. Disponível em: http://siagas.cprm.gov.br. Acesso em: dezembro de 2018

Custodio E, Llamas MR (2001) Hidrologia subterrânea. Omega, Barcelona

Domenico PA, Schwartz FW (eds) (1998) Physical and chemical hydrogeology. Wiley, New Jersey

Ford DC, Williams PW (2007) Karst Geomorphology and Hydrology. John Wiley & Sons Inc, West Sussex

Fleury S (2009) Land use policy and practice on karst terrains: Living on limestone. Springer Science and Business. Media B.V. https://doi.org/10.1007/978-1-4020-9670-9

Freeze RA, Cherry JA (1979) Groundwater. Prentice-Hall Inc. Prentice-Hall, Englewood Cliffs

Galvão P, Halihan T, Hirata R (2015a) Evaluating karst geotechnical risk in the urbanized area of Sete Lagoas, Minas Gerais, Brazil. Hydrogeol J 23(7):1499–1513

Galvão P, Halihan T, Hirata R (2015b) The Karst permeability scale effect of Sete Lagoas, MG, Brazil. J Hydrol. https://doi.org/10.1016/j.jhydrol.2015.11.02

Galvão P, Hirata R, Cordeiro A, Osório DB, Peñaranda J (2016) Geologic conceptual model of the municipality of Sete Lagoas (MG, Brazil) and the surroundings. An Acad Bras Ciênc 88(1):35–53

Galvão PHF (2016) Hydrogeological conceptual model of Sete Lagoas (MG) and associated implications of urban development in Karst region. Universidade de São Paulo, Thesis

Galvão P, Hirata R, Halihan T, Terada R (2017) Recharge sources and hydrochemical evolution of an urban karst aquifer, Sete Lagoas, MG, Brazil. Environ Earth Sci 76:159

HIDROVIA (2016) Estudos de dinâmica hídrica e da zona de proteção do Monumento Natural Lapa Vermelha. Fashion City Brasil, Environmental Impact Assessment. Unpublished technical report

HIDROVIA (2017) Estudo Hidrogeológico dos Aquíferos Cársticos Cobertos da Região Aeroportuária de Confins, Bicho do Mato/BH Airport, Environmental impact assessment. Unpublished technical report

Instituto Brasileiro de Geografia e Estatística - IBGE (2016). https://cidades.ibge.gov.br/. Accessed 2 Nov 2016

Karfunkel J, Aranha PRA, Büchi A, Pimenta F, Pagung R, Silva Pinto JA, Hofmann M, Hoppe A (2006) Registros neotectônicos entre Vespasiano e Pedro Leopoldo ao norte de Belo Horizonte, Minas Gerais. Paper presented at the 43th Congresso Brasileiro de Geologia, Aracaju, 3–8 Sept 2006

Kuchenbecker M, Babinski M, Pedrosa-Soares AC, Costa RD, Lopes-Silva L, Pimenta F (2013) Proveniência e análise sedimentar da porção basal do Grupo Bambuí em Arcos (MG). Geol USP Sér Cient 13(4):4–61

Kuchenbecker M, Reis HLS, Silva LC, Costa RD, Fragoso DGC, Knauer LG, Dussin IA, Pedrosa-Soares AC (2015) Age constraints for deposition and sedimentary provenance of Espinhaço

Supergroup and Bambui Group in Eastern São Francisco Craton. Geonomos 23(2):14–28. https://doi.org/10.18285/geonomos.v23i2.708

Magalhães L (1988) Análise Estrutural Qualitativa dos sedimentos do Grupo Bambuí, região da bacia do São Francisco (Faixa Sete Lagoas-Serra do Cipó). Dissertation, Universidade Federal de Ouro Preto

Martins M, Lemos VB (2007) Análise estratigráfica das sequências neoproterozoicas da Bacia do São Francisco. Revista Brasileira de Geociências 37(4):156–167

Paula-Santos GM, Babinski M, Kuchenbecker M, Caetano-Filho S, Trindade RI, Pedrosa-Soares AC (2015) New evidence of na Ediacaran age for the Bambuí Group in Southern São Francisco craton (eastern Brazil) from zircon U-Pb data and isotope chemostratigraphy. Gondwana Res 28(2):702–720

Peñaranda SJR (2016) Condicionamento estrutural e litológico da porosidade cárstica da Formação Sete Lagoas, Município de Sete Lagoas (MG). Dissertation, Universidade de São Paulo

Pessoa PFP (1996) Caracterização Hidrogeológica da Região Cárstica de Sete Lagoas, MG: Potencialidades e Riscos. Dissertation, Universidade de São Paulo

Pessoa PFP, Mourão MAA Levantamento Hidrogeológico (1998) In: CPRM – Companhia de Pesquisa de Recursos Minerais/IBAMA – Instituto Brasileiro do Meio Ambiente e dos Recursos Naturais Renováveis APA Carste de Lagoa Santa – Meio Físico/Volume I. CPRM/IBAMA Belo Horizonte p 1–39

Pessoa PFP (2005) Hidrogeologia dos Aquíferos Cársticos Cobertos de Lagoa Santa. Thesis, Universidade Federal de Minas Gerais, MG

Pessoa PFP, Auler AS (2015) Characterization and Dynamics of two Karst Springs in a Soil Covered Karst Area, Lagoa Santa, Southeastern Brazil. In: Andreo B et al (eds) Hydrogeological and environmental investigations in karst systems. Springer, Dordrecht

Piló LB (1998) Morfologia cárstica e materiais constituintes: dinâmica e evolução da depressão poligonal Macacos-Baú. PhD thesis, Universidade Federal de São Paulo

Pinto CP, Martins-Neto MA (eds) (2001) Bacia do São Francisco: Geologia e Recursos Naturais. Sociedade Brasileira de Geologia, Belo Horizonte

Rauch HW, White WB (1970) Lithologic controls on the development of solution porosity in carbonate aquifers. Water Resour Res 6:1175–1192

Reis HLS (2016) Neoproterozoic evolution of the São Francisco basin, SE Brazil: effects of tectonic inheritance on foreland sedimentation and deformation. Universidade Federal de Ouro Preto, Thesis

Reis HLS, Alkmin FF, Fonseca CS, Nascimento TC, Suss JF, Prevatti LD (2017) The São Francisco Basin. In: Alkmim FF, Cordani UG, Heilbron M (eds) São Francisco Craton, Eastern Brazil—Tectonic Genealogy of a Miniature Continent. Springer, Cham, pp 117–143

Ribeiro CG, Meireles CG, Lopes NHB, Arcos REC (2016) Levantamento Geológico Estrutural aplicado aos fluxos dos aquíferos cársticos-fissurais da região da APA Carste de Lagoa Santa, Minas Gerais. Final Graduation Report (?), Universidade Federal de Minas Gerais

Ribeiro JH, Tuller MP, Danderfer Filho A (2003) Mapeamento geológico da região de Sete lagoas, Pedro Leopoldo, Matozinhos, Lagoa Santa, Vespasiano, Capim Branco, Prudente de Morais, Confins e Funilândia, Minas Gerais (1:50.000) In: Projeto Vida. Companhia de Pesquisa de Recursos Minerais, Belo Horizonte

Sanchez EAM, Fairchild TR (2017) Reavaliação de fósseis do Grupo Bambuí: Implicações paleobiológicas para o neoproterozóico tardio do Brasil. Geonomos 25(2):1–11

Sawasato EY (1995) Estruturação da porção meridional da bacia Alto San-Fransciscana, Cretáceo do Oeste de Minas Gerais. Dissertation, Universidade Federal de Ouro Preto

Shinzato E (1998) O Carste da Área de Proteção Ambiental de Lagoa Santa (MG) e sua influência na formação dos solos. Universidade do Norte Fluminense, Thesis

Tayer TC (2016) Avaliação da vulnerabilidade intrínseca do aquífero cárstico da APA de Lagoa Santa, MG, utilizando o método COP. Dissertation, Universidade Federal de Minas Gerais

Teodoro MIP (2018) Interconexões e Hidrodinâmica do Sistema Aquífero Cárstico Bambuí, com uso de traçadores corantes, na região de Lagoa Santa, Minas Gerais. Dissertation, Universidade Federal de Minas Gerais

Tuller MP, Ribeiro JH, Signorelli N, Féboli WL, Pinho JMM (2010) Projeto Sete Lagoas—Abaeté (1:100.000). Convênio Secretária de Geologia, Mineração e Transformação Mineral/ Companhia de Pesquisa de Recursos Minerais - Serviço Geológico do Brasil, Belo Horizonte

Velásquez LNM, Vieira LCM, Teodoro MIP, Amaral GPA (2018) Projeto de Adequação e Implantação de uma Rede de Monitoramento de Águas Subterrâneas em Áreas com Cavidades Cársticas da Bacia do Rio São Francisco Aplicado à Área Piloto da APA Carste de Lagoa Santa. Minas Gerais - CPRM, IGAM, ICMBio, CDTN e CECAV

Veress M (2016) Covered Karsts. Springer Geology, Dordrecht

Vieira LC, Trindade RIF, Nogueira ACR, Ader M (2007) Identification of a Sturtian cap carbonate in the Neoproterozoic Sete Lagoas carbonate platform, Bambuí Group, Brazil. CR Geosci 339:240–258

Vieira LCM (2015) Características da Geologia e Geoquímica do Grupo Bambuí na APA Carste de Lagoa Santa, MG. Final Graduation Report (?), Universidade Federal de Minas Gerais

Vieira LCM (2018) Hidrogeoquímica dos Aquíferos da Região da APA Carste de Lagoa Santa, MG. Dissertation, Universidade Federal de Minas Gerais

Warren LV, Quaglio F, Riccomini C, Simões MG, Poiré DG, Strikis NM, Anelli LE, Strikis PC (2014) The puzzle assembled: Ediacaran guide fossil Cloudina reveals an old proto-gondwana seaway. Geology 42(5):391–394. https://doi.org/10.1130/G35304.1

White WB (1998) Geomorphology and hydrology of karst terrains. Oxford University Press, Thesis

Worthington SRH (1991) Karst hydrogeology of the Canadian rocky mountains. Mc Master University, Thesis

Caves and Speleogenesis in the Lagoa Santa Karst

Augusto S. Auler

Abstract

Nearly 1,400 caves have been registered in the Lagoa Santa Karst, the most-researched karst area in Brazil, as of December 2017. These caves developed in the Proterozoic carbonates of the Sete Lagoas Formation, mostly in the relatively pure upper Lagoa Santa Member. Much of our sample of 1,399 caves comprise short caves (<20 m in length), which were identified during detailed cave prospecting in environmental assessment studies for limestone quarrying projects. Due to the small area of the karst, which is limited by two rivers (the Velhas and Mata) that act as local base levels, and the limited noncarbonate areas, most of the caves are fed by autogenic (i.e., within the carbonate) water sources. The fragmented nature of the outcrops favors the development of mostly small caves; the longest one, Santo Antonio Cave, is 2.6 km long. As expected, the karst plateaus present the largest number of caves. Paragenetic caves are extremely common due to the availability of sediment and low hydraulic gradients, although syngenetic caves, mostly floodwater mazes, also exist. Floodwater mazes are favored by the existence of numerous sinkhole lakes located at the base of limestone cliffs. The smaller caves are usually fragmented sections of large caves, short vadose shafts, epikarst voids or, less frequently, talus caves. The Lagoa Santa Karst caves display abundant sedimentation and have been subjected to several paleoclimate-controlled episodes of sediment input and removal. Indurated mid to late Pleistocene sedimentation partially or totally fills many cave passages, some of which contain remarkable vertebrate remains. Chemical sedimentation is not frequent, with the exception of ubiquitous calcite layers that can cover significant portions of the cave passages. The ^{230}Th ages of the speleothems show that in some caves, older calcite precipitated before 650 kyr. The age of the initiation of speleogenesis is presently unknown but is likely to have occurred well within the Tertiary period, when groundwater flow started under the now largely removed metapelite cover.

Keywords

Lagoa Santa Karst • Speleogenesis • Syngenesis • Paragenesis • Floodwater maze

1 Introduction

The caves of the Lagoa Santa Karst have been used by humans since the early Holocene (Da-Glória et al. 2017), and since the late eighteenth and early nineteenth centuries, they have been subjected to intensive saltpeter excavations (Duarte 1899). In the early nineteenth century, Danish paleontologist Peter Wilhelm Lund brought the fossil content of these caves to the attention of an international audience (Lund 1837, 1841, 1845), establishing Lagoa Santa as a "scientific pilgrimage center" for many naturalists such as Burmeister, Reinhardt, and Liais, among many others (see chapter "History of Research in the Lagoa Santa Karst", this volume). Although Lund and Reinhardt provided an initial hypothesis on cave formation, the main focus of the cave research remained the rich fossil content. Over much of the twentieth century, these caves were researched for paleontological remains and later for archaeological findings (Walter 1948; Hurt 1960; Hurt and Blasi 1969; see a review in Da-Glória et al. 2017). Research that focused on the caves themselves, including their genesis and morphology, did not start until the late 1970s when a French scientific mission provided information on speleogenesis as part of a more comprehensive karst characterization project (Coutard et al. 1978). In the last decade of the twentieth Century, Auler (1995, 1998) discussed the speleogenesis, and Piló (1998) provided a link between cave sedimentation, cave genesis,

A. S. Auler (✉)
Instituto do Carste, Rua Barcelona 240/302, Belo Horizonte, MG 30360-260, Brazil
e-mail: aauler@gmail.com

A. S. Auler and P. Pessoa (eds.), *Lagoa Santa Karst: Brazil's Iconic Karst Region*, Cave and Karst Systems of the World, https://doi.org/10.1007/978-3-030-35940-9_9

and evolution in the small area associated with the Baú Cave. Auler (1999) provided a description of the cave's morphological features and how they relate to speleogenesis, aiming to produce a hypothesis of cave genesis related to paragenesis. Auler (1999) and Auler et al. (2009) discussed the role of paleoclimate cycles on cave sedimentation in the Lagoa Santa area. More detailed work on cave sedimentation was presented for Sumidouro Cave by Piló et al. (2005) and for Cuvieri Cave by Hubbe et al. (2011) and Haddad-Martim et al. (2017). Research by Galvão et al. (2016) and Peñaranda (2016) provided new insights into the permeability of the carbonate sequences. A summary of the karst and caves of the Lagoa Santa area was provided by Auler and Piló (2015a).

As of December 2017, the national official cave database (CANIE 2018) had registered 1,399 caves in the area (Fig. 1). Until approximately 2005, prior to the promulgation of cave-related laws that required detailed cave evaluations for mining projects (Auler and Piló 2015b), cave identification and surveys were performed by amateur caving clubs, which focused on the larger and more easily accessible caves. After that date, however, the bulk of cave studies was directed towards environmental assessment studies, which generally established a length of 5 m or even less as the lower limit for a void to be considered a cave (Auler and Piló 2015b; Auler 2016). This new approach resulted in a very large number of small caves (<20 m in length) being added to the database. The identification of caves in the Lagoa Santa Karst is quite exhaustive in areas owned by mining companies, mostly in the western portion of the area, while identification remains incomplete in the remaining areas.

This work will present a summary of the speleogenesis and the main morphological features of the Lagoa Santa Karst caves, including an overview of the cave initiation processes, the resulting morphology, and the chemical and clastic deposition.

2 Karst Geology and Porosity Development

Karst development in the Lagoa Santa area occurs in the two members (Pedro Leopoldo and Lagoa Santa) of the Sete Lagoas Formation of the Proterozoic Bambuí Group. Peñaranda (2016) presented data on the geochemistry of these two members (Table 1) at the northern limit of the area. The less-pure character of the lower Pedro Leopoldo Formation, with its higher silica content (for further analysis see also Piló (1998)), inhibits karst generation both at the surface and at the soil contact (epikarst) (Dutra et al. 1998). The same appears to hold true regarding accessible caves. Of the large dataset of the caves in the Lagoa Santa Karst, only 3% occur in the lower Pedro Leopoldo Member. However, well-developed conduit groundwater flow takes place in the contact between the two members, as attested by studies of the springs at the southern border of the area (Pessoa and Auler 2015).

Primary porosity is quite reduced due to the old age and marked recrystallization of the carbonate. Peñaranda (2016) reported values below 3% for both the Pedro Leopoldo and Lagoa Santa members, with pore sizes <0.03 mm. The primary porosity thus does not appear to have an important influence on speleogenesis in the area. Secondary porosity, represented by bedding planes and tectonic structures, represents the major speleogenetic control in the area. Bedding planes dip gently (<5°) towards the east (Auler 1994, 1999). This direction is generally coincident with the hydraulic gradient towards the Velhas River, which creates favorable conditions for speleogenesis along these geological horizons. Galvão et al. (2016), through the analysis of well logs in the Sete Lagoas area, points out the existence of two solution-enlarged bedding planes that occur over large areas: the upper one (now in the vadose zone due to groundwater withdrawal) is represented by a void up to 1–8 m high, and the lower one, located 10–20 m below, is between 20 cm and 1 m high. Field observations of outcrops in the Matozinhos and Pedro Leopoldo region also show that some bedding planes are preferentially enlarged, favoring the creation of an anastomotic network of channels (Fig. 2). The existence of favorable horizons aligned along the hydraulic gradient results in the strong directional control of conduit development. Subvertical joints are also important and exert control on many caves, especially the ones generated by lakes (Auler 1995). This control, however, tends to be limited to floodwater situations with reduced areal coverage.

3 Speleogenesis and Cave Morphology

The very long timespan since carbonate diagenesis (since c. 900 Myr) results in extra-long timescales for speleogenesis. Continuous denudation and the degree of fragmentation of the karst imply, however, that former ancient cave systems have already been removed from the landscape. Thus, present caves represent short and more recent snapshots in time of a much longer, ongoing process involving cave formation in the phreatic zone, including draining as well as removal from the landscape.

Evidence of initial processes of speleogenesis is widespread in the area and occurs as isolated cylindrical tubes, interpreted as fragmented protoconduits, and dissolution-enlarged bedding planes (Fig. 3). These protoconduits are seldom more than 15 cm in diameter and occur at various elevations, but they are sometimes concentrated at selected horizons. They are not to be confused with alveoles, also frequent, which are closed alveolar karren features

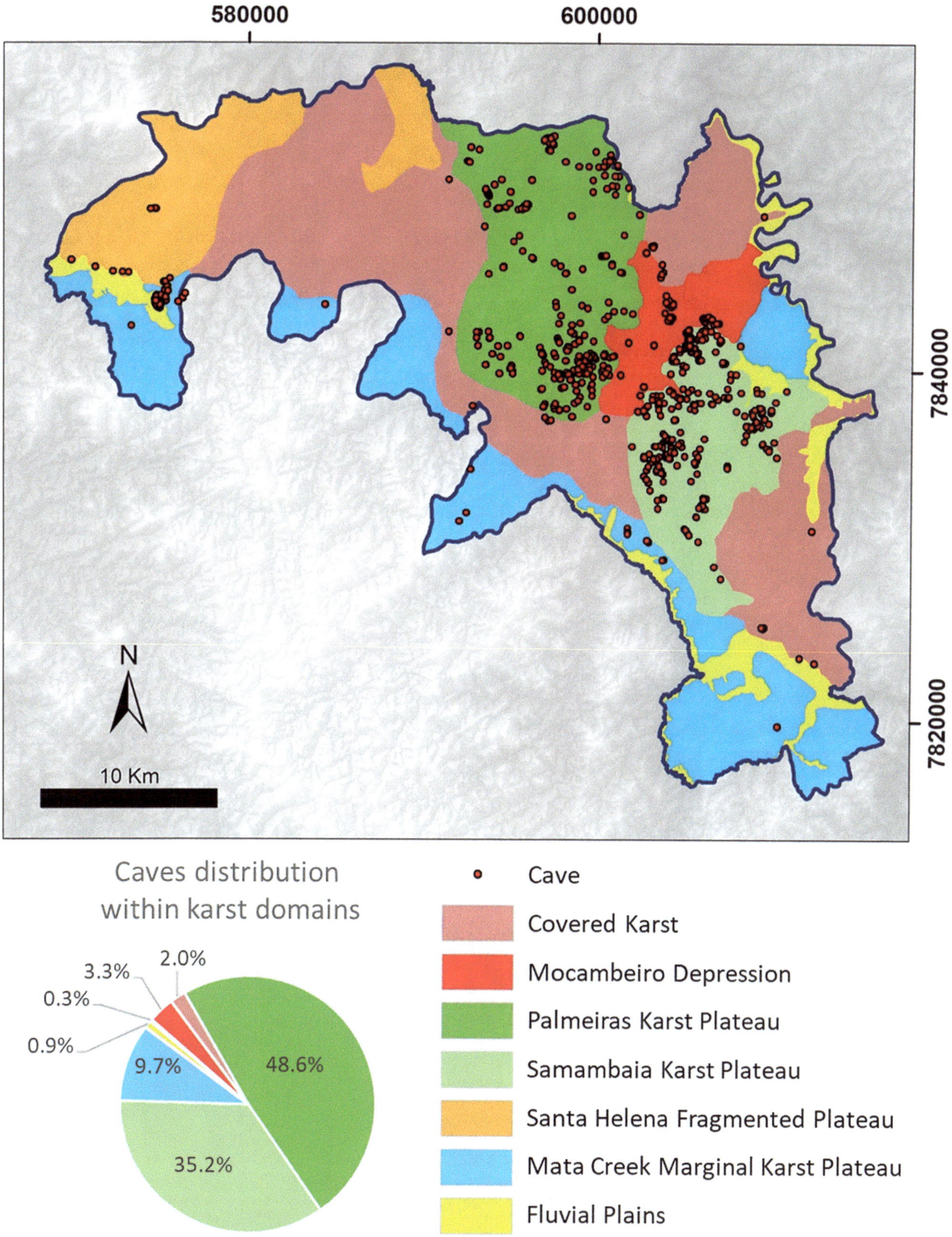

Fig. 1 Location of caves in the Lagoa Santa Karst in relation to karst domains. Due to the proximity of the caves, a single dot may represent several caves

Table 1 Geochemistry of the Pedro Leopoldo and Lagoa Santa members of the Sete Lagoas Formation. Data in percentage

Lagoa Santa Member					
Oxides	Mean	Max	Min	SD	CV
CaO	54.13	55.73	50.66	1.35	0.03
MgO	0.44	0.69	0.24	0.13	0.30
Al_2O_3	0.31	1.29	0.02	0.34	1.11
Fe_2O_3	0.19	0.60	0.03	0.17	0.87
MnO	0.01	0.03	0.00	0.01	0.60
SiO_2	1.58	5.90	0.08	1.56	0.99
Na_2O	0.04	0.12	0.01	0.04	0.87
K_2O	0.10	0.42	0.01	0.11	1.05
P_2O_5	0.12	0.33	0.06	0.05	0.45
TiO_2	0.02	0.07	0.00	0.02	1.10
Pedro Leopoldo Member					
CaO	44.90	55.19	37.11	4.24	0.09
MgO	3.83	10.35	0.41	2.58	0.67
Al_2O_3	1.62	2.72	0.09	0.57	0.35
Fe_2O_3	0.82	2.17	0.08	0.46	0.56
MnO	0.03	0.07	0.01	0.01	0.41
SiO_2	8.69	17.32	0.64	3.87	0.44
Na_2O	0.27	0.70	0.02	0.17	0.65
K_2O	0.34	0.61	0.03	0.18	0.54
P_2O_5	0.07	0.12	0.04	0.02	0.24
TiO_2	0.08	0.13	0.01	0.03	0.35

Data from Peñaranda (2016)
SD Standard deviation *CV* Coefficient of variation

Fig. 2 A subhorizontal bedding plane at Cuvieri Cave favors the development of anastomotic channels, with the largest passage (center) representing the cave entrance

Fig. 3 Outcrop in the Limeira area, showing numerous aborted protoconduits in the vicinity of a cave

(see chapter "Karst Landforms in the Lagoa Santa Area", this volume). Although there is no specific study addressing the spatial occurrence of these aborted protoconduits, their abundance is noteworthy and may indicate favorable speleogenetic conditions with the later exploitation of only a few preferred routes.

Bedding planes are expanded by subsoil and surface processes, generating joint karren. These enlarged bedding also play the role of initiation horizons. Dissolution-enlarged planes are observed at outcrops and quarry faces and sometimes represent the initial horizon for paragenetic cave development (see section on paragenesis). They are laterally extensive and can be traced over several meters in outcrops. Most caves in the area show initial control by bedding planes, especially anastomotic paragenetic caves as well as meandering syngenetic caves, such as Taboa Cave (Peñaranda 2016). An example of a cave developed at the contact between the carbonate and the crystalline basement is found at Irmãos Piriás Cave, where an active river passage runs alongside the contact.

Caves tend to be concentrated in the karst plateaus (Fig. 1), where outcrops and karst features are more developed. A list of the major caves in the Lagoa Santa Karst (in terms of length, depth, area, and volume) is presented in Table 2. A statistical analysis of the major parameters (length, depth, area, and volume) showed that the sample set is dominated by smaller caves (Fig. 4). The relatively small size of the caves in the area (Table 3) is due to the small area of the karst (implying relatively short groundwater routes), the highly fragmented nature of the outcrops and the dominant autogenic diffuse recharge. Furthermore, there is a sampling bias towards identifying every enterable cave during environmental assessment studies, distorting the analysis towards lower values. Interestingly, no specific speleogenetic mechanism is responsible for the large caves, since the three longest ones were generated mostly by distinct processes (Santo Antonio Cave—syngenetic, Escadas Cave—paragenetic and Lapa Vermelha I Cave—floodwater maze). Classic syngenetic caves appear to be more common in the western and eastern portions of the Lagoa Santa Karst, probably related to the existence of streams reaching the limestone from noncarbonate areas. In these areas, there is no concordance between the direction of the bedding planes and the hydraulic gradient. Floodwater mazes are particularly conspicuous in higher recharge areas in the Samambaia and Palmeiras karst plateaus and are associated with well-developed dolines at the base of limestone cliffs. Paragenetic caves occur throughout the area. The general direction of the caves follows approximately E-W, N-S joints (Dutra et al. 1998). In areas controlled by the major base level, the Velhas River, the trend in the nonnetwork caves tends to be towards the east. Isolated caves, such as those at the limits of the area (Taboa, Cascata or Irmãos Piriás caves), have orientations controlled by local base levels.

The depth distribution (Fig. 4) is strongly controlled by the relatively shallow nature of the karst. The proximity to the crystalline basement in the western portion of the area

Table 2 List of major caves in the Lagoa Santa Karst and their length, depth, area, and volume

Cave	Length (m)	Cave	Area (m^2)
Santo Antônio	2,615	Rei do Mato	21,160
Vermelha I	1,870	Santo Antônio	16,589
Escada	1,828	Vermelha I	10,755
Lemniskos	1,394	Taboa	6,027
Túneis	1,250	Boi	5,761
Cerca Grande	1,055	Buraco do Medo	5,226
Rei do Mato	998	Baú	4,708
Morro Redondo	960	Irmãos Piriás	3,739
Irmãos Piriás	898	Túneis	3,360
Gruta do Baú	885	Milagres	3,275
Cave	Depth (m)	Cave	Volume (m^3)
Morro Redondo	75	Rei do Mato	193,248
Encanação	68	Santo Antônio	113,365
Santo Antônio	66	Vermelha I	81,465
Estudantes	63	Buraco do Medo	28,025
Caveira	59	Baú	25,837
Pia	54	Milagres	25,703
Irmãos Piriás	50	Irmãos Piriás	25,549
Tombo	47	Boi	24,837
Timbau	45	Taboa	22,851
Lagoa Bonita	44	Lemniskos	21,584

and the low thickness of the limestone above the water table (seldom over 100 m) considerably limit the occurrence of deep caves.

3.1 Paragenesis

Paragenetic caves occur throughout the area and represent the most common cave-forming process. Kruger (1965) and Coutard et al. (1978) were the first to describe paragenesis as a significant mechanism of cave genesis in the Lagoa Santa Karst. Paragenetic caves usually occur as truncated (by outcrop fragmentation or doline deepening) segments of once-larger caves that may have extended for over one kilometer but that are now represented by small caves. The most common cave pattern is the anastomotic maze (Auler 1999). Two long representative caves, Escadas and Baú (Fig. 5), are segments of larger caves exposed at a cliff face (Escadas) or restricted to an isolated outcrop (Baú). The passage morphology varies but usually shows a triangular shape; heights reach up to 15 m but are usually less than 10 m, with the ratio of the height/width of the passages usually above 3 (Auler 1999). The floor is nearly always covered by sediment. The thickness of the sediment, where observed (for example, at Lavoura Cave), shows that the exposed (above the floor) portion is just a fraction of a much larger passage, presently mostly sediment-filled. Meandering canyons also occur, such as in Lemniskos Cave, in which the meander migration vectors, following Lauritzen and Lauritsen (1995), demonstrate paragenetic development. Other paragenetic features, such as pendants, parasitic wall tubes, and ascending meander junctions (Auler 1999) are common (Fig. 6). An interesting feature of the Lagoa Santa paragenetic caves is the absence of scallops in the cave walls (Fig. 7). The "smooth" character of the cave walls has been noted since the nineteenth century (Lund 1841; Reinhardt 1867). Two explanations were put forward by Auler (1999): (i) obliteration of scallops due to later events of sediment input (Auler et al. 2009) or (ii) slow-moving water flow, producing scallop lengths that blend with the cave sinuosity and become indistinguishable from the cave meander length.

At these caves, the ceiling meanders show no guiding joint at the ceiling level and cut across the bedding planes (Figs. 6d and 7), suggesting that they did not initiate at the ceiling (Auler 1999). Observations at quarry faces and in outcrops show that some major bedding planes play the role of a favorable initiation horizon and that competing paragenetic channels grow upwards; some develop faster and form higher passages. The availability of soil coupled with the low hydraulic gradients results in groundwater flow velocities that are too slow to entrain sediments, favoring paragenesis. Paleoclimate-controlled cycles of sediment

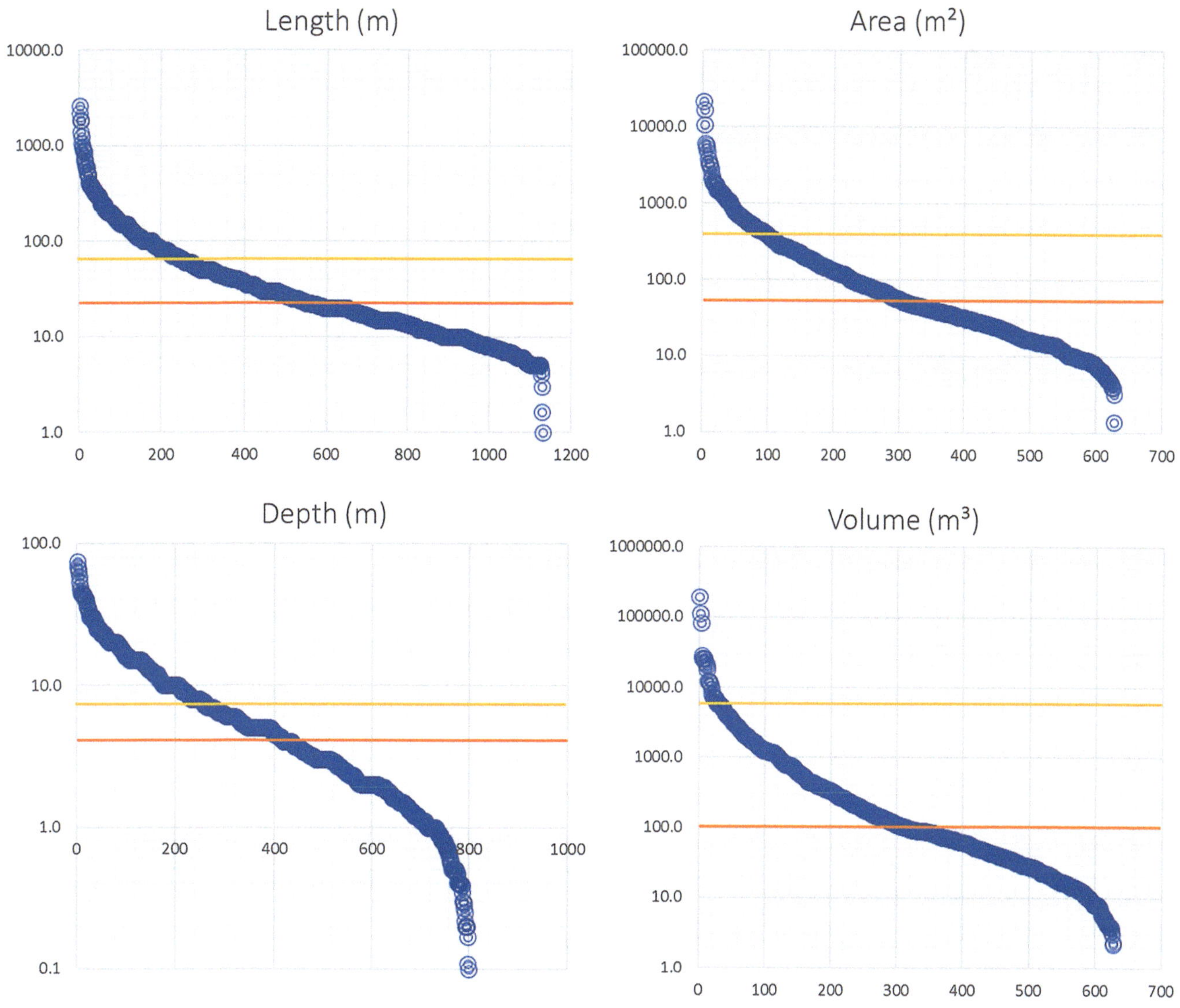

Fig. 4 Distribution of caves in the Lagoa Santa Karst in terms of length, depth, area, and volume. Red line = median, yellow line = mean (see Table 3 for details)

Table 3 Sample size, mean and median of the length, depth, area, and volume of the Lagoa Santa Karst caves. The sample size varies because some of the parameters (especially area and volume) are calculated only during professional environmental studies. The low median value shows the dominance of small caves

Parameter	Sample size	Median (m)	Mean (m)
Length	1129	22.1	66.4
Depth	861	4.0	7.3
Area	628	49.1	319.7
Volume	628	100.7	1648.4

input and erosion (Auler et al. 2009) may be responsible for later paragenetic processes, resulting in the expansion of passages or the generation of localized paragenetic features. At Morro Redondo Cave, the upper level comprises a floodwater maze that was, at a later stage, filled with sediment. Paragenesis took place and generated a large paragenetic meander that cuts across the network maze (Fig. 6d). Anastomoses and pendants located at the midlevel of the

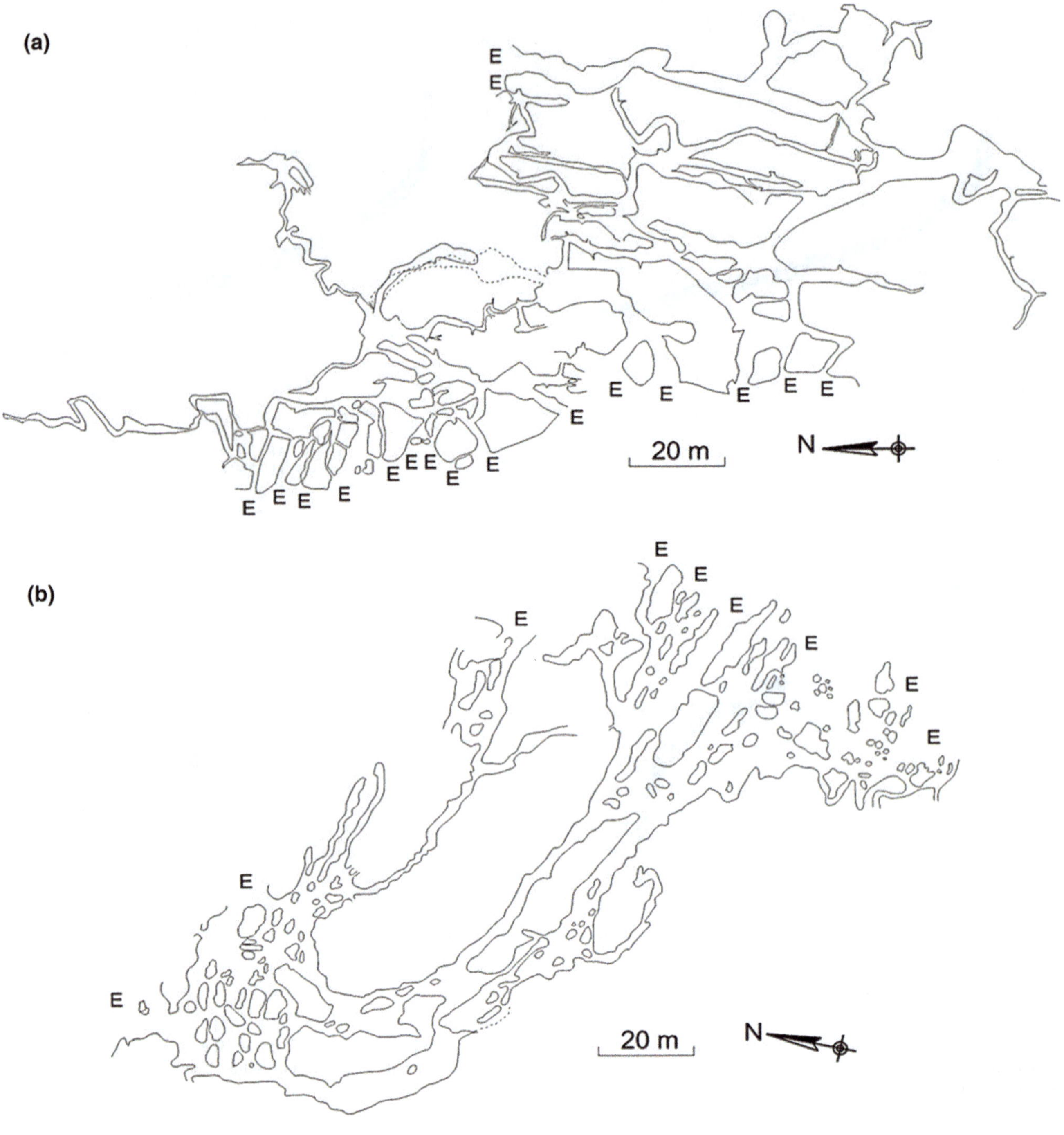

Fig. 5 Plan map of **a** Escadas and **b** Baú caves. E—Entrance. Map of Escadas Cave by Sociedade Excursionista e Espeleológica. Map of Baú Cave from Piló (1998)

Fig. 6 Paragenetic features. **a** and **b** Pendants in caves in Sete Lagoas and Pedro Leopoldo, respectively. **c** Wall grooves at Santo Cave. **d** Paragenetic meander at Morro Redondo Cave. **e** Ceiling half tube at Ouro Cave

Fig. 7 The ceiling of this passage in Baú Cave shows no control by guiding joints. Fluvial scallops are absent, but vertical wall grooves can be observed

passages may be the result of paragenetic processes in situations where new sediment input did not reach the ceiling, such as in Túneis Cave.

3.2 Syngenesis

Two main types of syngenetic caves occur in the area: floodwater mazes and dendritic river caves. Many of the major caves in the area are syngenetic. Syngenetic caves occur in specific geomorphic conditions in which sediment input is restricted or flow velocities are beyond the entrainment threshold. Such situations take place in two hydrodynamic extremes: slow-moving water (generating floodwater mazes) and fast-flowing streams and/or limited sediment supply (forming syngenetic river caves).

3.2.1 Floodwater Mazes

Auler (1995) described the characteristic floodwater network mazes in the area. The common occurrence of this style of cave is due to the existence of lakes associated with dolines at the base of limestone cliffs. These lakes show irregular level variations, probably related to the heterogeneous sealing of the sinking point by impermeable sediment (Auler 1995). Water that feeds the lake usually comes from rainfall and runoff from the doline slope. In some situations, such as in the Lapa Vermelha and Caetano cliffs, water discharging from a spring can be observed during low flow situations inside the lake perimeter. Water levels vary considerably, and presently few of these lakes are perennial. Some dolines, such as Escrivânia, may remain dry for several years, while others, such as Lapa Vermelha (Fig. 8), although holding water for several months, display network passages well above the maximum present level. Floodwater speleogenesis appears to be a largely inactive speleogenetic process at present, and relict network mazes can serve as indicators of former lake basins (Auler 1995). Considering the highly seasonal climate of the area, unsaturated water would feed these lakes during the wet season (austral summer), favoring episodic speleogenesis. Larger caves may bear relation to more stable/prolonged lakes.

The enlargement of single joints may result in a series of rectilinear passages, such as in Escrivânia (Fig. 9a). These joints may represent single passages or join as a network, as observed in Lapa Vermelha I Cave (Fig. 9b). In this cave, a syngenetically large meandering passage was subject to floods, creating a lateral network maze. An additional example of an ancient floodwater network maze, now well above the water table and isolated in a residual hill, is the Morro Redondo Cave upper level.

Floodwater mazes are characterized by the presence of nondirectional scallops, associated with existing or former lakes. These scallops are usually large in dimension, denoting slow-moving water without a defined direction, and tend to occur throughout the caves as well as in cliffs subject to flooding (Fig. 10). Passages in floodwater mazes comprise rectilinear rifts crossing at approximately 90° angles. They display a cross-section narrowing upwards and tend to pinch out away from the source of water.

3.2.2 Dendritic River Caves

This style of cave is not very common in the karst plateaus, where paragenetic caves predominate but instead tends to occur at the borders of the karst area. Perhaps the best

Fig. 8 Lake at the entrance of Lapa Vermelha Cave I, the longest floodwater maze in the Lagoa Santa Karst

example is in the Santa Helena Ridge, where a small stream issuing from metapelites generates two dendritic river caves, Cascata I and Cascata II. Another example is the bedding plane-controlled Taboa Cave (Peñaranda 2016). A keyhole cross-section showing the transition between phreatic and vadose conditions can be observed in some caves (Fig. 11). These caves contain small perennial drainages that meander in stream beds with gravel. Tributaries are not frequent but do occur in most of the caves. The drainages may represent perched streams, such as the dry downstream passage of Taboa Cave, which lies approximately 6 m above the present level of the spring. Upper levels occur in the Cascata II, Buraco do Medo, and Irmãos Piriás caves and are represented by large chambers in the two latter caves. It is possible that some of the caves presently recognized as syngenetic are, in fact, former paragenetic caves later excavated by a syngenetic drainage. The existence of scallops in the lower portion of the cave walls that are distinct from the scallops that occur in the higher wall sections indicates a change in the hydrological regime. The masking or obliteration of the original speleogens may prevent the recognition of the original speleogenetic process.

Due to the characteristic fragmentation of the Lagoa Santa Karst, these caves tend to represent short tunnels, such as the sequence of caves generated by Palmeiras Creek. In these caves, there are sometimes distinct former upper dry routes located approximately above the active ones. Some of the caves end in sumps or narrow passages or are plugged with sediment.

3.2.3 Vadose Shafts, Fragmentary Caves, Epikarst Caves, and Talus Caves

The Lagoa Santa Karst contains a very large number of caves that do not clearly fit into any of the above-mentioned speleogenetic types. The majority are represented by small caves (see Fig. 4) identified through detailed environmental consulting work. These caves fall into four general categories: (i) vadose shafts; (ii) fragmentary caves that represent shorter sections of once much longer caves; (iii) epikarst caves, and (iv) talus caves.

Vadose shafts occur mostly in the karst plateaus and are associated with doline bottoms or deeper portions of grykes/corridors in the inner sectors of outcrops. Most of them are related to the dissolutional action of runoff from the surface. These vertical caves may contain a few vertical sections and inclined passages and are usually not deeper than a few tens of meters. The deepest dissolutional shaft in the area is the 53 m-deep pit in Morro Redondo Cave that connects the upper network/paragenetic section with the lower fluvial dendritic passage. Shafts generated by collapse may also occur, and the best example is the 44 m-deep pit at the entrance of Estudantes Cave, which provides the only access to a horizontal passage.

Fragmentary caves are extremely common in most outcrops and can occur as clustered caves. An analysis of 10 detailed cave prospecting studies in the Lagoa Santa Karst showed an average of 33 caves/km^2, with some areas presenting as many as 234 caves/km^2 (Auler 2019). This very high density includes mostly fragmentary caves but also a

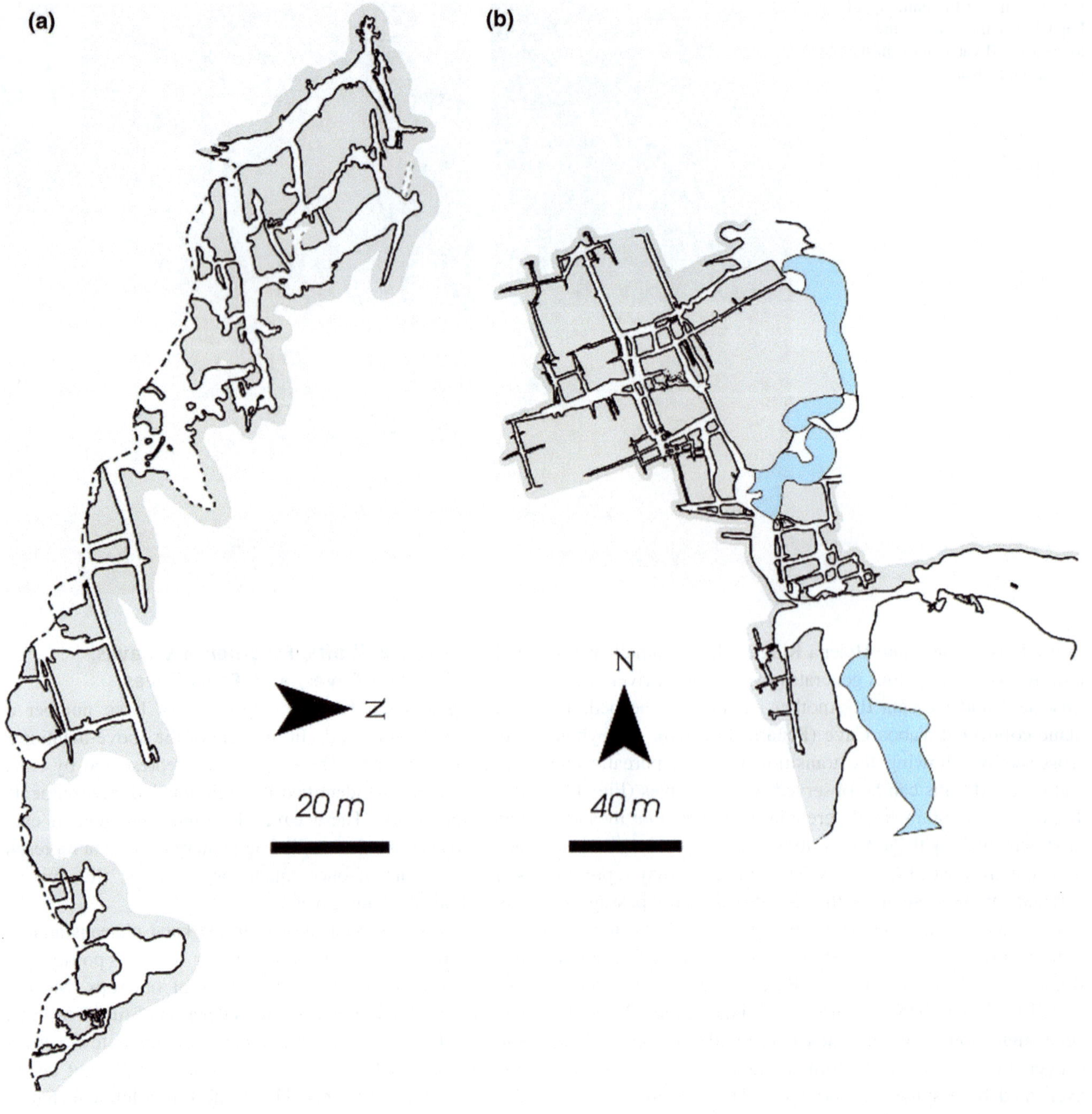

Fig. 9 Plan map of Escrivânia Cave (**a**) and Lapa Vermelha I Cave (**b**). Maps courtesy, respectively, of the Meandros Espeleo Clube and Grupo Bambuí de Pesquisas Espeleológicas

considerable number of epikarst caves. Most fragmentary caves are short sections of paragenetic mazes, usually plugged by speleothems or sediments or simply exposed as short tunnels due to outcrop fragmentation. The Lapinha outcrop displays both large caves, such as the paragenetic Túneis Cave (1.2 km long), and dozens of smallish caves a few meters long, as noted by Knez et al. (2012). As pointed out by Auler (1998), some of these caves are not related to present local base levels but represent former (now relict) routes of groundwater flow. A special type of fragmentary cave is represented by entranceless caves exposed by quarry blasting. A well-known case is represented by Cristais Cave, a short rift with walls lined with calcite crystals that were partially exposed in the mid-twentieth century.

Epikarst cave is a generic term to designate short sections of enlarged joints, vertical rifts or expanded joint karren or

Fig. 10 Nondirectional scallops in a cliff near the entrance of Vargem Cave. Water level mark (lighter color at middle of the cliff) is about 1.5 m above the floor

Fig. 11 Keyhole cross-sections in **a** a cave in Pedro Leopoldo and **b** a cave in Prudente de Morais

alveoles that probably evolved initially under soil cover. As denudation progresses, these caves are exposed and may reach the required official length criteria to be classified as a cave by the Brazilian government. Rarely, these caves surpass a few meters in length and tend to be devoid of significance. Shapes vary but are controlled by either bedding planes or joints.

Talus caves are not common in the Lagoa Santa Karst but are observed in a few situations. They tend to be associated with the inner portions of outcrops, where the interconnected network of towers and corridors favor breakdown. They are usually small and have been identified during environmental consulting work.

4 Cave Sediments

Since the late eighteenth century, cave sediments have been a magnet for attracting miners and scientists to the Lagoa Santa area. Initially, saltpeter diggers worked in hundreds of caves, removing vast amounts of sediments, and from 1835 onwards, the Dane Peter Wilhelm Lund drew international attention to the remarkable fossil content of the caves.

The abundant sedimentation intrigued most of the early scientists. Peter Lund went to great lengths to explain the mechanisms of sediment input and calcite precipitation and how they relate to the fossils. In one of his final papers, Lund (1845) stressed that as the caves became enlarged, sediments from the soil above would fill the caves; later, runoff from streams would have the sediments removed. These two processes, according to Lund, could occur simultaneously. Additionally, the extensive calcite layers that, according to Lund (1841), constituted "a carpet that marks the limit separating the past and the present" were used as a temporal marker and thus were considered crucial for understanding the bones' antiquity. Lund (1845) recognized the complexity of the sediment deposits, in which an inverted stratigraphy could occur, placing older sediment above younger ones. Claussen (1841), Reinhardt (1867), Liais (1872) and many twentieth century scientists also commented on the sediments. A more quantitative approach, with geochemical and geochronological analyses, was not applied until the late twentieth century (Piló 1998; Auler 1999), and detailed studies about cave sedimentation have now been performed (Piló et al. 2005; Auler et al. 2006, 2009; Hubbe et al. 2011; Haddad-Martim et al. 2017). Processes related to clastic and chemical sedimentation show strong paleoclimate control, while fossil emplacement is taphonomically controlled. Those processes are closely interrelated with each other and should be addressed by taking into consideration the dynamics of the surface and the cave environment.

4.1 Clastic Sediments

Most caves of the Lagoa Santa Karst display a sediment-covered floor, and the original bedrock floor is rarely observed. Remnants of sediment can also be observed attached to the ceilings or preserved in pockets higher in the passages (Fig. 12a), suggesting that the caves were totally or partially filled with sediment in the past. Auler et al. (2009) list some common characteristics of older cave sediments: they normally have a very indurated nature, commonly with calcite overgrowth within pores and voids (Fig. 12b); they are predominantly fine-grained but sometimes have a brecciated nature; they are commonly associated with the presence of fragmented fossil bones; they lack a laminated structure denoting the limited influence of static water; and they have a yellowish/reddish color and are frequently associated with calcite layers.

Although sediments can be brought from the surface and bear resemblance to the soil, sediment emplacement can also

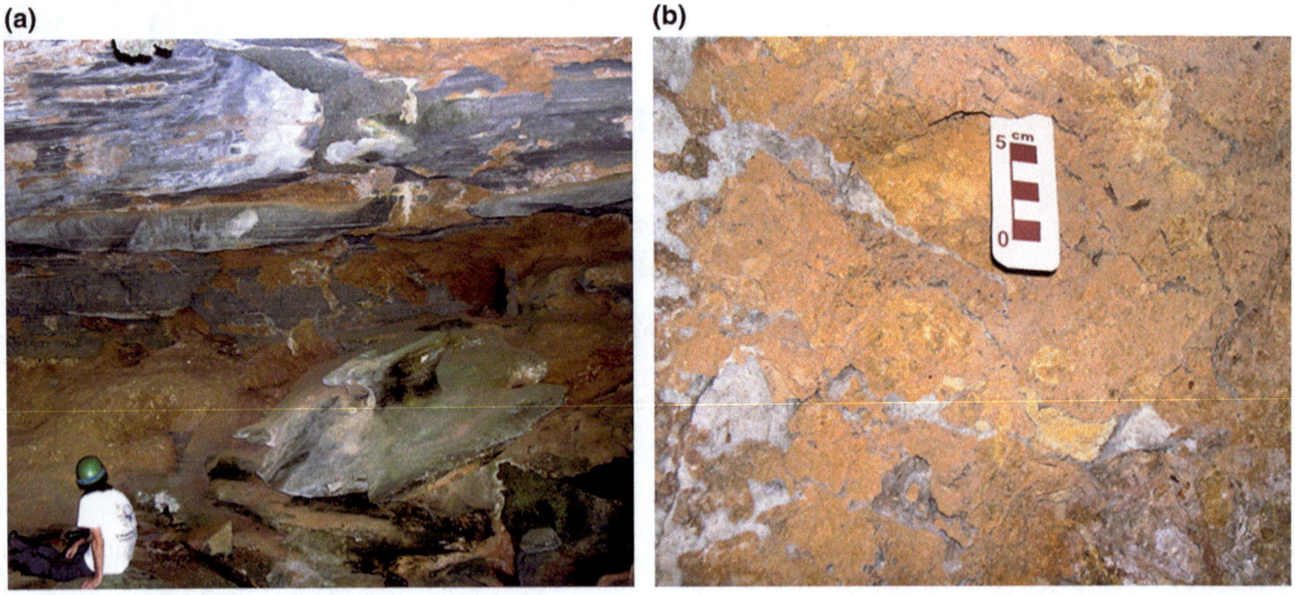

Fig. 12 **a** Remnants of sediment attached to the wall and ceiling in a cave in the Cochos area. **b** Indurated clastic sediment attached to a cave in the Palmeiras Karst Plateau, showing intercalation with calcite

occur through streams and from doline slopes and can result from human occupation. The latter type can be exemplified by Santo Cave, a large rock shelter with a widespread archaeological occupation, in which the sediments are mostly comprised of ash of human origin (Villagran et al. 2017).

Cave sediments are subject to a complex post-depositional history, including disturbance by water action, later infilling and removal, and precipitation or accumulation of chemical compounds. The high intersite variation that commonly occurs in caves (Campy and Chaline 1993) results in a highly complex assemblage that normally cannot be correlated between caves or even between sectors within the same cave (Auler et al. 2006). In many caves and outcrops, the brecciated and very indurated sediment fragments that fill higher pockets have been shown to be considerably old (>300 kyr) (see section on speleothems). Haddad-Martim et al. (2017) attribute an epikarst origin for this older sediment.

In addition to archaeological research at cave entrances, detailed studies on sediments in the internal passages of specific caves have been performed at Sumidouro Cave (Lund 1845; Piló et al. 2005) and Cuvieri Cave (Hubbe et al. 2011; Haddad-Martim et al. 2017). In the former cave, sediments are related to a nearby swallet and to vertical input from fissures in the ceiling (Piló et al. 2005). In Cuvieri Cave, older accumulations are related to the influx from the doline slope associated with the horizontal entrance, while more recent sediment originated from the ceiling (Hubbe et al. 2011). The fossil content in each of these situations varies considerably because not all processes responsible for sediment accumulation favor the emplacement of fossils. It appears, thus, that the sediment characteristics are controlled by several factors, such as the influence of water, the existence and position of openings (at ceiling level or associated with horizontal entrances) and the geomorphological situation (entrance associated with a doline slope, with swallets, at base of shafts, etc.). These highly variable controls do not allow for the establishment of an all-encompassing model of cave sedimentation for the area. Further control relates to paleoclimate conditions. For doline-fed sediments, Auler et al. (2009) developed a model in which the three main processes, i.e., sediment input, sediment erosion, and speleothem deposition, are correlated to paleoclimate conditions. Using semiarid northeastern Brazil, an area in which sediment input is actively occurring, as a model, sediment input in the Lagoa Santa Karst was interpreted as being associated with dry conditions, in which the lack of vegetation on the slopes favors sediment erosion and transport towards the caves (Auler et al. 2009). Sediment removal, a situation that prevails at present, was linked with intermediate conditions, in which vegetation on the slopes prevents erosion and runoff removes previously emplaced sediments (Auler et al. 2009), while speleothem deposition was unequivocally linked to wetter conditions (Auler et al. 2009).

Another mechanism responsible for clastic sedimentation is the partial collapse of the cave ceiling. The crystalline and thick-bedded nature of the limestone does not favor ceiling collapse, although it occurs in many caves and is usually represented by large tabular blocks controlled by bedding planes. Another type of cave sediment typical of the area, cave saltpeter, was an important economic asset before the focus shifted to sediment fossil content. Many caves were extensively excavated, and some were virtually emptied of their sediment for saltpeter extraction. There is no modern research on cave saltpeter, but Lund (1841) claimed that the saltpeter was due to the migration of chemical compounds that originated from the surface soil and later concentrated in the cave sediment, an explanation that would be adopted 150 years later (Hill and Forti 1997).

4.2 Vertebrate Fossil Deposits

The importance of the fossils found within the cave sediments was recognized by Peter Lund, who made pioneering contributions related mainly to the taxonomy and mode of emplacement (Lund 1845, 1846). The number of vertebrate remains varies widely between caves, and in only approximately 60 out of 1,000 caves was Lund able to find fossil bones (Reinhardt 1888). This number may be related to the high degree of saltpeter extraction, which removed and destroyed several fossil cave sites. Auler et al. (2006) presented a review of the chronology and processes of fossil emplacement at the Lagoa Santa caves, and Piló et al. (2005) attempted to reinterpret the heavily excavated deposits at Sumidouro Cave. Controlled excavations have been performed during archaeological studies at cave entrances, and only at Cuvieri Cave has a thorough study of the chronology and taphonomy of fossil deposits been performed (Hubbe et al. 2011; Haddad-Martim et al. 2017).

Fossils can accumulate in the local caves by at least five different mechanisms, according to Lund (1845): (i) animals that fall into vertical caves or vertical sections inside caves, the most common mechanism according to Lund; (ii) animals that enter the caves in search of water or saltpeter and cannot exit; (iii) animals carried into the caves by predators; (iv) animals that use caves regularly during their lives, such as bats and (v) fossils brought into the caves by running water. These distinct mechanisms can and do occur together and exert taphonomical control on the number, timing, and type of fossils. Thus, the fauna found in cave deposits does not necessarily mirror the surface vertebrate fauna (Auler et al. 2006).

The ages of fossil accumulation have been determined by both direct radiocarbon dating in bones and by U-series dating in associated calcite layers (see review in Neves and Piló 2003; Auler et al. 2006). Minimum ages are spread from the mid-Pleistocene (360 kyr) to the early Holocene. The youngest ages (9.1 kyr on Smilodon remains) suggest contemporaneity between the extinct fauna and early Americans (Neves and Piló 2003). The large spread of ages confirms that the fossil deposits are neither contemporaneous nor represent a single episode related to the Pleistocene/Holocene transition, as previously believed by Cartelle (1999).

4.3 Speleothems

Caves in the Lagoa Santa Karst are not known for their diversity and abundance of speleothems. The various cycles of sediment input and removal (Auler et al. 2009) may have prevented many of the cave passages from depositing or preserving speleothems. The main reason, however, appears to be related to the paragenesis model of speleogenesis, in which there is no guiding joint at the cave ceilings, not allowing for water infiltration and speleothem precipitation. Furthermore, the seasonal climate, with rainfall concentrated in the few austral summer months, results in annual floods in some caves.

Despite the general absence of chemical deposits, a few caves stand out. Rei do Mato Cave presents a series of long cylindrical stalagmites up to 8 m high. This cave is also profusely decorated and contains helictites. Helictites Cave at the Lapinha outcrop is known for its beautiful helictites. Salão Brilhante Cave is a small cave filled with very pure crystalline flowstone and helictites; it was formerly a closed chamber exposed by blasting. The same happened at Cristais Cave, a short enlarged rift also exposed by quarry blasting, where both walls are lined with subaqueous calcite spar. The network maze of Lapa Vermelha Cave also contains a rich variety of speleothems, mostly coralloids. Taboa Cave presents an impressive sequence of rimstone dams, some over 5 m in height. It is interesting to note that more common speleothems, such as stalactites and flowstones, are largely absent from the Lagoa Santa Karst caves, an effect probably related to the abovementioned lack of joints at the ceiling level. The mineralogy of the Lagoa Santa caves speleothems has not been studied in detail, but the clear majority appear to comprise calcite, with minor aragonite. Sulfates are largely absent.

The most characteristic speleothem type in the Lagoa Santa Karst is the calcite layers (Fig. 13). These layers occur within sediment sequences and can extend for considerable lengths, covering areas of hundreds of square meters. Their thicknesses can exceed 50 cm but are usually less than 10 cm (Auler et al. 2009). The speleothem layers attracted the attention of the early cave scientists because they represented an easy-to-recognize stratigraphic marker by which to separate old and recent faunal remains (Lund 1841). The abundance of speleothem layers and the absence of a dripping source intrigued Lund (1841), who hypothesized that a faster dripping rate would not allow the precipitation of calcite at the source point but instead would favor runoff and deposition over large areas. Auler (1999) and Auler et al. (2009) recognized three different mechanisms that can promote calcite layer precipitation: (i) precipitation from bicarbonate-rich percolation water that flows laterally and precipitates over sediment from a few input points in the walls or ceiling, as suggested by Lund (1841); (ii) per ascensum formation through evaporative concentration at the upper horizons of sediment sequences, resulting in non-stratified, detrital-rich layers that show an increased calcite content towards the top and (iii) formation within sediment, generating thin, irregular calcite layers, or by the filling of voids (small geodes), resulting in an irregular array of layers within the sediment. These three processes are not mutually exclusive and can occur together, rendering considerable complexity to the morphology and genesis of the calcite layers. Another type of calcite layer occurs over the sediment at the contact with the ceiling. It comprises acicular calcite following the contours of the cave ceiling, as observed in Ballet Cave (Fig. 13d). Its genesis has not been studied but may involve both per ascensum and per descensum processes.

Several U-series dates have been produced from the Lagoa Santa Karst speleothems, and most have been reported in Piló et al. (2005), Auler et al. (2006) and Auler et al. (2009). Figure 14 presents a synthesis of the 72 U-series ages available for the area. Speleothem deposition at present is largely absent.

5 Age and Evolution of Caves

Early authors speculated about the age of draining (removal from the phreatic zone) of the Lagoa Santa Karst caves, but without the benefit of absolute ages. King (1956) assigned cave draining to the Pleistocene, while Kohler (1989) placed both the initiation of surface karstification and cave draining in the mid-Pleistocene. Constraining the age of the Lagoa Santa Karst caves is difficult due to the lack of dates that reach beyond the Pleistocene. Several ^{230}Th dates (see Fig. 14) on calcite layers are beyond the ^{230}Th dating method, i.e., older than 650 kyr. This is the case for Ballet Cave, where nearly all calcite dates are older than 650 kyr. Morro Redondo Cave is at least 75 m above the local water table. If the surface lowering rates of 30 $\pm$ 10 m/Myr reported for the São Francisco craton by Harman et al. (1998) are applied to the above cave, minimum ages of draining between 3.7 and 1.9 Myr are obtained (Auler 1999). These ages demonstrate that cave development

Fig. 13 Calcite layers in the Lagoa Santa Karst. **a** and **b**, respectively, in a cave in Sete Lagoas and in Macumba Cave. **c** Detail of a calcite layer showing terrigenous intercalations in Salão Brilhante Cave. **d** Calcite layer over sediment (now removed) along the ceiling in Ballet Cave

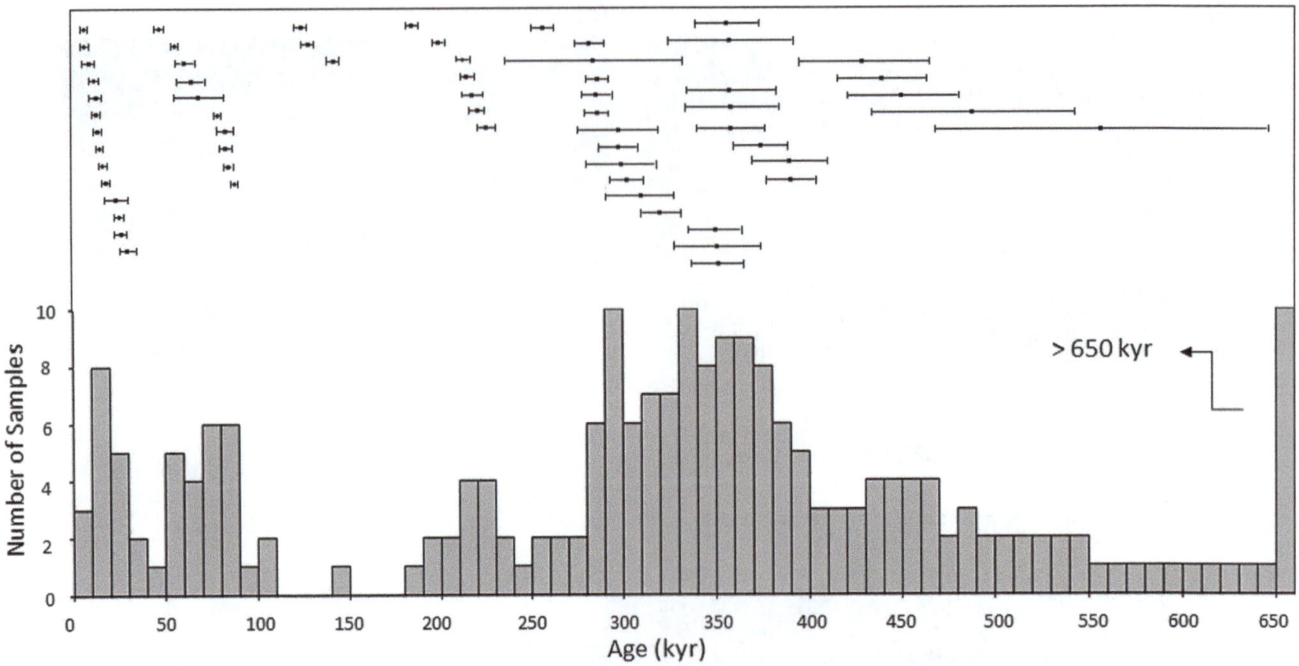

Fig. 14 Frequency plot of calcite U-series ages (including errors) for the Lagoa Santa Karst in 10,000 yrs intervals

started well within the Tertiary, and the beginning of cave draining and karstification occurred prior to the Pleistocene. The antiquity of the Lagoa Santa caves is further demonstrated by the fact that many caves are now mere remnants in karst towers or represent decoupled hydrologic routes that do not bear any relationship with current groundwater flow paths (Auler 1998).

As regional uplift progresses and doline floor deepening takes place, former cave entrances previously connected to doline bottoms become isolated at higher elevations and are no longer subject to water or sediment input. These "hanging entrances" are commonly observed in cliffs. Coutard et al. (1978) claimed that there are two levels of cave passages visible high above cliffs, such as in Cerca Grande. These cave levels (interpreted as former swallets) occur at 30–35 m and 12–15 m above the main base level, the Velhas River (Coutard et al. 1978). The higher passages were tentatively correlated with the Velhas River fluvial terraces. However, this interesting hypothesis has not yet been quantitatively tested. The analysis of cave scallops in several caves led Auler (1998) to identify past flow directions that are distinct from those at present, demonstrating that there were base level changes in the two major drainage basins, the Palmeiras/Mocambo and Samambaia. The timescales for such changes, however, are not known.

The evolution of the caves is closely related to the evolution of the karst. The short distances between the recharge zone and the base levels do not allow the development of extensive phreatic systems. As outcrop fragmentation proceeds and uplift elevates the formerly phreatic caves above the water table, the caves will tend to become segmented. The higher the degree of outcrop fragmentation, the larger the number of short cave sections. The recharge area of the karst plateaus to the east of Pedro Leopoldo and Matozinhos shows several hundreds of short paragenetic caves.

The transition of paragenetic caves from the phreatic zone towards the vadose zone has left many caves with total or partial sediment infilling. Some of these caves have been subject to sediment removal and later paragenetic (vadose) processes, named by Auler (1999) as "secondary paragenesis", in contrast to the initial "primary" paragenetic formation process. Vadose invasion streams commonly intercept these caves, reactivating speleogenesis or just promoting a heterogeneous removal of sediments. In the later stages of karst evolution, denuded outcrops favor the processes of breakdown and further fragmentation, generating many small caves and gradually unroofing and removing the caves from the landscape.

6 Conclusions

The Lagoa Santa Karst is home to 1,399 caves as of December 2017 and is presently the karst area in Brazil with the largest number of registered caves. This is not only due to the long history of cave research, starting in the eighteenth century with saltpeter miners, but also to the detailed work performed by environmental consulting companies. Caves occur mostly in the upper Lagoa Santa Member of the Sete Lagoas Formation. Caves tend to be small due to the relatively short groundwater routes and the degree of fragmentation of the karst outcrops. The cave depth is limited by the shallow (<100 m) limestone thickness above the crystalline basement.

Most caves show paragenetic development favored by the low hydraulic gradient (which prevents sediment removal once the sediment reaches the cave passages) and the availability of soil. Paragenetic anastomoses are the most common cave pattern, comprising meandering intertwining passages with no guiding joint at the ceiling level. Passages display scallop-free walls and abundant sedimentation, related to the primary paragenetic phase or to later input/removal due to paleoclimate cycles. Caves generated by lakes are also common. Vertebrate fossil-rich sediment occurs in several caves and has been extensively excavated since the early nineteenth century. The fossil concentration is related to the mechanism of emplacement. Speleothem deposition occurs in only a few caves. The most characteristic speleothem type is the ubiquitous calcite layer, which precipitates over/within the sediment.

Cave initiation started when the upper layers of the Sete Lagoas Formation limestone were subjected to groundwater flow. The thick sequence of limestone now denuded over most of the area suggests that the speleogenesis had a long history, starting probably well into the Tertiary if one assumes regional incision/denudation rates. Uplift and outcrop fragmentation resulted in many drained caves, with the degree of cave segmentation increasing as the outcrops became increasingly subdued until their final unroofing and removal from the landscape. Due to their economic and scientific potential, the Lagoa Santa Karst caves have acquired international relevance since the nineteenth century and presently lie in an area where significant portions of the landscape have been permanently preserved, although threats related to urban expansion, industrial development, groundwater pollution, and quarrying remain.

Acknowledgements This work results from three decades of research in the Lagoa Santa Karst. I acknowledge the help of many individuals, especially the members of the Grupo Bambuí de Pesquisas Espeleológicas. Local farmers and mining companies graciously allowed access to their properties. The staff at Carste Ciência e Meio Ambiente helped immensely, especially Bárbara Zambelli, Fernanda Pizani, Rafael Cruz, Gabriela Teixeira, and Caroline Reis. Thanks also to Tatiana Souza for reviewing the manuscript.

References

Auler AS (1994) Hydrogeological and hydrochemical characterization of the Matozinhos-Pedro Leopoldo Karst, Brazil. MSc thesis, Western Kentucky University

Auler AS (1995) Lakes as a speleogenetic agent in the karst of Lagoa Santa, Brazil. Cave Karst Sci 21:105–110

Auler AS (1998) Base-level changes inferred from cave paleoflow analysis in the Lagoa Santa Karst, Brazil. J Caves Karst Stud 60:58–62

Auler AS (1999) Karst evolution and palaeoclimate in eastern Brazil. PhD thesis, University of Bristol

Auler AS, Piló LB, Smart PL, Wang X, Hoffmann D, Richards DA, Edwards RL, Neves WA, Cheng H (2006) U-series dating and taphonomy of quaternary vertebrates from Brazilian caves. Palaeogeogr Palaeoclimatol Palaeoecol 240:508–522

Auler AS (2016) Cave protection as a karst conservation tool in the environmentally sensitive Lagoa Santa Karst, southeastern Brazil. Acta Carsologica 45:131–145

Auler AS (2019) Histórico, ocorrência e potencial de cavernas no Brasil. In: Rubbioli et al (eds) Cavernas. Atlas do Brasil subterrâneo. Grupo Bambuí de Pesquisas Espeleológicas, Belo Horizonte. In press

Auler AS, Piló LB (2015a) Lagoa Santa Karst: cradle of Brazilian cave studies. In: Vieira BC et al (eds) Landscapes and landforms of Brazil. Springer, Dordrecht

Auler AS, Piló LB (2015b) Caves and mining in Brazil: the dilemma of cave preservation within a mining context. In: Andreo B et al (eds) Hydrogeological and environmental investigations in karst systems. Springer, Dordrecht

Auler AS, Smart PL, Wang X, Piló LB, Edwards RL, Cheng H (2009) Cyclic sedimentation in Brazilian caves: mechanisms and palaeoenvironmental significance. Geomorphology 106:142–153

Campy M, Chaline J (1993) Missing records and depositional breaks in French late Pleistocene cave sediments. Quatern Res 40:318–331

CANIE (2018) Centro Nacional de Informações Espeleológicas—Centro Nacional de Pesquisa e Conservação de Cavernas. www.icmbio.org.br/cecav/canie. Accessed 20 May 2018

Cartelle C (1999) Pleistocene mammals of the cerrado and caatinga of Brazil. In: Ensenberg JB, Redford KH (eds) Mammals of the neotropics: the central neotropics. University of Chicago Press, Chicago

Claussen P (1841) Notes géologiques sur la province de Minas Geraes au Brésil. Bulletin de la Academie Royale de Bruxelles 8:322–343

Coutard JP, Kohler HC, Journaux A (1978) Carte du Karst. Region de Pedro Leopoldo—Lagoa Santa, Minas Gerais, Brasil. University of Caen, Caen

Da-Glória P, Neves WA, Hubbe M (2017) Archaeological and paleontological research in Lagoa Santa. Springer

Duarte MB (1899) Relassão das salitreiras que no prezente se trabalhão no Destricto de Matozinhos. Revista do Archivo Publico Mineiro 4:304–305

Dutra GM, Horta LS, Berbert-Born ML (1998) Levantamento espeleológico. In: APA Carste de Lagoa Santa, vol 3, 71 pp. Patrimônio espeleológico, histórico e cultural, CPRM

Galvão P, Halihan T, Hirata R (2016) The karst permeability scale effect of Sete Lagoas, MG, Brazil. J Hydrol 532:149–162

Haddad-Martim PM, Hubbe A, Giannini PCF, Auler AS, Piló LB, Hubbe M, Mayer E, Wang X, Cheng H, Edwards RL (2017) Quaternary depositional facies in cave entrances and their relation to landscape evolution: the example of Cuvieri Cave, eastern Brazil. CATENA 157:372–387

Harman R, Gallagher K, Brown R, Raza A, Bizzi L (1998) Accelerated denudation and tectonic/geomorphic reactivation of the cratons of northeastern Brazil during the Late Cretaceous. J Geophys Res 103:27091–27105

Hill C, Forti P (1997) Cave minerals of the world. National Speleological Society, Huntsville

Hubbe A, Haddad-Martim PM, Hubbe M, Mayer EL, Strauss A, Auler AS, Piló LB, Neves WA (2011) Identification and importance of critical depositional gaps in pitfall cave environments: the fossiliferous deposit of Cuvieri Cave, eastern Brazil. Palaeogeogr Palaeoclimatol Palaeoecol 312:66–78

Hurt WR (1960) The cultural complexes from the Lagoa Santa region. Am Anthropol 62:569–585

Hurt WR, Blasi O (1969) O Projeto Arqueológico "Lagoa Santa" – Minas Gerais, Brasil. Arquivos do Museu Paranaense 4:1–63

King LC (1956) A geomorfologia do Brasil oriental. Revista Brasileira de Geografia 18:147–265

Knez M, Slabe T, Travassos LEP (2012) Karren on laminar calcarenitic rock of Lagoa Santa (Minas Gerais, Brazil). Acta Carsologica 40:357–367

Kohler HC (1989) Geomorfologia cárstica da região de Lagoa Santa, MG. Dissertation, Universidade de São Paulo

Kruger FL (1965) A gruta do Baú. Revista da Escola de Minas 24:103–107

Lauritzen SE, Lauritsen A (1995) Differential diagnosis of paragenetic and vadose canyons. Cave Karst Sci 21:55–59

Liais E (1872) Climats, geologie, faune et geographie botanique du Brésil. Garnier Freres, Paris

Lund PW (1837) Om huler i kalksteen i det indre af Brasilien, der tildeels indeholde fossile knokler. Anden Afhandling. Det Kongelige danske Videnskabernes Selskabs Skrifter 6:309–330

Lund PW (1841) Blik paa Brasiliens dyreverden för sidste jordomvæltning. Første Afhandling. Det Kongelige Danske Videnskabernes Selskabs Naturvidenskabelige og Mathematiske Afhandlinger 8:29–60

Lund PW (1845) Notice sur des ossements humains fossiles, trouvés dans une caverne du Brésil. Mémoires de la Societé Royale des Antiquaires du Nord 49–77

Lund PW (1846) Meddelelse af det Udbytte de I 1844 undersogte Knoglehuler have afgivet til Kundskaben om Brasiliens Dyreverden for sidste Jordomvaeltning. Det Kongelige Danske Videnskabernes Selskabs Naturvidenskabelige og Mathematiske Afhandlinger 12:60–92

Neves WA, Piló LB (2003) Solving Lund's dilemma: new AMS dates confirm that humans and megafauna coexisted at Lagoa Santa. Curr Res Pleistocene 20:57–60

Peñaranda JR (2016) Condicionamento estrutural e litológico da porosidade cárstica da Formação Sete Lagoas, município de Sete Lagoas (MG). Dissertation, Universidade de São Paulo

Pessoa PFP, Auler AS (2015) Characterization and dynamics of two karst springs in a soil-covered karst area, Lagoa Santa, southeastern Brazil. In: Andreo B et al (eds) Hydrogeological and environmental investigations in karst systems. Springer, Dordrecht

Piló LB (1998) Morfologia cárstica e materiais constituintes: Dinâmica e evolução da depressão poligonal Macacos-Baú – Carste de Lagoa Santa, MG. Thesis, Universidade de São Paulo

Piló LB, Auler AS, Neves WA, Wang X, Cheng H, Edwards RL (2005) Geochronology, sediment provenance, and fossil emplacement at Sumidouro Cave, a classic late pleistocene/early holocene paleoanthropological site in eastern Brazil. Geoarchaeology 20:751–764

Reinhardt JT (1867) De brasilianske Knoglehuler og de i dem forekommende Dyrelevninger. Tidsskrift for populaere Fremstillingen af Naturvidenskaben 4(193–207):291–353

Reinhardt JT (1888) De Brasilianske knogelhuler og de i dem forekommende dyrelevninger. E Museo Lundii 1:1–56

Ruellan A (1950) Notas sobre a Gruta da Lapinha. Anais da Associação de Geógrafos Brasileiros 4:108–113

Villagran XS, Strauss A, Miller C, Ligouis B, Oliveira R (2017) Buried in ashes: site formation processes at Lapa do Santo rockshelter, east-central Brazil. J Archaeol Sci 77:10–34

Walter HV (1948) The Pre-history of the Lagoa Santa Region (Minas Gerais). Tipografia Brasil, Belo Horizonte

Biospeleology of the Lagoa Santa Karst

Rodrigo Lopes Ferreira, Marcus Paulo Alves de Oliveira, and Marconi Souza-Silva

Abstract

The region of Lagoa Santa stands out as the cradle of studies on epigean and subterranean ecology in Brazil. The subterranean habitats in the Lagoa Santa Karst area are structurally and biologically diverse, especially due to the wide variety of macro and microhabitats and food resources availability for the fauna. The observed biodiversity is a consequence of the distinct conditions of isolation, atmospheric stability, food availability, and the presence of dry and wet subterranean habitats. In terms of obligate cave fauna, the Lagoa Santa karst tends to present a moderate richness in absolute numbers (41 spp.) but low richness in relative terms (0.48 species per cave). However, such numbers were counted from inventories conducted in approximately 80 out of the 900 caves, so that many other species may still be discovered. The four described troglobites are *Brasilomma enigmatica, Tisentnops onix, Charinus taboa,* and *Spelunconiscus castroi*, but 37 other sampled troglomorphic species are still not described. Moreover, there are many human activities causing impacts to the Lagoa Santa Karst, which come from agriculture, cattle ranching, water exploitation, urbanization, deforestation, industries, mining, road construction, etc. The urgency for conservation actions is evident since studies on the population size and ecology of cave fauna are incipient and the low number of conservation units has led these species to an extreme degree of vulnerability and threats. In this perspective, the intensification of biological studies in caves of the Lagoa Santa karst is highly recommended to avoid species loss.

Keywords

Subterranean biology • Karst fauna • Cave ecology • Troglobites • Impacts

1 Subterranean Environments in the Lagoa Santa Karst

The high age of the limestone at the Lagoa Santa Karst (Neoproterozoic) and its long exposition to surface weathering contributed to the modeling of landscape in plateaus with different landforms (Berbert-Born 2000; Auler and Piló 2015). Such features influenced directly the hydrological dynamics, the availability of habitats and organic debris availability for the fauna associated with caves and other subterranean environments. Furthermore, rainfall and seasonality favor heterogeneous and peculiar vegetation (Brazilian Savannah) and primarily autogenic hydrology, which forms vertically dissected limestone outcrops with paragenetic caves and frequently short subterranean drainages (Auler and Piló 2015). The elevation ranging from 650 to 900 m asl represents a catchment divide between the basins of Mata Creek and Velhas River (Berbert-Born et al. 1998). Such dynamics and interaction of landscape elements have generated different habitats throughout the time, capable to shelter a specialized fauna occupying subsurface (epikarst) and deep subterranean habitats (dry caves, deep cracks, subterranean rivers, flooded travertines, water table, among others) from millimetric to decametric volumes.

R. L. Ferreira (✉) · M. Souza-Silva
Centro de Estudos em Biologia Subterrânea, Departamento de Biologia, Universidade Federal de Lavras, Lavras, Brazil
e-mail: drops@dbi.ufla.br

M. Souza-Silva
e-mail: marconisilva@dbi.ufla.br

R. L. Ferreira · M. P. A. de Oliveira · M. Souza-Silva
Programa de Pós-graduação em Ecologia Aplicada, Departamento de Biologia, Universidade Federal de Lavras, Lavras, Brazil
e-mail: marcuspbr@gmail.com

R. L. Ferreira · M. Souza-Silva
Setor de Zoologia Geral, Departamento de Biologia, Universidade Federal de Lavras, Campus Universitário, Caixa Postal 3037, Lavras, Minas Gerais CEP 37200-000, Brazil

A. S. Auler and P. Pessoa (eds.), *Lagoa Santa Karst: Brazil's Iconic Karst Region*, Cave and Karst Systems of the World, https://doi.org/10.1007/978-3-030-35940-9_10

The epikarst can be characterized as the uppermost weathered zone of carbonate rocks with substantially enhanced and more homogeneously distributed permeability and porosity when compared to the bulk rock mass below it (Klimchouk 2004). Despite the fact that this environment is poorly studied for caves of this region, it is plausible to assume that they represent important habitats for several species (e.g., copepods), as happens in several other karstic areas around the world (Culver and Pipan 2009). The meteoric water infiltrating the soil is partially retained in the epikarst and slowly percolates reaching caves via active speleothems even in dry periods of the year. Such waters may transport organic matter dissolved in the soil, therefore representing an important organic resource for microorganisms and small invertebrates, both in epikarstic zones and caves. In this sense, the great variety of mineral deposits (speleothems) present in caves of the region, besides improving the availability of microhabitats for fauna, are good indicators of the intense current and previous hydric dynamics.

Lotic and lentic groundwater bodies certainly comprise essential habitats for several organisms. However, more than 900 caves recorded for the region comprise dry passages with entrances in the lower portion of depressions or in the basis of limestone outcrops. Caves of different extensions that may be associated with the water table represent a minor percentage in this region (Auler 1998), although the occurrence of periodic swallets supplying the phreatic zone is not rare (Berbert-Born et al. 1998).

In these different subterranean habitats, the dynamics of food resource availability for the fauna tend to be complex due to the large territorial extension of the karst submitted to variations on the density of fractures (and caves) at different elevations. Thus, there are several and distinct possibilities of resource input into the caves due to biological (bats, rodents, accidentals) or physical elements (percolation water, rivers, flash flood) (Ferreira and Pompeu 1997; Berbert-Born et al. 1998; Ferreira et al. 2000; Souza-Silva et al. 2005). The input of organic resources (mainly plant debris) may occur gravitationally (via ceiling openings), from the deciduous vegetation present in the top of outcrops, by means of the water movement of flash flood and allogenic drainages, or by organic matter dissolved in percolation water. The vegetation associated with the limestone outcrops of Lagoa Santa karst present many deciduous plant species, which provides a greater accumulation of leaves in the litter during dry periods of the year (see chapter "The Vegetation of Lagoa Santa Karst", this volume). Part of this material may be carried to the interior of caves mainly during rainy periods, as described for other Brazilian karst regions (Souza-Silva et al. 2011). On the other hand, in permanently dry caves bat guano seems to be predominant trophic resource, comprising an important food source for several groups of invertebrates (Berbert-Born et al. 1998; Ferreira et al. 2000a) and microorganisms (Ferreira et al. 2000b).

In summary, caves in the Lagoa Santa Karst area are structurally and biologically too variable and complex, especially due to the wide variety of macro and microhabitats and food resources for the fauna. Thus, the invertebrates and vertebrates' communities have variable composition and diversity and were structured based on distinct conditions of isolation, atmospheric stability, food availability, and terrestrial and aquatic habitats.

2 Species Composition

Caves of the karst region of Lagoa Santa shelter a great variety of organisms that have been described since the nineteenth century. Paleontological sites sampled by the naturalist Peter W. Lund in caves of the region revealed living species that possibly used such environments as shelter, besides the extraordinary Pleistocene fauna (see chapter "Cave Paleontology in the Lagoa Santa Karst", this volume). Among such discoveries, the canids *Cerdocyon thous* Hamilton Smith 1839 (Berta 1982) and *Speothos venaticus* Lund 1842 (DeMatteo and Loiselle 2008) stand out, besides some rodent species like *Phyllomys brasiliensis* Lund 1840 (Emmons et al. 2002), *Blarinomys breviceps* Winge 1888 (Silva et al. 2003, Machado et al. 2016) and *Pseudoryzomys simplex* Winge 1887 (Machado et al. 2016). In the same century, members of the expedition Thayer (1865–1866) collected mite specimens, although they were described only 100 years after (*Whartonia W. pachywhartoni* Vercammen—Grandjean 1966, recently reviewed by Silveira et al. 2015). Although they had sampled components of cave fauna, these first studies involved themes non-related to subterranean biology (e.g., paleontology, photographic inventory).

Later, descriptions of new species sampled in caves of the region were published (ex. Costa-Lima 1940; Wygodzinsky 1950; Schubart 1956) and from 1990 on many works took into account the occurrence of subterranean fauna in the area (ex. Berbert-Born and Horta 1994; Pinto-da-Rocha 1995; Fontanetti 1996; Hermann et al. 1998; Trajano et al. 2000). Currently there are several studies reporting species of Lagoa Santa caves regarding different areas of biology, like ecology (Ferreira and Martins 1999; Gomes et al. 2000; Iniesta et al. 2012; Pellegrini et al. 2016), parasitology (Oliveira et al. 2000; Filho et al. 2007; Coelho-Finamore et al. 2011; Sábio et al. 2015), taxonomy (Gil-Santana et al. 1999; Pena et al. 2005; Mews and Sperber 2008; Campos-Filho et al. 2016), biochemistry (Dantas et al. 2014, 2016), among others. As a result, despite the volume of information being still incipient, such karst region may be considered one of the better sampled areas of Brazil in terms of subterranean fauna, what does not mean it is sufficiently known regarding bioespeleology.

Subterranean arthropods recorded for this karst are beyond those species considered restricted to underground habitats (as indicated in the topic "Troglobites"). Several non-troglobitic species (belonging to different taxonomic groups) were described for those caves, like Amphipoda (Hyalellidae: *Hyalella warmingi* Stebbing 1899), Gastropoda (Bulimulidae: *Thaumastus T. lundi* Pena et al. 2005), Pseudoscorpiones (Olpiidae: *Progarypus gracilis* Mahnert 2001), Diplopoda (Pseudonannolenidae: *Pseudonannolene microzoporus* Mauries 1987 and *Pseudonannolene chaimowiczi* Fontanetti 1996; Chelodesmidae: *Obiricodesmus rupestris* and *Sandalodesmus stramineus* Schubart 1956), Orthoptera (Phalangopsidae: *Endecous cavernicolus* Costa-Lima 1940, reviewed by Mews and Sperber 2008), Heteroptera (Reduviidae: *Mayemesa lapinhaensis* Wygodzinsky 1950, reviewed by Gil-Santana et al. 1999) and Isopoda (*Trichorhina pataxosi* Campos-Filho et al. 2016). Other species of wide distribution were sampled in some caves, among which some dipterans with medical importance stand out (Psychodidae: *Lutzomyia longipalpis* Lutz and Neiva 1912, in Oliveira et al. 2000 and Coelho-Finamore et al. 2011; *Lutzomyia L. renei* Martins, Falcão and Silva 1957, in Sábio et al. 2015; *Nyssomyia intermedia* Lutz and Neiva 1912 and *Nyssomyia neivai* Pinto, 1926, in Filho et al. 2007). Besides Diptera, other groups were also recorded, as spiders (Araneidae: *Alpaida septemmammata* Octavius Pickard-Cambridge 1889, in Hermann et al. 1998; Ctenidae: *Enoploctenus maculipes* Strand 1909, in Pinto-da-Rocha 1995; Oecobidae: *Oecobius annulipes* Lucas 1846, in Hermann et al. 1998; Theridiidae: *Nesticodes rufipes* Lucas 1846, in Hermann et al. 1998; Sicariidae: *Loxosceles similis* Moenkhaus 1898, in several studies like Pinto-da-Rocha 1995; Hermann et al. 1998; Gomes et al. 2000; Dantas et al. 2014, 2016; Trechaleidae: *Trechaleoides keyserlingi* Octavius Pickard-Cambridge 1903, in Hermann et al. 1998), scorpions (Buthidae: *Tityus serrulatus* Lutz and Mello 1922, in Hermann et al. 1998), pseudoscorpions (Chthoniidae: *Pseudochthonius strinatii* Mahnert 2001, Chernetidae: *Spelaeochernes dubius, Spelaeochernes eleonorae* and *Spelaeochernes gracilipalpus* Mahnert 2001), mites (Anystidae: *Erythracarus nasutus* Otto 1999, in Iniesta et al. 2012; Argasidae: *Ornithodoros talaje* Guerin-Méneville 1849, in Hermann et al. 1998), gastropods (Subulinidae: *Subulina octona* Bruguière 1798, in Hermann et al. 1998), bettles (Leiodidae: *Dissochaetus murrayi* Reitter 1884, in Pinto-da-Rocha 1995), Psocoptera (Psyllipsocidae: *Psyllipsocus ramburii* Selys-Longchamps 1872, in Lienhard and Ferreira 2015) and Hemiptera (Reduviidae: *Zelurus variegatus* Costa-Lima 1940, in Hermann et al. 1998). Even components of microbiota were identified in sand flies sheltered in subterranean habitats of the region, like bacteria and protozoa (*Stenonotrophomonas maltophilia, Enterobacter cloacae, Pseudomonas putida, Acinetobacter lwoffii, Flavimonas orizihabitans, Klebsiella p. ozaenae, Bacillus thuringiensis* in Oliveira et al. 2000; and *Leishmania infantum* Coelho-Finamore et al. 2011).

These records represent only a small part of specimens observed and/or collected, since many of them were not identified to the species level, and consequently they were not formally described (as indicated in Pinto-da-Rocha 1995; Hermann et al. 1998; Ferreira and Martins 1999; Gomes et al. 2000; Trajano et al. 2000; Mahnert 2001; Iniesta et al. 2012; Pellegrini et al. 2016 and in several other abstracts of meetings and thesis conducted in this karst region). Some species from the caves in the Lagoa Santa Karst are shown in Fig. 1.

Besides recent (living) species of mammals revealed by excavations of Peter W. Lund, several other vertebrates were recorded in caves of the Lagoa Santa karst. Many representatives of birdlife from Cerrado eventually use the caves entrances to build nests, like João-de-Barro (*Furnarius rufus badius* Lichtenstein 1823, in Pinto-da-Rocha 1995), Juriti (*Leptotila verreauxi decipiens* Salvadori 1871, in Pinto-da-Rocha 1995), Saí-Andorinha (*Tersina viridis* Illiger 1811, in Herman et al. 1998), Graúna (*Gnorimopsar chopi* Vieillot 1819, in Herman et al. 1998), owls (*Tyto furcata* Temminck 1827, in Herman et al. 1998), parrots and parakeets (*Amazona aestiva* Linnaeus 1758, *Eupsittula aurea* Gmelin 1789; in Pinto-da-Rocha 1995). In relation to anurans, most of the records do not have a taxonomic refinement to the species level, except for the species *Rhinella schneideri* Werner 1894 (Herman et al. 1998). However, amphibians are frequently found in Brazilian caves and there are occurrences in many studies in the region of Lagoa Santa (ex. Berbert-Born and Horta 1994; Pinto-da-Rocha 1995; Hermann et al. 1998; Matavelli et al. 2015). In relation to fishes, the only identified species comprises the Characidae *Astyanax bimaculatus* Linnaeus 1758 (Herman et al. 1998), although, as for other groups, there are records of other species identified at higher taxonomic levels (Pinto-da-Rocha 1995; Herman et al. 1998). Among mammals, bats may be considered the most efficient in using caves as shelters (Kunz 1982). Besides performing several ecological functions, they are responsible for the resource input to the hypogean environment (see the topic *Ecology*). In Lagoa Santa area, studies have indicated the presence of at least 23 species of bats inhabiting caves or cracks in the exokarst (Table 1). Besides the bats, other mammals are also frequently found, like raccoon (*Procyon cancrivorous* Cuvier 1798) and skunks (*Didelphis albiventris* Lund 1840).

In this sense, the faunistic composition of Lagoa Santa Karst comprises a great diversity of taxonomic groups that are not exclusive of subterranean environments, although there are species found exclusively in caves. Their adequate

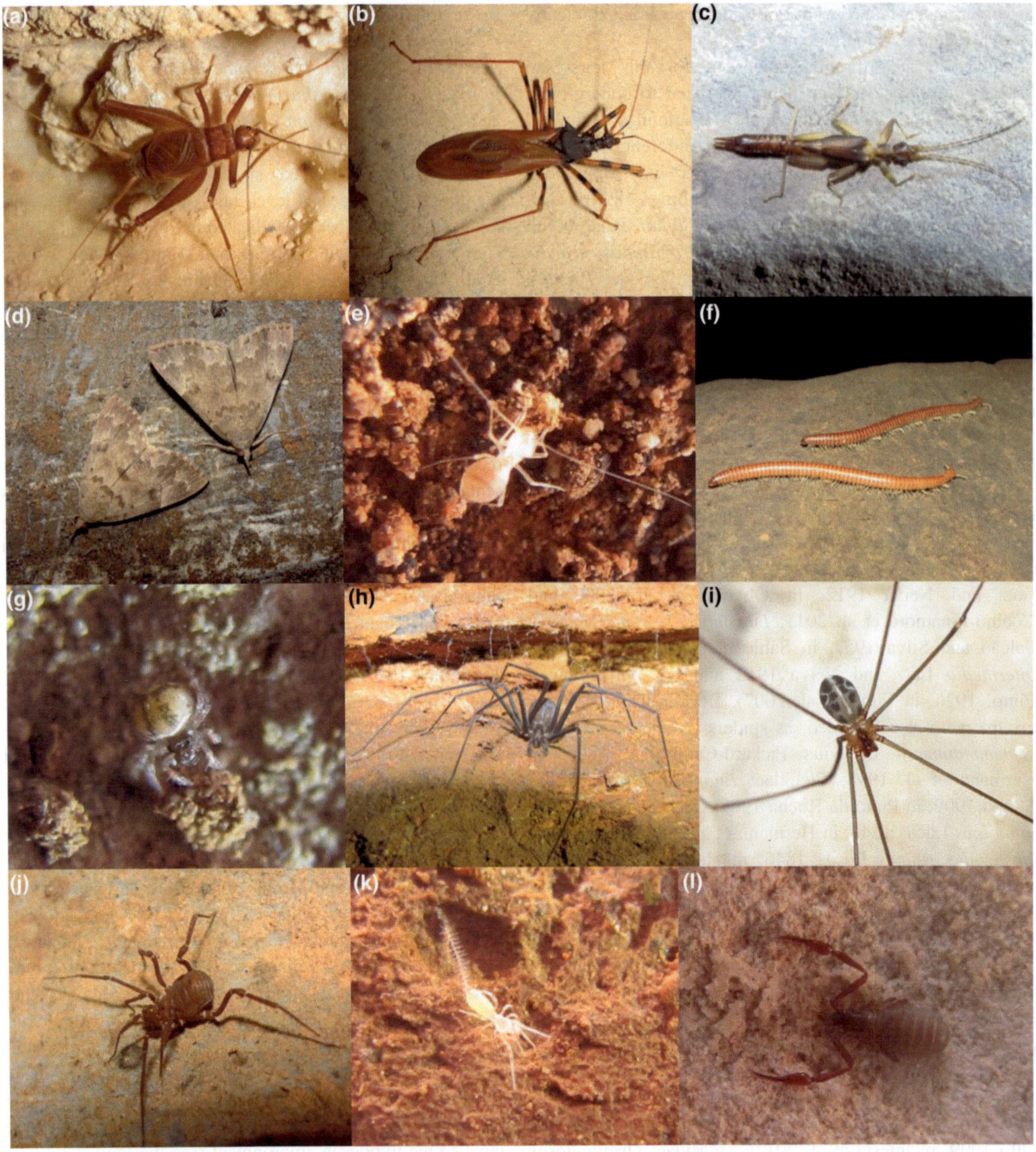

Fig. 1 Some troglophile and trogloxene species found in the caves of the Lagoa Santa Karst: **a** *Endecous cavernicolus* (Phalangopsidae); **b** *Zelurus zikani* (Reduviidae); **c** Diplatyidae earwig (Dermaptera); **d** *Hypena* sp. (Noctuidae); **e** *Psyllipsocus falcifer* (Psyllipsocidae); **f** *Pseudonannolene taboa* (Pseudonannolenidae); **g** *Symphytognatha carstica* (Symphytognathidae); **h** *Loxosceles similis* (Sicariidae); **i** *Mesabolivar* sp. (Pholcidae); **j** *Eusarcus* sp. (Gonyleptidae); **k** *Eukoenenia florenciae* (Eukoeneniidae); **l** *Progarypus* sp. (Olpiidae)

management depends on the knowledge on levels of adaptation and uses for different species that compose such communities. Moreover, the knowledge regarding all the present fauna allows distinguishing invasive species in these systems (Mazza et al. 2014). Then, it is increasingly necessary to study communities present in caves of this region

Table 1 Bat species recorded for the karst region of Lagoa Santa. The references indicate studies in which the individuals were sampled. Only occurrences in caves and shelters were considered

Family/Species	References
Emballonuridae	
Peropteryx macrotis Wagner 1843	Tavares et al. (2010), Talamoni et al. (2013), Guimarães and Ferreira (2014)
Phyllostomidae	
Anoura caudifer E. Geoffroy 1818	Guimarães and Ferreira (2014)
Anoura geoffroyi Gray 1838	Guimarães and Ferreira (2014)
Artibeus literatus Olfers 1818	Grelle et al. (1997), Guimarães and Ferreira (2014)
Artibeus obscurus Schinz, 1821	Talamoni et al. (2013)
Artibeus planirostris Apix 1823	Grelle et al. (1997), Talamoni et al. (2013)
Carollia perspicillata Linnaeus 1758	Grelle et al. (1997), Tavares et al. (2010), Talamoni et al. (2013), Guimarães and Ferreira (2014)
Chiroderma villosum Peters 1860	Talamoni et al. (2013)
Chrotopterus auritus Peters 1865	Hermann et al. (1998), Guimarães and Ferreira (2014)
Desmodus rotundus E. Geoffroy 1810	Grelle et al. (1997), Hermann et al. (1998), Tavares et al. (2010), Torquetti et al. (2013), Talamoni et al. (2013), Guimarães and Ferreira (2014)
Diaemus youngi Jentink 1893	Torquetti et al. (2013)
Diphylla ecaudata Spix 1823	Tavares et al. (2010), Talamoni et al. (2013), Guimarães and Ferreira (2014)
Glossophaga soricina Pallas 1766	Grelle et al. (1997), Torquetti et al. (2013), Talamoni et al. (2013), Guimarães and Ferreira (2014)
Micronycteris megalotis Gray 1842	Talamoni et al. (2013), Guimarães and Ferreira (2014)
Mimon bennettii Gray 1838	Talamoni et al. (2013)
Phyllostomus hastatus Pallas 1767	Tavares et al. (2010), Torquetti et al. (2013)
Phyllostomus discolor Wagner 1843	Guimarães and Ferreira (2014)
Platyrrhinus lineatus E. Geoffroy 1810	Grelle et al. (1997), Talamoni et al. (2013)
Vampyressa pusilla Thomas 1900	Grelle et al. (1997), Tavares et al. (2010)
Vespertilionidae	
Histiotus velatus I. Geoffroy, 1824	Talamoni et al. (2013)
Myotis nigricans Schinz 1821	Grelle et al. (1997), Talamoni et al. (2013)
Molossidae	
Molossus molossus Pallas 1766	Grelle et al. (1997)
Nyctinomops laticaudatus E. Geoffroy 1805	Talamoni et al. (2013)

for conservation purposes. As the number of inventories increases new species certainly will be described and/or new occurrences will be published for the Lagoa Santa Karst.

3 Troglobites

Different karst areas in Cerrado (Brazilian Savannah) were inventoried and revealed distinct richness of troglobitic species. The Lagoa Santa Karst, when compared to other areas, tends to present a moderate richness of troglobites (in absolute terms), although this richness, in relative terms, may be considered low. Until the present, there are 41 species with troglomorphic traits (potentially troglobitic) recorded for the Lagoa Santa caves. This number was taken from inventories performed in approximately 80 caves, as the mean of troglobites per cave is considerably low (mean of 0.48 troglomorphic species per cave). However, it is important to mention that such inventories were made in a small number of the caves recorded in the area, and then many other species may still be found, as happens for most of the Brazilian karst areas.

For comparison between the troglobites richness of Lagoa Santa Karst and other regions, some studies will be mentioned. A study performed in 15 caves in the karst region of Cordisburgo (adjacent to Lagoa Santa Karst area) revealed 39 troglomorphic species (mean of 2.6 species per cave) (Souza 2012). In the limestone speleological province of Arcos-Pains-Doresópolis, 294 caves were inventoried and 79 troglomorphic species were recorded (mean of 0.27 species per cave) (Zampaulo 2010). Simões (2013) studied 55 caves located in the karst region of Northwest Minas Gerais and found 35 troglomorphic species (mean of 0.84 species per cave). Finally, one study performed in 51 limestone caves from Central-North Minas Gerais revealed 94 troglomorphic species (mean of 2.75 species per cave) (Rabelo 2016).

It is noticeable that the absolute number of troglomorphic species observed for the region of Lagoa Santa Karst (41) is comparable to those found in other regions (such as Cordisburgo—39 species—and Northwest Minas Gerais—35 species). However, the relative number of troglobitic species (0.48 species/cave) is low and only higher to be observed for the region of Arcos-Pains-Doresópolis (0.27 species/cave). It is important to cite that these two previously mentioned areas stand out due to the predominance of small and dry caves, factors that may have contributed to the low relative richness of troglobitic species. Simões et al. (2015) demonstrated that the presence of water bodies and the length of caves are determining factors for the occurrence and richness of troglomorphic species in caves of some karst areas of the Cerrado.

More than 40 troglomorphic (potentially troglobitic) species have been found in the Lagoa Santa Karst, but only 4 have been described: the spiders *Brasilomma enigmatica* (Prodidomidae—Fig. 2a) and *Tisentnops onix* (Caponiidae—Fig. 2b), the tailless whip scorpion *Charinus taboa* (Charinidae—Fig. 2d) and the isopod *Spelunconiscus castroi* (Styloniscidae). However, there is a larger number of troglomorphic species already identified from caves of this area (37 species). Great part of these species is deposited in the reference collection of the Study Center on Subterranean Biology at the Federal University of Lavras (CEBS/UFLA 2016), and some of them are under the process of description. Such species were collected in the municipality of Confins (palpigrades, isopods, and spiders—Prodidomidae and Oonopidae), Lagoa Santa (mites Labidostomatidae, spiders *Nesticus* sp and isopods *Trichorhina*), Matozinhos (pseudoscorpions Chthoniidae, springtails Entomobryiidae, isopods *Trichorhina* and Styloniscidae and bettles Rizophagidae), Prudente de Morais (spiders Oonopidae, springtails *Arrhopalites*, millipedes Cryptodesmidae and isopods *Trichorhina* and Styloniscidae) and in Sete Lagoas (springtails Entomobryomorpha, isopods *Trichorhina* and *Spelunconiscus*, harvestmen *Spelaeoleptes* and millipedes Pyrgodesmidae). Some troglobitic species from the caves in the Lagoa Santa Karst are shown in Fig. 2.

Until the present, no exclusively aquatic species has been found in the Lagoa Santa Karst. Considering the 41 known species, 78% are terrestrial and the others are amphibious Styloniscidae (which favor the water but can move along terrestrial environments). Furthermore, most of the troglobitic species present a restricted distribution (occurring in only one cave) and only 5% of them occur in two or more caves. The caves that stand out in the region for their number of troglobitic species are Taboa Cave (6 species), Rei do Mato Cave (4), Lapa Vermelha Cave (4), and Meandro Abismante Cave (3).

In relation to the described species, *Charinus taboa* (Charinidae) is found in only two relatively close caves (Taboa Cave, 800 m long and BR24 cave, 33.8 m long), both in the municipality of Sete Lagoas, Minas Gerais state. The adults of *C. taboa* are found in the deeper portions of caves, mainly among speleothems, wall, and ceiling, while juveniles shelter in spaces under the rocks. This species may be considered vulnerable since only 31 specimens were observed in the two caves over years of research. Moreover, the epigean environment around the cave has been altered by human activities. Other species described for Taboa Cave is *Tisentnops onix* (Caponiidae) through the unique female collected under a rock in aphotic zone (at 15–20 m from the entrance). The occurrence of *T. onix* in other cave close to Taboa (Cave 64, 58.2 m long) is a strong indicator of the subterranean connectivity of these caves.

Fig. 2 Some troglobitic species found in the caves of the Lagoa Santa Karst: **a** *Brasilomma enigmatica* (Prodidomidae); **b** *Tisentnops onix* (Caponiidae—photo: Ubirajara Oliveira); **c** *Lygromma* sp. (Prodidomidae); **d** *Charinus taboa* (Charinidae); **e** Labidostomatidae mite; **f** *Spaeleoleptes* sp. (Escadabiidae); **g** Pyrgodesmidae millipede; **h** *Trichorhina* sp. (Platyarthridae); **i** *Spelunconiscus castroi* (Styloniscidae)

The isopod *Spelunconiscus castroi* (Styloniscidae) was described for the cave MOC32 (Campos-Filho et al. 2014), although nearly no information about this species was presented by the authors (except mentioning that it is an amphibious species). On the other hand, the spider *Brasilomma enigmatica* occurred in only one cave from Lagoa Santa karst (MOC131) but was also found in other areas (Brescovit et al. 2012). This species presents a wide distribution beyond the Lagoa Santa Karst, also occurring in ferruginous and quartzite caves along with an extension of more than 300 km without apparent subsurface connection with caves of the Lagoa Santa Karst (in this scenario the species was named "*enigmatica*" alluding to its distribution). The only specimen collected in MOC31 was collected 30 m below the surface level (Brescovit et al. 2012).

Information on the population size and ecology of the described troglobites and the other undescribed troglomorphic species does not exist until the present. Such lack of information associated with the intense human activities in the region and the reduced number of conservation units has led these species to an extreme degree of vulnerability and threats. In this perspective, the intensification of biological studies in caves of the Lagoa Santa karst is highly recommended in order to avoid the species extinction (mainly the troglobites) even before they are formally described (see the topic "Threats").

4 Ecology

The Lagoa Santa Karst certainly presents an historical relevance regarding biospeleology in Brazil, not only due to the species described from caves (as previously cited) but also especially considering that the first studies on ecology of Brazilian cave communities were performed in this region (e.g., Ferreira and Pompeu 1997; Ferreira et al. 2000b), among which we highlight those regarding ecology of guano communities.

Guano piles are widely considered as an important source of organic matter in caves (Harris 1970; Decu 1986; Ferreira and Martins 1999; Gnaspini 2012). In tropical caves, especially those permanently dry, bat guano may represent the most important source of organic matter, forming the trophic base for the structure of many invertebrate communities (Decu 1986; Ferreira and Martins 1998, 1999). The diversity of bats feeding habits in the tropics is remarkable (Herzig-Straschil and Robinson 1978). The three most common guano types found in the caves of the Lagoa Santa Karst are produced by frugivorous bats (with small undigested seeds), hematophagous bats (with a pasty consistency and reddish color when fresh, becoming black and often powdery when older), and by insectivorous bats (containing small chitinous parts of insects or other arthropods). Although other types also may occur, the most commonly found piles are produced by hematophagous bats (especially *Desmodus rotundus*), and this certainly has been caused by a historical replacement of the natural forests by pastures (see the topic "Threats").

Bat guano communities in caves are extremely interesting for ecological studies, especially due to their physical and chemical conditions (that are easily measurable and quite distinguishable when comparing different piles) and the great number of easily defined interactions with the overall cave environment (Fig. 3). However, during decades such communities were considered "isolated" from the cave environment as a whole, but in fact a high number of cave invertebrate groups associate and interact with them (Ferreira et al. 2000a).

Several guano piles traits act structuring the associated communities in the Lagoa Santa Karst (Ferreira et al. 2000b). Ferreira and Pompeu (1997) showed that the distance from the entrance was an important parameter determining the richness of guano communities in Taboa Cave (Sete Lagoas municipality) (Fig. 4b). Other studies performed later found similar (Ferreira et al. 2007) but also distinct results (Ferreira et al. 2000), indicating that the distance from the cave entrance is not always an important factor for the bat guano communities. Since most invertebrates occurring in guano piles are troglophiles (with some rare troglobitic species), the colonization of each deposit may occur from any place inside the cave, and not only from the entrance. On the other hand, the area of guano piles is always important, so that the richest and most diverse communities are always associated to biggest piles (Ferreira and Pompeu 1997; Ferreira and Martins 1998; Ferreira et al. 2000a; Bahia and Ferreira 2005; Ferreira et al. 2007) (Fig. 4c, d). The richness or abundance of species that use faecal deposits tends to be proportional to the availability of this resource (Doube 1986). Furthermore, one would expect some area effect of greater potential microhabitat variability over a larger surface. Hence, in caves of the Lagoa Santa Karst, larger deposits with a greater diversity of microhabitats (like small holes and cracks) have more associated species.

It is also important to mention that guano communities are structured over some physicochemical parameters, like pH and organic content. The reduction of guano pH over time is well known, since fresh guano is alkaline and becomes acidic due to ammoniac fermentation (Hutchinson 1950). However, pH does not always reflect guano age, since the piles are open systems subject to be affected by chemical (percolation water, floods, etc.) and physical (landslides, sedimentation, etc.) processes occurring in the cave environment (Ferreira et al. 2000a). In the caves of the Lagoa Santa Karst, the richest piles generally present moderate pH (Fig. 4c, d). The relationship between guano pH and richness of the associated communities is probably the result of a preference (by most species of the community) for an optimum and intermediate value of pH (Ferreira et al. 2000a). However, the same authors postulated that if guano pH reflects the piles age, fresh piles (more basic) may not have had enough time to be colonized, while old piles (more acidic) may have been exhausted as a food resource due to consumption or mineralization (Ferreira et al. 2000a). The organic content is also important, as shown by Ferreira et al. (2000a) in Lavoura Cave (Matozinhos municipality) located in the Lagoa Santa Karst. Such authors observed that richer communities were associated with guano piles with higher organic contents, demonstrating the importance of such parameters for these communities.

Finally, there are some evidence that larger piles can act as sources of colonization for smaller ones. In Lavoura Cave (Matozinhos municipality), the community similarities between one big guano pile and smaller piles surrounding it (unpublished data) were correlated. It was observed that smaller piles closer to the biggest ones are more similar to it than piles located far from it (Fig. 4a). This suggests that guano communities in a given cave may act as metacommunities. The concept of metacommunity was first proposed

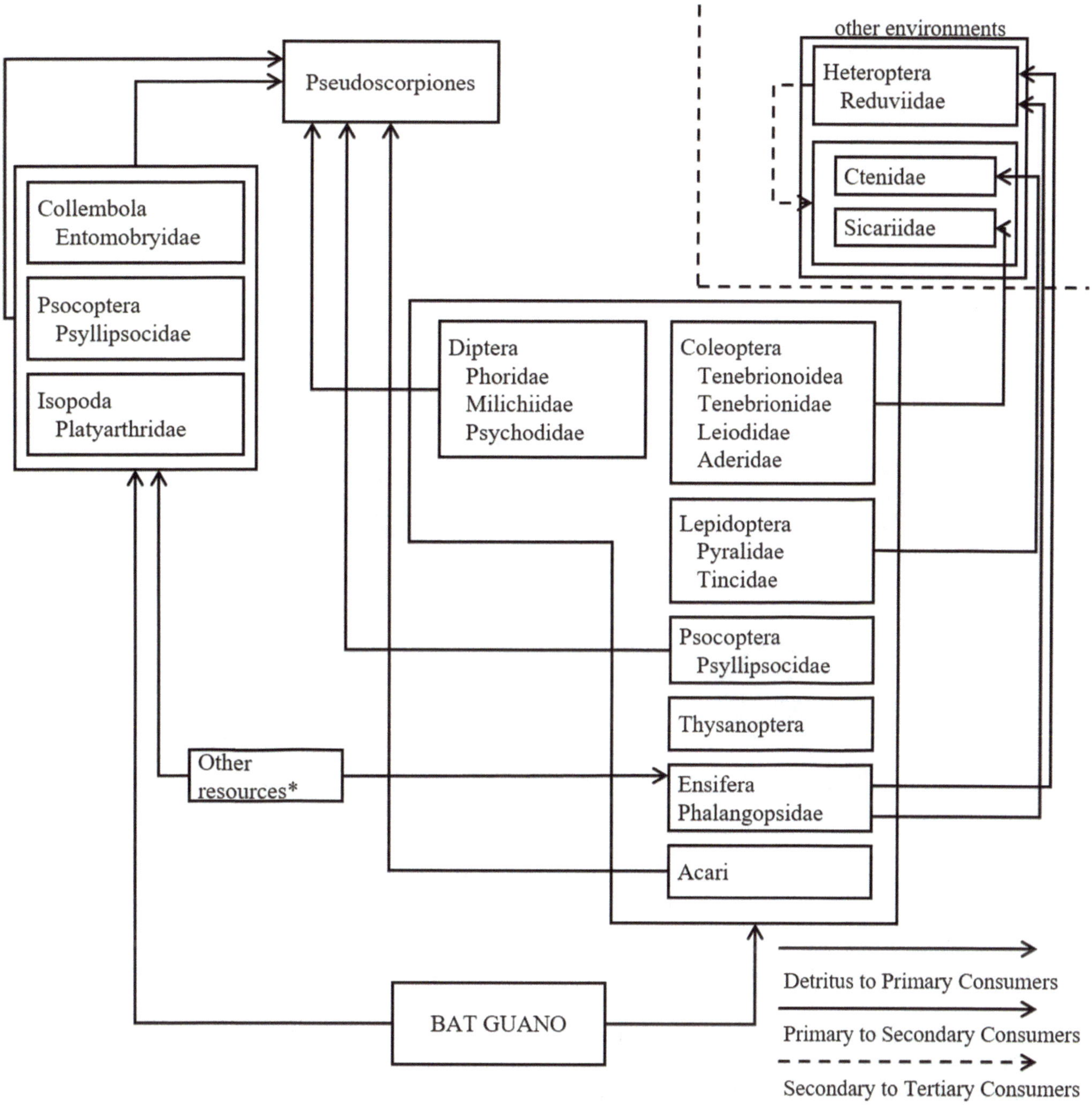

Fig. 3 Food web of a bat guano community in Lavoura Cave (Matozinhos municipality)

by Gilpin and Hanski (1991) to represent a set of local communities from different sites connected by dispersion of one or more of its components. A metacommunity follows the same dynamics of a metapopulation, and therefore is subject to local extinctions and recolonizations. The low abundance of organisms in some old guano piles enhances local extinctions. As the guano becomes older, its quality as an organic resource decreases, thus the community progressively leaves the pile and colonize others in order to maintain their populations. However, the dynamics of movements between guano piles remain largely unknown, certainly deserving further investigations.

Some studies performed in caves of the Lagoa Santa Karst focused on aspects of population ecology of some species. Ferreira et al. (2005) monitored a cave population of the brown recluse spider (*Loxosceles similis*) in the Lavoura Cave (located in Matozinhos municipality) and found some patterns especially regarding the displacement of such spiders inside the cave. *Loxosceles* species are usually considered sedentary having a life area restricted to the vicinities of

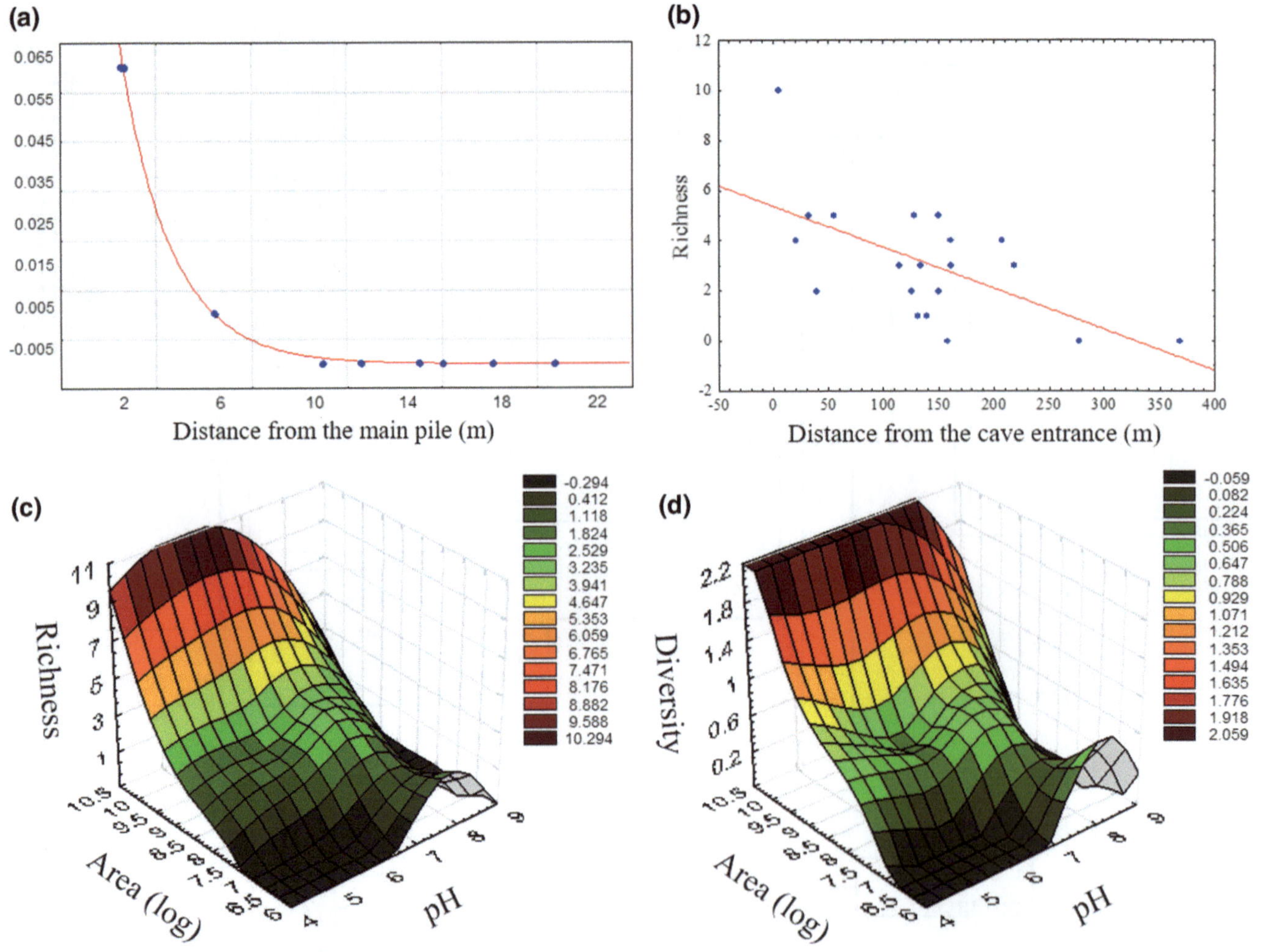

Fig. 4 **a** Similarity values (Renkonen index) between a larger pile (species "source") and small peripheral piles (species "sink") according to their distance in relation to the main (larger) pile in Lavoura Cave (Matozinhos municipality); **b** relation between the distance from the cave entrance and the richness of guano communities in Taboa Cave (Sete Lagoas municipality); **c** relation between the area, the pH of the guano piles and the richness of guano communities in Lavoura Cave (Matozinhos municipality); **d** relation between the area, the pH of the guano piles and the diversity of guano communities in Lavoura Cave (Matozinhos municipality)

their webs (Bücherl 1961; Foelix 1987). However, Ferreira et al. (2005) registered a great mobility of such spiders. Many individuals of *L. similis* were recaptured at distances from 20 to 80 m from the place where they were initially marked (a movement of 40 m was registered in only one week). These distances are much longer than those recorded for some species like *L. rufipes* (Lucas 1834) and *L. intermedia* Mello-Leitão, 1934 (around 2 m in one week, what was considered a significant mobility) (Delgado 1966; Fischer 1996). However, the cave environment presents different conditions from those found in external or anthropogenic environments. The scarcity and irregular distribution of food resources in a cave may result in lower prey availability, which may have led the spiders (in the Lavoura Cave) to move longer distances looking for preys.

Another important study performed in the Lagoa Santa Karst presented, for the first time for the tropical region, data regarding the responses of a cave population (of the moth *Hypena* sp.—Lepidoptera: Noctuidae) to seasonal variations in the epigean environments (Ferreira et al. 2015). Several studies demonstrated the effects of external climatic fluctuations on subterranean populations and communities in the temperate region (Bourne and Cherix 1978; Carchini et al. 1994; Novak et al. 2004; Tobin et al. 2013). Although such effects are expected in areas where external climatic fluctuations are more intense (Mitchell 1969), no studies were historically performed in tropical regions, maybe due to the false impression that both small temperature and humidity fluctuations (typical of the tropics) have little or no effect over the subterranean fauna. In this sense, Ferreira et al.

(2015) demonstrated that a cave population responds to seasonal external variations (in this case the Taboa Cave, municipality of Sete Lagoas). During the rainy seasons the population of *Hypena* sp. was closer to the cave entrance, while during the dry periods, the population peaks occurred slightly deeper inside the cave (Fig. 5). This study is important especially to demonstrate that patterns occurring in temperate caves can also be observed in the tropics. However, further studies are needed to actually understand how cave populations or communities from temperate and tropical regions differ or not, in terms of function and structure.

Although only few studies regarding general cave communities in the Lagoa Santa Karst were performed (Iniesta et al. 2012), some "patterns" were found, indicating that communities in this region present some similarities. In the Irmãos Piriás Cave (located in Matozinhos municipality), four collection events were conducted between 1999 and 2000. In all events, a reduction of invertebrate richness was observed from the entrance to deeper portions of the cave (Fig. 6a). This entrance-mediated effect was expected, especially in tropical regions, due to the remarkable richness of external communities, which represent an important source of organisms prone to colonize caves. Therefore, cave entrances in the tropics tend to present richer communities when compared to inner portions of the caves. However, although some species can establish themselves in the hypogean habitats, most of them cannot, due to their lack of preadaptive traits. Cave entrances are considered ecotones between epigean and hypogean environments (Prous et al. 2004, 2015). Hence, such regions tend to present intermediate conditions between those habitats. As an example, entrances are more stable than epigean environments (favoring the presence of many species), although they present less organic resources (preventing other species to colonize them). Prous et al. (2015) argued that ecotones in the cave entrances may function as selective permeability membranes acting differently over distinct species. According to Strayer et al. (2003), this "membrane" can act on matter, energy, or species in different ways and can partially transmit, transform, absorb, reflect, amplify them, or act neutrally. Moreover, filters present in these membranes can be biological or physical and the combination of both filters is responsible for the presence and distribution of species in the cave entrances and also within the caves (Prous et al. 2015).

However, although there is a tendency toward the reduction of richness from the entrance to deeper areas in caves, each species inside a cave can present a distinct pattern of distribution. In the Irmãos Piriás Cave, some species presents higher densities in regions closer to the entrance (as the spider *Plato* sp.—Theridiosomatidae) while for others the higher densities are observed on inner portions of the cave (e.g., the cricket *Endecous* sp.—Phalangopsidae) (Fig. 6b, c). Furthermore, such patterns of distribution vary not only in space but also throughout time. The community from Irmãos Piriás Cave varied considerably from 1999 to 2001. Richer communities were observed during rainy periods (89 and 94 invertebrate species in January 2000 and January 2001, respectively), while in the dry periods the communities reduced the number of species (44 and 64 invertebrate species in July 1999 and July 2000, respectively). The species turnover between the four sampling events corresponded to 54.18% indicating a considerable species replacement over time. The turnover was calculated using data of presence and absence through the index of Harrison et al. (1992); modified by Whittaker (1960), in order to compare samples of different sizes. $\beta_{Harrison} = \{[(S/a) - 1]/(N - 1)\}9100)$, where S = total species richness, a = average richness values, and N = number of samples. This measure ranges from 0 (no turnover) to 100 (each sample has a unique set of species) (Koleff et al. 2003).

In the Taboa Cave (Sete Lagoas Municipality) it was observed a similar pattern to the observed in the Irmãos Piriás Cave. In the rainy periods, not only the richness but also the diversity was higher than in dry seasons. In this cave, in both rainy periods the richness and diversity corresponded to 101 species and $H' = 3.04$ (January 2000) and 126 species and $H' = 3.29$ (January 2001), while in the dry periods those values corresponded to 35 species and $H' = 2.23$ (July 1999) and 75 species and $H' = 2.73$ (July 2000). However, the community equitability did not change considerably along the sampling period, corresponding to 0.63 in both dry periods and to 0.66 (January 2000) and 0.68 (January 2001) in the rainy periods. This indicates that although the community can vary along the time in richness, diversity and even in composition, the average distribution of population size is stable, and such "stability" can be the result of the oligotrophic conditions, which represents a limiting pressure upon the distinct populations inhabiting the cave.

Finally, the epigean landscape can also be important to determine the structure and composition of the cave fauna. It is well known that patterns of biodiversity usually respond to habitat disturbances and different land uses (MacDonald et al. 2000). However, until recently no study had demonstrated spatial-scale influences in subterranean terrestrial communities. In a study conducted in the Lagoa Santa Karst, Pellegrini et al. (2016) analyzed how land use and cave physical structure could influence the terrestrial cave invertebrate species composition. Such authors also determined the influence of different spatial scales under analysis on the composition of invertebrate cave fauna. For that purpose, they determined land uses at three different landscape scales, by gathering data into circular areas of different sizes (50, 100 and 250 m) with centroids in the cave entrances (Fig. 7). The authors found that the best explanatory variable

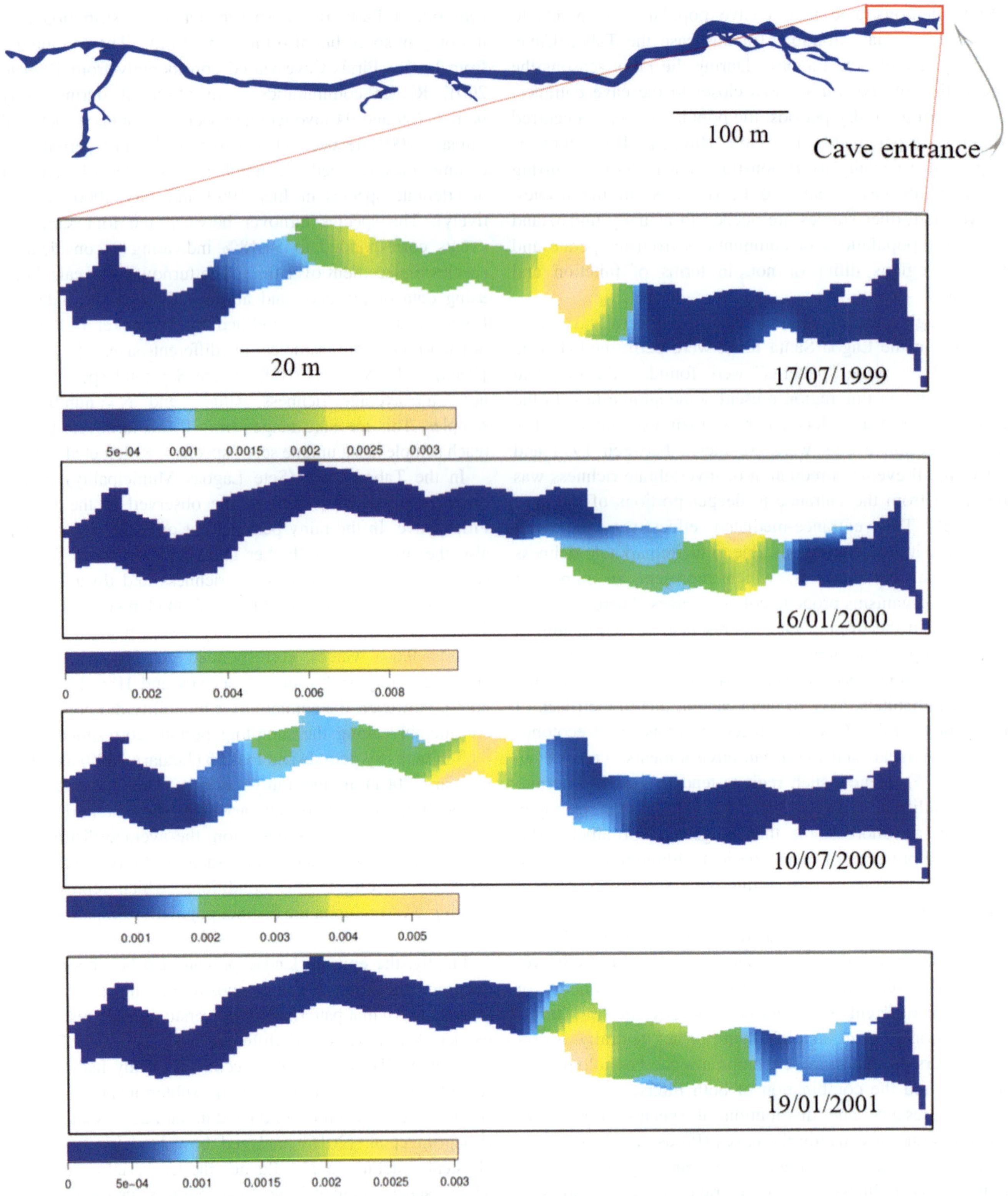

Fig. 5 Spatial distribution maps of *Hypena* sp. in Taboa Cave (Sete Lagoas municipality) demonstrating different densities between seasons. In the dry seasons (17/07/1999 and 10/07/2000) individuals are located far from the cave entrance, while during rainy seasons (16/01/2000 and 19/01/2001) individuals migrate to areas near the entrance. Blue colors indicate low densities while light yellow colors indicate high densities. Modified from Ferreira et al. (2015)

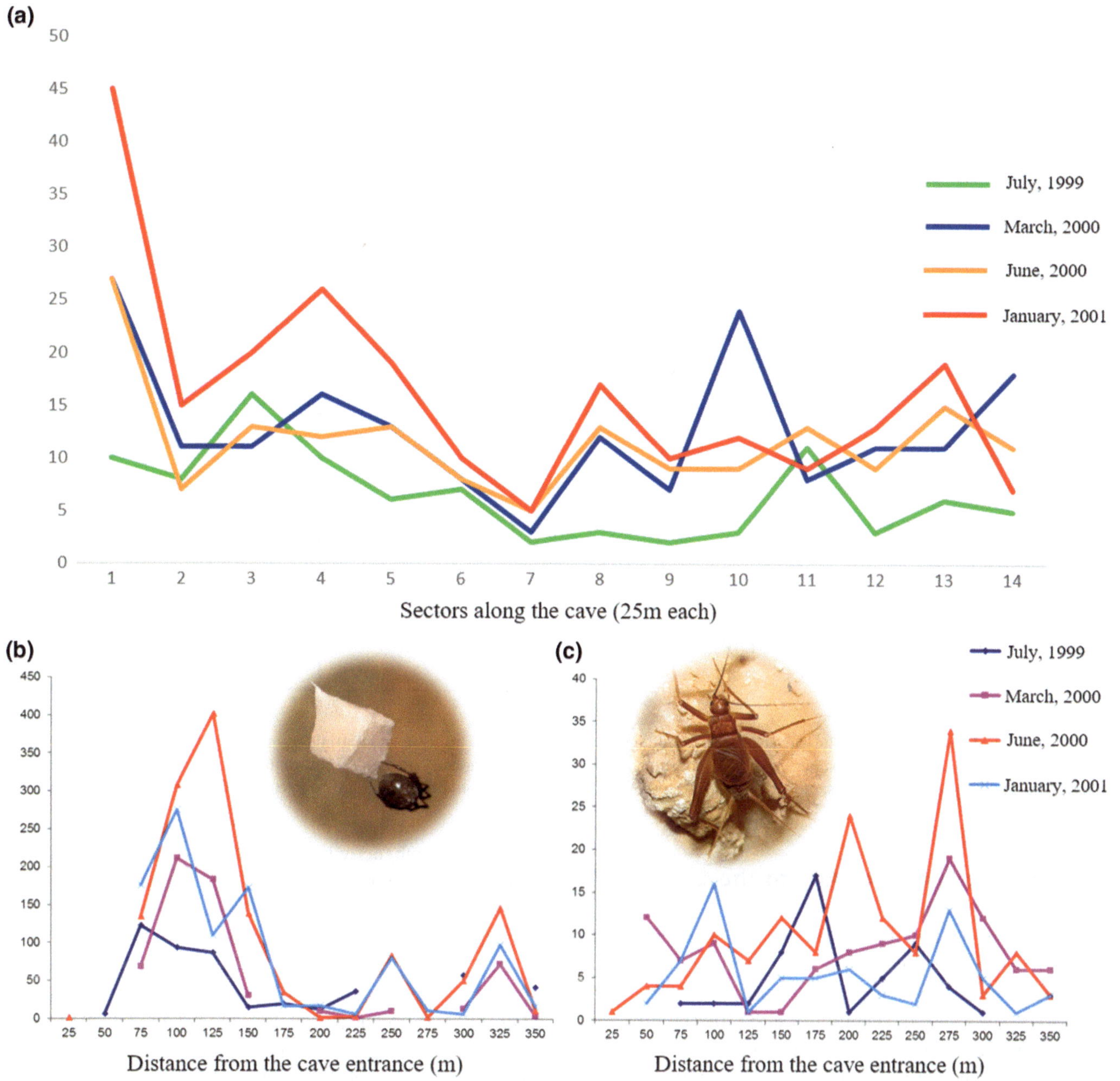

Fig. 6 **a** Relation between the distance from the cave entrance and the invertebrates richness (vertical axis) in different seasons in Irmãos Piriás Cave (Matozinhos municipality); **b** relation between the distance from the cave entrance and the density of the spider *Plato* sp. (Theridiosomatidae) in different seasons in Irmãos Piriás Cave; **c** relation between the distance from the cave entrance and the density of the cave cricket *Endecous* sp. (Phalangopsidae) in different seasons in Irmãos Piriás Cave

for cave invertebrate composition was the percentage of limestone outcrops covering the surface environments. Furthermore, they confirmed the scale-dependence hypothesis since the models generated during the study (by Distance Based Linear Modeling) became more precise as larger scales were analyzed to explain cave invertebrate composition. According to them, larger scales embrace more precisely more environmental traits that influence the cave fauna composition. In conclusion, Pellegrini et al. (2016) found a strong influence of limestone outcrops structuring cave communities independently of the scale under analysis (although the scale can enhance the precision of the analysis), demonstrating the major importance of preserving those outcrops for the maintenance of the cave fauna composition.

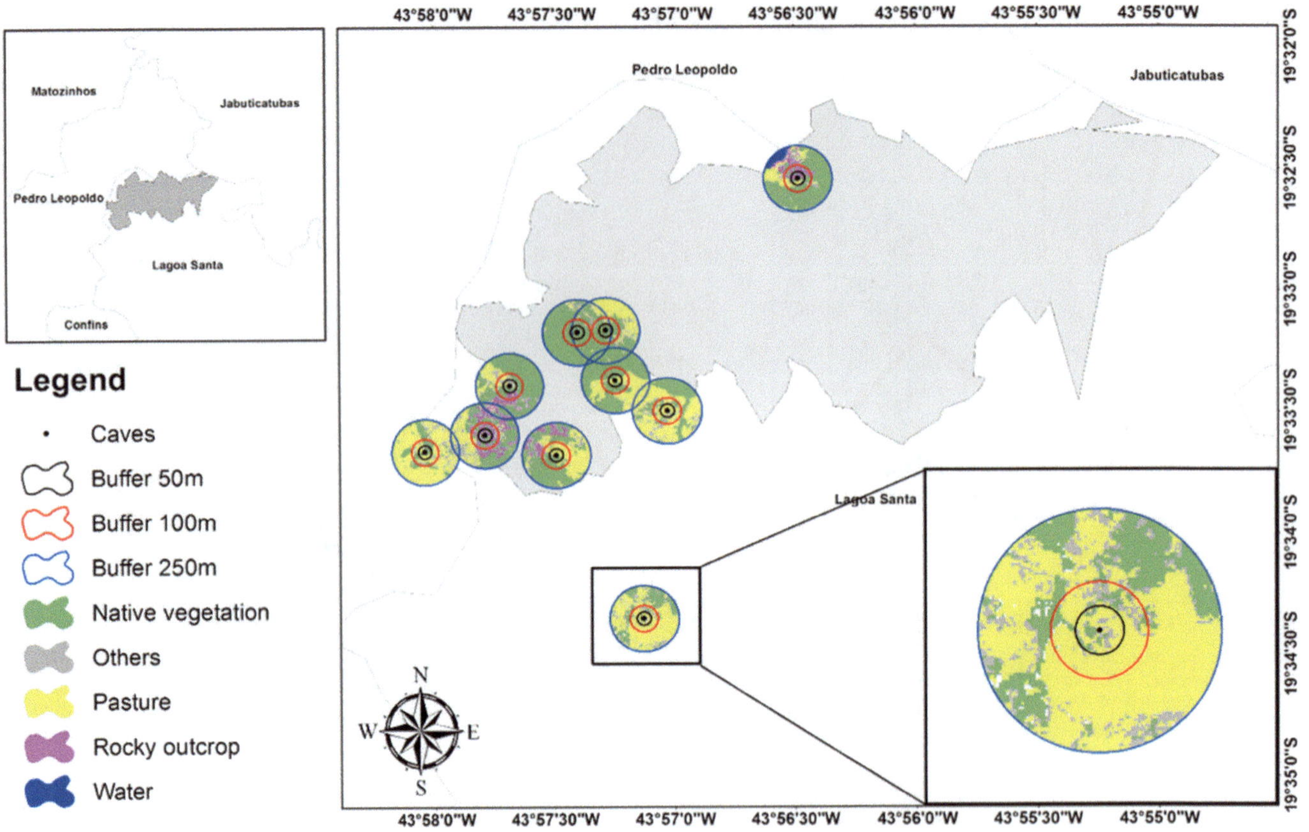

Fig. 7 Spatial characterization of the landscape at "Sumidouro State Park". Different colors represent distinct vegetation cover or land-use types. The black dots indicate the sampled caves, surrounded by the buffers of 50, 100, and 250 m for analyzing the effect of spatial scale on the explanatory power of environmental variables at the cave invertebrate communities. Modified from Pellegrini et al. (2016)

5 Tourism: Enemy or Ally to the Cave Life?

Several caves around the world present an extraordinary scenic beauty. Such caves historically attracted the attention of general public, which started to use them for touristic purposes. According to Hamilton-Smith (2004) tourist caves can be defined as those attracting the interest of the general people, thus generating some type of financial return, as charge for admission or sale of souvenirs. Several caves have been used worldwide for touristic purposes during decades. Many of these caves receive thousands of visitors every year. Culver and Sket (2002) attested that caves that receive large number of visitors could present a high diversity of species, despite the obvious alterations caused by tourism. Accordingly, the opening of a given cave to tourism does not mean that their biodiversity will necessarily be altered or lost, as in the cases of some show caves, as the Mammoth Cave (Kentucky, USA), Postojna-Planina Cave System (Slovenia), Baget-Sainte Catherine System (France), and Vjetrenica Cave (Bosnia and Herzegovina) (Culver and Sket 2002). However, one cannot assume that touristic use never has deleterious effects in caves, since alterations caused by such activities can affect distinct subterranean species in different ways.

In Brazil, there are many show caves with diverse infrastructure or use. In the Lagoa Santa Karst, two show caves can be highlighted, especially due to the large number of visitors they attract every year. Such caves are Rei do Mato Cave (located in Sete Lagoas municipality) and Lapinha Cave (Located in Lagoa Santa municipality). Both caves were inventoried regarding the fauna and at least in one of them (Lapinha Cave) it was possible to evaluate the effects of the touristic use on the cave communities.

In Rei do Mato Cave (Sete Lagoas municipality), a total of 204 invertebrate species were found during five sampling events (unpublished data from the authors). However, the species richness varied along the time, and richer communities were found during rainy periods, following the general pattern of the cave communities in the area (30 species on average in the dry periods contrasting with 67 species on average in the rainy periods). Most invertebrate species were troglophiles and trogloxenes, and only four species presented troglomorphic traits (two isopods, one springtail, and one harvestman). Furthermore, most species (more than 60%) were concentrated near the entrance during all

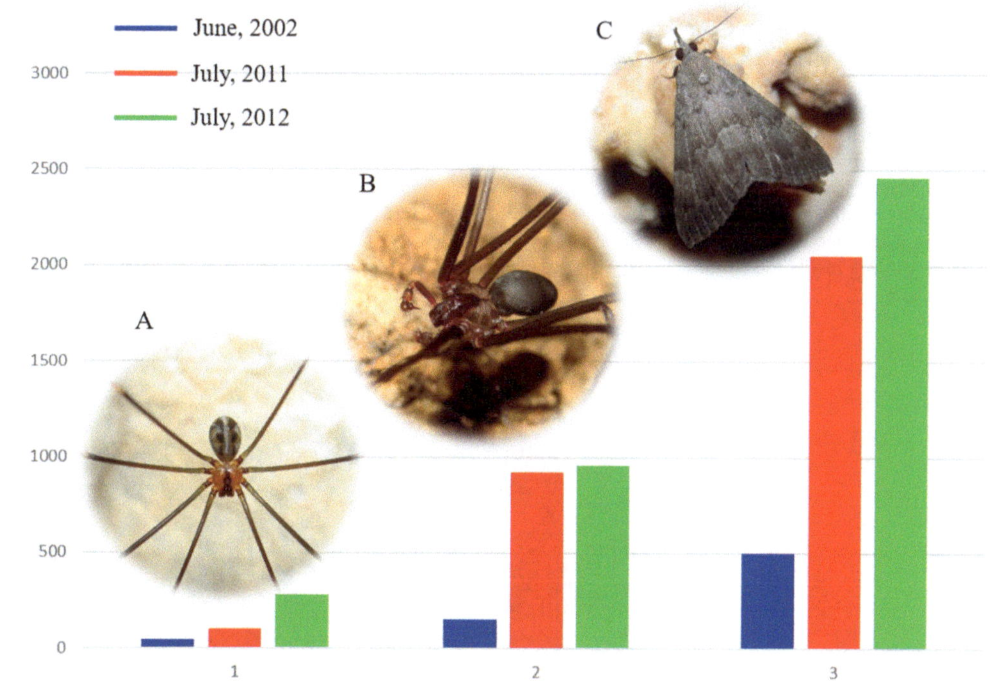

Fig. 8 Differences observed in population sizes of three species in sampling events conducted during touristic use (June 2002 and July 2012) and during non-touristic use (July 2011), in Lapinha Cave (Lagoa Santa municipality): **a** *Mesabolivar* sp. (Pholcidae); **b** *Loxosceles similis* (Sicariidae); **c** *Hypena* sp. (Noctuidae)

sampling events, indicating the importance of ecotone communities for this cave, or contrarily, the highly restrictive conditions observed in the inner portions of the cave (caused by natural and/or touristic factors).

In the Lapinha Cave (Lagoa Santa municipality), three sampling events were performed, all during dry periods (unpublished data from the authors). However, the sampling occurred in quite distinct situations. The first sampling (held in June 2002) occurred when the cave was under constant touristic use, since it was opened for this purpose in 1968. However, in 2009, touristic activities were interrupted due to the need of a management plan for the cave, and no tourists were allowed in. The second sampling event was performed in July 2011, 2 years after it was closed for tourism. After the second sampling, tourism was reestablished in the cave. The third sampling occurred in July 2012, one year after the reopening of the cave to tourists. Therefore, the samplings allowed a comparison regarding the direct effects of tourism on the cave communities of this cave. The most significant result observed was an increase in the invertebrate richness of the cave after the tourism ceased in the cave. The first sampling (during touristic use) revealed 58 invertebrate species, while the second event (no tourism in course) revealed 75. Hence, there was an increase of almost 30% in the richness during the period without tourism. The third collection (one year after the reestablishment of tourism) surprisingly revealed again 58 species, thus indicating that the tourism in this cave presented an obvious effect reducing the richness of invertebrate species.

However, diversity did not change considerably comparing the three events. In the first collection (during tourism), it corresponded to 2.3, in the second (after closing the cave) to 2.1, and in the third collection (after the reopening to tourism) it was 2.5. The increase in the diversity after the reestablishment of tourism resulted from a reduction in the dominance in this period ($D = 0.14$, compared to 0.21 and 0.22 from the first and second samplings, respectively). Furthermore, in general, the abundance tended to increase from the first to the last sampling events, and the highest values were observed after the reopening to tourism (8,141 individuals in the whole cave, compared to 4,966 and 1,210 individuals observed in the first and second samplings, respectively). Some common troglophilic or trogloxene species, as the cave moth *Hypena* sp. (Noctuidae) and the spiders *Mesabolivar* sp. (Pholcidae) and *Loxosceles* sp. (Sicariidae) presented the same pattern (Fig. 8).

However, it is important to highlight that the reopening of the cave to tourism occurred after the management plan, which indicated the necessity of preserving some areas (as the deeper chamber of the cave) for the fauna, especially due to the presence of a bat colony (*Phyllostomus* sp.). Accordingly, after the visitation ceased in this chamber, the bat colony increased (as the guano production), allowing more species to establish themselves and to grow their populations. Thus, the deleterious effects of tourism can be reduced by closing some areas to visitors, keeping them preserved and far from the direct effects of the visitation (such as lights, trampling, among others).

Fig. 9 **a** Algae growing on a speleothem in Rei do Mato Cave (Sete Lagoas municipality); **b** *Psyllipsocus falcifer* feeding on such algae (see the greenish color of the specimen's abdomen); **c** an angiosperm growing inside the Rei do Mato Cave; **d** fungi (especially *Fusarium solani*) growing associated with the electric cabling inside the Rei do Mato Cave

According to Hamilton-Smith (2004), the most severe impacts observed in tourist caves usually result from inappropriate use. However, the tourism, per se, can cause several modifications on the cave environment, most of them primarily produced by the introduction of artificial light, which creates a photoperiod in the pristine aphotic areas of the cave. Such photoperiod provides conditions for the growth of photosynthetic organisms (especially algae and moss, but sometimes ferns and other more complex plants) (Fig. 9a–c). In turn, those organisms represent organic resources that were absent in aphotic zones, thus attracting animals that would not be found in such areas. Furthermore, these organisms (Fig. 9c) can be consumed by cave species and the effects of such consumption on the populations (positive or negative) are unknown. Another important alteration caused by the touristic infrastructure in show caves of the Lagoa Santa Karst was the emergence of some fungi that started growing associated to the electrical wiring inside the caves (Fig. 9d). The most important species found in those cables were *Fusarium solani*, which are also commonly observed in soil. However, since there were no studies on the fungal community before the installation of those cables, it is not possible to evaluate whether those fungi already inhabited the cave soil or were introduced following the emplacement of such infrastructure. The concern lays over the fact that *Fusarium* species can cause opportunistic mycosis, thus offering risks to the visitors. Furthermore, hundreds of invertebrates (especially barklices and springtails) were observed feeding on those fungi colonies growing on the cabling. Accordingly, the fungi appearance may have caused a disproportional growth of some cave invertebrate populations, besides attracting those species to areas in which they are more exposed to trampling by visitors.

Finally, it is important to highlight the importance of tourism as a powerful tool for environmental education in the karst systems as a whole. Although such activities can produce deleterious effects on the caves and their fauna, such effects can be reduced by management plans and actions. On the other hand, one of the most effective approaches to promote the sustainable use of karst landscapes and caves is education. Show caves, in this context, are extremely important, since they can be used in instructive visits in order to provide a broader view of the environment for the cave visitors and incite more interest for the preservation of those unique and endangered habitats. Accordingly, show caves offer the opportunity to educate the general audience, since during visits technical information can be transmitted, thus providing a subtle inclusion of concepts, expanding the vision, as well as providing human interaction with the subterranean environment.

6 Threats

The Lagoa Santa Karst faces many environmental threats (see chapter "Environmental Problems in the Lagoa Santa Karst", this volume). The karst region of Lagoa Santa comprises one of the main areas of mineral exploitation of the state of Minas Gerais, with cement production as the main economic activity of the region. This mineral derivative represents an important economic pillar for the state, which is leader on the cement production of Brazil, with more than 14 million tons per year (CBIC 2013). According to Oss and Padovani (2003), for each ton of cement produced approximately 1.4 tons of limestone or similar rock is required. Thus, the accumulation of industries of such segment associated with the limited spatial distribution of Lagoa Santa karst constitutes relevant threats to the local subterranean environment.

The impacts of such economic activity over subterranean environments lead to the complete removal of bedrock for the production of commercial derivatives, alterations in groundwater routes of the local aquifer to access the mineral deposits and water supply of the industrial plant, besides the emission of gases and particulate sediments. The first mentioned impact promotes the complete alteration of landscape in function of the removal and processing of the rock mass. This process may cause the total or partial destruction of cavities and indirectly may interrupt the hydric flow responsible by the genesis and contribution of epigean trophic resources in the karst system. As it has been already demonstrated for other lithologies, the use of water during mining activities alters the balance between the aquifer recharge and the residence time of subterranean waters, limiting the capability of hydric resilience of the system (Gama and Matias 2015). In this sense, the lack of adequate management of the water resources may cause the extinction of many troglobitic amphibious species endemic of this karst area (Trajano and Bichuette 2010; Campos-Filho et al. 2014).

In relation to the emission of gases, studies demonstrate that this industrial sector is responsible for 5–7% of the total anthropic CO_2 emitted to the atmosphere (Chen et al. 2010). Additionally, the dust resulting from the process of cement production may accumulate in leaves of different vegetal strata, thus promoting necrosis and chlorosis in plants (Kumar et al. 2008). Considering the Lagoa Santa Karst, this is worrisome since the main trophic resources found in caves are allochthonous (Gomes et al. 2000; Pellegrini et al. 2016; Iniesta et al. 2012). Therefore, alterations on the quality and/or quantity of leaf litter carried into the caves may affect the fauna community, such as the local exclusion of species by inter or intraspecific competition.

The proximity of the Lagoa Santa Karst to a large metropolitan center makes the urban expansion and other anthropic activities possible threats to its conservation. The metropolitan region of Belo Horizonte, in which all the municipalities that compose the karst are included, is the third largest urban agglomeration of Brazil with population over 2.5 million people (IBGE 2016). Environmental impacts of the urban expansion vary from the excessive consumption and alterations of water quality, for modifications of the biodiversity and landscapes at different scales (Larson et al. 2009). According to White et al. (1986), the effects of urbanization over the karst are even more severe. The presence of urban centers affects the natural cycle of water infiltration in the soil, thus generating karst collapse through the increase of hydraulic gradient due to the lowering of the water table and alterations of the rainfall. The contamination of the water table by slurry and other anthropic wastes also occurs in some caves, given the proximity of dumps. One example occurred in a cave located in the municipality of Matozinhos (Meandro Abismante Cave). A faunal inventory conducted in 2002 revealed the presence of troglobitic fauna including an amphibious species of Styloniscidae (probably of the genus *Spelunconiscus* —Fig. 10b). After this inventory, a technical report was elaborated, which indicated a strong potential of contamination of the aquifer inside the cave given its proximity in relation to the municipal landfill of Matozinhos (Fig. 10a), what would have irreversible consequences for the fauna (especially for the troglobitic isopod that inhabited the small stream in the cave). This report also indicated the need to monitor such population in function of this potential impact. However, unfortunately, no attitude was taken by environmental agencies. In 2005, in a visit to the same cave, the presence of a dense biofilm was observed covering the entire

Fig. 10 **a** Landfill of the Matozinhos municipality (red arrows) located closely to the entrance of the Meandro Abismante Cave (white arrow); **b** *Spelunconiscus* sp. isopods present in the same cave before contamination; **c** dense biofilm covering the drainage bottom, as well as blackened fragments apparently from the contamination by the landfill

drainage bottom, as well as blackened fragments apparently from the contamination by the landfill (Fig. 10c). No specimen of the troglobitic Styloniscidae was observed during this visit. It is important to highlight that in 2002 dozens of individuals were seen along the accessible stretch of the drainage. Up to now, there is no more information about individuals of this species, and so it is possible that this important species, endemic and new to science, has been extinct even before it was described.

Furthermore, the economic and social development of the region results in a disordered exploration of natural resources and the substitution of important forest areas of Cerrado by grassland areas for livestock. This is the main threat to this biome, considered one of the hotspots for the conservation of world biodiversity (Myers et al. 2000). Nowadays, pastures for extensive livestock occupy almost 42% of the native area of Cerrado, and deforestation occurs at 30,000 km^2/year, a rate superior to those found for the Amazon region (Klink and Machado 2005). A recent study on different land use and its influence on the composition of terrestrial invertebrates from the Lagoa Santa Karst, pastures were present around all the studied caves, representing up to 76% of the buffer area (Pellegrini et al. 2016). In this work, the authors indicate the presence of limestone outcrops as the most determining factor for the composition of the subterranean community, as previously mentioned. Among other factors, this occurs due to the low anthropic influence, since the karst depressions in areas adjacent to the caves and outcrops are frequently used for crops and livestock (Pellegrini et al. 2016). The substitution of the native forests by monocultures generates a series of environmental problems like fragmentation, biodiversity loss, introduction of invasive species, erosion and soil degradation, water pollution, changes in the fire regime (natural impact of this biome), imbalances in the carbon cycle and probable modification of the regional climate (Klink and Machado 2005). Such alterations of epigean landscape may affect directly or indirectly the caves, thus hampering the maintenance of the environmental stability characteristic of subterranean environments.

Since the Lagoa Santa Karst lies close to large metropolitan areas, it presents a high density of roads and highways that makes possible the flow of people and services. Studies demonstrate that the establishment of a road network is among the main anthropic activities that affect directly or indirectly the dynamics of ecological processes, altering the physical, chemical, and biological properties of the environment (Coffin 2007; Forman and Alexander 1998). Direct impacts comprise the death of animals by trampling (Clevenger et al. 2003; Coelho et al. 2008) and isolation of populations (Shepard et al. 2008). Indirectly the roads favor the anthropic access and occupation in contiguous areas (Freitas et al. 2010) and alter the surface water runoff, thus promoting erosion in the soil and carrying the particulate material to other areas. Additionally, the facility of access makes marginal areas more susceptible to the fire (Laurance et al. 2009). The existence of these structures fractioning natural habitats interrupts the genetic flow, what may alter the permanence of populations in fragmented areas (Epps et al. 2005). Despite the absence of studies correlating the presence of roads and the effects of them over the hypogean fauna, it is likely that these anthropic actions may modify the structure of cave communities. Among the possible impacts, one may cite the alteration of the dynamics of trophic resources for the cave, what may severely influence the preservation and maintenance of the fauna. The proximity of roads may still generate constant vibrations in the caves, facilitate the accumulation of garbage and other anthropic wastes, besides altering the adjacent external landscape.

The presented threats represent only part of the risks to the maintenance of fauna and stability of subterranean environments. Since the Lagoa Santa Karst is inserted in an important Brazilian economic region and within the biome most used by agricultural activities, several other impacts may affect the caves and their surroundings (for example, forest fires, land invasion, pesticide contamination, vandalism and graffiti in caves, among others). Due to the social and economic plurality, preventing and mitigating the mentioned impacts is beyond the application of public policies and delimitation of areas for conservation. The awareness of the society and the assignment of value to the subterranean environment through environmental education are essential to guarantee the sustainable management of this karst.

References

Auler AS (1998) Base-level changes inferred from cave paleoflow analysis in the Lagoa Santa Karst, Brazil. J Cave Karst Stud 60 (1):58–62

Auler AS and Piló LB (2015) Lagoa Santa Karst: cradle of Brazilian cave studies. In: Vieira BC et al (eds) Landscapes and Landforms of Brazil. World geomorphological landscapes, pp 183–190

Bahia GR, Ferreira RL (2005) Influência das características físico-químicas e da matéria orgânica de depósitos recentes de guano de morcegos na riqueza e diversidade de invertebrados de uma caverna calcária. Revista Brasileira de Zoociências 1(7):165–180

Berbert-Born MLC (2000) The Lagoa Santa karst. Sítios Geológicos e Paleontológicos do Brasil. http://www.unb.br/ig/sigep/sitio015/sitio015.htm. Accessed 25 Oct 2016

Berbert-Born MLC, Horta LCS (1994) Informações básicas para gestão territorial: espeleologia – inventário de cavidades naturais da região Matozinhos-Mocambeiro. Companhia de Pesquisa de Recursos Minerais, Belo Horizonte

Berbert-Born MLC, Horta LCS, Dutra GM (1998) APA Carste de Lagoa Santa: Projeto vida - volume III Levantamento espeleológico. Companhia de Pesquisa de Recursos Minerais. Belo Horizonte

Berta A (1982) *Cerdocyon thous*. Mamm Species Arch 186:1–4

Bourne JD, Cherix D (1978) Note sur l'écophase souterraine de *Triphosa dubitata* L. (Lep. Geometridae) et *Scoliopteryx libatrix* L. (Lep. Noctuidae). Bulletin de la Société Vaudoise des Sciences Naturelles 354(74):147–156. https://doi.org/10.5169/seals-277411

Brescovit AD, Ferreira RL, Souza-Silva M, Rheims C (2012) *Brasilomma* gen. nov., a new prodidomid genus from Brazil (Araneae, Prodidomidae). Zootaxa (Auckland. Print) (3572):23–32

Bücherl W (1961) Aranhas do gênero *Loxosceles* e o Loxoscelismo na América. Ciência e Cultura 13(4):213–224

CamposFilho IS, Araujo PB, Bichuette ME, Trajano E, Taiti S (2014) Terrestrial isopods (Crustacea: Isopoda: Oniscidea) from Brazilian caves. Zool J Linn Soc 172(2):360–425

CamposFilho IS, Araujo PB, Bichuette ME, Taiti S (2016) Three new species of terrestrial isopods (Crustacea, Isopoda, Oniscidea) from Brazilian caves. Nauplius 24:e2016001

Carchini G, Rampini M, Sbordoni V (1994) Life cycle and population ecology of the cave cricket *Dolichopoda geniculata* (Costa) from Valmarino cave (Central Italy). Int J Speleol 23:203–218. https://doi.org/10.5038/1827-806X.23.3.6

CBIC (2013) Câmara Brasileira da Indústria da Construção http://www.cbicdados.com.br/menu/materiais-de-construcao/cimento. Accessed 26 Oct 2016

CEBS-UFLA (2016) Center of studies on subterranean biology from Federal University of Lavras, Minas Gerais, Brazil (www.biologiasubterranea.com.br)

Chen C, Habert G, Bouzidi Y, Jullien A (2010) Environmental impact of cement production: detail of the different processes and cement plant variability evaluation. J Clean Prod 18(5):478–485

Clevenger AP, Chruszcz B, Gunson KE (2003) Spatial patterns and factors influencing small vertebrate fauna road-kill aggregations. Biol Conserv 109(1):15–26

Coelho IP, Kindel A, Coelho AVP (2008) Roadkills of vertebrate species on two highways through the Atlantic Forest Biosphere Reserve, southern Brazil. Eur J Wildl Res 54:689–699

Coelho-Finamore JM, Freitas VC, Assis RR, Melo MN, Novozhilova N, Secundino NF, Pimenta PF, Turco SJ, Soares RP (2011) *Leishmania infantum*: Lipophosphoglycan intraspecific variation and interaction with vertebrate and invertebrate hosts. Int J Parasitol 41(3):333–342

Coffin AW (2007) From roadkill to road ecology: A review of the ecological effects of roads. J Transp Geogr 15:396–406

Costa-Lima A (1940) Um novo grilo cavernícola de Minas Gerais (Orthoptera: Grylloidea: Phalangositidae). Papéis Avuls Zool 6:43–50

Culver DC, Pipan T (2009) The biology of caves and other subterranean habitats. Oxford University Press, Oxford

Culver DC, Sket B (2002) Biological monitoring in caves. Acta Carsologica 31(1):55–64

Dantas AE, Horta CCR, Martins TM, Carmo AO, Oliveira-Mendes BBR, Goes AM, Kalapothakis E, Gomes DA (2014) Whole venom of *Loxosceles similis* activates caspases-3,-6,-7, and-9 in human primary skin fibroblasts. Toxicon 84:56–64

Dantas AE, Carmo AO, Horta CCR, Leal HG, Oliveira-Mendes BBR, Martins APV, Chavez-Olortegui C, Kalapothakis E (2016) Description of Loxtox protein family and identification of a new group of Phospholipases D from *Loxosceles similis* venom gland. Toxicon 120:97–106

Decu V (1986) Some considerations on the bat guano synusia. Travaux de l'Institute de Spéologie "Emile Racovitza" (25):41–51

Delgado A (1966) Investigacion ecológica sobre *Loxosceles rufipes* (Lucas) en la region costera del Perú. Mem Inst Butantan 33 (3):683–688

DeMatteo KE, Loiselle BA (2008) New data on the status and distribution of the bush dog (*Speothos venaticus*): evaluating its quality of protection and directing research efforts. Biol Conserv 141(10):2494–2505

Doube BM (1986) Spatial and temporal organization in communities associated with dung pads and carcasses. In: Gee JHR, Giller PS (eds) Organization of communities past and present. Blackwell Scientific Publications, Oxford, pp 255–280

Emmons LH, Leite YL, Kock D, Costa LP (2002) A review of the named forms of *Phyllomys* (Rodentia: Echimyidae) with the description of a new species from coastal Brazil. Am Mus Novit 1–40

Epps CW, Palsbøll PJ, Wehausen JD, Roderick GK, Ramey RR, Mccullough DR (2005) Highways block gene flow and cause a rapid decline in genetic diversity of desert bighorn sheep. Ecol Lett 8(10):1029–1038

Ferreira RL, Martins RP (1998) Diversity and distribution of spiders associated with bat guano piles in Morrinho cave (Bahia State, Brazil). Divers Distrib 4:235–241

Ferreira RL, Martins RP (1999) Trophic structure and natural history of bat guano invertebrate communities, with special reference to Brazilian caves. Trop Zool 12(2):231–252. https://doi.org/10.1080/03946975.1999.10539391

Ferreira RL, Pompeu PS (1997) Fatores que influenciam a riqueza e a diversidade da fauna associada a depósitos de guano na gruta Taboa, Sete Lagoas, Minas Gerais, Brasil. O Carste 2:30–33

Ferreira RL, Prous X, Machado SF, Martins RP (2005) Population dynamics os *Loxosceles similis* (Moenkhaus, 1898) in a Brazilian dry cave: a new method for evaluation of population size. Revista Brasileira de Zoociências 7(1):129–141

Ferreira RL (2000) Lixeiras de formigueiros: recursos adicionais em sistemas cavernícolas?. O Carste 3(12):154–158

Ferreira RL, Martins RP, Yanega D (2000a) Ecology of bat guano arthropod communities in a Brazilian dry cave. Ecotropica 6:105–116

Ferreira RL, Nonaka E, Rosa C (2000b) Riqueza e abundância de fungos associados a guano de morcegos na gruta da Lavoura. O Carste 12:46–51

Ferreira RL, Prous X, Martins RP (2007) Structure of bat guano communities in a Brazilian dry cave. Trop Zool 20:55–74

Ferreira RL, Martins VM, Paixão EA, Silva MS (2015) Spatial and temporal fluctuations of the abundance of Neotropical cave-dwelling moth *Hypena* sp. (Noctuidae, Lepidoptera) influenced by temperature and humidity. Subterr Biol 16:47–60

Filho JDA, Galati EAB, Falcão AL (2007) Nyssomyia intermedia (Lutz & Neiva, 1912) and Nyssomyia neivai (Pinto, 1926) (Diptera: Psychodidae: Phlebotominae) geographical distribution and epidemiological importance. Memórias do Instituto Oswaldo Cruz 102 (4):481–487

Fischer ML (1996) Biologia e ecologia de *Loxosceles intermedia* (Mello-Leitão, 1934) (Aranae, Sicariidae), no município de Curitiba, PR. Dissertation, Universidade Federal do Paraná

Foelix RF (1987) Biology of spiders. Harvard University Press, Cambridge

Fontanetti CS (1996) Description of three cave diplopods of *Pseudonannolene* Silvestri (Diplopoda, Pseudonannolenida, Pseudonannolenidae). Revista Brasileira de Zoologia 13(2):427–433

Forman RT, Alexander LE (1998) Roads and their major ecological effects. Annu Rev Ecol Syst 29:207–231

Freitas SR, Hawbaker TJ, Metzger JP (2010) Effects of roads, topography, and land use on forest cover dynamics in the Brazilian Atlantic Forest. For Ecol Manag 259(3):410–417

Gama EM, Matias GP (2015) Hidrogeologia e os geossistemas ferruginosos. In. Carmo FF, Kamino LHY (eds) Geossistemas ferruginosos do Brasil: Áreas prioritárias para conservação da diversidade geológica e biológica, patrimônio cultural e serviços ambientais. 3i Editora, Belo Horizonte, pp 103–124

Gilpin ME, Hanski IA (1991) Metapopulation dynamics: empirical and theoretical investigations. Academic Press, London

Gil-Santana HR, Zeraik SO, Costa LAA (1999) Sinonimização dos gêneros *Amilcaria* Wygodzinsky, 1950 e *Mayemesa* Wygodzinsky, 1945 e descrição do macho de M. lapinhaensis (Wygodzinsky, 1950), Combinação Nova (Hemiptera, Reduviidae, Emesinae) Boletim do Museu Nacional. Zoologia 400:1–10

Gnaspini P (2012) Guano communities. In: White WB, Culver DC (Org.) Encyclopedia of caves, 2 edn. Academic Press/Elsevier, Oxford, Waltham, pp 357–364

Gomes FTMC, Ferreira RL, Jacobi CM (2000) Comunidade de artrópodes de uma caverna calcária em área de mineração: composição e estrutura. Revista Brasileira de Zoociências 3(2): 77–96

Grelle CE, Fonseca MT, Moura RT, Aguiar L (1997) Bats from karstic area on Lagoa Santa, Minas Gerais: a preliminary survey. Chiropt Neotrop 3(1):68–70

Guimarães MM, Ferreira RL (2014) Morcegos cavernícolas do Brasil: novos registros e desafios para conservação. Revista Brasileira de Espeleologia 2(4):1–33

Hamilton-Smith E (2004) Tourist caves. In: Gunn J (ed) Encyclopedia of caves and karst science. Taylor and Francis, New York, pp 1554–1561

Harris JA (1970) Bat-guano cave environment. Science 169:1342–1343

Harrison S, Ross SJ, Lawton JH (1992) Beta diversity on geographic gradients in Britain. J Anim Ecol 61:151–158

Hermann G, Kohler HC, Duarte JC Carvalho PGS (1998) APA Carste de Lagoa Santa - Volume III: meio biótico. IBAMA, Belo Horizonte

Herzig-Straschil B, Robinson GA (1978) On the ecology of the fruit bat, *Rousettus aegyptiacus leachi* (A. Smith, 1829) in the Tsitsikama coastal National Park. Koedoe 21:101–110

Hutchinson P (1950) Vertebrate excretion (cave guano). Bull Am MusM Nat Hist 96:381–485

IBGE - Instituto Brasileiro de Geografia e Estatística (2016) Estimativas populacionais para os municípios brasileiros em 01/07/2016. http://www.ibge.gov.br/home/estatistica/populacao/estimativa2016/estimativa_tcu.shtm. Accessed 25 Apr 2017

Iniesta LFM, Ázara LN, Souza-Silva M, Ferreira RL (2012) Biodiversidade em sete cavernas no Parque Estadual do Sumidouro (Lagoa Santa, MG). Revista Brasileira de Espeleologia 2(2):18–37

Klimchouk A (2004) Towards defining, delimiting and classifying epikarst: its origin, processes and variants of geomorphic evolution. Speleogenesis Evol Karst Aquifers 2(1):1–13

Klink CA, Machado RB (2005) Conservation of the Brazilian Cerrado. Conserv Biol 19(3):707–713

Koleff P, Gaston KJ, Lennon JJ (2003) Measuring beta diversity for presence–absence data. J Anim Ecol 72:367–382

Kumar SS, Singh NA, Kumar V, Sunisha B, Preeti S, Deepali S, Nath SR (2008) Impact of dust emission on plant vegetation in the vicinity of cement plant. Environ Eng Manag J 7(1):31–35

Kunz TH (1982) Roosting ecology of bats. In: Kunz TH (ed) ecology of bats. Plenum Press, New York, pp 1–55

Larson KL, Casagrande D, Harlan SL, Yabiku ST (2009) Residents' yard choices and rationales in a desert city: social priorities, ecological impacts, and decision tradeoffs. Environ Manag 44 (5):921–937

Laurance WF, Goosem M, Laurance SG (2009) Impacts of roads and linear clearings on tropical forests. Trends Ecol Evol 24(12):659–669

Lienhard C, Ferreira RL (2015) Review of Brazilian cave psocids of the families Psyllipsocidae and Prionoglarididae (Psocodea: Psocoptera: Trogiomorpha) with a key to the South American species of these families. Rev Suisse Zool 122(1):121–142

MacDonald D, Crabtree J, Wiesinger G (2000) Agricultural abandonment in mountain areas of Europe: Environmental consequences and policy response. J Environ Manag 59(1):47–69. https://doi.org/10.1006/jema.1999.0335

Machado FS, Lima IJ, Lopes APM, Moura AS, Abreu TCK (2016) New occurrences and biological aspects to four species of rodents (Mammalia: Cricetidae) from Brazil. Revista Agrogeoambiental 8 (2):35–51. http://dx.doi.org/10.18406/2316-1817v8n22016800

Mahnert V (2001) Cave-dwelling pseudoscorpions (Arachnida, Pseudoscorpiones). Rev Suisse Zool 108(1):95–148

Matavelli R, Campos AM, Feio RN, Ferreira RL (2015) Occurrence of Anurans in Brazilian caves. Acta Carsologica 44(1):107–120

Mazza G, Reboleira ASPS, Gonçalves F, Aquiloni L, Inghilesi AF, Spigoli D, Stoch F, Taiti S, Gherardi F, Tricarico E (2014) A new threat to groundwater ecosystems: first occurrences of the invasive crayfish *Procambarus clarkii* (Girard, 1852) in European caves. J Cave Karst Stud 76(1):62–65

Mews CM, Sperber CF (2008) A new species of *Endecous* Saussure, 1878 and redescription of *Endecous cavernicolus* Costa-Lima, 1940 (Orthoptera: Grylloidea: Phalangopsidae). Stud Neotrop Fauna Environ 43(2):159–167

Mitchell RW (1969) A comparison of temperate and tropical cave communities. Southwest Nat 14(1):73–88. https://doi.org/10.2307/3669249

Myers N, Mittermeier RA, Mittermeier CG, Fonseca GA, Kent J (2000) Biodiversity hotspots for conservation priorities. Nature 403 (6772):853–858

Novak T, Sambol J, Janžekovič F (2004) Faunal dynamics in the Železna Jama cave. Acta Carsologica 33(2):249–267

Oliveira SMPD, Moraes BAD, Gonçalves CA, Giordano-Dias CM, d'Almeida JM, Asensi MD, Mello RP, Brazil RP (2000) Prevalence of the microbiota in the digestive tract of wild-caught females of *Lutzomyia longipalpis* (Lutz & Neiva) (Diptera: Psychodidae). Rev Soc Bras Med Trop 33(3):319–322

Oss HG, Padovani AC (2003) Cement manufacture and the environment part II: environmental challenges and opportunities. J Ind Ecol 7(1):93–126

Pellegrini T, Sales LP, Aguiar P, Ferreira RL (2016) Linking spatial scale dependence of land-use descriptors and invertebrate cave community composition. Subterr Biol 18:17–38

Pena MS, Salgado NC, Coelho AC (2005) New species of *Thaumastus* from Lagoa Santa region, Minas Gerais State, Brazil (Mollusca, Gastropoda, Pulmonata, Bulimuloidea). Iheringia Série Zoologia 95 (1):21–24

Pinto-da-Rocha R (1995) Sinopse da fauna cavernícola do Brasil (1907–1994). Papéis avulsos de Zoologia 39:61–73

Prous X, Ferreira RL, Martins RP (2004) Ecotone delimitation: epigean-hypogean transition in cave ecosystems. Austral Ecol 29:374–382

Prous X, Ferreira RL, Jacobi CM (2015) The entrance as a complex ecotone in a Neotropical cave. Int J Speleol 44:177–189

Rabelo LM (2016) Ecologia e conservação de cavernas no eixo Centro-norte de Minas Gerais, Dissertation, Universidade Federal de Lavras

Sábio PB, Andrade AJ Galati EA (2015) Redescription of *Lutzomyia (Lutzomyia) renei* Martins, Falcão & Silva, 1957 (Diptera: Psychodidae: Phlebotominae). Zootaxa, 3999(4):589–599

Schubart O (1956) "Leptodesmidae" brasileiras II. O gênero *Obiricodesmus* (Diplopoda, Proterospermophora). Revta Bras Biol 16 (3):341–348

Shepard DB, Kuhns AR, Dreslik MJ, Phillips CA (2008) Roads as barriers to animal movement in fragmented landscapes. Anim Conserv 11(4):288–296

Silva CR, Percequillo AR, Ximenes GL, Vivo M (2003) New distributional records of Blarinomys breviceps (Winge, 1888) (Sigmodontinae, Rodentia). Mammalia 67(1):147–152

Silveira PSA, Bernardi, LFO, Pepato AR (2015) New records of the genus *Whartonia* (Acari, Leeuwenhoekiidae) associated with the bat *Carollia perspicillata* from southeastern Brazil. Check List 11 (6):1793. http://dx.doi.org/10.15560/11.6.1793

Simões MH (2013) Invertebrados cavernícolas: subsídios para a determinação de cavernas e áreas prioritárias para a conservação no noroeste de Minas Gerais. Dissertation, Universidade Federal de Lavras

Simões MH, Souza-Silva M, Ferreira RL (2015) Cave physical attributes influencing the structure of terrestrial invertebrate communities in Neotropics. Subterr Biol 16:103–121

Souza MFVR (2012) Diversidade de invertebrados subterrâneos da região de Cordisburgo, Minas Gerais: subsídios para a definição de cavernas prioritárias para a conservação e para o manejo biológico de cavidades turisticas. Dissertation, Universidade Federal de Lavras

Souza-Silva M, Bernardi LFO, Ferreira RL (2005) Caracterização sistêmica da gruta da Lavoura (Matozinhos, MG): aspectos topoclimáticos, tróficos e biológicos. In: XXVIII Congresso Brasileiro de Espeleologia, Campinas, pp 109–115

Souza-Silva M, Martins RP, Ferreira RL (2011) Trophic dynamics in a neotropical limestone cave. Subterr Biol 9:127–138

Strayer DL, Power ME, Fagan WF, Pickett STA, Belnap J (2003) A classification of ecological boundaries. Bioscience 53(8):723–729

Talamoni SA, Coelho DA, Dias-Silva LH, Amaral AS (2013) Bat assemblages in conservation areas of a metropolitan region in Southeastern Brazil, including an important karst habitat. Braz J Biol 73(2):309–319

Tavares VC, Aguiar LM, Perini FA, Falcão FC, Gregorin R (2010) Bats of the state of Minas Gerais, southeastern Brasil. Chiropt Neotrop 16(1):675–705

Tobin BW, Hutchins BT, Schwartz BF (2013) Spatial and temporal changes in invertebrate assemblage structure from the entrance to deep-cave zone of a temperate marble cave. Int J Speleol 42(3):203–214. https://doi.org/10.5038/1827-806X.42.3.4

Torquetti CG, Assis MAC, Carmo SSA, Talamoni SA (2013) Observations of a diurnal roost of the white-winged vampire bat *Diaemus youngi* in a karstic area of southeastern Brazil. Chiropt Neotrop 19(1):1147–1150

Trajano E, Bichuette ME (2010) Diversity of Brazilian subterranean invertebrates, with a list of troglomorphic taxa. Subterr Biol 7:1–16

Trajano E, Golovatch SI, Geoffroy JJ, Pinto-da-Rocha R, Fontanetti CS (2000) Synopsis of Brazilian cave-dwelling millipedes (Diplopoda). Papéis Avulsos de Zoologia 41(18):259–287

White EL, Aron G, White WB (1986) The influence of urbanization of sinkhole development in central Pennsylvania. Environ Geol Water Sci 8(1–2):91–97

Whittaker RH (1960) Vegetation of the Siskiyou Mountains, Oregon and California. Ecol Monogr 30:279–338

Wygodzinsky P (1950) Sobre alguns "Emesinae" (Reduviidae, Hemiptera). Revta bras Biol 10(1):73–78

Zampaulo RA (2010) Diversidade de invertebrados cavernícolas na província espeleológica de Arcos, Pains e Doresópolis (MG): subsídios para a determinação de áreas prioritárias para a conservação. Dissertation, Universidade Federal de Lavras

Cave Paleontology in the Lagoa Santa Karst

Cástor Cartelle

Abstract

In Brazil, the majority of fossils of Quaternary tetrapods were found in caves. Among the fossil sites, the Lagoa Santa Karst stands out, especially regarding mammal paleontology. This chapter aims to present a review of the paleontological research in the area, which started in the beginning of the nineteenth century with Danish naturalist Peter Wilhelm Lund. Besides having discovered 22 extinct mammal species, Lund also described eight living species. Lund also performed studies about other vertebrates such as birds. He was also the first to discover remains of humans associated with the extinct megafauna. It took nearly one hundred years before systematic paleontological research in Lagoa Santa was resumed, through the work of members of the Academia Mineira de Ciências (Minas Gerais Academy of Sciences) and National Museum at Rio de Janeiro. Among the scientists, the names of Aníbal Mattos, H.V. Walter, Fausto Souza Cunha, and Carlos de Paula Couto stand out. Up to now, the life and works of Peter Lund are still being researched. In the 2000s, more sophisticated techniques were applied to Lagoa Santa fossil sites, including radiometric dating and a taphonomical approach, allowing the confirmation of the contemporaneity between humans and the extinct fauna. Efforts were also made toward understanding the mechanisms associated with the formation of the fossiliferous cave deposits in the area.

Keywords

Quaternary mammals • Lagoa Santa Karst • Peter Wilhelm Lund • Cave paleontology • Minas Gerais

C. Cartelle (✉)
Pontifícia Universidade Católica de Minas Gerais, Dom José Gaspar St., 500, Coração Eucarístico, Belo Horizonte, MG CEP 30535-610, Brazil
e-mail: cartelle@pucminas.br

1 Lund's Discoveries in Lagoa Santa

Peter Wilhelm Lund (Fig. 1) was born in Copenhagen, Denmark in 1801 and died in Lagoa Santa (state of Minas Gerais, Brazil) in 1880. He went to medical school at the University of Copenhagen for 6 years. At that time, Natural Sciences were studied as a complementary program for students in the Medical and the Theology programs. After graduating, he exchanged the hospital for the Botanical Garden and The Natural History Museum of Denmark. Subsequently, he received a grant to gather zoological and botanical collections for the aforementioned institutions and decided to travel to Brazil where he arrived in 1825. His work was mainly in the vicinity of Rio de Janeiro, Niterói, and Nova Friburgo, in the state of Rio de Janeiro. He collected botanical material, fish, skulls, and skeletons of mammals, insects, and shells, for the future king Prince Christian's (1786–1848) collection.

He returned to Europe 4 years later. After defending his doctoral dissertation at the University of Kiel (Germany) in November 1829, he traveled to several European countries: Italy, Austria, Germany, and Switzerland, contacting renowned scientists of the time. After that, he arrived in Paris in 1830, remaining there for 9 months. There he contacted several famous naturalists such as von Humboldt (1769–1850), Milne-Edwards (1800–1855), and Audouin (1793–1841). The latest introduced Lund to Cuvier (1769–1832); he then attended the famous scientist's History of Science Course as a listener at the Collège de France. Cuvier influenced Lund's way of interpreting the history of living beings (Hoch 1982).

As a student Lund read Cuvier's first book of Zoology: *Tableau élementaire d'histoire naturelle des animaux*, published in 1798 and translated into Danish in 1801. In 1824, Lund wrote a thorough essay on physiology based on experimentation, as opposed to writing from philosophical speculations, which was common at the time. The essay was published in German the following year and was used as

A. S. Auler and P. Pessoa (eds.), *Lagoa Santa Karst: Brazil's Iconic Karst Region*, Cave and Karst Systems of the World, https://doi.org/10.1007/978-3-030-35940-9_11

Fig. 1 Peter W. Lund (1801–1880). Photo by E. Warming around 1865. Collection of the Natural History Museum of Copenhagen

textbook in several European universities (*Physiologische Resultate der Vivisectionen neuerer Zeit*). In the Preface, Lund mentions Cuvier as a renowned person in the field of zoology.

During his stay in Paris, in addition to Cuvier's course, he attended courses in astronomy, physics, and chemistry with Professors Biot (1774–1862), Ampère (1775–1836), and Thénard (1776–1857), respectively. In 1831, the *Annales des Sciences Naturelles* published a paper by Lund: "*Lettre sur les Habitudes de Quelques Fourmis du Brésil, adressée to M. Audouin.*" Lund brought from Brazil a collection of ants with 30 different species, only three had been previously described.

The death of his mother, the monotony of the European fauna and flora, and the climate, made Lund decline the invitation of several universities and return to Brazil, where he arrived in 1833, a year after Charles Darwin (1809–1882) was in Rio de Janeiro, and stood there until his death.

He immediately resumed his studies in Rio de Janeiro, where he was in touch with members of the Langsdorff Expedition, in special the naturalist Olfers (1793–1872) and the botanist Riedel (1790–1861), both Germans. Olfers narrated the findings of two giant sloths in Rio Grande do Sul which were sent to Germany, while Riedel, who in 1829 finished his work on the expedition mentioned, would become the Director of the Botanical Garden of Rio de Janeiro. Riedel and Lund decided to carry out an expedition for collecting samples of the Brazilian Cerrado flora. The project included traveling through the States of Rio de Janeiro, São Paulo, and Minas Gerais, ending in the capital of the state of Goiás. The expedition departed in October 1833.

The obvious difficulties of the trip along with a disease that struck Riedel were responsible for aborting the expedition when they reached the area currently named Uberaba, a city in the Center-West of Minas Gerais. On the way back, they crossed the São Francisco River in Paracatu (Minas Gerais). They went to Ouro Preto where they took the road Estrada Real, leading to Rio de Janeiro.

In Santo Antônio de Curvelo, a city in the center of Minas Gerais, they spent the night on a ranch. And by chance the Danish naturalist–botanist had his life changed. The fortuitous meeting with an also Danish guest at the ranch, Peter Claussen (1801–1872), resulted in a radical change in the scientific life of P. W. Lund.

Like Riedel, Claussen during the Langsdorff Expedition worked with the German paleontologist Sellow gathering fossils in the state of Rio Grande do Sul and in Uruguay. Sellow drowned in the Doce River, in Minas Gerais, and Claussen eventually acquired a small farm near Santo Antônio de Curvelo, where there were numerous caves in which he obtained saltpeter and mammal fossils he traded with European museums. Most certainly his encounter with Claussen made him relive his contact with Cuvier and with paleontology, the new science. This new science became very important in the early nineteenth century, mostly due to the work of two exceptional scientists, the French Georges Cuvier and the English Richard Owen (1804–1892).

The trip continued to Ouro Preto, which was at that time the capital of the "Province" Minas Gerais. Recovered from his illness, Riedel continued his trip to Rio de Janeiro but Lund returned to the vicinities of Santo Antônio de Curvelo. On his way back, he went through the small town of Lagoa Santa, which had approximately five thousand inhabitants. There were hundreds of cave openings in the vicinity where he chose to live.

The decision was made: his interest in botany shifted to paleontology. His first excavation took place in the Maquiné Cave in 1835, which had been discovered shortly before. Impressed by the odd beauty of the cave, he described it in his First Memoir (Lund 1837). In addition to the narration of his first paleontological findings, he published the wonderful cave map (Fig. 2) made by his recently hired secretary and draftsman Peter Andreas Brandt (1792–1862), Norwegian, who worked with him until his death.

Lund's numerous publications proved his magnificent intellectual readiness of a genius. In addition, they showed a progressive evolution in the accuracy of his taxonomic

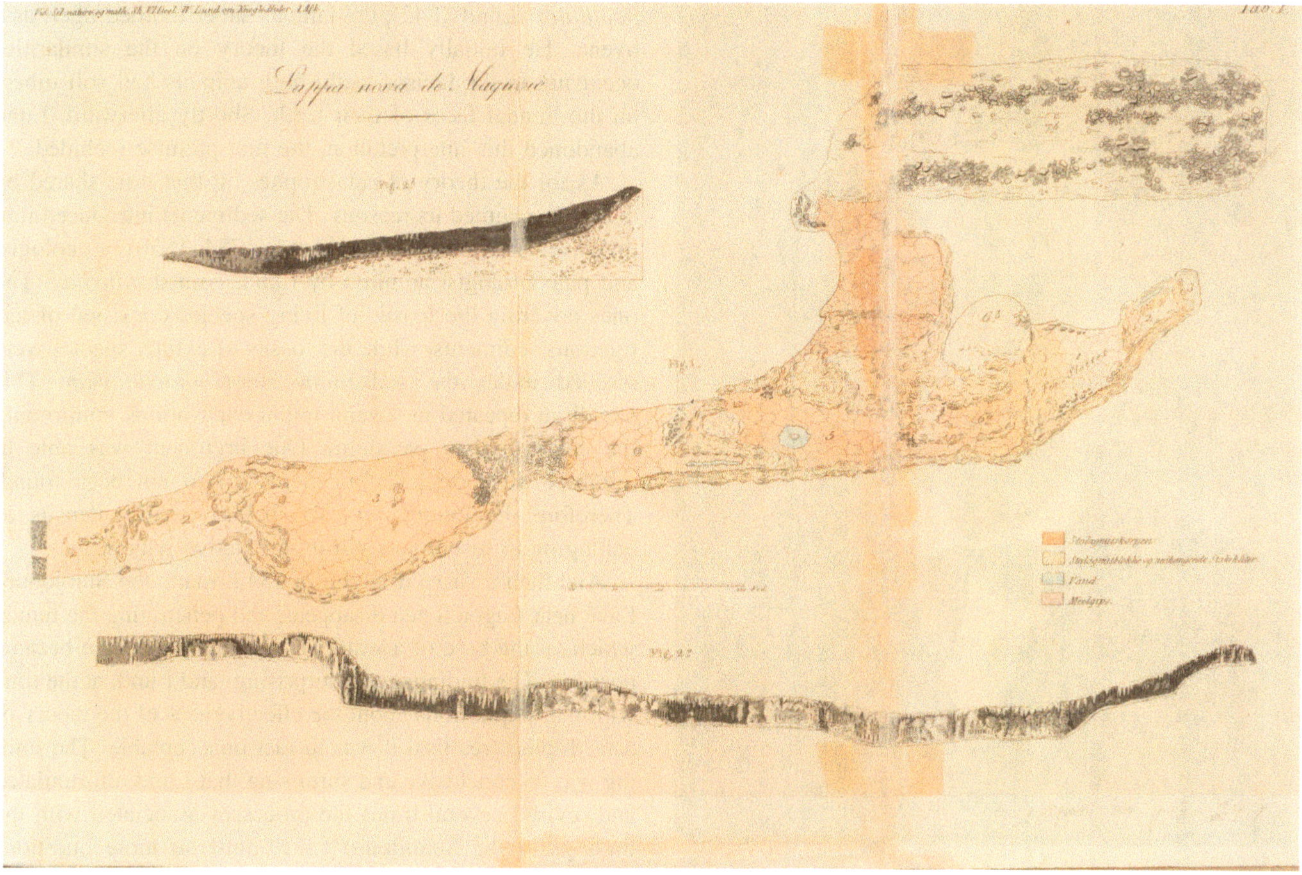

Fig. 2 Map of Maquiné Cave (Lapa Nova de Maquiné). Survey and drawing by P. A. Brandt. From Lund (1837)

determinations. Lund pioneered the survey of the Quaternary faunal composition of a region, identifying numerous species of mammals. Showing an encyclopedic knowledge in his isolation in Lagoa Santa, over the years he modified his taxonomic determinations being able to identify among the recovered fossils the extinct species and the living ones. Throughout his memoirs, he offered ever more accurate lists of the extinct and the living species whose fossils he encountered during expeditions to the enormous number of eight hundred caves. He has thus reached an unprecedented achievement: a broad and comprehensive survey of the species of a region in two distant historical moments separated by ten thousand years. The exactness of his taxonomic determinations shows his vast knowledge. The mistakes made by him can be considered to be almost "logical."

The repetition of his findings and the high costs incurred in the excavations were decisive for Lund's termination of the project, donating the collection consisting of over 12,000 fossil pieces—most of the items being mammals—to King Christian VIII in Denmark, in 1845. In his letter to the King that same year, Lund explained: "This collection must be used by science as soon and as completely as possible due to interest and the scientific value of it" (Lund and Paula Couto 1950: 10). The material was sent along with a manuscript Catalog with numerical record, morphological and specific identifications, as well as the location of each finding. In addition, he has also sent a collection of breccias from several caves, as well as a huge collection of micro mammal fossils collected from a chamber in Escrivania Cave. These small bones were the result of regurgitations of successive generations of owls on the site, which left such record at the end of the Pleistocene—there will be further development on this finding.

The collection sent to Denmark became a reference in Europe and placed Lagoa Santa in the world scientific circuit. The discoveries were, in addition to being original, broad, and diverse, and they became the subject for several researchers' studies.

The monumental piece of work entitled *E Museo Lundii* deserves special emphasis. It started in 1888 and ended in 1915, consisting of five volumes with high-quality printing and images. 1,368 pages were published, 1,188 of which were written by Herluf Winge (Fig. 3). These works were enriched by 81 slides containing 564 images. Recently, an *Addendum to E Museo Lundii* was published with 103 pages and 41 images (Hansen 2012). Winge's studies cover practically all the

Fig. 3 Herluf Winge (1857–1923). Winge published several works based on Lund's collection, mainly related by mammals. Collection of the Natural History Museum of Copenhagen

species discovered by Lund. He is rightly considered one of the leading mastozoologists of the twentieth century, being compared to P. W. Lund. His works dedicated to the study of the material collected by Lund in Lagoa Santa took place in Winge (1887, 1892, 1893, 1895a, b, 1906, 1915).

2 The Mammal Fossils Found by Lund

When beginning his paleontological work in Lagoa Santa, Lund worked with two premises guiding his research: the composition of the fauna he discovered should be like a replica of the old-world fauna, including Africa; and Cuvier's theory of catastrophes should ground the interpretation of the findings through the new scientific path undertaken by Lund.

Thus, during his first excavation, which was in the Maquiné Cave, about 50 km from Lagoa Santa, the neophyte paleontologist identified a typical South American deer, *Mazama*, as an antelope, a species from the Old Continent that had never occurred in America. In addition, in 1842 Lund would first identify the feline *Smilodon populator* (Lund 1842), the famous saber-toothed tiger, as a hyena. He initially based the theory on the similarities occurring in the incisor teeth; both animals had soft tubers on the lingual faces of their teeth. Shortly afterward, Lund abandoned this interpretation, the first premise included.

As for the theory of catastrophes, at that time shared by many, he claimed its reasons. The sediments introduced into the caves showed what Buckland (1784–1856), a geologist and paleontologist admired by Lund, coined *diluvium*. The ones covering the fossils of living species consisted of terrigenous sediments, while the fossils of extinct species were rescued under the stalagmitic floor, *antediluvian*. This paradigm repeated in several initial excavations, reinforcing the catastrophism paradigm. One argument was able to reinforce this point: A human fossil had not been found. Therefore, he interpreted cave paintings and utensils as belonging to recent and still-living Native Americans.

And then a sharp drought, in 1843, made the Sumidouro Lake near Lagoa Santa disappear, and penetrating the tunnel which, at the base of a wall, was the lagoon's drain became possible. The findings were surprising, and Lund, at the time with many questions about the effectiveness of the theory of catastrophes, realized it was in fact unacceptable. The finding was so conclusive and surprising that Lund reformulated and revised several times the processes associated with the findings in the Sumidouro Lake, until no more questions were left. In the same sediment, an "impossible" association according to the Cuvierian theory occurred: human fossils along with remains from extinct and still-living species. Lund carried out a detailed and accurate description of the findings, justifying his conclusions (Piló et al. 2004, 2005). The theory of catastrophes was finally refuted and Lund's second premise guiding his research could not be maintained.

Following the publication of these discoveries, Lund considered his paleontological research almost done. The scientific isolation and the high costs of the excavations were confluent causes for this decision. From then on, he dedicated himself mainly to what he identified as the "sweet idleness of the countryside." The discoveries in the Sumidouro Lake were the last straw, making him realize that his mentor's theory was incorrect. In addition to the above-mentioned reasons, the fear of criticism which could arise in Europe contributed to his behavior. He justified himself by stating that the findings had become repetitive. In that particular aspect, he was quite right. After Lund's findings, so far, only one extinct mammal species has been discovered in Lagoa Santa.

In his works, Lund continuously published lists containing extinct species and living species whose fossils he recovered, along with numerous references to the fauna of mammals existing in the region. According to Lund and

Paula Couto (1950), Lund recorded 149 species of mammals. Thirty-one were first described by Lund, 22 of which were extinct species and nine living species (Cartelle 2012). It was a unique, absolutely original wide survey in such a geographically restricted area, including species which were distant in time.

The discoveries which prove Lund's exceptional knowledge are now highlighted. Over time, some species have been synonymized while others have been modified. But comprehensiveness and identifications attest to the genius of the most Brazilian Danish person.

The first list was elaborated by Lund and Paula Couto (1950); then, it was updated by Cartelle (2012). And this is the list we use hereafter—with only minor modifications. The rapid dynamics of Systematics allows the interpretation or correction of the nomenclature previously used. In bold, we indicate the extinct species and in italics the still-living species, both as Lund has established.

Among the Primates, he discovered the extinct species ***Protopithecus brasiliensis*** (Lund 1839). All extinct species of Carnivorous in Lagoa Santa were determined by Lund: the Canidae ***Protocyon troglodytes*** (Lund 1838), ***Speothos pacivorus*** (Lund 1839) (Fig. 4), the Felidae ***S. populator***, and the Ursidae ***Arctotherium brasiliense*** (Lund 1839).

Among the Xenarthrans, besides fossils of extant species of anteaters and armadillos, he determined a group of extinct species. In Pilosa, with material from his first excavation, he coined the Nothrotheriidae ***Nothrotherium maquinense*** (Lund 1839), and later the Megatheridae ***Eremotherium laurillardi*** (Lund 1842), the Mylodontidae ***Ocnotherium giganteum*** (Lund 1839), ***Catonyx cuvieri*** (Lund 1839), and ***Valgipes bucklandi*** (Lund 1844). Among the Cingulata are the Dasipodidae ***Cabassous squamicaudis*** (Lund 1844), ***Propraopus punctatus*** (Lund 1839), and ***Propraopus sulcatus*** (Lund 1842), and also, the Pampatheriidae ***Pampatherium humboldti*** (Lund 1839) and finally, the Glyptodontidae ***Hoplophorus euphractus*** (Lund 1839).

Among the Perissodactyl, he identified ***Equus neogeus*** (Lund 1840), and ***Hippidion principale*** (Lund 1845). Among the Artiodactyla, Lund identified the extinct ***Brasiliochoerus stenocephallus*** (Lund in Gervais 1867–69) and the Laminae ***Palaeolama major*** (Lund in Liais 1872).

As for the Rodents, Lund identified the hedgehog ***Coendou magnus*** (Lund 1839), the giant capybara ***Neochoerus sulcidens*** (Lund 1839), and the giant ***Cuniculus major*** (Lund 1839) which, according to Mayer et al. (2016), should be referred to as ***Cuniculus rugiceps*** (Lund 1837).

Among the still-living species, Lund identified the Didelphomorphia *Didelphis alviventris* (Lund 1840) and *Marmosops incanus* (Lund 1840); as for Carnivora, the *Lycalopex vetulus* (Lund 1842) and the *Speothos venaticus* (Lund 1842); and Rodents: *Rhipidomys mastacalis* (Lund 1840), *Carterodon sulcidens* (Lund 1839), *Phyllomys brasiliensis* (Lund 1840), and *Tricomys apereoides* (Lund 1839).

Winge, in the material sent by Lund, identified two other extinct species: the Perissodatyls *Tapirus cristatellus* (Winge 1906) and the Rodent *Juscelinomys vulpinus* (Winge 1887). Among the fossils recovered by Lund, Winge (1906), holding a few cranial remains, identified the Gomphothere *Mastodon andium* (Cuvier 1824) (= *Stegomastodon waringi* (Holland 1920)). With seven cranial fragments, he also identified the Toxodontinae *Toxodon platensis* (Owen 1837). However, due to the existence in sympatry in the Brazilian intertropical territory of this species and the Toxodontinae *Piauhytherium capivarae* (Guérin and Faure 2013), Guérin and Faure (2013) such identification is uncertain. The material recovered by Lund could belong to this species. Even less numerous was the material Winge attributed to Litopterna *Macrauchenia patachonica* (Owen 1838): vertebral fragments and a metapodial distal epiphysis. It is possible that such material belongs to the species *Xenorhinotherium bahiense* (Cartelle and Lessa 1988), since there are no records in intertropical territories of *M. patachonica*.

Lund named 22 of the 27 extinct species he identified. Lund determined eight of the 123 living species he collected. Few naturalists were able to make that many discoveries and taxonomic determinations. Some of the species, mentioned above, are worth being further discussed because of their exceptional importance.

3 Some of Lund's Emblematic Findings

Lund began his publications, as we have already pointed out, in 1825. He then wrote a few articles on invertebrates, the stomach of a bird, and botany. His first publication concerning his discoveries in Lagoa Santa was in 1837, and the last ones in 1846. Lund's complete bibliography about his findings in Lagoa Santa was published by Mones (1986). This bibliographic survey—the most exhaustive so far—includes original works, abstracts, and some translations of his works. In this list, Mones has entered 60 publications, 38 of which are in Danish, nine are in German, 7 are in French, 5 are in the Portuguese, and 2 are in English.

We have already pointed out that the first excavation carried out by Lund was in the Maquiné Cave, on the outskirts of Lagoa Santa. There was little material found, and it does not differ from the findings made later in Lagoa Santa Caves. *N. maquinense* was his first discovery and the first species identified by him as extinct. Later, he would find further subjects of that species in caves in the area of Lagoa

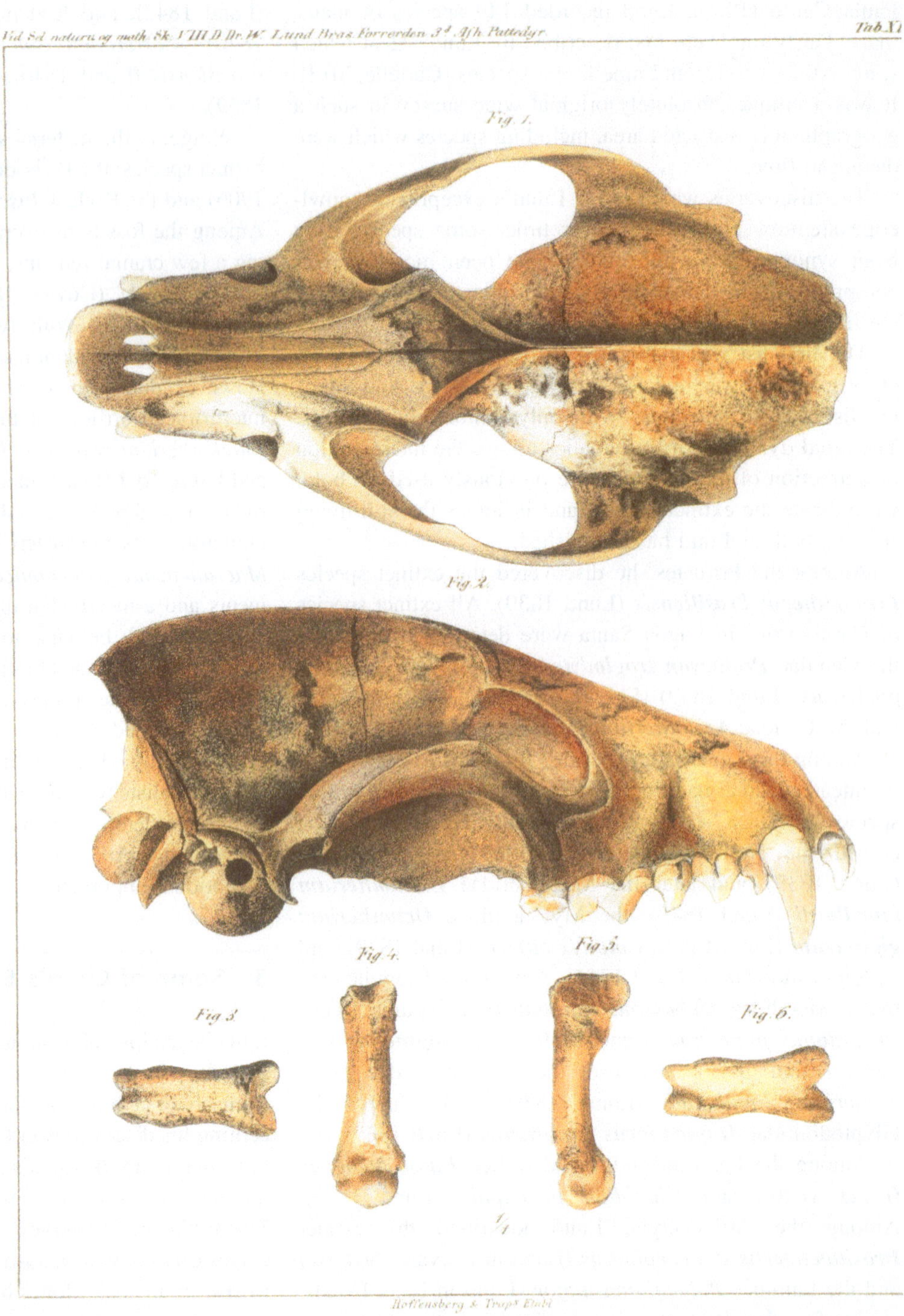

Fig. 4 Skull and foot bones of the canid *Speothos pacivorus*. Hand-colored illustration in Lund (1840)

Santa. In his first report, he tells about the finding that which today we understand is the symbol species of the Brazilian paleontology.

Initially, he identified the species he had found in the Maquiné Cave as a *Megatherium* of the size of a tapir. This interpretation was heavily influenced by events that had happened earlier. The discovery, in 1788, of *Megatherium americanum* (Cuvier 1796) was emblematic for the South American paleontology. A very complete skeleton found by a priest in the area where Luján (Argentina) is settled was sent to Spain and kept in the Madrid's Royal Natural History Cabinet. The Institution's researchers identified it as a South American elephant and set it up with palmar and plantar supports—currently, it is still exhibited this way. Being an exceptional and historical finding, it was kept as original, even if erroneous and with the reconstruction of some missing pieces. During his first stay in Brazil, Lund became acquainted with the findings of Megatherians through Olfers, but he most likely extended his knowledge when in contact with Cuvier. He obtained drawings of the Madrid skeleton and concluded that the South American "elephant" was, in fact, a giant extinct ground sloth. And the *Megatherium* became a consumer's dream for more than a few European museums. *N. maquinense* was much smaller than *Megatherium* and would eventually be identified as a different family.

The Xenarthra had Lund's special attention. In addition to the work by Cuvier, he learned about the sloths discovered in North America, later expanding his knowledge with some of Owen's works, among others. He gradually changed concepts in such a way that in his final works the identifications were very consistent. He rightly attributed ossicles shaped like small nodules to osteoderms, bones located in the dermis of some extinct sloths. From the skeletal remains, he inferred the habits and behaviors of these animals.

Lund found two molariforms belonging to young sloths in two separate caves. Even with so little material, the determination at the time was accurate: *Megatherium laurillardi* (Lund 1842). With such limited material, he was able to identify a large sloth. Only in the second half of the twentieth century did the exact name of the species come to be known as *E. laurillardi*. At the end of this chapter, we will mention this species again.

Also noteworthy is the determination of several extinct Cingulata, the Xenarthra with exoskeleton. Aware of Sellow's research, probably through Claussen, who had described a Glyptodont, Lund identified a species of these Cingulata, the *H. euphracthus*, which was large and had rigid carapace and tail. He also identified *P. humboldti*, a pampathere, as well as three extinct armadillos *D. punctatus, P. sulcatus,* and *C. squamicaudis*. His confidence in identifying those species is admirable, which shows the deep knowledge of the current fauna of the region he acquired in a short period.

Unfortunately, none of Lund's possessions in Lagoa Santa was preserved. His house was put down to build a school on the site; his furniture and especially his library also disappeared. Rumor says his books were destroyed with the pages being used to make fireworks. But it is obvious that during the early years of his research Lund had access to North American publications about Xenarthra Pilosa more than to the ones published in Europe. His findings of medium-sized sloths must then be interpreted with the generic name *Megalonyx* following the identifications in North America. He subsequently abandoned such identifications, renaming the specimens he found as *Scelidotherium*, probably because of the works received from Europe, notably those by Owen. That denomination identified medium-sized sloths species in South America.

One of the greatest discoveries made in Lagoa Santa occurred in 1837. Lund realized the importance of it and did not take long to write the memoir with an account of that discovery (Lund 1839). For the first time in history, Primate fossils were found and identified as *Protopithecus brasiliensis* (which translates as "first monkey from Brazil"). This determination, once again, shows Lund's confidence and great knowledge. The material recovered in two caves was minimal: a pair of phalanges, a metatarsal, a cervical and a caudal vertebra, proximal half of a femur and distal half of a humerus. It was enough for him not only to define the species but also to consider it extinct.

The human findings lingered and happened only in 1840, 5 years after Lund's arrival in Lagoa Santa. Lund realized that cranial morphological features did not match the modern Native Americans. He ended up comparing them to the skulls of Egyptian civilization. At present, archaeologist Walter Neves beckons the possibility of those being cranial features of former ancient dwellers of Lagoa Santa with Negroid skulls. Lund got it almost right.

In Carnivores, he was precise when separating, with great skill, the fossils belonging to extinct and still-living species. The extinct *S. populator*, as has already been mentioned, initially considered being a hyena was recognized as Felidae. Prior to this species, Lund had identified species *A. brasiliense* (Ursidae) and *P. troglodytes* (Canidae).

But there are two species of Canidae that deserve attention. Once again it shows Lund's confidence in his knowledge. In 1839, he determined, by means of a very well-preserved skull found in the Cerca Grande complex, the extinct species *S. pacivorus*. When finding numerous remains of paca in that place, he reached the conclusion that the carnivore—"a devourer of pacas"—was responsible for the accumulation of fossils of that species, which translated into the specific denomination.

At that time, Lund had heard frequently about a small, wild carnivore occurring in the area. The descriptions provided by a few who saw the animal waved the possibility of it being a species still unknown. After offering a reward, hunters located two subjects in the woods: one was killed, and the other one was captured. Lund kept it alive in his yard for over a year, watching his behavior. The skeleton of the dead animal was prepared. He had previously encountered a fossilized skull of the same species, concluding that the unknown animal, the bushdog, belonged to the same genus as the extinct *S. pacivorus*. It was coined *S. venaticus*. For the first time, an extinct fossil species was recognized earlier than a current species belonging to the same genus.

Another finding that showed Lund's consistency was the discovery of Equids. In a cave of uncertain location named Horse Cave, Lund identified a metacarpal. The anatomy of horses, zebras, and donkeys is very similar. Lund initially raised the hypothesis that he might have had in hands a horse-type metacarpus originally introduced by the Portuguese settlers. After numerous comparisons, he came to the conclusion that the piece belonged to an extinct horse: "*It is noticeably wider and flatter than those of all the living Horses with which I have been able to compare it*" (Lund and Paula Couto 1950: 315).

In 1833, Darwin found on the surroundings of the Paraná basin two upper molars of Equidae together with the remains of extinct animals. Owen described the finding. As a result, the South American horses were extinguished before the arrival of the colonizers. Owen (1840: 109) realized the importance of the discovery: "*...is not one of the least interesting fruits of Mr. Darwin's paleontological discoveries.*"

Lund in turn wrote an article in Lagoa Santa in April 1840. That same year it was published in France (Lund 1840). In the isolation of Lagoa Santa, Lund, surprisingly, had already learned of Darwin's discovery: "*How he found in the temperate parts of America (in the United States and in the Republic of Uruguay) fossilized horse bones*" and concluded that having disappeared completely it would only reappear in America introduced by the colonizers (Lund and Paula Couto 1950: 315–316). Later findings by Lund confirmed the correctness of the identification. In Lagoa Santa, there occurred two species of extinct Equidae *E. neogeus* and *H. principale*.

In the Artiodactyla Tayassuidae, in addition to the fossil remains of the two extant species, the "cateto" and the "queixada," Lund identified a rare species *Brasiliochoerus stenocephalus* of which only two skulls are known, one found by Lund and the other by researchers to be further mentioned. Both findings occurred in the region of Lagoa Santa. Still among the Artiodactyls, we highlight the findings of *P. major*; many bones of that species were recovered by Lund. He correctly identified the specimens as belonging to Camelids. But in his publications he did not determine the species and only published the image of a molar of a young animal. In the absence of comparative material, he preferred omission to the possibility of error.

4 An Unusual Find

In one of the caves of the Escrivania complex, the "wise man of Lagoa Santa" discovered what today we consider to be of great importance; nevertheless, it has not received due attention. The following account presents data acquired from one of Lund's memoirs (Lund 1846).

During three and a half months, the work of removing fragments was carried out, mostly of micromammals coming from the regurgitations of owls probably during a great sequence of years. 6,552 barrels of sediment were withdrawn from inside a cave. Using teeth and long bones, predominantly from rodents and marsupials, Lund estimated there had been 6,881,500 individuals there.

In his house, Lund had a pair of cave owls, *Tyto furcata* (Temminck 1827), which allowed him to observe its habits. On average, they ate four small animals per meal. From this assumption, the deposition in the passage of the cave where he had made the finding would have occurred during five thousand years, the time spent by couples of owls in the cave to accumulate so many bones.

We do not know of any study with this material. Lund and Reinhardt made intensive collections of small mammals of the Lagoa Santa region in the middle of the nineteenth century, which, like the previous findings, are also deposited in the Museum of Zoology in Copenhagen. By combining data obtained from these two moments with data from the present situation, it would be possible to compare three scenarios, one possibly from the late Pleistocene (Escrivania), data from 150 years ago and the current material. The confluence of these circumstances, hardly replicated elsewhere, can provide data and information on possible environmental changes over a long period of time.

5 The Paleontological Findings After Lund

After Lund, paleontological research in the Lagoa Santa region was not immediately carried on. Virtually, a century passed by before some research took place.

Between 1935 and 1960, members of the now-disbanded Minas Gerais Academy of Sciences resumed researches in the field of paleontology and archeology. Among others, Harold V. Walter (1889–1969), English consul in Belo Horizonte, and Aníbal Mattos (1897–1976), professor and artist, excavated in several municipalities around Lagoa Santa, such as Confins and Pedro Leopoldo. Their main interest was archeology, but over the years they carried out paleontological collections today housed in the Museum of Natural History of the Federal University of Minas Gerais. They published their works with a general scope, with emphasis on archaeological findings (Mattos 1934, 1938, 1939; Walter 1940, 1943, 1949).

Some of these paleontological findings were later used for studies by several paleontologists, among which Carlos de Paula Couto (Fig. 5) stands out. He warned about the importance of Harold Walter's private collection, mentioning he was willing to sell it. It was bought by the Federal University of Minas Gerais in the 1960s.

Based on the material found by H. Walter, Paula Couto (1947) described an incomplete skull, molariforms, the posterior half of a carapace, some limb bones, and the caudal tube as belonging to the Glyptodont *H. euphractus*.

Fig. 5 Carlos de Paula Couto (1910–1982) was an important Brazilian paleontologist who translated into Portuguese Lund's paleontological works. Collection of the Fundação Zoo-botânica do Rio Grande do Sul

The fossils were found in Borges Cave (Pedro Leopoldo). Paula Couto confirmed Lund was right in identifying only few carapace fragments as *H. euphractus* (1839). Paula Couto (1957a) resumed studying this species considering the previous article as a preliminary note. In addition to the description and figuration, he reconstructed the species defined by Lund. After Lund, only one extinct species was described in Lagoa Santa with material he had not found: *Tetrastylus walteri* (Paula Couto 1951), a Cavioidea rodent. Species of that genus were only known from the Pliocene of Argentina. The finding consisted of incomplete right teeth with Incisive, PM4 and M1 to M3. The piece was part of the Walter Collection but, at present, it is deposited in the collection of the National Department of Mineral Production in Rio de Janeiro (Paula Couto 1951).

Later, the illustrious paleontologist (Paula Couto 1960) studied the South American Tremarctine species, dedicated to the "*Pararctotherium brasiliensis*" (= *A. brasiliense*). He increased the restricted knowledge about the species by studying a very complete skull with dentition and left ulna found in Lagoa Funda Cave, in the municipality of Pedro Leopoldo, also included in Walter's collection.

Fonseca (1979) in an unpublished Master's degree thesis studied the material of Tayassuidae belonging to the Museum of Natural History of the Federal University of Minas Gerais. The pieces studied also belonged to Walter's collection. In this work, he came to the conclusion that a rare extinct species, *B. stenocephalus*, of which two skulls are known, one discovered and studied by Lund, and the other discovered by H. Walter and studied by Fonseca, is valid. Paula Couto (1981) ratified Fonseca's conclusion.

In 1956, a joint expedition of the National Museum of Rio de Janeiro and of the University of South Dakota occurred in Cerca Grande (Paula Couto 1958). The main purpose of the expedition was the search for archaeological elements and to prove the coexistence of human beings and the extinct fauna. Paula Couto (1958, 1970) did not believe this was possible. Excavations were made in Caetano Cave and in the cliff of Cerca Grande, both located in the municipality of Matozinhos. Subsequently, with the new data, Paula Couto changed his mind. The expedition collected a few pieces of Tayassuidae and fragments of the carapace of the extinct armadillo "*Propraopus punctatus*" (= *Dasypus punctatus* (Lund 1839)), which is currently in the collection of the Museum of Natural Sciences of the Zoo-botanical Foundation of Rio Grande do Sul (Paula Couto 1980).

Souza Cunha (1960, 1964) described the teeth of *H. principale* collected in the Mortuária de Confins Cave (Confins municipality), deposited in the National Museum of Rio de Janeiro and collected by Padberg-Drenkpohl in 1926, an employee at the Museum, according to unpublished reports.

In the 1970s, archaeologists Annette Laming-Emperaire and André Prous performed an excavation at Lapa Vermelha IV Cave, located in the region of Lagoa Santa, in the municipality of Pedro Leopoldo, which culminated in the finding of skull of "Luzia, the oldest Brazilian." Based on material collected in this excavation, Souza Cunha and Guimarães (1978) cite the findings of a *Glossotherium* nail phalange (such identification is questionable), as well as a list of small mammals at a generic level that Souza Cunha and Guimarães (1978), Magalhães and Locks (1983), and Locks and Montenegro (1985) also pointed out.

André Prous, an archaeologist at the Museum of Natural History at the Federal University, found in the 1980s a very complete post-cranial skeleton of *C. cuvieri* in the Cuvieri

Cave (Matozinhos). This specimen is deposited in that institution and about this nothing has been published.

Souza Cunha and Magalhães (1986) pointed out new Cervidae material which Magalhães (1987), in a Master's degree thesis, later defined more accurately. In his work, he reviewed the Cervidae fossil/subfossil found in several Brazilian regions, emphasizing the findings in Lagoa Santa: *Mazama gouazoubira* (Fisher 1814), *Mazama sp.*, *Ozotoceros bezoarticus* (Linnaeus 1758), *Blastocerus dichotomus* (Illiger 1815), *Antifer sp.*, and, doubtfully, *Morenelaphus sp.* He also pointed out an indeterminate Odocoilini from Lapa Vermelha.

C. Cartelle in sporadic excavations in caves in the Jaguara Farm (Matozinhos), Baú Cave (Pedro Leopoldo), Cerca Grande (Matozinhos), and Escrivânia (Prudente de Morais) in the 1980s collected several pieces of Carnivores, Xenarthra, Rodents, and Tayassuids of known species. This unpublished material is deposited in the Museum of Natural Sciences of the Catholic University of Minas Gerais.

The Archaeological Project *Origins and Microevolution of Men in America* started in the beginning of the year 2000 and is still ongoing. Directed by archaeologist Walter Neves, from the University of São Paulo, it reached spectacular results in the archeology field (see Da-Glória et al. 2017). At the same time, his team developed a paleontological excavation in the Cuvieri Cave where, years before, the archaeologist André Prous found the skeleton mentioned before. This excavation was an attempt to find extinct fauna and men in association. No human record took place in this cave. Mammalian findings repeated some of those identified by Lund. The material is deposited in the Human Evolutionary Studies Laboratory of the University of São Paulo.

According to Hubbe (2008), Mayer (2011), and Mayer et al. (2016), the mammal remains recovered from Cuvieri cave were *Tapirus terrestris* (Linnaeus 1758), *Mazama sp.*, *C. cuvieri, V. bucklandi, Panthera onca* (Linnaeus 1758), *S. populator, Cuniculus paca* (Linnaeus 1766), *Cuniculus rugiceps*, (Lund 1837) *Dasyprocta sp.*, *Euphracthus sexcinctus* (Linnaeus 1758), *Sylvilagus brasiliensis* (Linnaeus 1758), *Didelphis sp.*, *Tayassu pecari* (Link 1795), *Tayassu tajacu* (Linnaeus 1758), *Cavia sp.*, *Akodon sp.*, *Necromys sp.*, *Oxymicterus sp.*, and *Oryzomys sp.*

Perônico and Araújo (2002) report a excavation in Bauzinho dos Ossos Cave (Pedro Leopoldo municipality) and findings of mammal at the Order level (Pilosa and Artiodactyla). Vasconcelos et al. (2015) carried out taphonomic and taxonomic analysis of his findings in a small cave located in the municipality of Prudente de Morais. Common specimens of *T. pecari*, *Dasypus novemcinctus* (Linnaeus 1758), *Speothos sp.*, *Eira barbara* Linnaeus 1758, *Sylvilagus* sp., *V. bucklandi*, and a rare occurrence: the rather complete jaw of a *C. rugiceps* were found.

6 Publications of Mammal Fossil Related to Lagoa Santa

After Lund's publications, many authors turned their attention to Lagoa Santa. The main ones are here summarized. The related work mentioned before which had direct or more explicit links with the Lund's findings will not be listed hereafter, such as the work by Winge in *E Museo Lundii* as previously noted or the aforementioned works by Paula Couto.

The German Burmeister (1871, 1885), who came to visit Lund in Lagoa Santa around 1852, wrote articles about the species *H. euphractus* and the "*Coelodon maquinense*" (= *N. maquinense*). Boas (1881) once again addressed the subject of Equines discovered by Lund claiming the possibility that *H. principale* could be a zebra.

By 1845, Lund had already sent thirty boxes of fossils to Denmark. That year the Galathea corvette, the Danish version of Darwin's Beagle voyage, departed from Copenhagen. Lund invited Reinhardt (Fig. 6), one of his mentor's

Fig. 6 Johannes T. Reinhardt (1816–1882). Reinhardt worked as Lund's assistant in Lagoa Santa and he would become the first Curator of Lund's paleontology collection. Collection of the Natural History Museum of Copenhagen

son, to be the first curator of the collection. The Danish government ordered the young man to interrupt his voyage in the Galatea to visit Lund in Lagoa Santa. Reinhardt would return two more times to the small town (1852–54 and 1856–58).

On his second stay in Lagoa Santa, Reinhardt received the book with Brandt's drawings as a gift. On the previous voyage, he had received Lund's notes and the Manuscript Catalog with the record of the specimens collected as well as the journal with a detailed record of each excavation. Therefore, we know the dates and the excavated caves. In a cross-checking of data, it is possible to infer when and where Lund collected each one of his pieces. In addition to the records of the excavations carried out by Lund (Reinhardt 1888), he wrote some articles (Reinhardt 1868, 1870, 1875, 1877, 1878, 1880) dealing with birds, armadillos, glyptodonts, tayassuids, *N. maquinense,* and Lund's general findings.

Works by Paula Couto have already been mentioned in this chapter; they were prepared with material from the Walter collection composed of findings which succeeded Lund's and extended the knowledge about several species. But there are other works by this author which are essential to the knowledge of the discoveries which took place in Lagoa Santa. The update of Winge's nomenclature is a highlight (Paula Couto 1946), as well as an article on *S. populator* in which he reviewed the material from Minas Gerais deposited in several Museums (Copenhagen, Paris, and London). In the Walter collection, there are no remains of this species (Paula Couto 1955).

The publication of Lund's works in Portuguese is particularly noteworthy: Memories on Brazilian Paleontology (Lund and Paula Couto 1950). In addition to an introduction and the translation of Lund's scientific works, on a large number of notes Paula Couto updates the nomenclature and observations with the most exact concepts at the time. Images from Lund's works and an index make this all-important 592-page publication, containing 55 images a complete edition.

Four of the five Memoirs published by Lund in Danish were translated into French by Henri Gorceix (1842–1919), a French geologist, physicist, and mathematician who was hired by Emperor Dom Pedro II to found the Ouro Preto Mining School (MG). The Emperor, who admired Lund's work, paid for the costs.

In 1979, the magnificent Treaty of Paleomastozoology in which Paula Couto (1979) indicates all the species discovered by Lund was published. Previously, in several articles (1957a, b, 1970, 1971), he had already written about all the mammal fossils Lagoa Santa.

Hoffstetter (1908–1999), a renowned French paleontologist, in several works, and notably in his Ph.D. thesis (Hoffstetter 1952) made frequent allusions and comparisons with species of Megatheriidae, Pampatheridae, Equidae, Scelidotheriinae, and Lamini, which Lund and Winge defined for Lagoa Santa. Subsequently (Hoffstetter 1954) with the material deposited in Denmark, from Lagoa Santa, came to a series of conclusions about some Xenarthra that were accepted for a long time, influencing the erroneous analysis of Brazilian species by several authors. To a large extent, such conclusions were refuted by Cartelle et al. (2009).

As previously mentioned, in 1837, Lund made a very important discovery. He hastened to send a work to Europe (Lund 1839) in which he reported the discovery of *P. brasiliensis*. For the first time, a fossil of monkey was recorded. Lund's discovery was almost forgotten after the publication of Winge (1895a), who illustrated the remains Lund discovered. The revival of the importance of this finding is due to Hartwig (1995a). The title of the article is a summary of the content: "*Protopithecus: rediscovering the first fossil primate.*" Later, Hartwig (1995b, 2002) described the remains stored in the Museum of Zoology, in Copenhagen.

Cartelle and Fonseca (1983) improved the somewhat restricted knowledge about the *N. maquinense* sloth, a symbol of Lund's paleontological discoveries. Czaplewski and Cartelle (1998) did a survey on the Chiroptera fossil found in several regions of Brazil, including Lagoa Santa. Earlier, Winge (1892) identified 27 bat species in the material Lund had collected. Czaplewski and Cartelle (1998) reached almost the same conclusion: they identified 26 species in Lagoa Santa. Berta (1984) improved the knowledge of the Carnivora *S. pacivorus*. In the same sense, Cartelle and Langguth (1999) extended the knowledge about another species of Carnivora determined by Lund: *P. troglodytes*.

Alberdi et al. (2003) reviewed the Equidae Lund had discovered. Cartelle and Bohórquez (1982) revalidated the giant sloth species *E. laurillardi*. Cartelle and De Iuliis (1995) reinforced this revalidation in an important article. Also, Cartelle et al. (2009) revalidated the species *V. bucklandi*. These denominations are now in disuse.

7 Other Views

Besides the already mentioned work, there were other approaches that moved away from those reported, with an emphasis on taphonomy. We relate Perônico and Araújo (2002), Auler et al. (2006), Hubbe (2008), Hubbe et al. (2011), Mayer et al. (2009a, b), Mayer (2011), and Vasconcelos et al. (2015).

Another approach was that of the publications that carried out a survey focusing on the fossil fauna of Lagoa Santa (Cartelle 1994). Cartelle et al. (1998), in a comprehensive report justifying the creation of the Environmental Protection Area of Carste in Lagoa Santa (and which until today has not been fully implemented), plotted and listed caves and the findings therein. Most of them, with rare exceptions, are those explored by Lund.

The analysis of the extinct Caatinga and Cerrado fauna by Cartelle (1999), including Lagoa Santa, can be referred to in this item. In such work, the author defended the hypothesis that the main cause of the extinction of the Late Quaternary was climatic, reaching especially the megafauna of herbivorous mammals from open field. Examples of the fauna discovered by Lund were the basis for this hypothesis.

Cartelle (2012) broadened and deepened the work produced in 1994. He placed special emphasis on the area of Lagoa Santa, on Lund, and obviously on the fauna he discovered. Special attention was given to the illustrations and the original descriptions made by P. W. Lund. A list of species of late Quaternary mammals of Minas Gerais was included in this publication.

We have finally and rightly to point out some publications which refer to the life of P. W. Lund or to aspects of it or to its discoveries. Reinhardt (1880) summed up Lund's scientific life in the year of his death. Shortly thereafter, a niece (Lund 1885) wrote a biography that was used by Nereo dos Santos, Lund's adopted son, as a basis to write a short booklet about his father's life (Santos 1923). Degerböl (1945) provided the essential data for organizing a biography used by Paula Couto as an introduction to the publication of the Memoirs of Brazilian Paleontology (Lund and Paula Couto 1950: 5–26). Schøllhammer (2002) briefly pointed out several facets of Lund's life in Brazil.

It is worth highlighting Marchesotti's (2005) Master's degree dissertation, and the Luna (2007) Ph.D. thesis, both carried out in-depth analyzes with solid foundations. Recently, two excellent books were translated from Danish to Portuguese. In the first one, Holten and Sterll (2011), supported by numerous documents, carried out in-depth research work portraying Lund's life and its discoveries. In the second one, Holten et al. (2012) depicted the life and work of Lund's main assistant, his secretary, and artist Peter Andreas Brandt. Both Luna (2007) and Holten and Sterll (2011) used primary information contained in the extensive correspondence (family and scientific) Lund kept throughout the years he was in Lagoa Santa, and which is preserved in the Royal Library of Copenhagen.

8 Chronological Data Related to the Lagoa Santa Fauna

At the end of the nineteenth century and in the beginning of the twentieth century, issues pertaining to humans' coexistence with the extinct fauna identified by Lund as a result of his discoveries in Sumidouro Cave were the subject of several controversies. Piló et al. (2004, 2005), after detailed topographic, sedimentological, and taphonomic studies, provided data that, I believe, undoubtedly clarified the question: human beings and the extinct fauna were contemporaneous in the Lagoa Santa region. Lund was right. By means of dating, they obtained the minimum ages of human bones and of eight-thousand-year extinct fauna. The contemporaneity is also supported by absolute dates.

Previously, Neves and Piló (2003) recorded new dates with material collected by Lund and deposited in the Museum of Copenhagen and the Museum of Natural History of UFMG. Collagen from bones belonging to extinct species was used—*E. (A.) neogaeus, C. cuvieri, S. populator,* and *Homo sapiens* (Linnaeus 1758). In addition, they used tooth collagen of *E. (A.) neogaeus* and charcoal. They were aged between 9,260 ± 150 (*S. populator*) and 16,900 ± 70 years (*E. neogaeus*). The date for *H. sapiens* was 9,780 ± 70 BP (Calibrated: 10,790–11,270) which overlaps the confidence interval of the calibrated age of the *S. populator* mentioned above (calibrated 9920–11,060) and also of a *C. cuvieri* from Cuvieri cave which was dated at 9,990 ± 150 (calibrated 10,900–12,020), and was also reported in Neves and Piló (2003). Hubbe et al. (2009) offered new dates with material collected in the previously cited Cuvieri cave. For *S. populator,* they indicate conventional, 10,790 ± 60 (calibrated 12,560–12,750), for *V. bucklandi*: conventional, 11,020 ± 40 (calibrated 12,730–12,980), and for Tayassuidae: conventional 10,470 ± 40 (calibrated: 12,050–12,540). Additional ages are provided by Auler et al. (2006) and Hubbe et al. (2011).

9 Epilogue

This chapter ends with a final observation emphasizing Peter Lund's scientific figure. We consider that he finished his publications on the extinct fauna of Lagoa Santa in 1846. Now, 170 years later, 22 species determined by Lund (Table 1) are still valid, as previously indicated. During this long period, from the Brazilian intertropical Pleistocene, 17 extinct species were either established or identified (Table 2). Two were designated with material collected by Lund in Lagoa Santa (*J. vulpinus* and *T. cristatellus*). A species of Chiroptera, *Desmodus draculae* (Morgan, Linares and Ray 1988), discovered in Venezuela and also identified in Bahia and São Paulo (Czaplewski and Cartelle 1998) can be added to the list.

Mammals were Lund's main interest. But he also found fossils of other vertebrates and some of them were recorded in his "Katalog," a manuscript he sent to Denmark along with his collection. At the end of his brief but astonishing paleontological career, he recorded 12,622 specimens. For

Table 1 List of extinct mammals of Lagoa Santa named by Lund (22 species)

Pilosa
Nothrotherium maquinense (Lund 1839)
Eremotherium laurillardi (Lund 1842)
Catonyx cuvieri (Lund 1839)
Ocnotherium giganteum (Lund 1839)
Valgipes bucklandi (Lund 1839)
Cingulata
Cabassous squamicaudis (Lund 1844)
Propraopus punctatus (Lund 1839)
Propraopus sulcatus (Lund 1842)
Pampatherium humboldti (Lund 1839)
Hoplophorus euphractus Lund 1839
Primate
Protopithecus brasiliensis Lund 1839
Carnivore
Protocyon troglodytes (Lund 1840)
Speothos pacivorus (Lund 1839)
Arctotherium brasilienses (Lund 1839)
Smilodon populator Lund 1842
Perissodactyla
Equus neogeus (Lund 1840)
Hippidion principale (Lund 1845)
Artiodactyla
Brasiliochoerus stenocephalus (Lund in Gervais 1867–69)
Palaeolama maior (Lund in Liais 1872)
Rodent
Coendou magnus (Lund 1839)
Neochoerus sulcidens (Lund 1839)
Cuniculus rugiceps (= major) (Lund 1839)

Table 2 List of extinct mammals of intertropical Brazil named after Lund's works (17 species). The ones that occurred in the region of Lagoa Santa are written in bold

Perissodactyla
Tapirus cristatellus (Winge 1906)
Rodent
Juscelinomys vulpinus (Winge 1887)
Tetrastylus walteri, Paula Couto 1951
Proboscidea
Stegomastodon waringi (Holland 1920)
Cingulata
Panochthus greslebini (Castellanos 1941)
Panochthus jaguaribensis (Moreira 1965)
Pampatherium paulacoutoi, Cartelle and Bohórquez 1983
Pachyarmatherium brasiliense, Porpino, Fernicola and Bergqvist 2009
Glossotherium phoenesis Cartelle, De Iuliis, Boscccini and Pujos, 2019
Pilosa
Mylodonopsis ibseni Cartelle, 1991
Aytherium aureum Cartelle, De Iuliis & Pujos, 2008
Australonyx aquaae De Iuliis, Pujos and Cartelle, 2009
Litopterna
Xenorhinotherium bahiense, Cartelle and Lessa 1988
Toxodontia
Piauitherium. capivarae, Guérin and Faure 2013
Primate
Caipora bambuiorum Cartelle and Hartwig, 1996
Alouatta mauroi Tejedor, Rosenberg and Cartelle, 2008
Cartelles coimbrafilhoi Halenar and Rosenberg, 2013

each one of them, he indicated number, morphological identification, specific identification, and location of the finding.

Lund's work in Brazil was broader than previously reported. In addition to the paleontological works and their implications, the zoological collections and the innumerable botany findings must be highlighted. This is due to him advising students from Denmark who lived with him: Reinhardt, who would become the first Curator of Lund's paleontology collection, and E. Warming who with his studies on interrelations in the Cerrado should be considered the founder of Ecology. Lund's material concerning botany is deposited at the Botanical Museum in Copenhagen. In addition to photographs taken by Warming during his stay in Lagoa Santa (1863–1866) and his notes, the correspondence between him and Lund has been preserved. In the herbarium of the Botanical Museum, sixteen thousand exsiccatae (Holten and Sterll 2011) collected by Lund and Warming are deposited. We further summarize other findings of vertebrate fossils also collected by Lund. He did, in some of his works, restricted allusions to these specimens.

Reinhardt, directed by Lund, carried on an important collection of fish in the Velhas River, which he sent to Denmark. This collection was studied by Lütken (1875, 2001) (Fig. 7), which identified 55 species. It seems that 14 of these are not found in the ichthyofauna in its currently polluted waters (Alves and Pompeu 2010). Lund found, mainly in the Sumidouro Cave, fossils belonging to eight fish species (Hansen 2012) no longer found in the lake: *Oseudoplatystoma corruscans* (Spix and Agassiz 1829), *Pimelodus maculatus* (Lacèpède 1803), *Bergiaria westermanni* (Lütken 1874), *Pseudopimelodus charus* (Valenciennes 1840), *Rhynel*

Fig. 7 Christian F. Lütken (1827–1901) studied the collection of fish of the Velhas River, which Reinhardt sent to Denmark. Collection of the Natural History Museum of Copenhagen

quelen (Quoy and Gaimard 1824), *Hoplias malabaricus* (Bloch 1974), *Prochilodus costatus* (Valenciennes 1860), and *Astyanax lacustris* (Lütken 1875).

In several caves, Lund also collected parts of the amphibian *Ceratophrys aurita* (Raddi 1823), the toad-totan or ox-toad. Non-avian reptile findings were broader (Hansen 2012). They took place in 13 caves: two species of turtles—*Phrynops hilarii* (Duméril and Bibron 1835)—and *Platemys platycephala* (Scheneider 1792), a Crocodylia—*Caiman latirostris* (Daudin 1802)—an Amphisbaena—*Amphisbaena alba* Linnaeus 1758)—four Lacertidae: *Ploychrus acutirrostris* Spix 1825, *Tropidurus torquatus* (Wied 1820), *Ameiva ameiva* (Linnaeus 1758), and *Tupinambis teguixin* (Linnaeus 1758), and finally, four snakes: *Boa constrictor* (Linnaeus 1758), *Mastigodryas bifossatus* (Raddi 1820), *Crotalus durissus* (Linnaeus 1758), and *Lachesis muta* (Linnaeus 1766).

Lund collected more than just a few fossils of different bird specimens. And it is worth noticing that, in addition to the fossils, excellent collections of the current vertebrate fauna (fish, amphibians, reptiles, birds, and mammals) were also sent to Denmark, as we have already mentioned. But it was mainly Reinhardt, during his visits to Lagoa Santa, who carried out the surveys of the regional bird fauna, entering 393 species to his list. Among them, 198 species of non-Passeriformes (Reinhardt 1878) were described in a remarkable and long publication (Krabbe 2007).

Winge (1887), on the other hand, indicated that Lund found in 19 caves fossil remains belonging to 126 species of non-passerines and a large quantity of remains of these birds. Twenty-two species identified by Winge are not found among the species recorded by Reinhardt and most of the species found as fossils in the caves still survive in the region. The greatest number of findings occurred in the cave Escrivania 5 (Prudente de Morais). Winge (1887) identified, in the material collected by Lund, the new species *Chenalopex pugil* (Winge 1887) as previously related.

There were no explicit references to fossil findings of fish, amphibians nor non-avian reptiles in the Lagoa Santa area after the records previously mentioned. In the material recovered in the Cuvieri Cave, a new extinct bird species was identified: *Pleistovultur nevesi* (Alvarenga et al. 2008). With remains from Lapa Vermelha IV, the cave where "Luzia" was found, Alvarenga (2007) identified the presence of *Vultur gryphus* (Linnaeus 1758) and of *Anodorhynchus glaucus* (Vieillot 1816) or *Anodorhynchus leari* (Bonaparte 1856). These two species are almost identical in their osteology so it was not possible to define the exact identity of the Psittacidae.

Concerning the birds, it is important to point out what has already been said about mammalian micro-fossils: the possibility of comparative studies between three historical moments of which there are enough study elements belonging to the Late Quaternary, second half of the nineteenth century and contemporary collections. In the material found by Lund, there are elements that allow inferring that the ornithofauna in the region he has studied would indicate, in the Late Quaternary, a predominant environment of forest. Such a paleoenvironment could explain a strange fact observed in the fauna composition of extinct megamammals in Lagoa Santa. In most of the Pleistocene deposits of the Brazilian intertropical region, numerous findings consist of specimens of *S. waringi* and, notably, *E. laurillardi*. Possibly the remains of this species are found in greater numbers and sites in this area. But, in Lagoa Santa, Lund's findings of these two species were very scarce. For the species *E. laurillardi,* only two molariforms were recovered. This fact, at the least, is exceptional since the collections were extensive and intensive. It is perhaps associated with a type of regional environment that is not in agreement with the most common in other intertropical areas. A plausible explanation for this finding could be the environment translated by the fossils of the birds identified previously indicated. Large animals, such as those mentioned, would hardly inhabit a distinctly predominantly forested environment in which large areas of pasture, necessary to such massive species, would be nearly absent.

Acknowledgements I acknowledge the help of André Gomide Vasconcelos in revising the original manuscript.

References

Alberdi MT, Cartelle C, Prado JL (2003) El registro Pleistoceno de *Equus (Amerhippus)* e *Hippidion* (Mammalia, Perissodactyla) del Brasil: consideraciones paleoecologicas y biogeograficas. Ameg 40 (2):173–196

Alvarenga H (2007) *Anodorhynchus glaucus* e *A. leari* (Psittaciformes, Psittacidae): osteologia, registros fósseis e antiga distribuição geográfica. Rev Bras Ornit 15(3):427–432

Alvarenga H, Brito G, Migotto R, Hubbe A et al (2008) *Pleistovultur nevesi* gen. et sp. nov. (Aves: Vulturidae) and the diversity of condors and vultures in the South American Pleistocene. Ameg 45 (3):613–618

Alves CBM, Pompeu PS (2010) A fauna de peixes da Bacia do Rio das Velhas no final do século XX. In: Alves CBM, Pompeu PS (eds) Peixes do Rio das Velhas: passado e presente. 2ed. argvmentvm, Belo Horizonte, pp 167–189

Auler AS, Piló LB, Smart PL et al (2006) U-series dating and taphonomy of quaternary vertebrates from Brazilian caves. Palaeogeog, Palaeoclim, Palaeoecol 240:508–522. https://doi.org/10.1016/j.palaeo.2006.03.002

Berta A (1984) The Pleistocene bush dog *Speothos pacivorus* (Canidae) from the Lagoa Santa caves, Brazil. J Mam 65(4):549–559. https://doi.org/10.2307/1380837

Boas JEV (1881) Om en fossil Zebra-Form fra Brasilienes Campos. Med. et Tillaeg om to Arter af Salaegten Hippidion. Det Kongelige Danske Videnskabernes Selskabs Skrifte Naturvidenskabelig og Mathematisk Afdeling ser 6, 1(5):305–330

Burmeister HG (1871) Ueber *Hoplophorus euphracthus*. Archiv. für Anat, Phys und Wissensc Med 164–179

Burmeister HG (1885) Berichtigung zu Coelodon. Sitzungsberichte der königlich preussischen. Akad der Wissensc zu Berlin 2:567–573

Cartelle C (1994) Tempo passado: mamíferos do Pleistoceno em Minas Gerais. Palco/Acesita, Belo Horizonte

Cartelle C (1999) Pleistocene mammals of the Cerrado and Caatinga of Brazil. In: Eisenberg JF, Redford KH (eds) Mammals of the neoprotropics. The University of Chicago Press, Chicago, pp 27–46

Cartelle C (2012) Das grutas à luz: os mamíferos pleistocênicos de Minas Gerais (From the caves to the light: the Pleistocene mammals of Minas Gerais). Bicho do Mato, Belo Horizonte

Cartelle C, Abuhid V, Ferreira MA et al (1998) Levantamento Paleontológico. In: CPRM, IBAMA, GERIDE (eds) APA Carste de Lagoa Santa (Patrimônio Espeléólogico, Histórico e Cultural) CPRM, IBAMA e GERIDE, Belo Horizonte, pp 177–198

Cartelle C, Bohórquez G (1982) *Eremotherium laurillardi* Lund, 1842. Parte I. Determinação específica e dimorfismo sexual. Iher sér Geol 7:45–63

Cartelle C, De Iuliis G (1995) *Eremotherium laurillardi* the panamerican late Pleistocene megatheriid sloth. J Vert Paleo 15(4):830–841

Cartelle C, De Iuliis G, Ferreira RL (2009) Systematic revision of intertropical Scelidotheriinae (Xenarthra, Mylodontidae). J Vert Paleo 29(2):555–566. https://doi.org/10.1671/039.029.0231

Cartelle C, Fonseca JS (1983) Contribuição ao melhor conhecimento da pequena preguiça terrícola *Nothrotherium maquinense* (Lund) Lydekker, 1889. Lund 2:11–48

Cartelle C, Langguth A (1999) *Protocyon Troglodytes* (Lund): um canídeo intertropical extinto. An Acad Bras Ciên 71(3):371–384

Czaplewski NJ, Cartelle C (1998) New records of Pleistocene bats from cave deposits in Bahia, Brazil. J Mam 79(3):784–803

Da-Glória P, Neves WA, Hubbe Mark (eds) (2017) Archaeological and paleontological research in Lagoa Santa: the quest for the first Americans. Springer, Cambridge

Degerböl M (1945) P.W. Lund og hans udforskning af Brasiliens Forhistoriske Dyreverden. Dir I Nat og Mus 111–134

Fonseca JS (1979) Taiassuídeos do pleistoceno de cavernas calcárias de Minas Gerais. Dissertation, Universidade Federal do Rio Grande do Sul

Guérin C, Faure M (2013) Un nouveau Toxodontidae (Mammalia, Notoungulata) du Pléistocène supérieur du Nordeste du Brésil. Geod 35(1):155–205. https://doi.org/10.5252/g2013n1a7

Hansen L (2012) E Museum Lundii addendum. Natural History Museum of Denmark, Copenhagen

Hartwig WC (1995a) *Protopithecus*: rediscovering the first fossil primate. Hist Ph Lif Sci 17(3):447–460

Hartwig WC (1995b) A giant new world monkey from the Pleistocene of Brazil. J Hum Evol 28:189–196. https://doi.org/10.1006/jhev.1995.1013

Hartwig WC (2002) A possible Giant Muriqui from the Pleistocene deposits of Lagoa Santa, Minas Gerais, Brazil. Field Museum Publications/National Zoological Park, Chicago

Hoch E (1982) The influence of Georges Cuvier on the danish naturalist Peter Lund, "Father of Brazilian Paleontology". In: Buffetaut E, Mazin JM, Salmon E (eds) Symposium Paleontologique Georges Cuvier, Montbéliard, October 1982. Actes du Symposium Paleontologique Georges Cuvier, La Ville, Montbéliard, pp 273–287

Hoffstetter R (1952) Les mammiferes pléistocénes de la République del'Equateur. Mém Soc Géol Fr, Nou Sér 31:1–391

Hoffstetter R (1954) Les gravigrades (Edentés Xénarthres) des cavernes de Lagoa Santa (Minas Gerais, Brasil). An Sci Nat—Zoo Biol Ani 16:741–764

Holten B, Sterll M (2011) Peter Lund e as Grutas com ossos em Lagoa Santa. Editora UFMG, Belo Horizonte

Holten B, Sterll M, Fjeldsa J (2012) O artista desaparecido. P.W. Lund e P. A. Brandt no Brasil. Editora UFMG, Belo Horizonte

Hubbe A (2008) Contextualização taxonômica, tafonômica e morfométrica dos remanescentes ósseos da megamastofauna da gruta Cuvieri (MG), um sítio paleontológico do Pleistoceno tardio. Dissertation, Universidade de São Paulo

Hubbe A, Haddad-Martim PM, Hubbe M et al (2011) Identification and importance of critical depositional gaps in pitfall cave environments: The fossiliferous deposit of Cuvieri Cave, eastern Brazil. Palaeogeog, Palaeoclim, Palaeoecol 312:66–78. https://doi.org/10.1016/j.palaeo.2011.09.010

Hubbe A, Hubbe M, Neves W (2009) New Late-Pleistocene dates for the extinct megafauna of Lagoa Santa, Brazil. Cur Res Pleist 26:154–156

Krabbe N (2007) Birds collected by P. W. Lund and J. T. Reinhardt in South-eastern Brazil between1825 and 1855, with notes on P. W. Lund's travels in Rio de Janeiro. Rev Bras Ornitol 15 (3):331–357

Locks M, Montenegro E (1985) Contribuição à paleontologia do "Grande abrigo da Lapa Vermelha Emperaire", Pedro Leopoldo, MG, Brasil. No 1 – um novo cavíineo quaternário (Mammalia-Rodentia). In: Campos DA, Ferreira CS, Brito IM et al (eds) Coletânea de Trabalhos Paleontológicos. Brasília, DNPM, pp 81–86

Luna PE (2007) Peter Wilhelm Lund: o auge das suas investigações científicas e a razão para o término das suas pesquisas. Thesis, Universidade de São Paulo

Lund H (1885) Naturforskeren Peter Wilhelm Lund: en biografisk skizze. C.A. Reitzel, Copenhagen

Lund PW (1837) Om Huler i Kalksteen i det indre af Brasilien, der tildeels inheholde fossile knokler- Förste Afhandling. Kong Dan Vidensk Sel Naturvid og Math Afh 6:207–249

Lund PW (1839) Blik paa Brasiliens Dyreverden för Sidste Jordomvaeltning. Anden Afhandling: Pattedyrene. Kong Dan Vidensk Sel Naturvid og Math Afh 8:61–144

Lund PW (1840) Nouvelles recherches sur la faune fossile du Brésil. An Sc Nat 2(13):310–319

Lund PW (1846) Meddelelse af Det Udbytte de i 1844 undersögte knoglehuler have afgivet til hundskaben om Brasiliens dyreverden för sidste jordomvaeltning. Kong Dan Vidensk Sel Naturvid og Math Afh 12:57–94

Lund PW, Paula Couto C (1950) Memórias sobre a paleontologia brasileira. Instituto Nacional do Livro, Rio de Janeiro

Lütken C (1875) Flodens Fiske. Et Bidrag til Brasiliens Ichtyologi efter Professor J. Reinhardtts Indsamlinger og Optegnelser. Copenhagen

Lütken C (2001) Peixes do Rio das Velhas: uma contribuição para a ictiologia do Brasil. In: Alves CBM, Pompeu PS (eds) Peixes do Rio das Velhas: passado e presente. 1ed. Segrac, Belo Horizonte, pp 23–164

Magalhães RM (1987) Os cervideos quaternários das Regiões Nordeste, Sudeste e Sul do Brasil (Cervidae, Odocoileinae). Dissertation, Universidade Federal do Rio de Janeiro

Magalhães RM, Locks M (1983) Contribuição à paleontologia do "Grande abrigo da Lapa Vermelha Emperaire", Pedro Leopoldo, MG. N.2 - sobre os cervídeos quaternários (Mammalia Artiodactyla). In: Campos DA, Ferreira CS, Brito IM et al (eds) Resumos das Comunicações Congresso Brasileiro de Paleontologia. DNPM, Rio de Janeiro, pp 87–92

Marchesotti AAP (2005) Peter Wilhelm Lund (1801–1880): O naturalista, sua rede de relações e sua obra, no seu tempo. Dissertation, Universidade Federal de Minas Gerais

Mattos A (1934) O sábio Dr. Lund e estudos sobre a prehistória brasileira. Edicões Apollo, Belo Horizonte

Mattos A (1938) A Pré-história brasileira. Cia Editora Nacional, São Paulo

Mattos A (1939) Peter Wilhelm Lund no Brasil: problemas de paleontologia brasileira. Cia Editora Nacional, São Paulo

Mayer EL (2011) Processo de formação de um depósito fossilífero em abismo na Gruta Cuvieri (MG): taxonomia, tafonomia e distribuição espacial de fósseis de mamíferos do Pleistoceno. Dissertation, Universidade de São Paulo

Mayer EL et al (2009a) Novo registro de pacas extintas (Rodentia, Cuniculidae) do Pleistoceno Terminal de Lagoa Santa, Minas Gerais. In: Gallo V, Silva HMA (eds) XXI Congresso Brasileiro de Paleontologia, Belém, September 2009. Boletim de resumos do XXI Congresso Brasileiro de Paleontologia, SBP, p 196

Mayer EL et al (2009b) Processos de acumulação óssea da mesomastofauna do Pleistoceno Terminal da Gruta Cuvieri, Lagoa Santa, Minas Gerais. In: Gallo V, Silva HMA (eds) XXI Congresso Brasileiro de Paleontologia, Belém, September 2009. Boletim de resumos do XXI Congresso Brasileiro de Paleontologia, SBP, p 197

Mayer EL, Hubbe A, Kerber L et al (2016) Taxonomic, biogeographic, and taphonomic reassessment of a large extinct species of paca from the Pleistocene of Brazil. Act Palaeo Pol 61:743–758. https://doi.org/10.4202/app.00236.2016

Mones A (1986) Palaeovertebrata Sudamericana. Catálogo Sistemático de los Vertebrados Fósiles de América del Sur. Parte I. Lista preliminar y Bibliografia. Courier Forschungsinstitut Senckenberg, Frankfurt

Neves WA, Piló LB (2003) Solving Lund's dilemma: new AMS confirm that humans and megafauna coexisted at Lagoa Santa. Cur Res Pleist 20:57–60

Owen, R (1840) Fossil Mammalia. In: Darwin C (ed) Fossil Mammalia Part 1 No. 4 of the zoology of the voyage of H.M.S. Beagle. Smith, Elder and Co., London, pp 81–111

Paula Couto C (1946) Atualização da nomenclatura genérica e específica usada por Herluf Winge em "E Museo Lundii". Est Bras Geol 1:59–80

Paula Couto C (1947) Contribuição para o estudo de *Hoplophorus euphractus* Lund, 1839. Sum Bras Geol 1:33–53

Paula Couto C (1951) Uma espécie do gênero *Tetrastylus* Ameghino, 1886, em Lagoa Santa. Bol Mus Nac 15:1–16

Paula Couto C (1955) O "Tigre dente de sabre" no Brasil. Bol Cons Nac Pesq 1:1–30

Paula Couto C (1957a) Sobre um Gliptodonte do Brasil. Bol Div de Geol e Min 165:1–37

Paula Couto C (1957b) Um urso fóssil de Minas Gerais. An Acad Bras Ciênc 29(2):3

Paula Couto C (1958) Notas à margem de uma expedição científica a Minas Gerais. Krit - Rev Fac Fil da U.M.G. 11(45–46):401–423

Paula Couto C (1960) Um urso extinto do Brasil. Bol Soc Bras Geol 9:5–27

Paula Couto C (1970) Lund e a paleontologia do Brasil. Bol Inf Cen Inf Arqu 2:5–13

Paula Couto C (1971) Mamíferos fósseis das cavernas de Minas Gerais. Esp 3:1–14

Paula Couto C (1979) Tratado de paleomastozoologia. Academia Brasileira de Ciências, Rio de Janeiro

Paula Couto C (1980) *Propraopus punctatus* (Lund, 1840) no Pleistoceno de Cerca Grande, Minas Gerais. An Acad Bras Ciênc 52:323–325

Paula Couto C (1981) On a extinct peccary from the Pleistocene of Minas Gerais. Iher (Geol) 6:75–78

Perônico C, Araújo ACS (2002) Diversidade faunística baseada em escavação tafonomicamente orientada de material subrecente preservado em caverna da região cárstica de Lagoa Santa, Minas Gerais. Arq Mus Nac 6:195–198

Piló LB, Auler AS, Neves WA et al (2004) Revisitando a Lapa do Sumidouro: marco paleo-antropológico do Quaternário americano. Rev Bras Paleo 7:337–348

Piló LB, Auler AS, Neves WA et al (2005) Geocronology, sediment provenance and fossil emplacement at Sumidouro Cave, a classiclate Pleistocene/early Holocene paleoantropologial site in eastern Brazil. Geoarc 20:751–764

Reinhardt J (1868) The bones caves of Brazil and their animal remains. Amer J Sci Arts 96:264–265

Reinhardt J (1870) Bidrag til Kundskab om Fuglefaunaen i Brasiliens Campos. Videnskabelige Meddelelser fra den Naturhistoriske Forening i København, Bianco Lunos Book Printing (F. S. Muhle), Copenhagen

Reinhardt J (1875) De i Brasil ens knoglehuler fundne Glyptodont – Levninger og em ny, til de gravigrade Edentater hörende slægt. Vindensk Medd Naturh Foren Kjob 9–15:165–236

Reinhardt J (1877) Tandforholdene hos baelte-dyr Slægten Dasypus. Vindensk Medd Naturh Foren Kjob 1877–1878 35–40:1–12

Reinhardt J (1878) Kaempedovendyr-Slægten Coelodon Vidensk. Kong Dan Vidensk Sel Naturvid og Math Afh 12(3):253–350

Reinhardt J (1880) De i Brasil ens knoglehule fundne navlesvin-arter. Vindensk Medd Naturh Foren Kjob 1879–1880:271–301

Reinhardt J (1888) De brasilianske Knoglehuler og de i dem forekommende Dyrelevninger. E Museo Lundii 1:1–49

Santos NC (1923) O Naturalista: Biographia do Dr. Pedro Guilherme Lund. Imprensa Official, Belo Horizonte

Schøllhammer EK (2002) The story of Peter Wilhelm Lund: between life and work. Lund 3(1):5–7

Souza Cunha FL (1960) Sobre o *Hippidion* da Lapa Mortuária de Confins, Lagoa Santa, Minas Gerais. Editorial Sul-Americana, Rio de Janeiro

Souza Cunha FL (1964) A Lapa Mortuária de Confins - Lagoa Santa. Rev Bras Arq 1:30–41

Souza Cunha FL, Guimarães L (1978) A fauna sub-recente de vertebrados do "Grande abrigo da Lapa Vermelha" de Pedro Leopoldo, Minas Gerais. Arq M His Nat, Belo Horizonte 3:201–240

Souza Cunha FL, Magalhães RM (1986) A fauna de vertebrados sub-recentes de Cerca Grande, Matozinhos, MG. Pub Av Mus Nac 65:119–131

Vasconcelos AG, Meyer KEB, Campello MS (2015) Mamíferos quaternários da cavidade ES-08, Município de Prudente de Morais, Minas Gerais: análises tafonômica e taxonômica. Rev Bras Paleo 18:171–190

Walter HV (1940) Urso extinto *Arctotherium brasiliensis* da caverna de Lagoa Funda, Minas Gerais. An Ac Ciê Minas Gerais 1–36

Walter HV (1943) Sobre alguns dos últimos achados de Mastodontes na região de Pedro Leopoldo. An Ac Ciê Minas Gerais 3–6

Walter HV (1949) The pre-history of the Lagoa Santa, Minas Gerais, Brazil. Velloso & Cia, Belo Horizonte

Winge H (1887) Jordfundne og nulevende Gnavere (Rodentia) fra Lagoa Santa, Minas Geraes, Brasilien: med udsigt over gnavernes indbyrdes slægtskab. E Museo Lundii 1(3):1–178

Winge H (1892) Jordfundne og nulevende Flagermus (Chiroptera) fra Lagoa Santa, Minas Geraes, Brasilien: med udsigt over flagermusenes indbyrdes slægtskab. E Museo Lundii 2(1):1–65

Winge H (1893) Jordfundne og nulevende Pungdyr (Marsupialia) fra Lagoa Santa, Minas Gerais, Brasilien: med udsigt over gnavares pungdyrdes slægtskab. E Museo Lundii 2(2):1–133

Winge H (1895a) Jordfundne og nulevende Aber (Primates) fra Lagoa Santa, Minas Gerais, Brasilien: med udsigt over abernes indyrdes slægtskab. E Museo Lundii 2(3):1–45

Winge H (1895b) Jordfundne og nulevende Rovdyr (Carnivora) fra Lagoa Santa, Minas Gerais, Brasilien: med udsigt over rovdyreues indyrdes slægtskab. E Museo Lundii 2(4):1–103

Winge H (1906) Jordfundne og nulevende Hovdyr (Ungulata) fra Lagoa Santa, Minas Gerais, Brasilien: med udsigt over hovdyrenes indbyrdes slægtskab. E Museo Lundii 3(1):1–239

Winge H (1915) Jordfundne og nulevende Gumlere (Edentata) fra Lagoa Santa, Minas Gerais, Brasilien: med udsigt over gumlernes indbyrdes slægtskab. E Museo Lundii 3(2):1–321

The Archaeological Record of Lagoa Santa (East-Central Brazil): From the Late Pleistocene to Historical Times

André Strauss, Igor M. Mariano Rodrigues, Alenice Baeta, Ximena S. Villagran, Marcony Alves, Francisco Pugliese Jr., Marcos Bissaro Jr., Rodrigo Elias de Oliveira, Gustavo Neves de Souza, Lucas Bueno, João Carlos Moreno de Sousa, Johnica J. Morrow, Karl J. Reinhard, Tiago Hermenegildo, Glauco Constantino Perez, Eliane Nunes Chim, Rafael de Oliveira dos Santos, Maurício de Paiva, Renato Kipnis, and Walter Neves

The original version of this chapter was revised: The author names have been updated. The correction to this chapter is available at https://doi.org/10.1007/978-3-030-35940-9_15

A. Strauss (✉) · E. N. Chim · M. de Paiva
Laboratório de Arqueologia e Antropologia Ambiental e Evolutiva, Museu de Arqueologia e Etnologia da, Universidade de São Paulo, Av. Prof. Almeida Prado, 1466, Butantã, São Paulo, SP 05508-070, Brazil
e-mail: strauss@usp.br

E. N. Chim
e-mail: chim@usp.br

M. de Paiva
e-mail: fotomauricio@gmail.com

I. M. Mariano Rodrigues
Laboratório de Estudos Interdisciplinares sobre Tecnologia e Território, Museu de Arqueologia e Etnologia, Universidade de São Paulo, Av. Prof. Almeida Prado, 1466, Butantã, São Paulo, SP 05508-070, Brazil
e-mail: igor_mmrodrigues@hotmail.com

M. Alves
Museu de Arqueologia e Etnologia, Universidade de São Paulo, Av. Prof. Almeida Prado, 1466, Butantã, São Paulo, SP 05508-070, Brazil
e-mail: marcony.alves@yahoo.com.br

A. Baeta
Museu de Arqueologia e Etnologia da Universidade de São Paulo, Av. Prof. Almeida Prado, 1466, Butantã, São Paulo, SP 05508-070, Brazil
e-mail: alenicebaeta@yahoo.com.br

X. S. Villagran
Laboratório de Microarqueologia, Museu de Arqueologia e Etnologia, Universidade de São Paulo, Av. Prof. Almeida Prado, 1466, Butantã, São Paulo, SP 05508-070, Brazil
e-mail: villagran@usp.br

F. Pugliese Jr.
Laboratório de Geocronologia, Instituto de Geociências, Universidade de Brasília, Brasília, Brazil
e-mail: pugliesefrancisco@yahoo.com.br

Laboratório de Arqueologia dos Trópicos, Museu de Arqueologia e Etnologia, Universidade de São Paulo, Asa Norte, Brasília, DF 70910-900, Brazil

M. Bissaro Jr.
Laboratório de Paleontologia de Ribeirão Preto, Faculdade de Ciências e Letras de Ribeirão Preto, Universidade de São Paulo, Av. Bandeirantes, 3900, Ribeirão Preto, SP 14040-901, Brazil
e-mail: marcosbissaro@gmail.com

R. E. de Oliveira
Laboratório de Arqueologia e Antropologia Ambiental e Evolutiva, Departamento de Genética e Biologia Evolutiva, Instituto de Biociências, Universidade de São Paulo, R. do Matão, 277, Butantã, São Paulo, SP 05508-090, Brazil
e-mail: eliaso@usp.br

G. N. de Souza
Universidade do Vale do São Francisco, Campus Serra da Capivara, Rua João Ferreira dos Santos, s/n. Bairro Campestre - São Raimundo Nonato, Petrolina, PI 64770-000, Brazil
e-mail: gustavo.souza@univasf.edu.br

L. Bueno
Laboratório de Estudos Interdisciplinares em Arqueologia, Departamento de História, Universidade Federal de Santa Catarina, Campus Universitário – UFSC, Bloco E, 6° andar, Bairro Trindade, Florianópolis, SC 88040-970, Brazil
e-mail: lucasreisbueno@gmail.com

A. S. Auler and P. Pessoa (eds.), *Lagoa Santa Karst: Brazil's Iconic Karst Region*, Cave and Karst Systems of the World, https://doi.org/10.1007/978-3-030-35940-9_12

Abstract
Lagoa Santa recorded ~12,500 years of non-continuous human occupation in cavities and open air sites. For the early- and middle Holocene occupations, lithic technology, zooarchaeology, osteological markers and multi-isotopic analyses indicate groups of foragers with low mobility and a subsistence strategy focused on gathering plant foods and hunting small and mid-sized animals but no megafauna. A high frequency of caries is observed among women. Lithics include small flakes and cores of quartz. Artefacts like projectile points and axe blades occur marginally. Funerary rituals included primary burials (10.3–10.6 cal kyBP), reduction of the body followed by secondary burial (9.4–9.6 cal kyBP) and pits filled with disarticulated and fragmented bones of a single individual (~8.2–8.6 cal kyBP). Rock art abounds, including the oldest securely dated case of South America (10.5 cal kyBP). Representations include animals, filiform anthropomorphs, geometric motifs, manioc's tubers and semi-lunar axes. Similar styles are found over a large area of Brazil. Ancient DNA extracted from several early Holocene skeletons indicates they are entirely nested within past and present Native American genetic diversity. Pottery arrives during the late Holocene with undecorated vessels of varying sizes (0.2–417 L), tubular pipes, spindle whorls, highly polished axes and ritual semi-lunar axes (Aratu-Sapucaí-Una Tradition). In the site 'Vereda III', more than 3,000 in situ sherds indicate repeated use for ritual feasts centred on fermented beverages: (maize?). Polychrome pottery occurring in low frequency testify to the late presence of Tupi-speaking groups. During the colonial period, Lagoa Santa housed large haciendas for cattle and agriculture, resulting in the formation of large farms with thousands of enslaved people. Stub-stemmed earthenware pipes from the nineteenth century, an important element of African diasporic identities, were produced in the region.

Keywords
American archaeology • American anthropology • Historical archaeology • Colonial Brazil • Pre-history • Slavery

1 Introduction

With approximately 170 registered sites (Table 1) the archaeological record of the Lagoa Santa region (Fig. 1) documents more than 12,000 years of human occupation in east-central Brazil (Araujo et al. 2012; Strauss et al. 2016; Da-Gloria et al. 2017, 2018). With a landscape typical of an evolved karst (Fig. 2 and chapter "Karst Landforms in the Lagoa Santa Area", this volume), the region has been well known among archaeologists and palaeontologists since the nineteenth century. The first human skeletons were found by the Danish naturalist Peter Lund between 1842 and 1844 in the Sumidouro cave in association with the bones of extinct megafauna (Lund 1844; Cartelle 1994; Piló and Auler 2002; Auler and Piló 2017; Luna 2007; Holten and Sterll 2017). Due to this putative coexistence of humans and megafauna, Lagoa Santa became the focus of many nineteenth century scholars (e.g., Kollman 1884; Ten Kate 1885; Hansen 1888; Hrdlička 1912). During the twentieth century, different teams pursued evidence that could confirm the coexistence hypothesis (Walter et al. 1937; Walter 1958; Hurt and Blasi 1969; Laming-Emperaire 1979a, b; Bányai 1997). As the result of more than 170 years of excavations, a large

J. C. M. de Sousa · G. C. Perez
Laboratório de Estudos Evolutivos Humanos, Departamento de Genética e Biologia Evolutiva, Instituto de Biociências, Universidade de São Paulo, R. do Matão, 277, Butantã, São Paulo, SP 05508-090, Brazil
e-mail: jcmorenodesousa@gmail.com

G. C. Perez
e-mail: glauco1113@hotmail.com

J. J. Morrow · K. J. Reinhard
School of Natural Resources, University of Nebraska-Lincoln, 719 South Hardin Hall, 3310 Holdrege Street, Lincoln, NE, Brazil
e-mail: jmorrow@csc.edu

K. J. Reinhard
e-mail: kreinhard1@mac.com

T. Hermenegildo
Department of Archaeology and Anthropology, University of Cambridge, Downing Street, Cambridge, CB2 3ER, UK
e-mail: thermenegildo@gmail.com

R. de Oliveira dos Santos
Fundação do Nacional do Índio, Coordenação técnica Local em Paraty, Rua Minas Gerais, 23 – Vila Residencial de Mambucaba, Paraty, RJ 29970-000, Brazil
e-mail: rafael.santos@funai.gov.br

R. Kipnis
Scientia Consultoria Científica, R. Prof. Campos Almeida, 75, Jardim Bonfiglioli, São Paulo, SP 05591-045, Brazil
e-mail: rkipnis@scientiaconsultoria.com.br

W. Neves
Instituto de Estudos Avançados, Universidade de São Paulo, R. do Anfiteatro, 513, Butantã, São Paulo, SP 05508-060, Brazil
e-mail: waneves@ib.usp.br

Table 1 Archaeological sites of the Lagoa Santa region

# map	Site	City	Rockshleter	Open air	Prehistoric	Historical	Pottery	Polished	Flaked	Rockart	Burials
1	A. do Angico	Matozinhos	x		x		x				
2	A. L. Vermelha I bis	Pedro Leopoldo	x		x					x	
3	A. Lapa Vermelha I	Pedro Leopoldo	x		x				x	x	
4	A. Lapa Vermelha II	Pedro Leopoldo	x		x					x	
5	A. Lapa Vermelha IV	Pedro Leopoldo	x		x		x		x	x	?
6	A. Pedra Esquecida	Matozinhos	x		x			x	x		
7	Abrigão das cerâmicas	Prudente de Morais	x		x		x		x		
8	A. Mata da Cauaia	Matozinhos	x		x					x	
9	Abrigo da Serrinha	Matozinhos	x		x		x				
10	Abrigo de Caieiras	Matozinhos	x		x			x	x	x	?
11	Abrigo do Bodão	Lagoa Santa	x		x					x	
12	Abrigo do Galinheiro	Confins	x		x		x	x	x	x	x
13	Abrigo dos Ideogramas	Matozinhos	x		x					x	
14	Abrigo Lapa Preta	Prudente de Morais	x		x				x		
15	Abrigo Leste	Pedro Leopoldo	x		x		x		x		
16	Abrigo Pedra Pesada	Matozinhos	x		x			x	x		
17	Abrigo Rei do Mato	Sete Lagoas	x		x				x	x	
18	Arruda	Lagoa Santa		x	x		x		x		
19	Ballet a céu aberto	Matozinhos		x	x		x				
20	Base	Pedro Leopoldo		x	x		x				
21	Baú	Pedro Leopoldo		x	x		x	x			x
22	Bebedouro	Matozinhos		x	x		x				
23	Bité	Lagoa Santa		x	x		x				
24	Busca Vida	Confins		x	x		x				
25	Cachorro Branco	Lagoa Santa		x	x		x				
26	Caetano	Matozinhos		x	x		x				
27	Cainhanga	Matozinhos	x		x		x				
28	Campinho	Pedro Leopoldo	x		x					x	
29	Capao das Eguas	Prudente de Morais	x		x				x	x	
30	Cauaia I	Matozinhos		x	x		x				
31	Cauaia II	Matozinhos		x	x		x				

(continued)

Table 1 (continued)

# map	Site	City	Rockshleter	Open air	Prehistoric	Historical	Pottery	Polished	Flaked	Rockart	Burials
32	Cerca Grande I	Matozinhos	x		x		x				
33	Cerca Grande II	Matozinhos	x		x		x	x	x	x	x
34	Cerca Grande III	Matozinhos	x		x			x	x	x	
35	Cerca Grande IV	Matozinhos	x		x			x	x		
36	Cerca Grande V	Matozinhos	x		x			x	x	x	x
37	Cerca Grande VI	Matozinhos	x		x		x	x	x		x
38	Cerca Grande VII	Matozinhos	x		x						x
39	Cerca Grande VIII	Matozinhos		x	x			x	x		
40	Coqueirinho	Lagoa Santa		x	x				x		
41	Criciúma I e II	Prudente de Morais	x		x					x	
42	Cruzeiro das Goiabeiras	Lagoa Santa		x		x					
43	Dolina da Bebida	Prudente de Morais	x		x					x	
44	Escrivania	Prudente de Morais	x		x				x	x	
45	Eucalipto	Pedro Leopoldo	x		x		x	x	x		x
46	Experiencia da Jaguara	Matozinhos	x	x	x		x		x	x	
47	Faz. Capao Grande Abrigo 1	Sete Lagoas	x		x			x	x	x	
48	Faz. Capao Grande Abrigo 3	Sete Lagoas									
49	Fazenda Alpercata	Lagoa Santa		x	x		x				
50	Fazenda da Lapa	Sete Lagoas		x	x		x				
51	Fazenda da Salmoura I e II	Pedro Leopoldo			x				x		
52	Fazenda da Salmoura III e IV	Pedro Leopoldo			x				x		
53	Fazenda Fidalgo	Lagoa Santa		x	x		x				
54	Fazenda Moinho	Lagoa Santa	x		x			x	x		
55	Fazenda Ribeira	Pedro Leopoldo		x	x		x				
56	Fazenda Tamboril	Pedro Leopoldo		x	x		x		x		
57	Ferradura	Matozinhos		x	x		x		x		
58	Gameleira	Matozinhos	x		x					x	
59	Gruta Cheirosa	Pedro Leopoldo	x		x		x				
60	Gruta da fazenda Paraíso	Prudente de Morais	x		x		x		x		
61	Gruta de Caieiras	Matozinhos	x		x					x	
62	Império da Jaragua	Matozinhos		x	x		x		x		

(continued)

Table 1 (continued)

# map	Site	City	Rockshleter	Open air	Prehistoric	Historical	Pottery	Polished	Flaked	Rockart	Burials
63	Jacques I	Lagoa Santa		x	x		x				
64	Jacques II	Lagoa Santa		x	x		x				
65	Janelas de Cerca Grande	Matozinhos	x		x					x	
66	Joao Bárbara	Matozinhos		x	x		x				
67	João Bosco	Lagoa Santa		x	x		x				
68	Julião	Matozinhos		x	x		x				
69	Lagoa dos Aracas	Matozinhos				x					
70	Lagoa Funda	Pedro Leopoldo	x		x						x
71	Lagoa Verde I, III e V	Prudente de Morais	x		x				x	x	
72	Lapa da Pia	Prudente de Morais	x		x					x	
73	Lapa da Ribeira	Pedro Leopoldo	x		x				x		
74	Lapa das Boleiras	Matozinhos	x		x		x	x	x		x
75	Lapa das Boleiras II	Matozinhos	x		x						x
76	Lapa do Ballet	Matozinhos	x		x		x		x	x	
77	Lapa do Baú	Pedro Leopoldo	x		x						x
78	Lapa do Caetano	Matozinhos	x		x					x	x
79	Lapa do Carrocao	Pedro Leopoldo	x		x		x		x		?
80	Lapa do Chapéu	Matozinhos	x		x				x		
81	Lapa do Jessé	Lagoa Santa	x		x				x		
82	Lapa do Ouro	Matozinhos	x		x				x		
83	Lapa do Salitre	Lagoa Santa	x		x		x				?
84	Lapa do Santo (ou Cinzenta)	Matozinhos	x		x		X	x	x	x	x
85	Lapa do Sumidouro I	Pedro Leopoldo	x		x		x	x	x	x	?
86	Lapa do Sumidouro II	Pedro Leopoldo	x		x						x
87	Lapa do Urubu	Confins	x		x			x	x		x
88	Lapa dos Micos	Lagoa Santa	x		x					x	
89	Lapa Limeira	Confins	x		x			x	x		x
90	Lapa Mortuária	Confins	x		x		x	x			x
91	Lapa Vermelha Soleil	Pedro Leopoldo		x	x		x				
92	Lapa Vermelha VI	Pedro Leopoldo	x		x		x	x	x		x
93	Lapinha I	Lagoa Santa	x		x		x		x		
94	Lapinha II Bis A	Lagoa Santa	x				x		x		x
95	Lapinha II Bis B	Lagoa Santa		x	x		x				

(continued)

Table 1 (continued)

# map	Site	City	Rockshleter	Open air	Prehistoric	Historical	Pottery	Polished	Flaked	Rockart	Burials
96	Lavarjão	Lagoa Santa		x	x		x		x		
97	Limeira	Prudente de Morais	x		x					x	
98	Lund	Lagoa Santa		x	x				x		
99	Macaúba	Lagoa Santa		x	x		x				
100	Mae Rosa	Pedro Leopoldo	x		x			x	x		x
101	Mandiocal	Matozinhos		x	x		x				
102	Marciano	Pedro Leopoldo	x		x	?	x		x		?
103	Mato Seco	Prudente de Morais	x		x				x	x	
104	Milharal do Macaco	Sete Lagoas		x	x		x		x		
105	Olhos d'água	Lagoa Santa		x	x		x	x			
106	Paredao do Trevo	Sete Lagoas	x		x					x	
107	Pastinho	Lagoa Santa		x	x		x	x	x		
108	Pasto do Topo	Matozinhos	x		x	x	x				
109	Pedra Falsa	Lagoa Santa		x	x		x		x		
110	Pedreira	Matozinhos		x	x		x				
111	Pedreira da Vitrina	Sete Lagoas		x	x		x		x		
112	Peri Peri	Matozinhos		x	x		x		x		x
113	Poço Azul	Lagoa Santa		x	x		x				
114	Poções I	Matozinhos	x		x				x	x	
115	Poçoes II (a e b)	Matozinhos	x	x	x		x		x		
116	Poções III	Matozinhos		x	x		x	x			
117	Porco Preto	Matozinhos		x	x		x				
118	Porco Preto	Matozinhos	x		x	x		x	x	x	
119	Quebra Cangalha	Lagoa Santa		x	x		x	x	x		
120	Quebra Prato	Lagoa Santa		x	x		x				
121	Quintalinho	Matozinhos		x	x		x				
122	Riacho Dantas	Matozinhos		x	x		x		x		
123	Rocinha	Pedro Leopoldo		x	x		x	x	x		
124	Ronaldo Nascimento	Pedro Leopoldo		x	x		x				
125	Ruínas Fazenda Bom Jardim	Matozinhos		x		x					
126	S. Açude do Barbosa	Matozinhos		x	x		x	x	x		
127	Salitre (Caianga)	Matozinhos	x		x		x				
128	Samambaia I	Pedro Leopoldo	x		x		x		x	x	x
129	Samambaia II	Pedro Leopoldo	x		x					x	
130	Samambaia III	Pedro Leopoldo	x		x	?			x	x	

(continued)

Table 1 (continued)

# map	Site	City	Rockshleter	Open air	Prehistoric	Historical	Pottery	Polished	Flaked	Rockart	Burials
131	Santo Antônio do mocambo	Matozinhos		x	x		x		x		
132	Santo Antonio II	Matozinhos		x	x		x				
133	Seu Nenê I	Matozinhos		x	x		x		x		
134	Seu Nenê II	Matozinhos		x	x		x				
135	Seu Nenê III	Matozinhos		x	x		x				
136	Sítio Aqueduto do Elmo	Lagoa Santa		x		x					
137	Sítio Aqueduto do Fidalgo	Lagoa Santa		x		x					
138	Sitio Ceramico do Bene	Lagoa Santa		x	x		x				
139	Sitio Ceramico Nilo Abreu	Lagoa Santa		x	x		x				
140	Sítio da Goiabeira	Prudente de Morais		x	x		x				
141	Sítio da Palhada	Sete Lagoas		x	x		x		x		
142	Sitio do Ari II	Sete Lagoas		x	x		x		x		
143	Sitio do Engenho	Pedro Leopoldo		x	x		x				
144	Sítio do Lameiro	Matozinhos	x		x			x			
145	Sitio do Maciel	Sete Lagoas		x	x		x				
146	Sitio do Padre	Lagoa Santa		x	x		x				
147	Sitio do Romero	Sete Lagoas		x	x		x				
148	Sitio do Valão	Sete Lagoas		x	x		x				
149	Sitio Federal	Sete Lagoas	x		x		x		x		
150	Sítio Lítico Formosa	Matozinhos		x	x				x		
151	Sitio Lontrinha	Sete Lagoas		x	x		x				
152	Sítio Sumidouro	Pedro Leopoldo		x	x		x		x		
153	Sobradinho	Lagoa Santa		x	x		x				
154	Sumidouro da V. da Cauaia	Matozinhos			x	x	x		x		
155	Tamboril II	Lagoa Santa		x	x		x		x		
156	Tiãozinho Fernandes	Lagoa Santa		x	x		x		x		
157	Vaca Prenha	Lagoa Santa		x	x		x				
158	Valetão	Pedro Leopoldo		x	x		x				
159	Vargem da Lapa	Lagoa Santa	x		x					x	
160	Vargem da Pedra	Matozinhos	x		x		x			x	x
161	Vargem Formosa	Matozinhos	x		x					x	
162	Vargem Grande	Sete Lagoas		x	x		x				
163	Vereda I	Prudente de Morais		x	x		x				
164	Vereda II	Prudente de Morais		x	x		x				

(continued)

Table 1 (continued)

# map	Site	City	Rockshleter	Open air	Prehistoric	Historical	Pottery	Polished	Flaked	Rockart	Burials
165	Vereda III	Prudente de Morais	x	x	x		x	x	x		
166	Vereda IV	Prudente de Morais	x		x		x				
167	Vereda V	Prudente de Morais		x	x		x				
168	Vereda VI	Prudente de Morais		x	x		x				
169	Viana	Lagoa Santa		x	x		x				
170	Vice Rei	Prudente de Morais	x		x					x	

collection of early Holocene skeletons was formed (Neves et al. 2013; Bernardo et al. 2011; Da-Gloria 2012; Costa 2017; Keuller 2017; Neto and Rodrigues-Carvalho 2017; Souza and Lyrio 2017).

Archaeological studies in the region were largely focused on the oldest periods and human skeletal remains. However, Lagoa Santa also provides one of the most complete diachronic sequences available in Brazil, registering both pre- and postcolonial history. In addition to the more commonly discussed early Holocene component, there is also evidence of foragers during the middle Holocene, different ceramist groups during the late Holocene (i.e., the Tupiguarani and Aratu-Sapucaí-Una) as well as archaeological evidence of the colonial and imperial periods. In this chapter, we provide a broad description of the archaeological record of the Lagoa Santa region.

2 Chronology

Lagoa Santa's chronology is based on approximately 185 radiocarbon dates and several optically stimulated luminescence (OSL) dates (Tables 2). The most thoroughly dated sites are the rock shelters of Lapa do Santo, Lapa das Boleiras, Lapa Vermelha IV and Lapa Grande de Taquaraçu and the open air sites of Sumidouro and Coqueirinho. Additionally, a few dates are available for Cerca Grande 6, Caieiras and Lagoa Funda. The first two radiocarbon dates for the region were obtained at the University of Pennsylvania from charcoals derived from the 1956 excavations of the American expedition led by Wesley Hurt and Oldemar Blasi (Hurt and Blasi 1969). Indicating an early Holocene occupation, the charcoals confirmed the antiquity of a human presence in the region—although not to Pleistocene times. During the 1970s, a total of 29 radiocarbon dates on charcoal were obtained for Lapa Vermelha IV by the French Mission coordinated by Annette Laming-Emperaire in the Gif-sur-Yvet laboratory (Laming-Emperaire et al. 1975; Delibrias et al. 1986; Prous 2017).

Since 2001, in the framework of the São Paulo Research Foundation (FAPESP) project 'Origins and microevolution of humans in America', coordinated by Walter Neves (Araujo et al. 2012; Bernardo et al. 2017), a large database of approximately 126 radiocarbon dates were provided by Beta Analytic on charcoal, plant remains, organic sediments and collagen extracted from the bones of humans and other animals (extinct megafauna included). Dozens of OSL dates from sediment samples—including single grain analysis—were generated by James Feathers at the University of Washington. From 2011 onwards, as part of the project 'Mortuary rituals of the first Americans' coordinated by André Strauss and Rodrigo Oliveira and funded by the Max Planck Institute for Evolutionary Anthropology (MPI-EVA), the first radiocarbon dates were produced based on collagen checked for quality parameters. Collagen extraction was performed at the Max Planck Institute for Evolutionary Anthropology by Sahra Talamo and dating was performed at the Klaus-Tschira-AMS facility of the Curt-Engelhorn Centre in Mannheim, Germany (Strauss et al. 2016). Additionally, 14 samples of charcoal were dated in the Oxford Radiocarbon Accelerator Unit (ORAU) (Villagran et al. 2017).

Lagoa Santa's chronology is compatible with a scenario of post-late glacial maximum (LGM) settlement of the New World (Bueno et al. 2013; Dias and Bueno 2013). A Bayesian chronological model based on Lapa do Santo, the better dated site of Lagoa Santa, indicates three discrete phases of occupation in the region. They roughly correspond to the early Holocene (12.7–11.7 kyr cal BP[1] to 8.3–8.0 kyr cal BP), middle Holocene (5.4–4.9 kyr cal BP to 4.3–3.9 kyr cal BP) and late Holocene (2.1–0.8 kyr cal BP to 0.9–0.2 kyr cal BP)

[1]'kyr cal BP': thousand years calibrated before the present. All ranges based on 95.4% confidence interval.

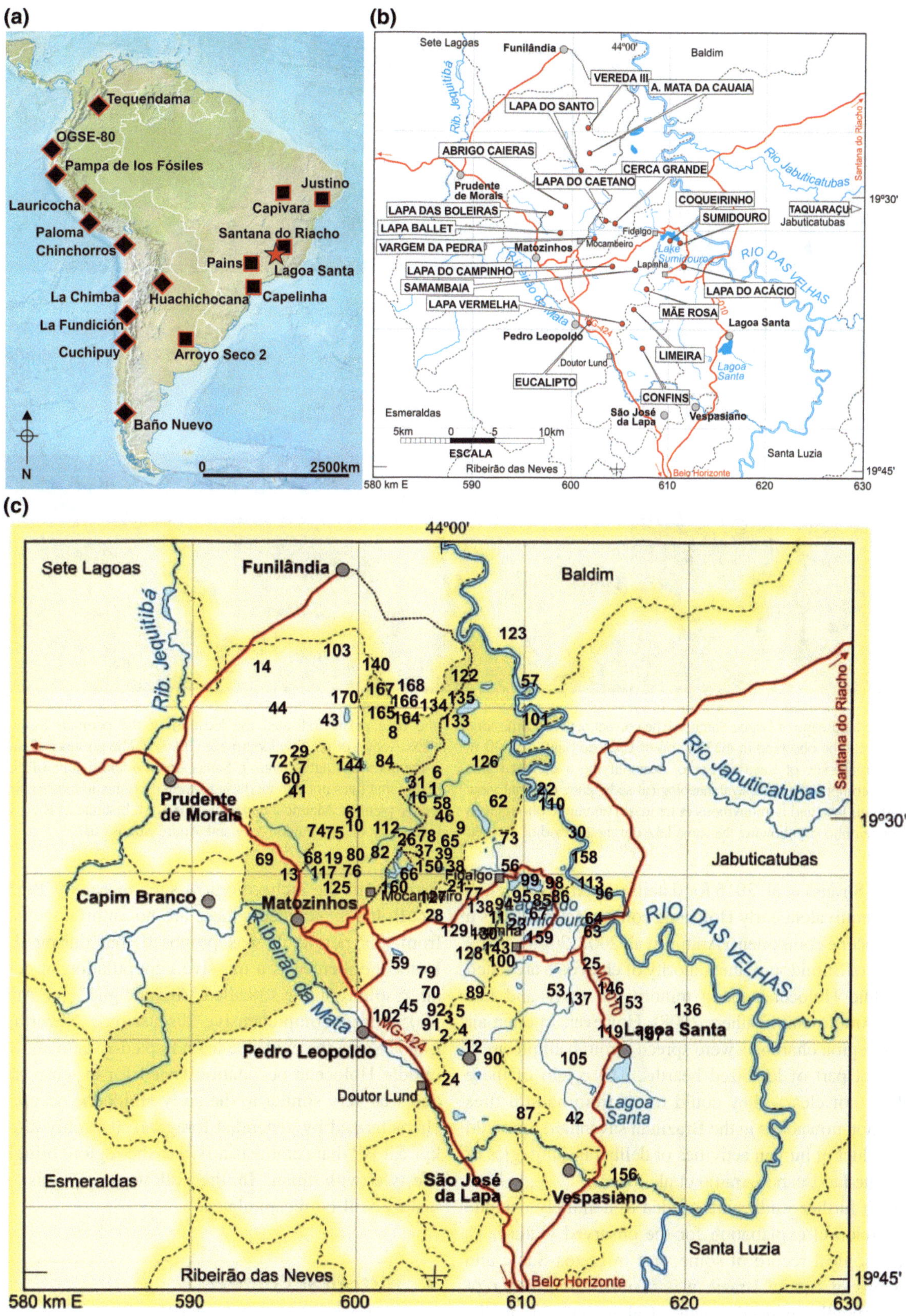

Fig. 1 **a** Map of South America showing the locations of Lagoa Santa (red star) and early Holocene sites with skeletal remains relevant for comparison; **b** map of the Lagoa Santa region, showing the most important archaeological sites mentioned in the text (red circles), city limits (dotted grey lines), major urban centres (grey circles), local communities (grey squares) and main roads (continuous red lines). Note that Taquaraçu is located outside the area shown in the map. **c** Map of the Lagoa Santa region showing the location of every registered archaeological sites (indicated by numbers from 1 to 170). Information concerning site type (open air or rockshelter), period (precolonial or historical) and associated material culture (pottery, polished stone tools, flaked stone tools, presence of rock art and presence of human skeletons) can be found on Table 1

Fig. 2 Karstic landscape of Lagoa Santa. **a** The contact between limestone and vegetation can be observed in the outcrop of Lapa do Santo (ca. 40 m height); **b** Internal view of Lapa do Santo, exemplifying a sheltered area whose ground comprises anthropic/archaeological sediments; **c** Aerial view of the Sumidouro lake and homonymous outcrop; **d** Panoramic photo taken from the Coqueirinho site showing the same lake during a period of drought (the blue arrow indicated the location of the open-air Sumidouro site); **e** Excavation of the Sumidouro site (2007); **f** The presence of subtle dolines is common in Lagoa Santa; **g** Santo Antonio lake is one of the few water bodies that does not dry out during severe droughts in modern times. Credits for the pictures: Maurício de Paiva (**a**, **b**, **f**, **g**), Institute of Biosciences of the University of São Paulo (**c**, **e**) and André Strauss (**d**)

periods (see Strauss et al. 2016 for details). Lapa das Boleiras presents an equivalent early Holocene component but lacks a middle Holocene component (Araujo et al. 2002, 2008). In the open air site of Sumidouro, the majority of charcoals are dated to the middle Holocene and a minority to early and late Holocene (Araujo and Feathers 2008). However, in open air sites of the region charcoals were spread through ample areas and were not part of localized hearths. The origin of these charcoals is not clear—they could result from natural fires (very common nowadays in the Brazilian savannah), or could be a by-product of human activities or deliberate strategies of managing the landscape (Araujo et al. 2013).

Although further studies are needed to definitely exclude a non-behavioural explanation for the observed hiatuses in the archaeological record of some sites in Lagoa Santa and other regions of central Brazil, it is assumed that they represent climatically induced events of abandonment or dispersion (Araujo et al. 2005; Bueno 2012). Actually, the middle Holocene anthropogenic component of Lapa do Santo and Lapa Vermelha IV is a rare finding. In the majority of the archaeological sites in central Brazil—Lapa das Boleiras included—there are no anthropogenic deposits from this period. It was proposed that during the middle Holocene there was a massive depopulation of central Brazil, resulting in the so-called 'Archaic gap' (see Araujo et al. 2005 for a comprehensive discussion). Nevertheless, for Lagoa Santa the evidence from Lapa do Santo shows that the middle Holocene occupation lasted for at least 1,000 years and was very similar to the early Holocene occupation (i.e., characterized by generalist foragers). It is only after 2.1–0.8 kyr cal BP that ceramists arrive to the region, bringing a new lifestyle with them. In the following sections, this long history will be recapitulated.

3 Historical Context

Portuguese colonization of South America started in 1500's, but it was only a century later that the interior of the continent started to be populated. During the seventeenth

Table 2 Complete list of radiocarbon dates for the Lagoa Santa region

Lab Id	Site	Material 2	Unit	X	Y	Z	% Coll	D13C	d15 N	%C	%N	C14 Age	Error
Gif - 2737	Caieiras	Wood charcoal	Sondage B Couche V	–	–	–	–	–	–	–	–	9600	200
Uni. Pennsylvania 519b	Cerca Grande 6	Wood charcoal	Nivel 2 e 3	–	–	–	–	n/a	–	–	–	9028	120
Uni. Pennsylvania 521b	Cerca Grande 6	Wood charcoal	Nivel 6 e 7	–	–	–	–	n/a	–	–	–	9720	128
Beta - 248889	Coqueirinho	Wood charcoal	S4	–	–	160–170	–	−27.4	–	–	–	100	40
Beta - 248887	Coqueirinho	Wood charcoal	S6	–	–	70–80	–	−25.8	–	–	–	2740	40
Beta - 248886	Coqueirinho	Wood charcoal	S6	–	–	70–80	–	−24.9	–	–	–	3620	40
Beta - 248885	Coqueirinho	Wood charcoal	S6	–	–	40–50	–	−25.9	–	–	–	3790	40
Beta - 248888	Coqueirinho	Wood charcoal	S6	–	–	80–90	–	−25.5	–	–	–	6800	50
Beta - 237346	Coqueirinho	Wood charcoal	S1	–	–	50–60	–	−25.1	–	–	–	10460	60
Gif - 2738	Fazenda Manguera	Wood charcoal	Puits a extraction d^argile, limite sediment jaune/sed.rouge	–	–	90–100	–	–	–	–	–	4670	130
Uni. Pennsylvania?	Lagoa Funda	Wood charcoal	–	–	–	–	–	–	–	–	–	3000	300
Beta - 162529	Lapa das Boleiras	Plant material	–	–	–	–	–	−28.3	–	–	–	140	40
Beta - 159237	Lapa das Boleiras	Wood charcoal	–	–	–	–	–	−25	–	–	–	160	70
Beta - 159231	Lapa das Boleiras	Wood charcoal	–	–	–	–	–	−26.3	–	–	–	180	40
Beta - 159234	Lapa das Boleiras	Wood charcoal	–	–	–	–	–	−25	–	–	–	210	60
Beta - 162530	Lapa das Boleiras	Organic Sediment	–	–	–	–	–	−16.4	–	–	–	810	40
Beta - 159235	Lapa das Boleiras	Wood charcoal	–	–	–	–	–	−25	–	–	–	3830	60
Beta - 159243	Lapa das Boleiras	Wood charcoal	–	–	–	–	–	−25	–	–	–	7560	110
Beta - 159232	Lapa das Boleiras	Wood charcoal	–	–	–	–	–	−25.3	–	–	–	8240	50
Beta - 159244	Lapa das Boleiras	Wood charcoal	–	–	–	–	–	−25.4	–	–	–	8360	50

(continued)

Table 2 (continued)

Lab Id	Site	Material 2	Unit	X	Y	Z	% Coll	D13C	d15 N	%C	%N	C14 Age	Error
Beta - 221456	Lapa das Boleiras	Wood charcoal	–	–	–	–	–	−26.7	–	–	–	8510	60
Beta - 221453	Lapa das Boleiras	Wood charcoal	–	–	–	–	–	−25.9	–	–	–	8530	60
Beta - 221454	Lapa das Boleiras	Wood charcoal	–	–	–	–	–	−27.2	–	–	–	8570	60
Beta - 159245	Lapa das Boleiras	Wood charcoal	–	–	–	–	–	−25	–	–	–	8730	110
Beta - 221452	Lapa das Boleiras	Wood charcoal	–	–	–	–	–	−23.5	–	–	–	8730	70
Beta - 183563	Lapa das Boleiras	Wood charcoal	–	–	–	–	–	−25.6	–	–	–	8750	50
Beta - 183564	Lapa das Boleiras	Wood charcoal	–	–	–	–	–	−25.9	–	–	–	8750	50
Beta - 221459	Lapa das Boleiras	Wood charcoal	–	–	–	–	–	−24.8	–	–	–	8810	60
Beta - 159242	Lapa das Boleiras	Wood charcoal	–	–	–	–	–	−25	–	–	–	8820	150
Beta - 221451	Lapa das Boleiras	Wood charcoal	–	–	–	–	–	−26.5	–	–	–	8980	60
Beta - 221457	Lapa das Boleiras	Wood charcoal	–	–	–	–	–	−26.9	–	–	–	9060	60
Beta - 159233	Lapa das Boleiras	Wood charcoal	–	–	–	–	–	−25	–	–	–	9210	130
Beta - 221460	Lapa das Boleiras	Wood charcoal	–	–	–	–	–	−26.7	–	–	–	9420	60
Beta - 159236	Lapa das Boleiras	Wood charcoal	–	–	–	–	–	−25.8	–	–	–	9600	60
Beta - 221458	Lapa das Boleiras	Wood charcoal	–	–	–	–	–	−26.4	–	–	–	9650	60
Beta - 168449	Lapa das Boleiras	Wood charcoal	–	–	–	–	–	−24.3	–	–	–	9850	40
Beta - 168451	Lapa das Boleiras	Wood charcoal	–	–	–	–	–	−25	–	–	–	10150	130
Beta - 168451	Lapa das Boleiras	Wood charcoal	–	–	–	–	–	−25	–	–	–	10150	130
Beta - 253497	Lapa do Santo	vertebrae (non-human?)	Burial 2	–	–	–	–	−19.2	4.2	–	–	790	40

(continued)

Table 2 (continued)

Lab Id	Site	Material 2	Unit	X	Y	Z	% Coll	D13C	d15 N	%C	%N	C14 Age	Error
Beta - 202764	Lapa do Santo	Wood charcoal	M05	12.630	4.860	1.097	–	−25.6	–	–	–	910	50
Beta - 183572	Lapa do Santo	Wood charcoal	M05	12.582	4.807	0.994	–	−25.0	–	–	–	960	70
Beta - 280491	Lapa do Santo	Wood charcoal	G13	6.280	13.082	−0.792	–	−26.6	–	–	–	1080	40
Beta - 202766	Lapa do Santo	Wood charcoal	M05	12.459	4.561	0.431	–	−26.0	–	–	–	3810	50
Beta - 202765	Lapa do Santo	Wood charcoal	M04	12.751	3.377	0.513	–	−25.7	–	–	–	3820	100
Beta - 183570	Lapa do Santo	Wood charcoal	M06	12.916	5.269	0.495	–	−25.5	–	–	–	3830	40
Beta - 216517	Lapa do Santo	Wood charcoal	M06	12.761	5.253	0.871	–	−24.9	–	–	–	3930	40
Beta - 243393	Lapa do Santo	Wood charcoal	M07	12.750	6.590	0.260	–	−23.4	–	–	–	3950	40
Beta - 214129	Lapa do Santo	Wood charcoal	M06	12.356	5.309	0.780	–	−26.7	–	–	–	3960	40
Beta - 183574	Lapa do Santo	Wood charcoal	L07	11.304	6.477	0.473	–	−25.0	–	–	–	4010	130
Beta - 216521	Lapa do Santo	Wood charcoal	M04	12.088	3.455	0.947	–	−25.0	–	–	–	4070	60
Oxford - 30686	Lapa do Santo	Wood charcoal	P11	15.954	10.967	−0.034	–	−27.04	–	–	–	4086	28
Beta - 243392	Lapa do Santo	Wood charcoal	M07	12.537	6.570	0.697	–	−26.3	–	–	–	4140	40
Beta - 202768	Lapa do Santo	Wood charcoal	M04	12.574	3.867	0.154	–	−25.7	–	–	–	4290	90
Beta - 248891	Lapa do Santo	Wood charcoal	AI1	34.099	0.630	−2.787	–	−24.5	–	–	–	4470	40
Beta - 215195	Lapa do Santo	bone	Burial 11	–	–	-0.120	–	-20.6	–	–	–	5990	40
Beta - 215194	Lapa do Santo	Bone	Burial 07	–	–	−0.271	–	-18.9	–	–	–	7400	40
Beta - 215200	Lapa do Santo	Bone (details not available)	Burial 19	–	–	−0.084	–	−18.6	–	–	–	7700	40
Beta - 214142	Lapa do Santo	Wood charcoal	I12	8.479	11.936	−0.606	–	−25.1	–	–	–	7890	40

(continued)

Table 2 (continued)

Lab Id	Site	Material 2	Unit	X	Y	Z	% Coll	D13C	d15 N	%C	%N	C14 Age	Error
Beta - 159246	Lapa do Santo	Wood charcoal	F13	–	–	−2.145	–	−26.5	–	–	–	7940	50
MAMS - 15242	Lapa do Santo	rib	Burial 2	–	–	–	1.2	−19.1	4.62	4.06	1.52	8062	36
Beta - 248893	Lapa do Santo	Wood charcoal	J08	8.980	7.637	0.415	–	−25.2	–	–	–	8170	50
Beta - 183573	Lapa do Santo	Wood charcoal	K08	10.789	7.002	0.577	–	−25.8	–	–	–	8230	50
Beta - 215196	Lapa do Santo	Bone (details not available)(adult)	Burial 14	–	–	0.140	–	−22.4	–	–	–	8230	40
MAMS - 15245	Lapa do Santo	bone	Burial 16	–	–	–	–	–	–	–	–	8290	35
MAMS - 29423	Lapa do Santo	Left petrous bone	Burial 30	–	–	–	0.3	−32.5	–	22.70	–	8299	29
MAMS - 16368	Lapa do Santo	Sphenoid fragment	Burial 26	–	–	−0.531	0.81	−19.0	5.86	3.00	1.17	8331	44
Beta - 265182	Lapa do Santo	Fragments of rib and long bone (adult)	Burial 17	–	–	−0.141	–	−19.0	–	–	–	8480	50
Oxford - 28666	Lapa do Santo	Wood charcoal	N11	13.994	10.225	−0.303	–	−27.61	–	–	–	8520	55
Beta - 202767	Lapa do Santo	Wood charcoal	G12	6.500	11.250	−1.303	–	−26.5	–	–	–	8530	40
Beta - 253511	Lapa do Santo	bone	Burial 26	–	–	−0.531	–	−19.8	7.50	–	–	8540	50
MAMS - 28706	Lapa do Santo	Tooth (18)	Burial 15	–	–	–	2.42	−19.8	9.40	30.90	10.40	8560	34
BETA 253507	Lapa do Santo	2 fragments of rib (sub-adult of ca. 1.5 years) and fragments of rib and long bone (adult)	Burial 17	–	–	-0.141	–	-19.0	8.50	–	–	8580	50
MAMS - 15246	Lapa do Santo	Long bone	Burial 21	–	–	-0.533	1.4	−18.6	6.40	7.31	2.75	8584	33
Oxford - 30690	Lapa do Santo	Wood charcoal	N11	13.980	10.752	−0.405	–	−24.71	–	–	–	8590	45
Beta - 183571	Lapa do Santo	Wood charcoal	M06	12.870	5.858	−0.513	–	−25.0	–	–	–	8600	160
Beta - 202763	Lapa do Santo	Wood charcoal	J08	9.908	7.737	0.034	–	−26.8	–	–	–	8600	50

(continued)

Table 2 (continued)

Lab Id	Site	Material 2	Unit	X	Y	Z	% Coll	D13C	d15 N	%C	%N	C14 Age	Error
MAMS - 28704	Lapa do Santo	Tooth (27)	Burial 23 Adulto #1	–	–	–	2.57	−18.3	6.60	37.20	13.20	8604	35
Beta - 216523	Lapa do Santo	Wood charcoal	L07	11.918	6.863	0.029	–	−26.0	–	–	–	8620	40
MAMS - 17190	Lapa do Santo	Bone (# 37 on drawing of Exposition 1)	Burial 32	–	–	–	–	–	–	–	–	8637	35
Oxford - 28665	Lapa do Santo	Wood charcoal	P11	15.238	10.841	−0.381	–	−26.86	–	–	–	8660	40
Beta - 214131	Lapa do Santo	Wood charcoal	L07	11.713	6.504	0.121	–	−26.8	–	–	–	8670	40
MAMS - 29425	Lapa do Santo	Right petrous bone	Burial 18 Subadult	–	–	–	0.3	−25.1	–	37.4	–	8677	27
Beta - 216519	Lapa do Santo	Wood charcoal	M06	12.763	5.059	−0.005	–	−26.2	–	–	–	8690	40
Beta - 214133	Lapa do Santo	Wood charcoal	M06	12.267	5.861	−0.763	–	−26.0	–	–	–	8700	40
MAMS - 28705	Lapa do Santo	Tooth (45)	Burial 34	–	–	–	3.2	−18.9	8.60	32.20	11.50	8709	35
Beta - 214136	Lapa do Santo	Wood charcoal	L11	11.598	10.503	−0.382	–	−25.8	–	–	–	8710	80
Beta - 216524	Lapa do Santo	Wood charcoal	K07	10.005	6.442	−0.145	–	−24.8	–	–	–	8710	40
Oxford - 30488	Lapa do Santo	Wood charcoal	N14	13.698	13.621	−0.864	–	−24.55	–	–	–	8723	37
Beta - 253505	Lapa do Santo	fragment of rib (subadult)	Burial 14	–	–	0.140	–	−19.6	8.50	–	–	8730	50
Beta - 214135	Lapa do Santo	Wood charcoal	J07	9.718	6.920	−0.093	–	−25.4	–	–	–	8750	40
Oxford - 30534	Lapa do Santo	wood charcoal	P11	15.499	10.750	−0.533	–	−26.42	–	–	–	8765	45
Beta - 214132	Lapa do Santo	Wood charcoal	L31	11.595	30.909	−6.303	–	−24.8	–	–	–	8790	40
Beta - 280490	Lapa do Santo	Wood charcoal	F13	5.467	12.881	−1.753	–	−26.2	–	–	–	8790	40
Beta - 216522	Lapa do Santo	Wood charcoal	M04	12.520	3.591	0.060	–	−26.3	–	–	–	8800	40
Oxford - 30687	Lapa do Santo	Wood charcoal	P11	15.921	10.857	−0.300	–	−24.59	–	–	–	8802	37

(continued)

Table 2 (continued)

Lab Id	Site	Material 2	Unit	X	Y	Z	% Coll	D13C	d15 N	%C	%N	C14 Age	Error
Beta - 216520	Lapa do Santo	Wood charcoal	G12	6.220	11.916	−0.928	–	−26.3	–	–	–	8810	90
Beta - 170723	Lapa do Santo	Wood charcoal	F13	–	–	−3.755	–	−26.4	–	–	–	8820	60
Beta - 214137	Lapa do Santo	Wood charcoal	L08	11.450	7.517	−0.067	–	−26.3	–	–	–	8820	40
Oxford - 30689	Lapa do Santo	Wood charcoal	O12	14.451	11.203	−0.498	–	−25.84	–	–	–	8820	38
Oxford - 31014	Lapa do Santo	Wood charcoal	N14	13.304	13.950	−0.751	–	−25.82	–	–	–	8830	40
Beta - 271249	Lapa do Santo	tooth	Burial 1	–	–	-0.806		-18.1	8.80	–	–	8840	29
Oxford - 30685	Lapa do Santo	Wood charcoal	N13	13.569	12.499	−0.555	–	−24.01	–	–	–	8853	36
Beta - 214134	Lapa do Santo	Wood charcoal	M06	12.795	5.591	−0.824	–	−24.1	–	–	–	8870	100
MAMS - 23435	Lapa do Santo	Tooth (28)	Burial 23 Adult #2	–	–	–	1.9	–	–	–	–	8878	29
Beta - 159247	Lapa do Santo	Wood charcoal	F13	–	–	−3.928	–	−26.4	–	–	–	8880	50
Beta - 280487	Lapa do Santo	Wood charcoal	F13	5.347	12.850	−2.567	–	−24.2	–	–	–	8890	40
Beta - 246246	Lapa do Santo	Wood charcoal	G12	6.398	11.918	−2.605	–	−24.9	–	–	–	8900	40
Beta - 214139	Lapa do Santo	Wood charcoal	G13	6.925	12.579	−1.900	–	−25.9	–	–	–	8930	40
Beta - 214140	Lapa do Santo	Wood charcoal	M06	12.070	5.503	−1.512	–	−25.2	–	–	–	8930	40
Oxford - 30692	Lapa do Santo	Wood charcoal	O13	14.898	12.039	−0.533	–	−25.42	–	–	–	8937	38
MAMS - 28703	Lapa do Santo	Tooth (34)	Burial 15	–	–	–	1.68	−19.7	8.80	16.70	5.90	8948	39
Oxford - 30691	Lapa do Santo	Wood charcoal	O13	14.898	12.039	−0.533	–	−24.71	–	–	–	8950	45
Oxford - 30489	Lapa do Santo	Wood charcoal	N11	13.896	10.525	−0.464	–	−25.34	–	–	–	8958	39
Oxford - 30688	Lapa do Santo	Wood charcoal	N11	13.915	10.510	−0.427	–	−25.27	–	–	–	8959	36

(continued)

Table 2 (continued)

Lab Id	Site	Material 2	Unit	X	Y	Z	% Coll	D13C	d15 N	%C	%N	C14 Age	Error
Beta - 214141	Lapa do Santo	Wood charcoal	G12	6.079	11.878	−2.072	–	−26.1	–	–	–	8980	40
Beta - 216518	Lapa do Santo	Wood charcoal	F13	5.818	12.142	−2.795	–	−24.1	–	–	–	9100	40
Beta - 214143	Lapa do Santo	Wood charcoal	G12	6.543	11.276	−2.101	–	−24.8	–	–	–	9150	40
MAMS - 15247	Lapa do Santo	Rib	Burial 27	–	–	-0.962	–	−18.8	6.53	6.29	2.35	9245	40
Beta - 263885	Lapa do Santo	Wood charcoal	G13	6.201	12.806	−3.881	–	–	–	–	–	9370	40
Beta - 263883	Lapa do Santo	Wood charcoal	G13	6.971	12.350	−3.740	–	–	–	–	–	9470	50
Beta - 256223	Lapa do Santo	Wood charcoal	AI1	34.611	0.359	−3.004	–	−24.9	–	–	–	9520	60
Beta - 263882	Lapa do Santo	Wood charcoal	G13	6.121	12.570	−4.255	–	−25.6	–	–	–	9650	50
Beta - 263884	Lapa do Santo	Wood charcoal	G13	6.422	12.794	−4.881	–	−23.1	–	–	–	9680	50
Beta - 280488	Lapa do Santo	Wood charcoal	G13	6.876	12.522	−4.043	–	−22.2	–	–	–	9720	40
Beta - 214130	Lapa do Santo	Wood charcoal	F13	5.741	12.125	−3.889	–	−23.7	–	–	–	9900	40
Beta - 214138	Lapa do Santo	Wood charcoal	L28	11.925	27.221	−6.579	–	−23.0	–	–	–	10070	100
Beta - 256224	Lapa do Santo	Wood charcoal	AI1	34.227	0.773	−3.015	–	−28.0	–	–	–	10130	60
Beta - 280489	Lapa do Santo	Wood charcoal	G13	6.203	12.805	−3.352	–	−23.2	–	–	–	10490	50
Beta - 216528	Lapa Grande de Taquaraçu	Wood charcoal	–	–	–	–	–	−29.9	–	–	–	1160	60
Beta - 216527	Lapa Grande de Taquaraçu	Wood charcoal	–	–	–	–	–	−25.1	–	–	–	8080	40
Beta - 183576	Lapa Grande de Taquaraçu	Wood charcoal	–	–	–	–	–	−23.9	–	–	–	8230	50

(continued)

Table 2 (continued)

Lab Id	Site	Material 2	Unit	X	Y	Z	% Coll	D13C	d15 N	%C	%N	C14 Age	Error
Beta - 216529	Lapa Grande de Taquaraçu	Wood charcoal	–	–	–	–	–	−25.8	–	–	–	8310	40
Beta - 183577	Lapa Grande de Taquaraçu	Wood charcoal	–	–	–	–	–	−26	–	–	–	8730	50
Beta - 216530	Lapa Grande de Taquaraçu	Wood charcoal	–	–	–	–	–	−23.7	–	–	–	8730	40
Beta - 216531	Lapa Grande de Taquaraçu	–	–	–	–	–	–	−25.8	–	–	–	8910	40
Beta - 216532	Lapa Grande de Taquaraçu	Wood charcoal	–	–	–	–	–	−24.9	–	–	–	9040	40
Beta - 216526	Lapa Grande de Taquaraçu	Wood charcoal	–	–	–	–	–	−25.5	–	–	–	9540	90
Beta - 183575	Lapa Grande de Taquaraçu	Wood charcoal	–	–	–	–	–	−24.3	–	–	–	9550	60
Beta - 216525	Lapa Grande de Taquaraçu	Wood charcoal	–	–	–	–	–	−26.3	–	–	–	9620	40
Beta - 242715	Lapa Grande de Taquaraçu	Wood charcoal	–	–	–	–	–	−27.1	–	–	–	9900	60
Beta - 242714	Lapa Grande de Taquaraçu	Wood charcoal	–	–	–	–	–	−23.1	–	–	–	9990	60
Gif - 2735	Lapa Vermelha IV	Wood charcoal	Sondage B (base de la couche I)	–	–	20	–	–	–	–	–	320	80
Gif - 3222	Lapa Vermelha IV	Wood charcoal	Base level B	–	–	–	–	–	–	–	–	1620	100
Gif - 3220	Lapa Vermelha IV	Wood charcoal	Surface	–	–	–	–	–	–	–	–	1880	140

(continued)

Table 2 (continued)

Lab Id	Site	Material 2	Unit	X	Y	Z	% Coll	D13C	d15 N	%C	%N	C14 Age	Error
Gif - 2732	Lapa Vermelha IV	Wood charcoal	Sondage B	–	–	115	–	–	–	–	–	3000	110
Gif - 3221	Lapa Vermelha IV	Wood charcoal	Level D	–	–	–	–	–	–	–	–	3070	110
Gif - 3211	Lapa Vermelha IV	Wood charcoal	–	–	–	–	–	–	–	–	–	3260	110
Gif - 3218	Lapa Vermelha IV	Wood charcoal	Base level D	–	–	–	–	–	–	–	–	3370	110
Gif - 3219	Lapa Vermelha IV	Wood charcoal	Base level C	–	–	–	–	–	–	–	–	3430	130
Gif - 3210	Lapa Vermelha IV	Wood charcoal	Level E	–	–	–	–	–	–	–	–	3580	130
Gif - 2734	Lapa Vermelha IV	Wood charcoal	Sondage B	–	–	210	–	–	–	–	–	3660	110
Gif - 2545	Lapa Vermelha IV	Wood charcoal	Sondage A	–	–	90	–	–	–	–	–	3720	120
Gif - 2733	Lapa Vermelha IV	Wood charcoal	Sondage B	–	–	150	–	–	–	–	–	3740	110
Gif - 3209	Lapa Vermelha IV	Wood charcoal	Level E	–	–	–	–	–	–	–	–	3750	110
Gif - 2543	Lapa Vermelha IV	Wood charcoal	Sondage C (1 m sous le F3)	–	–	435	–	–	–	–	–	4170	120
Gif - 3215	Lapa Vermelha IV	Wood charcoal	Secteur 25 C.D. Couche G1	–	–	–	–	–	–	–	–	4350	120
Gif - 2544	Lapa Vermelha IV	Wood charcoal	Sondage C (160 cm sous le F3)	–	–	500	–	–	–	–	–	4400	120

(continued)

Table 2 (continued)

Lab Id	Site	Material 2	Unit	X	Y	Z	% Coll	D13C	d15 N	%C	%N	C14 Age	Error
Gif - 3213	Lapa Vermelha IV	Wood charcoal	Level F	–	–	–	–	–	–	–	–	4550	130
Gif - 3214	Lapa Vermelha IV	Wood charcoal	Secteur 25 C.D. Couche G3	–	–	–	–	–	–	–	–	5120	130
Gif - 3907*	Lapa Vermelha IV	Wood charcoal	12.95–13.15 m	–	–	–	–	–	–	–	–	5400	500
Gif - 3207	Lapa Vermelha IV	wood charcoal	Secteur 33 D.	–	–	965	–	–	–	–	–	6830	150
Gif - 3217	Lapa Vermelha IV	Wood charcoal	Level I	–	–	–	–	–	–	–	–	6950	140
Gif - 3216	Lapa Vermelha IV	Wood charcoal	Level H	–	–	–	–	–	–	–	–	8490	160
Gif - 3208	Lapa Vermelha IV	Wood charcoal	Secteur 33 C.	–	–	1030–1080	–	–	–	–	–	9580	200
Gif - 3727	Lapa Vermelha IV	Wood charcoal	11.7–11.9 m	–	–	–	–	–	–	–	–	10200	220
Gif - 3726*	Lapa Vermelha IV	Wood charcoal	11.7 m	–	–	–	–	–	–	–	–	11680	500
Gif - 3906	Lapa Vermelha IV	Wood charcoal	12.6–12.8 m	–	–	–	–	–	–	–	–	12960	300
Gif - 3905	Lapa Vermelha IV	Wood charcoal	13.55–14.5 m	–	–	–	–	–	–	–	–	15300	400
Gif - 3908*	Lapa Vermelha IV	Wood charcoal	12.6–13.55 m	–	–	–	–	–	–	–	–	22410	400
Gif - 3725*	Lapa Vermelha IV	Wood charcoal	11.7–11.8 m	–	–	–	–	–	–	–	–	>25000	–

(continued)

Table 2 (continued)

Lab Id	Site	Material 2	Unit	X	Y	Z	% Coll	D13C	d15 N	%C	%N	C14 Age	Error
Beta - 191131	Sumidouro open-air	Wood charcoal	B2	–	–	115	–	−24.5	–	–	–	230	30
Beta - 234518	Sumidouro open-air	Wood charcoal	S4	–	–	25	–	−25.8	–	–	–	340	40
Beta - 234520	Sumidouro open-air	Wood charcoal	S4	–	–	31	–	−26.6	–	–	–	400	40
Beta - 234506	Sumidouro open-air	Wood charcoal	S4	–	–	235	–	−27.5	–	–	–	450	40
Beta - 234512	Sumidouro open-air	Wood charcoal	S4	–	–	26	–	−27.1	–	–	–	510	40
Beta - 256391	Sumidouro open-air	Wood charcoal	S7	–	–	200	–	−27.3	–	–	–	660	40
Beta - 234517	Sumidouro open-air	Wood charcoal	S4	–	–	46	–	−26.4	–	–	–	1350	40
Beta - 234510	Sumidouro open-air	Wood charcoal	S4	–	–	108	–	−24	–	–	–	2210	40
Beta - 234513	Sumidouro open-air	Wood charcoal	S4	–	–	130	–	−25.8	–	–	–	3450	40
Beta - 234511	Sumidouro open-air	Wood charcoal	S4	–	–	112	–	−25.7	–	–	–	4220	40
Beta - 234514	Sumidouro open-air	Wood charcoal	S4	–	–	146	–	−24.3	–	–	–	4460	40
Beta - 234508	Sumidouro open-air	Wood charcoal	S4	–	–	138	–	−26.4	–	–	–	4580	50
Beta - 234516	Sumidouro open-air	Wood charcoal	S4	–	–	135	–	−26.5	–	–	–	4640	40
Beta - 205350	Sumidouro open-air	Wood charcoal	B1	–	–	137	–	−25.6	–	–	–	5020	70
Beta - 234515	Sumidouro open-air	Wood charcoal	S4	–	–	154	–	−25.5	–	–	–	5130	40
Beta - 234509	Sumidouro open-air	Wood charcoal	S4	–	–	150	–	−25.8	–	–	–	5810	40
Beta - 234507	Sumidouro open-air	Wood charcoal	S4	–	–	225	–	−25.7	–	–	–	5860	50
Beta - 256402	Sumidouro open-air	Wood charcoal	S9	–	–	200	–	−24.8	–	–	–	6410	40

(continued)

Table 2 (continued)

Lab Id	Site	Material 2	Unit	X	Y	Z	% Coll	D13C	d15 N	%C	%N	C14 Age	Error
Beta - 205351	Sumidouro open-air	Wood charcoal	B1	–	–	160	–	−25.3	–	–	–	8310	40
OxA-2677-49	Sumidouro Cave	Petrous	ZMK1/1845:2352	–	–	–	–	–	–	–	–	8785	50
OxA-33337	Sumidouro Cave	Petrous	ZMK1/1845:2350 - SH10	–	–	–	–	–	–	–	–	8990	50
OxA-33336	Sumidouro Cave	Petrous	ZMK1/1845:15108 - SH16	–	–	–	–	–	–	–	–	9015	45
OxA-2663-56	Sumidouro Cave	Petrous	ZMK1/1845:15109 - SH09	–	–	–	–	–	–	–	–	9070	45
OxA-33335	Sumidouro Cave	Petrous	ZMK1/1845:15113 - SH04	–	–	–	–	–	–	–	–	9240	50

Fig. 3 Left: Jaguara Farm in 1868. Photographed by August Riedel (Housing institution: National Library of Rio de Janeiro). Right: Lagoa Santa village in the mid-19th century. Watercolour by Peter Andreas Brand (Housing institution: Zoologisk Museum, Copenhagen)

century, an expedition from São Paulo in pursuit of gold, gems and Amerindians to enslave arrived in Lagoa Santa (Monteiro 1994; Damasceno 2011). By the end of the seventeenth century, gold mines in the nearby Velhas River triggered a demographic expansion in the region (Moura 2016). During the early eighteenth century, the Kingdom of Portugal donated vast portions of land in Lagoa Santa for cattle and agriculture resulting in the formation of large farms with thousands of enslaved people (Revista do Arquivo Público Mineiro 1897:259–260, Haddad 2004) (Fig. 3). In 1853, Burmeister wrote: ‘Lagoa Santa is a poor village. It constituted a parish that has only one church made in wood and earth without spire. It does not have more than 60 or 80 houses and counts only 500 inhabitants, mulattos in majority’ (Burmeister 1853:231).

In 1822, independence was declared, and Brazil became more accessible to foreigners. The Danish naturalist Peter Lund, the German zoologist Hermann Burmeister and the English explorer Richard Burton studied the region during this period, establishing Lagoa Santa’s early relationship to science (see chapter “History of Research in the Lagoa Santa Karst”, this volume). During the nineteenth century, mining declined and the manual production of fabrics became the dominant economic activity. In 1896, after Brazil became a republic, the *Central do Brasil* Railway reached the region and by the beginning of the twentieth century villages like Lagoa Santa and Matozinhos were elevated to the condition of cities. As the economy diversified over the last century, the region played an active part in the industrialization of the country. Nevertheless, social inequality and poverty remain severe.

4 Historical Pots and Stub-Stemmed Earthenware Pipes

In spite of the great potential of Lagoa Santa for Historical Archaeology, few studies have so far been conducted in colonial sites. Nevertheless, historical pottery is known from local museums (Fig. 4a). Examples of this pottery show a mixing of different techniques, with the base and inferior waist moulded with a concave surface covered with fabric and a coiled body. Such a combination of techniques is still used today by potters of the neighbouring municipalities of Inhaúma and Santa Luzia (Snow and Abreu 1977; Machado 2011). Additionally, the production of ‘rude pottery, glazed with yellow and green’ in the village of Lagoa Santa (Burton 1869:33)—although not preserved in museums—is described in 19th chronicles and could be related to an African ancestry.

Stub-stemmed earthenware pipes, likely dating to the nineteenth century (Paiva 2015), are common in the collection of local museums as well (Fig. 4b). Although tobacco pipes made of kaolin were imported from England in the eighteenth and nineteenth centuries and earthenware pipes were produced in Germany, Italy and Turkey, the unique motifs of the archaeological specimens strongly suggest local production (Agostini 1997; Gaspar 2009; Souza 2012). Decoration was usually in high relief and the ‘two faces’ motif the most common—this motif involves a face and a body with breasts and navel on the bowl and an anthropomorphic figure on the shank (Fig. 4b, c) (Brancante 1981). The colonial pipes from Lagoa Santa were mostly used by enslaved blacks—especially women—and poor people in general, constituting an important element of diasporic identities (Wied-Neuwied 1820; Debret 1839). Interestingly, the motifs themselves were not African in their origin, as is true of motifs from other regions of Brazil. An important element of Lagoa Santa’s historical archaeology is the church at Jaguara Farm, as its construction involved the participation of Aleijadinho, a renowned Brazilian artist of the eighteenth century.

Studies of historical archaeology in the region must expand and intensify with the investigation of farms, slave houses, gold mines and ancient churches. This work will potentially shed light on the fate of the local Amerindians, the processes related to the African Diaspora[2] as well as the more recent intricate economic flux of foodstuffs and gold in the region.

[2]The forced migration of millions of enslaved Africans between the fifteenth and nineteenth centuries.

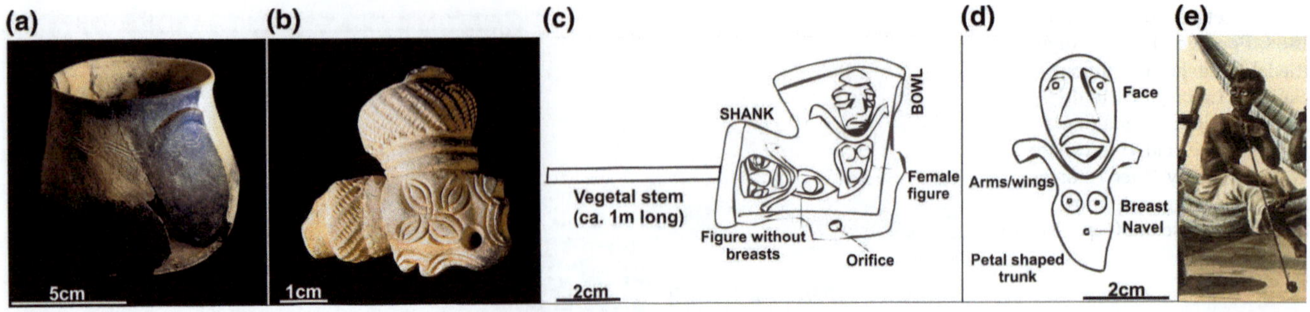

Fig. 4 **a** Terracotta pan made using a combination of moulding and coiling techniques. **b** Moulded earthenware pipe decorated with a handmade incised and stamped pattern. Credits for the pictures: Maurício de Paiva. **c** Schematic representation of a moulded earthenware pipe with a 'two-faces' motif. **d** Schematic representation of a female figure from the bowl of an earthenware pipe. **e** Painting by Joaquim Guilhobel in 1820s Rio de Janeiro showing an enslaved individual smoking a stub-stemmed pipe. Drawings by Marcony Alves. Housing institution: Archaeological Museum of Lagoa Santa-Lapinha (**a**, **b**), Historical Museum of Sete Lagoas (**c**), Natural History Museum and Botanical Garden of the Federal University of Minas Gerais (**d**) and Guita and José Mindlin Brasiliana Library of the University of São Paulo (*e*)

5 Precolonial Pottery

Precolonial archaeological contexts including pottery are very common in Brazil (Prous 1992). In Lagoa Santa alone, pottery remains are known from at least 106 archaeological sites (Table 1). Traditionally, pottery was not the focus of archaeological studies, and so the material largely lacks contextual information, a trend that has recently started to be reversed. In this section, we provide an overview of the pottery from the region and discuss its implications for broader debates surrounding the late Holocene ceramists of central Brazil.

In Lagoa Santa, there are basically two types of precolonial pottery: Tupiguarani and Aratu-Sapucaí-Una (ASU). The pottery from the Tupiguarani tradition (Fig. 5) occurs only marginally in Lagoa Santa and is associated with no proper archaeological site.[3] This pottery is normally decorated with red and black stripes painted over a white slip, referred to as 'polychromatic style'. A wide range of plastic decorations such as incised, fingernail marks, corrugated and punctuated decorations are often present in the external surface of the vessels (Prous 1992; Prous and Lima 2008). Geographically, this polychromatic pottery occurs over a vast area including the Amazon forest, most of the Atlantic coast and the southern Pampas, with known sites in Brazil, Bolivia, Paraguay, Uruguay and Argentina. Studies that articulate archaeological, ethnographic and linguistic data firmly support the link between this pottery and Tupi-speaking groups (Brochado 1989; Noelli 1996, 1998). The antiquity of the Tupiguarani pottery tradition is debatable. It originated either in central Brazil at 2.0 kyr cal BP or in Amazon around 4.0 kyr cal BP (Corrêa 2014). In Lagoa Santa, some isolated pots or sherds from this tradition were found both in open air sites (e.g., Sumidouro (Fig. 5g), Bité and Santo Antônio do Mocambo) and rock shelters (e.g., Boleiras and Lapinha I and II). There are no dates for the Tupiguarani material from the region, but it likely dates to ca. 2.0–0.5ky cal BP.[4]

The vast majority of the pottery found in Lagoa Santa, however, belongs to the ASU tradition. In comparison to the Tupiguarani tradition, the ASU is more discrete, usually lacking painted or plastic decorations[5] (Fig. 6). Twinned and vegetable-shaped (e.g., gourd-like) pots are known to exist, but 'diagnostic' vessels have a pear-shaped[6] morphology with generally smoothed—but not polished—walls (Prous 1992; Rodrigues 2011). Varying in size from very small pots to large receptacles (up to 400 L), they certainly served distinct functions and were in many instances reused as funerary urns (Fernandes 2003). Apparently, the smaller and larger vessels occur exclusively in rock shelters and open air sites, respectively. Based on this size and spatial distinction, archaeologists in the past reserved the term 'Una' for the smaller vessels and 'Aratu-Sapucaí' to the larger. However, recent studies advocate that this difference in size is of no cultural significance but reflects instead different activity patterns (Henriques Jr. 2006; Rodrigues 2014). Moreover, as more sites are properly excavated (e.g., Vereda III, see below) it is becoming clear that the size of the pots is independent from the type of the site.

[3]The closest Tupiguarani site (Santa Margarida II) is located in the margins of the Velhas River outside the limits of the karst.

[4]The oldest date associated with the Tupiguarani tradition in the state of Minas Gerais (where the Lagoa Santa region is located) is 1.75 ± 0.2 kyr cal BP (thermoluminescence date determined at FATEC on pottery from the site Córrego do Maranhão).

[5]Exceptions to this might be more common than it is usually assumed. Cases of red slip are known to occur (including sherds from the Lagoa Santa region) and decorated margins do exist in collections from north-east Brazil. In the coast of Rio de Janeiro and Espírito Santo (two states of Brazil), ASU pottery presents plastic decorations.

[6]In the Brazilian literature, this is often referred to as 'piriform'.

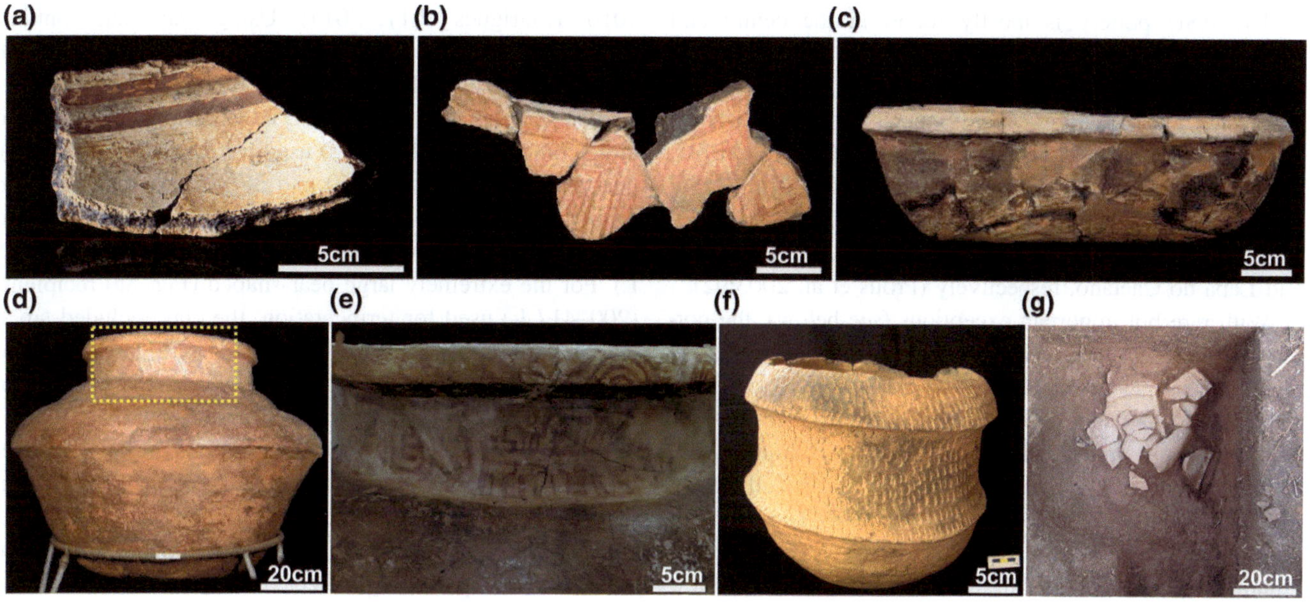

Fig. 5 Pottery of the Tupi-Guarani tradition from localities close to Lagoa Santa. **a–b** Sherd with a white slip and a red stripe; **c** Pot with a square mouth and reinforced rim (height: 15 cm); **d** Funerary urn (height: 73 cm) with a shoulder and decorated neck (see details of the area indicated by the yellow rectangle in '**e**'); **f** Pot with a 'fingernail' plastic decoration (height: 27.5 cm); **g** Fragmented vessel with a thickened rim - likely in the Tupi-Guarani style - found in unit S04 of the Sumidouro site in 2006. Credits for the pictures: Mauricio de Paiva (**a**), Marcos Gohm (**b–d**), Igor Rodrigues (**e**), Adriano Carvalho (**f**) and Institute of Biosciences of the University of São Paulo (**g**). Housing institutions: Archaeological Museum of Lagoa Santa-Lapinha (**a–e**) and Natural History Museum and Botanical Garden of the Federal University of Minas Gerais (**f**). The provenance of **a–e** is not known with certainty, but they likely come from the Tupi-Guarani site of Santa Margarida, located a few kilometres away from Lagoa Santa. Specimen F comes from the north of Minas Gerais (ca. 500 km away from Lagoa Santa)

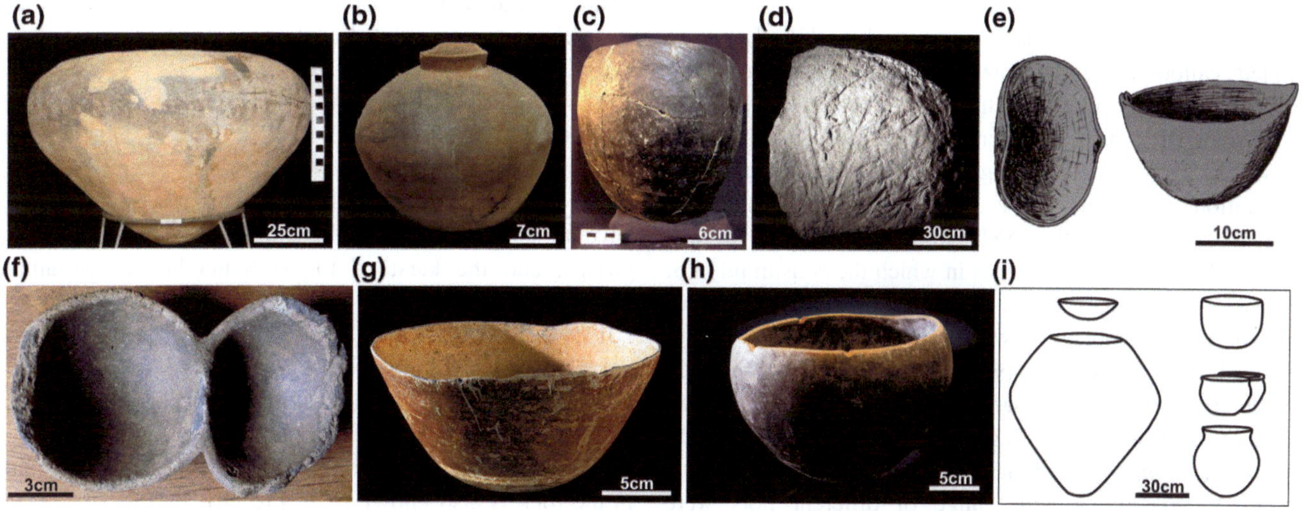

Fig. 6 Pottery of the Aratu-Sapucai-Una tradition from Lagoa Santa or neighbouring regions. **a** Pear-shape (i.e., piriform) funerary urn (height: ca. 80 cm) from Lagoa Santa; **b** Globular pot with neck from the Upper São Francisco region; **c** Ovaloid pot (height: 19.8 cm) from the north of Minas Gerais; **d** Base of a pot from the Lagoa Santa region, with a leaf print on the external surface; **e** Leaf-shaped pots from south-central Minas Gerais (approximate scale); **f** Twinned pot from south-central Minas Gerais; **g** Hemispheric pot with a red slip (height: 10 cm) from the Lagoa Santa region; **h** Globular pot (height: 18 cm) from the Lagoa Santa region; **i** Plate reproduced from Junqueira and Malta (1978) showing the diverse range of sizes and shapes of reconstructed pots from the Lagoa Santa region. Credits for the pictures: Mauricio de Paiva (**d**, **g**, **h**), Marcos Gohm (**a**), Igor Rodrigues (**b**), Adriano Carvalho (**c**), drawing adapted from Prous (1992:348) (**e**) and Gustavo Souza (**f**). Housing institutions: Archaeological Museum of Lagoa Santa-Lapinha (**a**, **d**, **g**, **h**), Natural History Museum and Botanical Garden of the Federal University of Minas Gerais (**c**, **e**, **i**), Archaeological Museum of the Carste do Alto São Francisco (**b**) and Museu e Arquivo Sacro-Histórico da Paróquia de Nossa Senhora do Carmo (**f**)

The ASU pottery is mostly found in the centre and north-east of the country. Linguistically, it is suggested to be associated with Macro-Jê speakers. The oldest date for the smallest (Una) and larger (Aratu-Sapucaí) pots in Brazil are 2.6 kyr cal BP and 1.1 kyr cal BP, respectively, (Prous 1992; Fernandes 2011). In Lagoa Santa, small pottery of the ASU tradition was found as part of funerary contexts dated to 2.0 kyrcal BP and 1.7 ky cal BP in Lapa Mortuária de Confins and Lapa do Caetano, respectively (Prous et al. 2003:92).

With rare but important exceptions (see below), the pottery recovered from Lagoa Santa was found on the surface and lacks proper archaeological or chronological context.[7] The first systematic description was based on material recovered from approximately 40 locations (Junqueira and Malta 1978). Although polychromatic and plastically decorated sherds were recognized (i.e., presumably Tupiguarani), the great majority were attributed to the ASU tradition[8] (Junqueira and Malta 1978; Prous et al. 2003). The few reconstructions showed a broad variety of forms, including pear-shaped, half-calotte, globular and hemispheric varieties (Fig. 6). Additionally, only materials such as quartz, hematite and phyllite were observed as antiplastic. The vessels were predominantly produced by the coiling technique and burnt in a reducing environment. The ASU pottery from Lagoa Santa also includes tubular pipes (Fig. 7a, b) and spindle whorls (Fig. 7c), indicating the consumption of tobacco and the production of cotton yarn. Finally, sherds presenting long and broad incisions that were not part of the plastic decoration are also common (Fig. 7d). They are called 'calibrators' in the Brazilian literature and were likely discarded pottery reused to straighten arrow shafts.

Vereda III is the only pottery-rich site in Lagoa Santa that has been excavated following modern techniques of documentation[9] (Rodrigues 2011, 2014). With more than 3,000 sherds assigned to the ASU tradition, Vereda III was repeatedly used for ritual feasts in which the consumption of fermented beverages was a central component (Rodrigues 2014). The assemblage is well preserved and can be categorized as de facto *refuse* (sensu Schiffer 1972). The study of starch grains indicates that the recipients were accustomed to process, at least, maize (*Zea mays sp.*) and sweet potato (*Ipomoea batatas*) (Gardiman et al. 2014).

In Vereda III, a wide range of different pots were employed for specific functions (Fig. 8) (Gardiman et al. 2016; Rodrigues 2011, 2014). Using clay that contains mineral non-plastic material and which was tempered both with small crushed sherds and large quantities of freshwater sponge spicules (known as *cauixi* in Brazil),[10] they produced large conical pots (Fig. 8b, c) used for cooking (70–180 L), pear-shaped recipients of 50 and 380 L that were used to ferment and stock, respectively, small hemispheric pots used to cook (3 L) and small open plates used to serve (0.4–1.5 L). For the extremely large pear-shaped (Fig. 8a) recipients (200–417 L) used for fermentation, the clay included small amounts of *cauxi* and was tempered with large crushed sherds. Finally, using a clay that contained only mineral non-plastic material, they produced multifunctional globular or hemispheric (Fig. 8d) recipients (2–10 L) that were decorated with red slip. The pottery of Lagoa Santa is unique in this region of Brazil because of the presence of *cauixi* (freshwater sponge spicules). In other regions of Brazil, the *cauixi* was independently collected and later mixed with the clay. In Lagoa Santa, on the other hand, *cauixi* was a natural constituent of the clay deposits (Rodrigues et al. 2017).

Beyond its immediate local relevance, the site of Vereda III shows the potential of Lagoa Santa for studies of pottery in central Brazil. It testifies that current traditions might be artificially partitioning the material culture that was produced by the same human groups. Traditionally, however, archaeologists have not focused their attention on this aspect of the archaeological record of the region. In the future, this tendency must be reversed and excavations and research on pottery-related contexts intensified.

6 Rock Art

Rock art provides one of the few windows archaeologists have to access the symbolic expressions of past human groups, and the karst of Lagoa Santa houses an ample repertoire (Figs. 9, 10, 11 and 12) (Prous 1977; Baeta 1991, 2011; Baeta and Prous 2017). Painting and low relief engraving by pecking (Fig. 11) were the most common techniques. Polychromatic paintings exist (Fig. 12b), but the majority are monochromatic in red or yellow. Additionally, a drawing technique based on subparallel fine incisions made in the rock is also known (e.g., Fig. 9t).

The most commonly depicted motifs are animals (Fig. 9 a–p). They were painted with simple contours and few anatomical details (Fig. 9e–p), varying in size from 0.1 to 1.5 m (the majority are between 20 and 45 cm). Deer and peccaries are more frequent, but jaguars, armadillos,

[7]No reliable date is available for the pottery from the Lagoa Santa region.

[8]In the original publication, Junqueira and Malta attributed the material strictly to the Sapucaí tradition. As mentioned before, some authors differentiate between Aratu, Sapucai and Una, while we adopt the position of grouping them into a single category (i.e., the ASU tradition).

[9]Surface samples were collected using a total station in 2003 and excavations took place in 2010.

[10]This type of sponge is abundant in neotropical regions. It is most commonly found in lakes or river margins (Volkmer-Ribeiro and Pauls 2000).

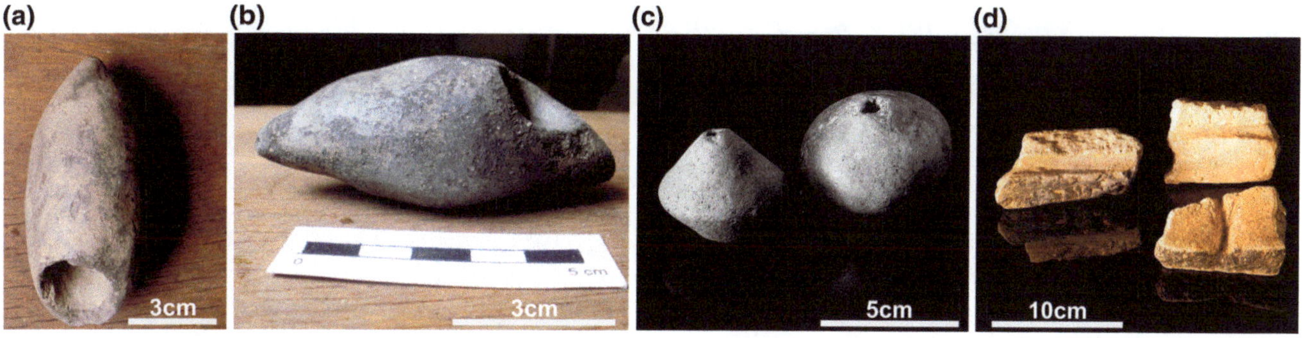

Fig. 7 **a–b** Tubular pipes of the Aratu-Sapucai-Unatradition (from southern Minas Gerais), which are very similar to specimens found in Lagoa Santa (presumably late Holocene); **c** Spindle whorls (ASU style) from the Lagoa Santa region; **d** Sherds reused as 'calibrators' for arrow shafts from Lagoa Santa.Credits for the pictures: Gustavo Souza (**a–b**) and Mauricio de Paiva (**c–d**). Housing institutions: Museu e Arquivo Sacro-Histórico da Paróquia de Nossa Senhora do Carmo and Archaeological Museum of Lagoa Santa-Lapinha (**c–d**)

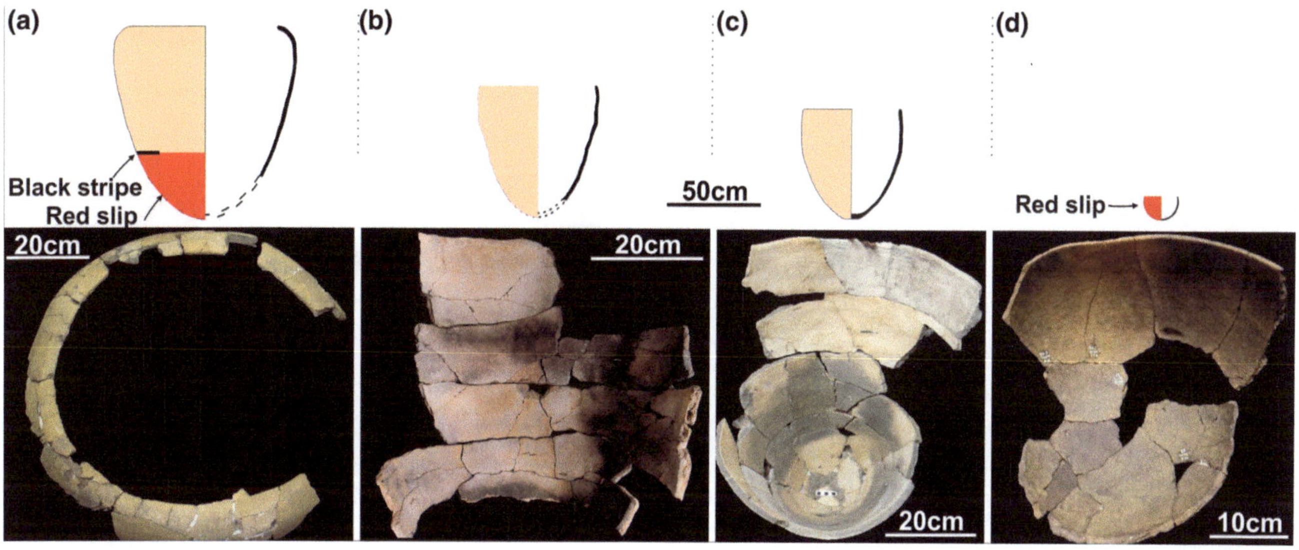

Fig. 8 Pottery from the Vereda III site (late Holocene, Lagoa Santa). **a** Pear-shaped pot used for beverage fermentation (417 litres); **b** Conical pot used for cooking (130 litres); **c** Conical pot used for cooking (80 litres); **d** Hemispheric multifunctional pot (three litres). Credits for the pictures and drawings: Igor Rodrigues. Housing Institution:), Natural History Museum and Botanical Garden of the Federal University of Minas Gerais

monkeys, lizards, birds and fishes appear as well. The contours are usually completely filled with pigments of the same colour, resulting in a flat figure (e.g., Fig. 9j–p, among many others). Less frequently, parallel stripes were used to fill the figures instead (e.g., Fig. 10z), a pattern more common in the nearby region of Serra do Cipó. Some of the deer have a 'comma-shaped' body (e.g., Fig. 10q), a style known to occur in archaeological sites in central and north-eastern Brazil.

Animals are normally depicted in isolation, but there are also composite scenes with other animals and/or humans. Examples include deer or peccaries being hunted with arrows (Fig. 10q, r, u) or ropes (Fig. 10v). Depictions of animals in other styles—with more details and different proportions—also exist in Lagoa Santa (Fig. 9a–d). When depicted together with humans, the images of animals are usually larger (Fig. 10r, t, u), leading to the suggestion that they depict extinct megafauna. However, based on anatomical features no undisputable case exists.

Different anthropomorphic motifs are found in Lagoa Santa. A common style is the filiform representation of the human body with a c-shaped head, tri-digit extremities, and well-marked sexual organ—elements that can appear together (Fig. 10a–e, g, j) or independently (Fig. 10m–o). The c-shaped head might represent the beak of a bird, as it is similar to avian depictions of a similar style (Fig. 10e, f). Therefore, these anthropomorphic representations might well evoke mythic entities instead of simple humans. In Lapa do Santo (Fig. 10d), a male anthropomorph of this style was dated to ca. 10.5 kyr cal BP (Neves et al. 2012).

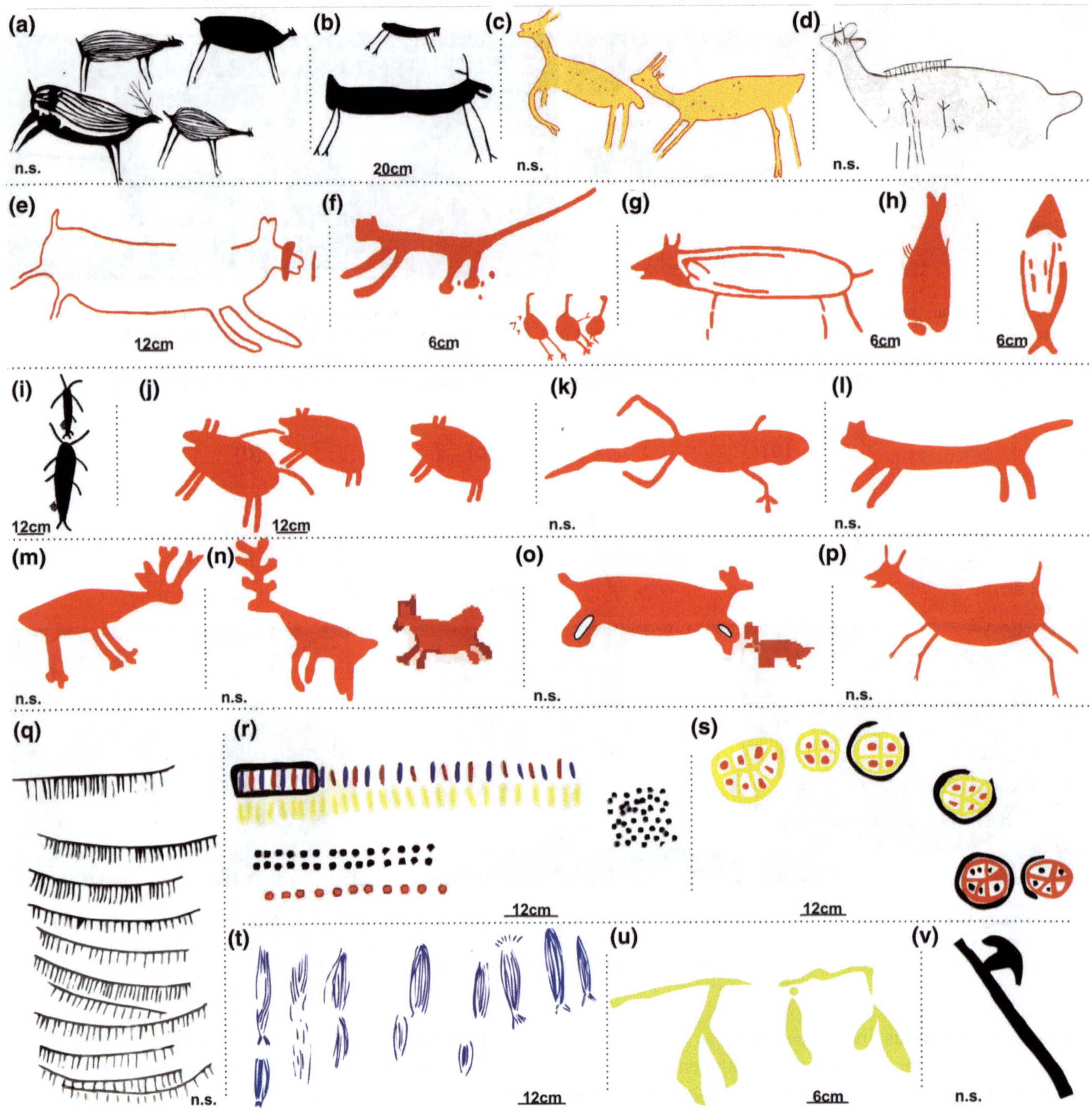

Fig. 9 Reproduction of rock art from Lagoa Santa, depicting zoomorphic (**a–p**), geometric (**q–t**) and horticulturally related (**u**, **v**) motifs. With the exception of T, which was produced using very fine incisions, all of the depictions were painted. The colours shown here correspond to the originals. The animals include deer (**a–e**, **m–p**), jaguars (**f**), peccaries (**g** and **j**), birds (e.g., Cariamidae in **f**) and fish (**h** and **i**). Rock art in Lagoa Santa is characterised by an ample variety of geometric motifs (**q–s**). The elliptical motifs that sometimes look like fish are produced by making thin incisions in the rock, and have been discovered in several sites in the region (**t**). Some motifs are associated with the most recent occupation of horticultural groups, such as tubers (likely cassava) (**u**) and semi-lunar axe (**v**). The depictions come from the following sites: Abrigo Mato Seco (**a**), Abrigo do Trevo (**h**), Abrigo Vargem da Pedra (**b**, **e**, **s**), Capão das Éguas (**f**, **g**), Cerca Grande (**k–o**), Gruta Três Porquinhos (**j**) (a.k.a Lapa Velha de Maquiné—this cave is located outside Lagoa Santa in the nearby region of Cordisburgo), Lapa do Caetano (**v**), Lapa da Cauaia (**c**), Lapa da Pia (**d**, **i**, **q**, **t**), Lapa da Samambaia (**p**), Lapa Vermelha (**v**) and Sumidouro (**r**). Credits of all reproductions to Alenice Baeta

Fig. 10 Reproduction of rock art from Lagoa Santa depicting anthropomorphic motifs. With the exception of **c**, **d**, **g**, **h** and **i**, which are low reliefs produced by pecking, the remaining depictions are paintings. The colours for the paintings correspond to the originals. The depictions are from the following sites: Abrigo da Serrinha (**i**), Abrigo Caieiras (**c**, **g**, **h**), Abrigo do Campinho (**f**), Abrigo Vargem da Pedra (**e**, **v**, **z**), Abrigo Vargem Formosa (**j**), Cerca Grande (**r**), Lapa do Capão das Éguas (**m**, **p**, **q**, **u**, **x**), Lapa da Criciúma (**k**), Lapa da Samambaia (**n**), Lapa da Sucupira* (**l**, **o**, **s**), Lapa do Ballet (**a**, **b**), Lapa do Santo (**d**) and Mata da Cauaia (**t**). *Indicates locality outside, but near to, Lagoa Santa. Credits for all reproductions: Alenice Baeta

This is the only absolute dating in the region for rock art of a recognizable style[11] and constitutes one of the oldest instances in the Americas. Very similar anthropomorphic depictions are found thousands of kilometres to the north in the state of Rio Grande do Norte (e.g., Seridó sub-tradition).

The anthropomorphs in Lagoa Santa can be found in isolation or in scenes suggestive of ritual-like congregations (Fig. 10y). Pregnancy (see the individual in the far right of Fig. 10a) and sex (Fig. 10e) scenes, as well as erected phalluses (Fig. 10d), are not common but do exist. Human representations are sometimes extremely schematic, composed of a few elementary lines or forming a simple 'X' (Fig. 10r–u, y). Representations that do not comply with the

[11] Painted limestone slabs felled from the ceiling/walls were found in Lapa Vermelha IV. Although there is associated chronology, it is not possible to define their stylistic identity.

Fig. 11 Reproduction of low relief rock art from Lagoa Santa, produced by the pecking technique (see photograph in Fig. 12c). The depictions are from the following sites: Abrigo Caieiras (**a**), Lapa do Capão das Éguas (**g**), Lapa da Cricíuma (**i**), Lapa Caieiras (**b**), Lapa da Samambaia (**h**), Lapa do Santo (**c**, **d**, **e**, **f**). Credits for all reproductions: Alenice Baeta

stylistic elements mentioned above have been documented as well (Fig. 10k, l).

Geometric motifs such as circles, parallel lines, and dots are also present (Fig. 9q–s). This geometric style is thought to represent an influence from the Upper São Francisco River, a region in Brazil where these are the predominant motifs. A local origin, however, cannot be discounted. Finally, paintings of manioc tubers (Fig. 9u) and semi-lunar axes (Fig. 9v) are associated with the more recent ceramist occupations. The latter is part of the ASU Tradition.

In Lagoa Santa, rock art spans more than 10,000 years. The ample diversity of forms and themes testifies to the great cultural diversity represented in the archaeological record of the region. Due to the large variety of motifs and styles and the lack of absolute chronology, the identification of synchronic and diachronic patterns is extremely challenging.

Fig. 12 Photographs of rock art from the Lagoa Santa region, exemplifying monochromatic (**a**) and polychromatic (**b**) painting and a pecked low relief (**c**). **a**, **b** and **c** are from Cerca Grande, Sumidouro and Caieiras, respectively. Credits for photographs: Maurício de Paiva (**a**), Artefacto/Orange (**b**) and Maurílio Craveiro da Costa (**c**). 'B' was digitally treated to enhance the paintings

Nevertheless, the motifs found in Lagoa Santa present a clear correspondence with other regions in Brazil, sometimes more than 1,000 km away, possibly indicating the existence of broad cultural networks during the Holocene.

7 Stone and Bone Technology

The physical hardness and chemical stability of rocks make them extremely resistant to the passage of time. Therefore, in the archaeological record, artefacts made from stones and the by-products generated in the process, known as 'lithics', are very common findings. The study of this material provides information on the technologies used in the past as well as information about which specific activities took place in a given locality. Additionally, as part of the material culture, they can also provide information on group identity and ethno-genetic processes.

In Lagoa Santa, lithics were identified in more than 100 sites, encompassing 12,500 years of human presence in the region. However, due to the general absence of formal artefacts in early contexts, the lithics from Lagoa Santa were not the focus of detailed studies until the twenty-first century (but see Walter 1958; Hurt and Blasi 1969). Here we provide a basic description of the lithics from Lagoa Santa region.

During the Pleistocene–Holocene transition and early Holocene period (ca. 12.5–8.0 cal kyr BP, also known as the 'early Archaic' period) there were two main conceptual schemes of stone tool production in sub-Amazonian Brazil (Araujo 2015): the Itaparica and Umbu industries (Fig. 13a, b). The former is characterized by elongated unifacial artefacts whose active edges result from the intersection of convex and planar surfaces (also termed 'limaces', 'lesmas' or 'plano-convex', see Fig. 13b) (Lourdeau 2015). This diagnostic artefact is usually associated with abundant flakes but rarely with bifacial artefacts.

The Umbu tradition, on the other hand, is broadly defined by the high frequency of bifacial stemmed points (e.g., arrowheads and spear points) (Fig. 13b) and the absence of limaces. Geographically, both Umbu and Itaparica are known in vast areas (>1 million square meters), the Umbu to the south of the country and Itaparica to the central/north-eastern region (Fig. 13b). In central Brazil, they occur relatively close to each other (ca. 50–500 km) but apparently have not frequently overlapped (Koole 2015). The lithic industry from Lagoa Santa during the early Archaic resembles neither of these more widespread traditions (Fig. 13c–e).

There are only four early Archaic sites that include chronological context, two rock shelters (Lapa do Santo, Lapa das Boleiras) and two open air sites located around the Sumidouro Lake (Coqueirinho and Sumidouro). Unless it is noted otherwise, the following descriptions are based on material from these sites. Additionally, there is a site outside the karst but sufficiently close to merit inclusion (Lapa Grande de Taquaraçu), as well as the material from excavations from the 1950s–1970s, such as Cerca Grande 6 and Lapa Vermelha IV (Bueno 2012).

The Lagoa Santa lithic industry is characterized by assemblages dominated by small flakes and small cores (25 ± 12 mm × 21 ± 10 mm × 9 ± 68 mm; 95% intervals) in which formal artefacts (e.g., arrow head, axes and scrapers) are rare (i.e., a *debitage* industry). Flakes and cores, obtained through unipolar and anvil-mediated knapping, present a low variability in forms, types and raw materials (Fig. 13c–e). The flakes are rarely retouched and sometimes present macroscopic use-wear traces in one edge. Preliminary experiments with crystal quartz flakes have shown that they quickly become dull, and since retouching is not observed, it is assumed that they were used only a few times before being discarded (Pugliese 2007). Small prismatic crystals (ca. 50–100 mm long and 10–40 mm thick; 95% intervals), commonly knapped upon exhaustion (Fig. 13d), were the most common support for *debitage* (i.e., cores). Together with the abundant presence of micro flakes,

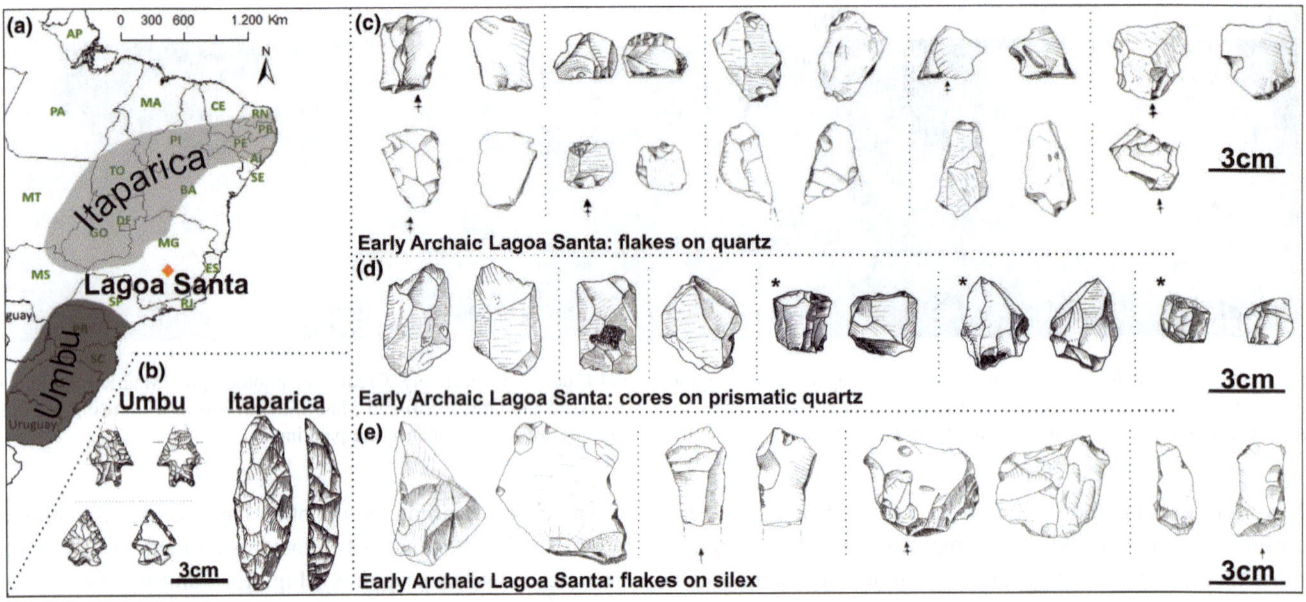

Fig. 13 Left: **a** Map of Brazil showing the approximate distribution of the Umbu and Itaparica traditions and the location of the Lagoa Santa region (orange diamond); **b** Examples of two Umbu points (from the Tunas site in the north of the state of Parana) and one Itaparica limace (from the Córrego do Ouro 19 (GO-CP-17) site in Goiás). Right: Selected specimens representing the most common lithics from the early Archaic period in Lagoa Santa, including examples of typical quartz flakes (**c**); Prismatic quartz cores (**d**); and Silicified sandstone flakes (**f**). The stars (*) in '**d**' indicate exhausted cores. Technological drawing by João Carlos Moreno de Sousa. All lithics on '**c**', '**d**' and '**e**' are from the early Holocene component of Lapa do Santo, and are housed in the Institute of Biosciences of the University of São Paulo

this suggests that these artefacts were produced and used in situ (Binford 1982; Parry and Kelly 1987; Panja 2003).

Lithic types remained very similar for at least 4,000 years—and perhaps even 8,000 years, if the middle Holocene component of Lapa do Santo is representative of the entire region—attesting to the stability of the related activities and the efficiency of this technology (Araujo et al. 2017). The raw materials used, on the other hand, varied through time (Fig. 14 shows the data for Lapa do Santo). Crystalline prismatic quartz was by far the dominant raw material at any moment during the early and middle Archaic, but from the beginning of the occupation (12.5 ky cal BP) until approximately 9.9 ± 0.5 cal kyBP, chert, quartzite, and silicified sandstone[12] also appeared as a minor component (ca. 10–50%). This pattern is observed both in rock shelters and open air sites (Pugliese 2007; Araujo and Pugliese 2009; Bueno 2012).

It has been suggested that crystalline quartz commonly occurs within the karstic region, while non-quartz material seldom does (Prous et al. 1998; Pugliese 2007; Araujo and Pugliese 2010). In this frame, at 9.9 ± 0.5 kyrcal BP, allochthonous raw material was abandoned, possibly indicating a heavier reliance on local items and decreased mobility. However, it has also been postulated that sources of quartz are also external to Lagoa Santa (Moreno de Sousa and Araujo 2018). Until studies of raw material provenance are undertaken for Lagoa Santa and nearby regions, this issue remains open. Only minor differences distinguish the artefacts produced with different raw materials (Pugliese 2007; but see also Moreno de Sousa 2014). Flakes, cores and artefacts made on chert are slightly smaller than those made on quartz. It has been suggested that thermal treatment was applied to chert but not to quartz (Pugliese 2007). However, the very existence of thermal treatment in Brazil is still debatable and further studies are needed to properly evaluate the issue (Moreno de Sousa 2014).

The density of the lithic assemblages also changes during the early Archaic. In Lapa do Santo (unit F13), for example, a very low density (ca. five pieces per level; i.e., 1.0 m × 1.0 m × 0.1 m) was found by the time humans first arrived in Lagoa Santa (ca. 12.0 ± 0.5 ky cal BP) and a gradual increase took place until a peak (ca. 270 pieces per level) is reached at 10.6 ± 0.5 kyr cal BP. After this time, lithic density reduces, increases and reduces once again, reaching corresponding minimums and maximums at 9.9 ± 0.5 kyr cal BP (ca. 50 pieces per level), 8.8 ± 0.5 kyr cal BP (ca. 240 pieces per level) and 8.0 ± 0.5 kyr calBP (ca. 10 pieces per level) (Fig. 14). The same pattern is present in other units of Lapa do Santo and Lapa das Boleiras. Additionally, in the middle Holocene deposits of Lapa do Santo, another peak of lithics is observed (ca. 70 pieces per level).

[12]The 'non-quartz' raw material of Lagoa Santa still awaits proper identification. We here refer to it as 'silicified sandstone', but there is a large variation.

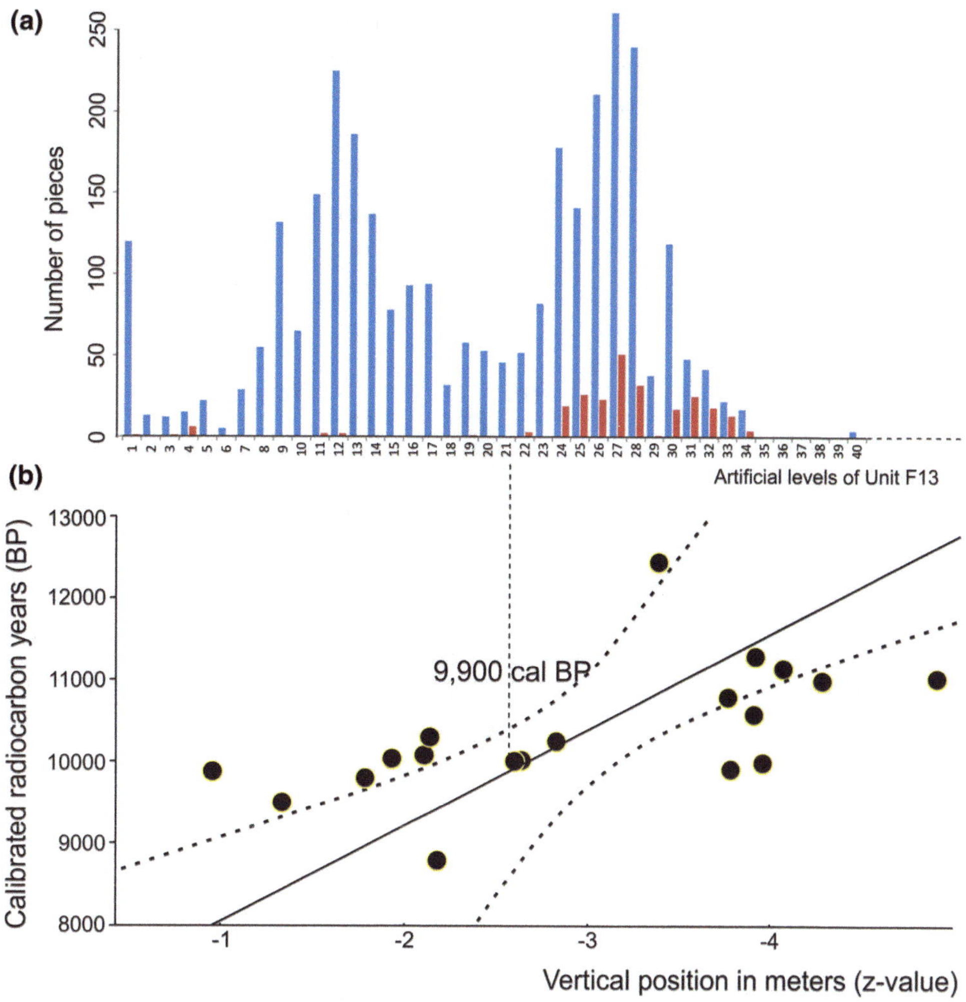

Fig. 14 Chronological behaviour of lithic abundance and the type of raw material in Lapa do Santo during the early Holocene. **a** Histogram of the abundance of quartz (blue bars) and silicified sandstone (red bars) along Lapa do Santo's unit F13 stratigraphy. Note that after level 22 (i.e., earlier in time), silicified sandstone is almost no longer used in the site; **b** the event identified in the histogram is dated by comparing its vertical position (z-value) with the calibrated radiocarbon dates of charcoals (black disks) from unit F13 and the adjacent unit G13 in the corresponding vertical position. The x-axes of both graphs are in alignment and on the same scale

As the first minimum of lithic abundance roughly coincides with the moment when non-quartz raw material stopped being used, as well as with the moment the rock shelters start being used as interment ground (i.e., 9.9 ± 0.5kyr cal BP), it is possible that a demic and/or cultural transformations took place in Lagoa Santa around this time. Alternatively, this could reflect changes in territoriality with a reduction in the average size of the human groups and an increase of foraging ranges.

Recent excavations around the Sumidouro Lake (the sites of Sumidouro and Coqueirinho) resulted in the first lithic assemblages for the region from dated open air sites (Strauss 2008; Bueno 2010; Araujo et al. 2013). Compared with rock shelters, the density of the lithics was much smaller (ca. one to 10 pieces per level) and no other type of archaeological remain was preserved. In the open air sites assemblages are also dominated by small flakes and cores (only one formal artefact was found in open air, see below). Size and shape of flakes are similar to those observed in the rock shelter and the same pattern of early use and later abandonment of non-quartz raw material was identified (Araujo et al. 2013). The relative proportions of lithic types, however, differ between open air sites and rock shelters (Table 4), indicating distinct activity patterns. In the rock shelters, there was an emphasis on the exploitation of the cores for the production and use of small flakes, while a more complex preparation of artefacts and retouching was more common in open air sites.

In addition to the tens of thousands of flakes and cores, some formal artefacts also exist (Fig. 15). This material mostly lacks chronological context or is known to belong to more recent ceramist occupations (e.g., polished blades, pestles or 'semi-lunar' axes, see below). Nevertheless, two stemmed arrowheads (Ls-5534 and Bl-5534) and a distal fragment of what was likely, based on its large size, a spear head (Co-195) were found in contexts securely dated to the early Holocene (Fig. 15).[13] Some consider the arrowheads as preforms (Pugliese 2007; Bueno 2012) while others

[13]A third preform of a steamed projectile point made on milky quartz was found in Lapa do Santo in 2009 on early Holocene levels. Unfortunately, we could not locate this specimen during preparation of this chapter.

Table 3 Abundance of lithic types in Lagoa Santa. Modified from Bueno (2012)

	Coqueirinho	Sumidouro	L. Santo	L. das Boleiras
Site type	Open air		Rockshelter	
Artifacts	46 (7%)	8 (6%)	20 (2%)	58 (3%)
Cores	6 (1%)	2 (1%)	98 (11%)	177 (10%)
Flakes	376 (57%)	81 (59%)	629 (69%)	1197 (67%)
Flakes frag.	138 (21%)	23 (17%)	93 (10%)	229 (13%)
Others	92 (14%)	23 (17%)	75 (8%)	118 (7%)

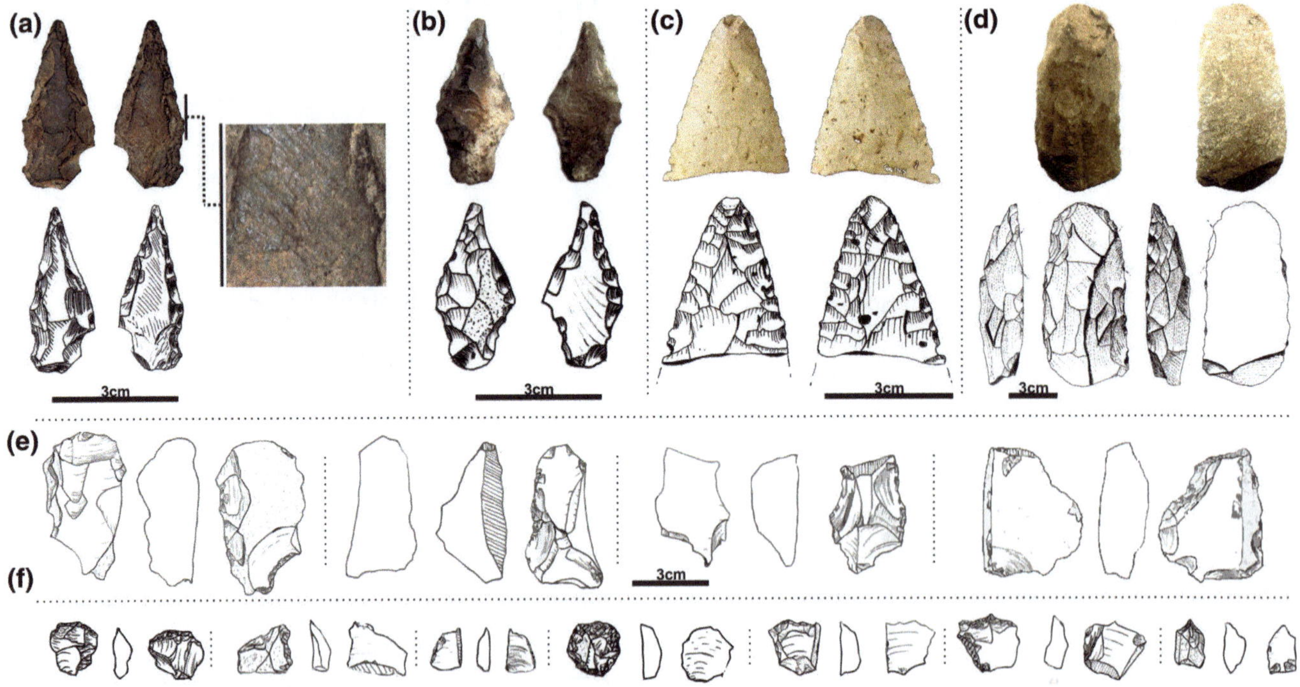

Fig. 15 Formal artefacts produced in Lagoa Santa during the early Archaic period. **a** Projectile point from Lapa do Santo (accession ID: Ls-6410). Grounded area shown in detail; **b** Projectile point from Lapa das Boleiras (accession ID: Bl-2179); **c** Distal fragment of a projectile point from the open-air site of Coqueirinho (accession ID: Coq-195); **d** Limace from Taquaraçu; **e** plano-convex artefacts over flakes from Cerca Grande 6; **f** Very small artefacts that were likely hafted into compound tools. Credits for pictures and technical drawings: **a–d** to João Carlos Moreno de Sousa and **e– f** to Lucas Reis Bueno. Housing institutions: Institute of Bioscience of the University of São Paulo and National Museum of of Rio de Janeiro

consider them fully functional artefacts (Moreno de Sousa 2014). The three projectile points are very different from each other, pointing to a high diversity of lithic types not recognizable when the *debitage* material alone is considered. The two arrowheads broadly recall those of the Umbu tradition, due to their similar size and the presence of a stemmed base. However, they were produced with very different techniques (e.g., bifacial flaking is absent and Ls-5534 presents signs of gridding) and are likely not directly related to the arrowheads found in southern Brazil. The larger specimen from Coqueirinho (Fig. 15c) finds no parallel in central Brazil. 'Plano-convex' artefacts (Fig. 15e) are also present in the early Holocene component of Lapa do Santo and a limace was found in the Pleistocene levels of Lapa Vermelha IV in Lagoa Santa and in the Taquaraçu rock shelter a few kilometres outside the region (Angeles Flores 2016) (Fig. 15d). In Cerca Grande, a few end-scrapers of silex were recovered (Prous, person. communication). Although specific studies are still lacking, it has been suggested that some of the small flakes were hafted as part of composite instruments (Pugliese 2007; Bueno 2012) (see Fig. 15f) and used for processing vegetables, cutting animal flesh and/or sectioning human corpses as part of funerary rituals (Pugliese 2007; Moreno de Sousa 2014; Bueno and Isnardis 2017; Strauss et al. 2016).

Axe blades,[14] locally known as 'coriscos', are the most common formal stone artefact found in the region. Sometimes used as paperweights or doorstops by local people, today they are known in the folklore as axes fallen from the sky inside of thunderbolts. There is a broad variation in size, morphology, degree of polishing and raw material. Although produced over more than 12,000 years of occupation, the majority of specimens were found on the surface, making it hard to establish a reliable typology. Nevertheless, it is possible to distinguish at least two very different groups. Highly polished—usually larger—specimens including axes, adzes, pestles and ritual semi-lunar axes[15] are associated with more recent ceramist occupations (Fig. 16a–f).

Smaller blades, usually asymmetric and rough in appearance, that present flaking scars over their surfaces which are not completely obliterated by pecking or polishing (sometimes referred to as 'cordiform type') seem to be associated with the early Holocene occupations—as far as inferences can be made from the two specimens found in context (Fig. 17a, b). Although the sample size is still small, hematite was likely the favoured raw material for the production of these early Holocene blades (although hematite pestles and blades are found in more recent occupations as well, see Fig. 16d). Other blades of the region cannot be easily assigned to specific traditions (e.g., Figs. 16g, h and 17c). If these blades were used to cut trees, they represent one of the earliest instances of forest management in the New World and a yet not fully discussed case of niche construction in central Brazil.

There are three possible explanations for the rare presence of formal artefacts (e.g., hematite axes and projectile points) during the early Holocene in Lagoa Santa. The first is that formal tools were commonly produced outside the rock shelters by the same groups that produced the flakes and cores, but were almost never discarded within the rock shelters. The second possibility is that groups with distinct lithic technology eventually passed through or coexisted in Lagoa Santa. The final possibility is that the groups from the karst were part of a broader socio-economic network and that these rare elements were obtained by some sort of interregional contact (Pugliese 2007). These are hypotheses deserving future investigation. It is worth noting that many artefacts found within rock shelters are entirely non-functional, as they are either broken (e.g., fishhooks, bone spatulas and axe blades) or represent preforms (e.g., arrowheads). Therefore, one cannot discard archaeological visibility as a main factor behind this overall absence of formal artefacts in the region.

Bone artefacts were common in early Holocene Lagoa Santa. Studies of Lapa do Santo and Lapa das Boleiras show strong similarities with other parts of central Brazil during this period. In Lapa do Santo, for example, a total of 198 bone artefacts or fragments of bone artefacts were found before 2009, including spatulas (71%) and burins (25%) (Fig. 18). Highly standardized and by far the most common type of artefact occurring in Lapa do Santo, the function of the so-called spatulas remains unknown. Manufacturing these artefacts did not require a considerable time investment and involved distinct types of techniques. The rare presence of elaborate J-shaped fishhooks made of animal bone—eight from Lapa do Santo, one from Lapa das Boleiras and one fragment from Lapa Grande de Taquaraçu (a.k.a. Lapa Niáctor)—indicates the importance of aquatic resources in the local diet. All fishhooks were part of the early Archaic component of the site representing the earliest fishhooks in South America. Of the bone artefacts, 57% were made of deer bone, which based on preliminary analysis based on the plotted faunistic remains was also by far the most commonly consumed animal in the site. A total of 49% of the artefacts were carved from metapodial deer bones (Santos 2011). Fishhooks made of bone were also recovered from Lapa Vermelha IV and Lapa do Urubu. A J-shaped fishhook made of wood was found in Santana do Riacho (Prous 1991). Finally, shell was also used as a raw material for tool production in the region, as testified by the archaeological record of Lapa do Eucalipto, São José de Confins and Lapa Vermelha IV.

8 Diet and Subsistence: Zooarchaeology, Isotopes and Caries

Zooarchaeological analyses are key to better understanding the relationship between humans and other animal populations, particularly for studies on diet and hunting strategies (Reitz and Wing 2008). In Lagoa Santa, such studies are available for the early Holocene components of Lapa das Boleiras, Lapa do Santo and Lapa Grande de Taquaraçu, and show that faunal remains include mammals, birds, reptiles (lizards and turtles), fish and mollusc shells (Kipnis et al. 2008; Mingatos and Okumura 2016; Mingatos 2017; Chim 2018). Mammals are the most representative group, comprising about 90% of all identified bones in Lapa das Boleiras, as well as in Lapa do Santo, and include the following taxa: *Mazama sp.*, *Tayassu sp.*, Carnivora, *Dasypus novemcinctus*, *Euphractus sexcinctus*, *Sylvilagus brasiliensis*, Didelphidae,

[14]The term 'axe blade' is used in the Brazilian literature as a descriptive term encompassing several kinds of artefacts, as depicted in Figs. 16 and 17. Although these artefacts were likely used as an axe or adze, their real function cannot be determined with certainty.

[15]These are assumed to be produced by the same groups that produced the ASU pottery.

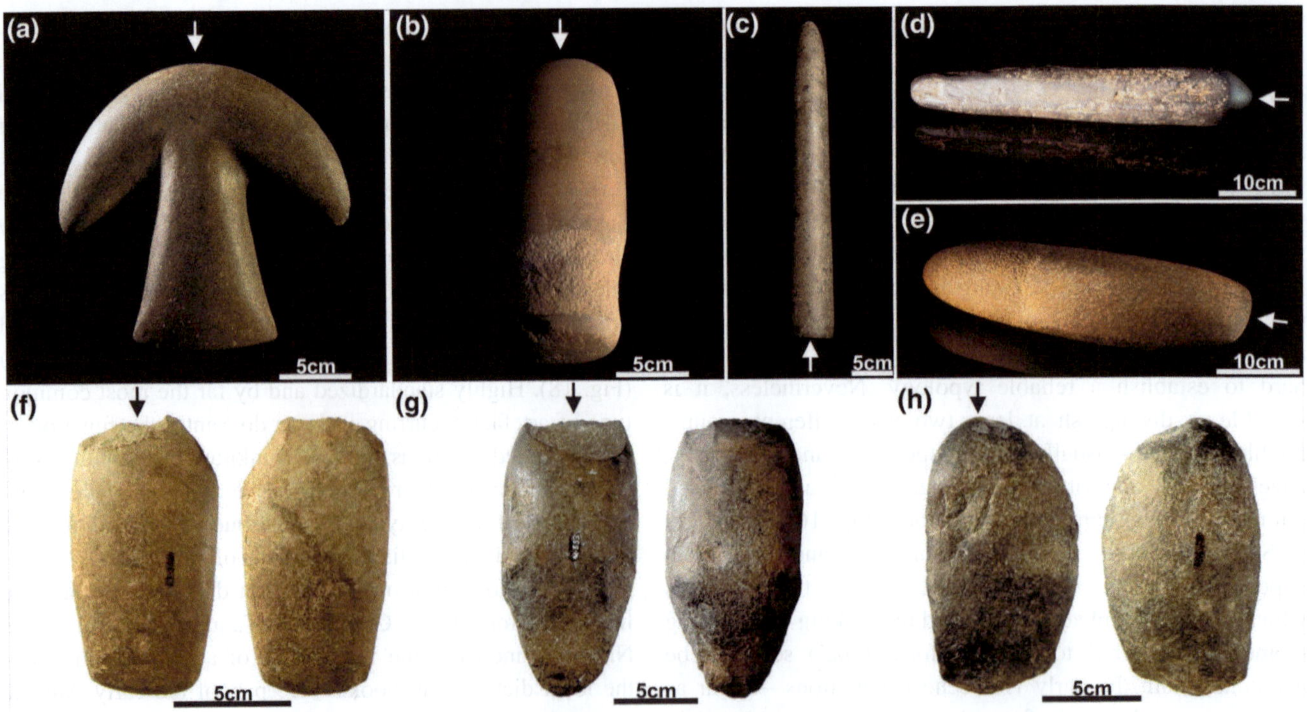

Fig. 16 **a** Polished semi-lunar axe blade, usually associated with pottery in the Aratu-Sapucaí-Una style and considered a ritual artefact; **b** Polished axe blade with a hafting neck; **c** Polished pestle made of igneous rock; **d** polished pestles over hematite; **e** Large axe blade with distinctive polishing and pecking. The provenance of **a**, **b**, **c**, **d** and **e** is not known. Credits for the pictures: Maurício de Paiva. Housing institution: Museu Arqueológico de Lagoa Santa; **f** Polished blade of unknown chronology from Lapa do Santo found outside the main excavation area (unit L29) (Ls-3560); **g** Polished blade from Lapa das Boleiras (Bl-833); **h** Semi-polished blade found on the surface of Lapa do Santo (Ls-9903). Arrows indicate the presumably active edge. Credits for the pictures: João Carlos Moreno de Souza. Housing institution: Archaeological Museum of Lagoa Santa-Lapinha **a–e**, Institute of Biosciences of the University of São Paulo **f–h**

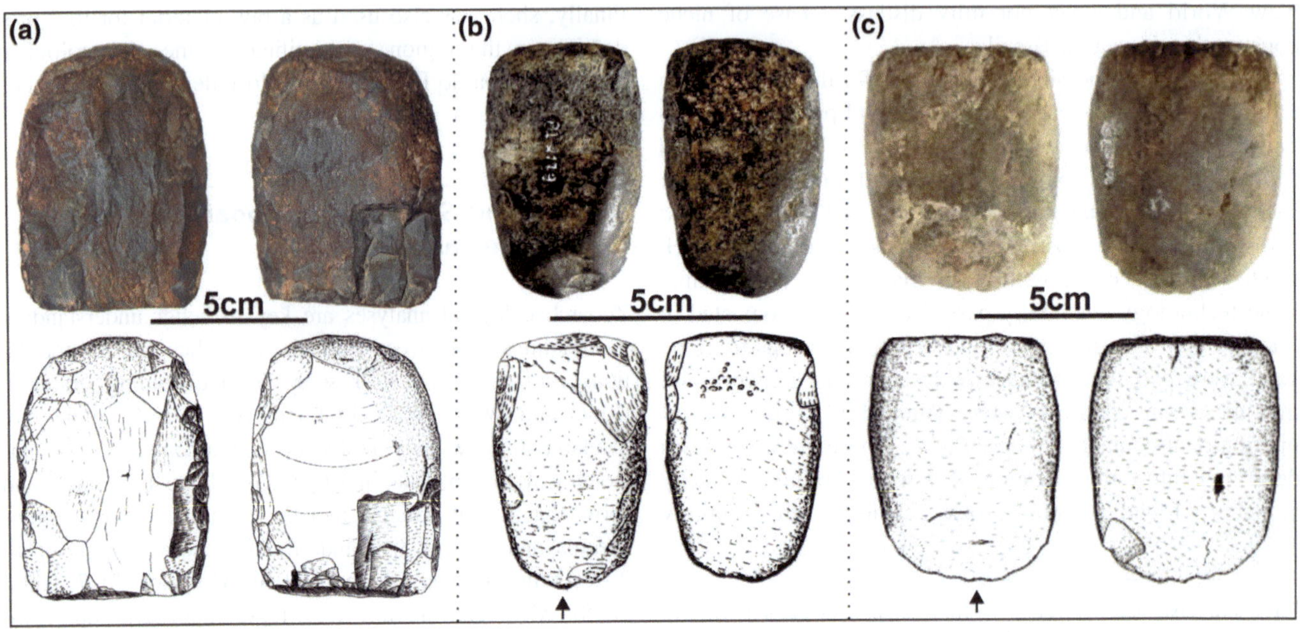

Fig. 17 Axe blades from Lagoa Santa. **a** Hematite blade (Ls-6410) found within the early Archaic component of Lapa do Santo; **b** Hematite polished blade from the early Holocene component of Lapa das Boleiras (Bl-2179); **c** Polished blade from Lapa do Santo, found in the transition from the early Holocene to the middle Holocene components of the site (Ls-2976). Credits for pictures and drawings: João Carlos Moreno de Sousa. Housing institution: Institute of Biosciences of the University of São Paulo

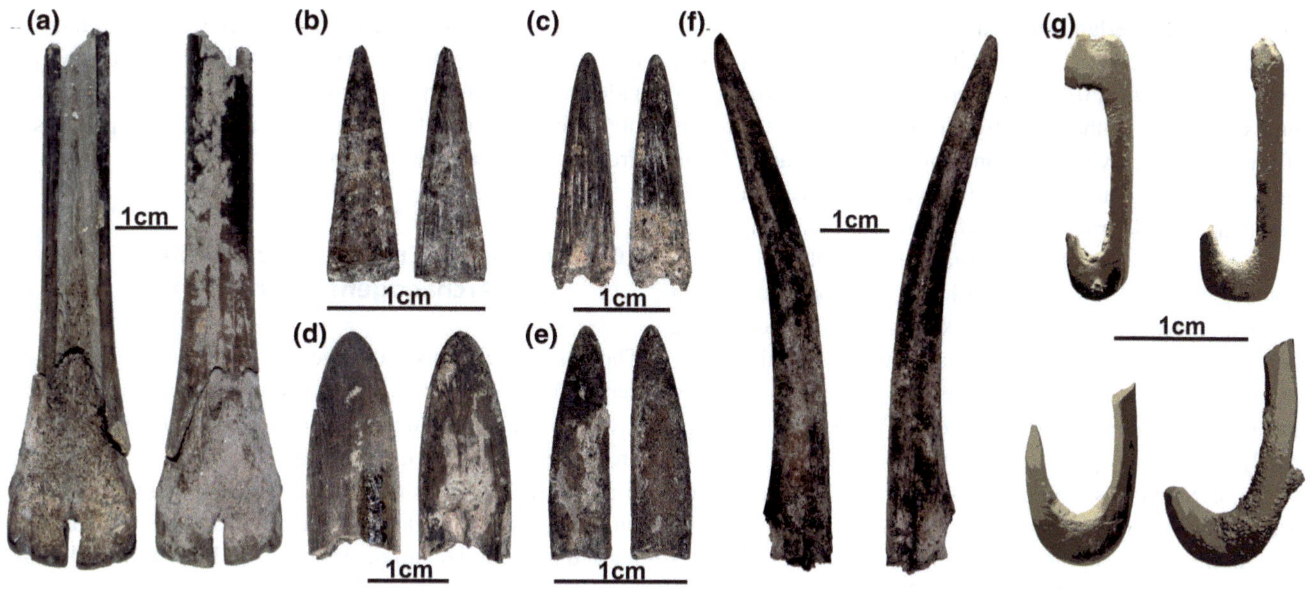

Fig. 18 Bone artefacts (**a**–**f**) and bone fishhooks (**g**) from the early Holocene levels of Lapa do Santo. Credits: André Strauss. Housing institution: Institute of Biosciences of the University of São Paulo

Primates, *Agouti paca* and Rodentia (see Kipnis et al. 2008; Strauss et al. 2016 for details).

In Lapa das Boleiras, medium to large-sized animals present with burning marks affecting between 16% (e.g., paca, cavies and peccaries) and 27% (deer) of their bones; on the other hand, such marks are virtually absent in small rodents (0.6%). In Lapa do Santo, damage from burning affects 21.6% of deer bones, including limbs, ribs and vertebrae (Bissaro Jr. 2008). According to ethnographic studies, roasting and boiling result in minimal damage (Kipnis 2002); therefore, although they were certainly cooked, burning marks on animal bones from Lagoa Santa are likely the result of them being disposed of near to hearths.

The distribution of cut marks in deer bones from Lapa das Boleiras is compatible with the butchering of medium-sized prey (Lyman et al. 1992; Kipnis et al. 2008). In Lapa do Santo, just 3.13% of deer bones have cut marks on them, including metatarsals, humeri, radium, tibia, talus and a cuboid-navicular joint. The butchering, processing and consumption of small and medium-sized animals have resulted in the introduction of nearly complete skeletons in Lapa do Santo and Lapa das Boleiras.

At Lapa Grande de Taquaraçu mammals represent 47% and fish represent 34% of the faunal remains (Chim 2018). Mammals include small and medium-size animals and fish identified in Taquaraçu include the familiae Erythrinidae, Serrasalmidae and Pimelodidae, but 97% of fish bones are unidentified. In Lapa do Santo the fishes identified include the taxa Siluriformes and *Hoplias* sp.

Dental pathologies are used as a dietary proxy in bioarchaeological studies. As expected for generalist foragers living in a tropical environment (Ströhler and Hahn 2011; Humphrey et al. 2014), the elevated frequency of dental abscesses and caries has been reported for early Holocene Lagoa Santa (Neves and Cornero 1997; Neves and Kipnis 2004; Cornero 2005), particularly among females (Da-Gloria and Larsen 2014, 2017; Da-Gloria and Oliveira 2017; Da-Gloria et al. 2017; Da-Gloria and Bueno 2018). In an environment with a wide variety of tubers and fruits such as pequi (*Caryocar brasiliense*), jatobá (*Hymenaea sp.*) and araticum (*Annona classiflora*), it is expected that humans would have had a diverse diet instead of focusing exclusively on meat (Kipnis 2002). Accordingly, dietary stable isotope data from adult individuals show slightly elevated δ^{13}C levels, which is consistent with C_3 sources and low δ^{15} N values, similar to local herbivores. The combined carbon and nitrogen isotope results suggest that early Holocene people from Lagoa Santa had diets mostly based around C_3 plants, with only a minor faunal component (Kipnis 2002; Hermenegildo 2009, see Fig. 5 in Strauss et al. 2016). Ethnographic work among local dwellers of the region shows that overall this subsistence strategy is still practiced today (Oliveira et al. 2018).

9 Coexistence with Extinct Megafauna

The possibility of coexistence of humans and megafauna in Lagoa Santa was originally proposed by Peter Lund. However, although it has been proposed that there was a chronological overlap between humans and megafauna in Lagoa Santa[16] (Neves and Piló 2003; Piló and Neves 2003; Hubbe et al. 2011, 2013a, b), there is no direct evidence for the exploitation of the latter by the former. Recent analysis of stratified microvestiges from two lakes in Lagoa Santa region spanning the last 23,000 years showed reduced abundances of *Sporormiella*—a fungus associated with large herbivores—and an inferred megafaunal population decline that began 18.0 kyr cal BP, with the functional extinction occurring between 12.0 and 11.5 kyr cal BP. The results are suggested to be compatible with climate causing the population collapse, with humans preventing population recovery and inducing extinction (Raczka et al. 2016, 2017; Raczka 2017).

Lagoa Santa—as well as many other localities in Central Brazil—lacks any evidence of megafauna bones associated with human occupation or which present anthropic marks (Prous and Fogaça 1999; Kipnis 2002). With the application of new dating methods to megafaunal bones that control for collagen contamination (e.g., ultrafiltration, C/N ratios), the aforementioned chronological overlap might be revealed to be artificial. If not, this would imply either an intentional avoidance of megafauna (e.g., a dietary taboo or the favouring of other game) or that humans and megafauna were never present in the region at the same time. Finally, one cannot discard the possibility that the consumption/exploitation of megafauna was sporadic or only took place outside of rock shelters, resulting in a lack of accessible archaeological records. Of relevance to this matter is the continuous presence of tapir (*Tapirus terrestris*) in the region during the Holocene. This is the largest extant mammal in South America (weighing between 150 and 320 kg), but its remains have never been found in archaeological sites in Lagoa Santa. On the other hand, even in the ethnoarchaeological collection of the Guajá, one of the last hunter-gatherer societies in the South American lowlands (Forline 1997; Prado 2009; Politis et al. 2013), there are few skeletal remains of tapir—although it is known that they did eventually hunt this species.

10 Human Skeletal Remains

As a result of more than 170 years of excavations in the Lagoa Santa region, a large collection of approximately 155 human skeletons has been formed (Da-Gloria 2012; Souza and Lyrio 2017). Based on a few dozen direct dates on bone collagen (Table 3), it is assumed that the majority of this material dates back to the early Holocene, although there are also skeletons from later periods. Skeletal remains can provide information on a population's history (i.e., ancestry), diet, health, mobility, interpersonal violence, patterns of growth and mortuary practices.

10.1 Ancestry: Craniometric Studies and Archaeogenetic Prospects

The genetic relationship between Native Americans and the earlier inhabitants of the continent has always been a topic of great interest and debate. Before the recent advent of archaeogenetics, comparative anatomy was the only proxy for ancestry (Strauss and Hubbe 2010), with studies focusing on the morphological comparison of skeletal remains—particularly cranial and dental remains. Peter Lund, although noting 'ethnographic peculiarities of the skulls' from Lagoa Santa, was of the opinion that they nevertheless conformed to the 'American race'. Others, however, suggested the existence of early and late American crania. While the latter (Amerindians) were characterized by short and wide neurocrania as well as orthognathic high faces with relatively high and narrow orbits and noses, crania found in Lagoa Santa (Paleoamericans) presented the opposite pattern (Figs. 19 and 20). Accordingly, the existence of an earlier 'primitive race' was postulated by nineteenth century scholars, an idea that would find both supporters (Lacerda and Peixoto 1876; Reinhardt 1888; Imbelloni 1938; Rivet 1942) and detractors (Blake 1864; Kollman 1884; Ten Kate 1885; Hansen 1888; Hrdlička 1912) in the following decades.

More recently, this seminal observation was confirmed by typological studies (Melo and Alvim 1977), multivariate statistical analysis (e.g., Neves and Hubbe 2005), two (González-José et al. 2008) and three-dimensional geometric morphometrics (Galland and Friess 2016; von Cramon-Taubadel et al. 2017), and was interpreted as reflecting the arrival of two distinct biological populations on the continent coming from north-east Asia via Beringia: the Paleoamericans and the Amerindians (a.k.a. Model of the Two Main Biological Components) (see Hubbe et al. 2015 for a review). Alternatively, differences in cranial shape could be a result of in situ differentiation, with the Amerindian morphology resulting from the population's adaptation to regular plant cultivation and consumption from the middle Holocene onwards (Perez and Monteiro 2009; Perez et al. 2011). Moreover, recent studies show that the Paleoamerican morphology was also present during postcolonial times (Strauss et al. 2015a). Finally, the very difference between Paleoamerican and Amerindian morphologies has being disputed (Raghavan et al. 2015).

[16]This overlap is based on an indirect date for a *Catonix cuvieri* at Lapa Vermelha IV (9580 ± 200) and two direct dates (i.e., bone collagen) on a *C. Cuvieri* (11.6–11.3 kyr cal BP) and a *Smilodon populator* (12.9–12.7 kyr cal BP), both from Cuvieri Cave.

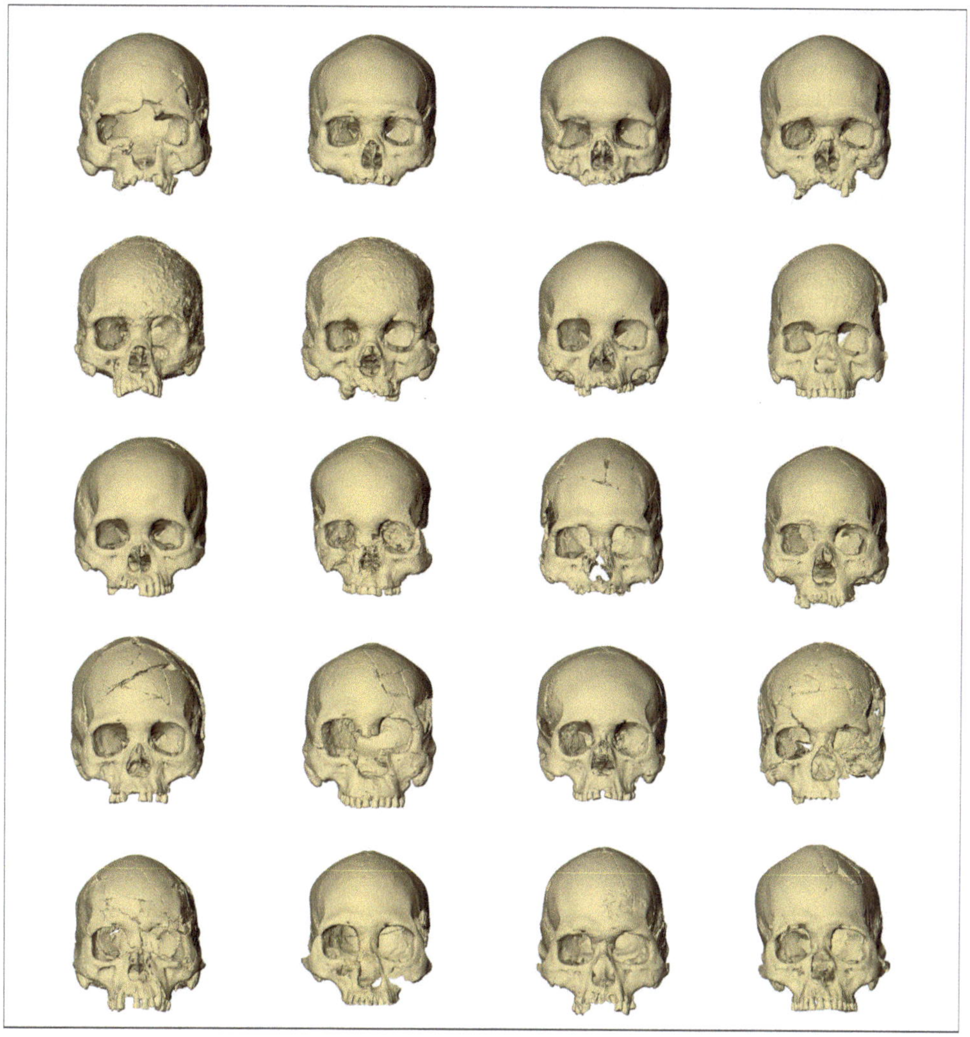

Fig. 19 Examples of early Holocene crania from Lagoa Santa, frontal view. The following specimens are shown (from top to bottom and from left to right): Row 1: Lapa do Acácio 2, Coleção Hélio Diniz 200, Coleção Hélio Diniz Sumidouro 2, Coleção Hélio Diniz Sumidouro 3; Row 2: Lapa da Samambaia—Adult Female, Lapa do Samambaia—Adult Male, Coleção Harold Walter 1, Coleção Harold Walter 4; Row 3: Coleção Harold Walter 5, Coleção Harold Walter 6, Coleção Harold Walter 9, Coleção Harold Walter 10; Row 4: Coleção Harold Walter 291, Coleção Harold Walter —'Confins' Man'; Santana do Riacho Burial 10, Lapa do Santo Burial 11; Row 5: Lapa do Santo Burial 15, Lapa do Santo Burial 17, Lapa do Santo Burial 21, Lapa do Santo Burial 26. Not to scale. Three-dimensional rendering carried out by André Strauss based on CT scans produced by Rodrigo de Oliveira

DNA extracted from early Holocene skeletons from Lapa do Santo (Posth et al. 2018) and Lapa do Sumidouro (Moreno-Mayar et al., 2018) have shown that these populations derive from a relatively early branch in the main radiation of Native American populations (Fig. 21). Mitochondrial DNA was extracted from 11 individuals from Lapa do Santo (A2 = 5, B2 = 3, C1d1 = 1, and D4h3a = 2) and six individuals from Lapa do Sumidouro (D1 = 3 and D4h3a = 3) indicating typical Amerindian mitochondrial DNA haplogroups. The presence of haplogroup D4h3a in high frequency in Lagoa Santa ($\sim$30%) is challenging to reconcile with previous suggestions that this was related to a Pacific coastal expansion (Perego et al. 2009).

Nuclear DNA was extracted from 7 individuals from Lapa do Santo and 5 individuals from Lapa do Sumidouro. Three males from Lapa do Santo have the Q1a2a Y-chromosome haplogroup, more specifically two individuals carry Q-M3 and one Q-CTS1780 haplogroups. Moreover, one Lapa do Santo male has the C2b Y-chromosome haplogroup that is extremely rare in present-day Native Americans, and was proposed to represent a post 6,000 years arrival to the continent—a hypothesis disproved by this Lapa do Santo individual directly dated to 10160-9600 yrcalBP. All five males from Lapa do Sumidouro are instead associated to the Y-chromosome haplogroup Q-M3. The presence of the today dominant Q-M3 and the rare Q-CTS1780 Y-chromosome haplogroups in Lagoa Santa during the Early Holocene suggests that both haplogroups were involved in the initial settlement of the region.

Genomic analysis of both Lapa do Santo and Lapa do Sumidouro individuals do not indicate an excess of extra-continental genetic affinities in relation to any modern or past Amerindian populations (Fig. 20). The unique exception is individual 'Sumidouro 5'—the only high coverage genome ($\sim$15x) available for Lagoa Santa—in which a putative $\sim$2–3% ancestry related to Australasian populations (i.e., Andaman, Onge, Papuans and Australians) was identified. However, the demographic implication—if any—of this signal is not clear at this point. Genomic data also indicates that Central and South American populations older than $\sim$9,000 yrcal BP have a distinctive genetic affinity to the Anzick-1 genome, a skeleton likely

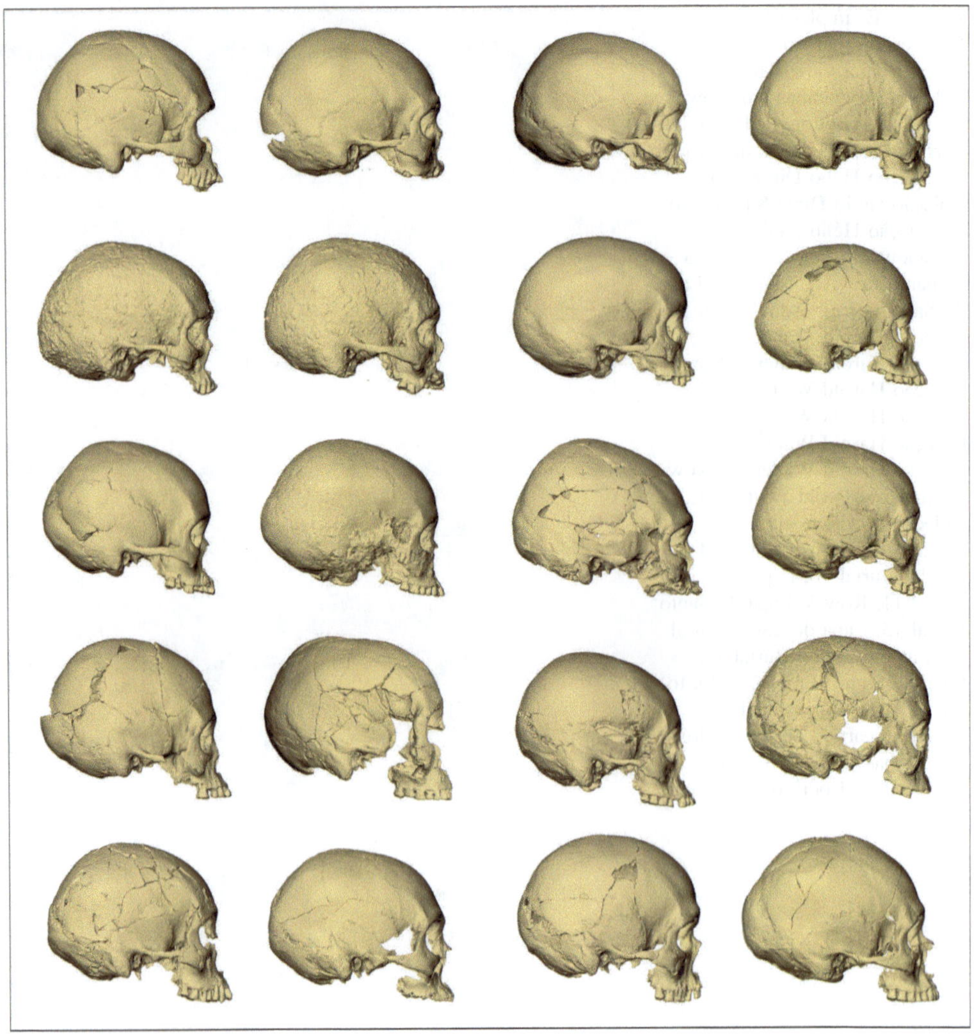

Fig. 20 Examples of early Holocene crania from Lagoa Santa, right lateral view. The following specimens are shown (from top to bottom and from left to right): Row 1: Lapa do Acácio 2, Coleção Hélio Diniz 200, Coleção Hélio Diniz Sumidouro 2, Coleção Hélio Diniz Sumidouro 3; Row 2: Lapa da Samambaia—Adult Female, Lapa do Samambaia—Adult Male, Coleção Harold Walter 1, Coleção Harold Walter 4; Row 3: Coleção Harold Walter 5, Coleção Harold Walter 6, Coleção Harold Walter 9, Coleção Harold Walter 10; Row 4: Coleção Harold Walter 291, Coleção Harold Walter —'Confins' Man'; Santana do Riacho Burial 10, Lapa do Santo Burial 11; Row 5: Lapa do Santo Burial 15, Lapa do Santo Burial 17, Lapa do Santo Burial 21, Lapa do Santo Burial 26. Not to scale. Three-dimensional rendering carried out by André Strauss based on CT scans produced by Rodrigo de Oliveira

belonging to the Clovis Culture. This ancestral component is no longer present in modern-day Native Americans indicating the occurrence of a substantial population turnover in their population history. Finally, the analysis of phenotypically related genes revealed that one individual from Lapa do Santo has the ancestral allele for the EDAR gene (rs3827760) showing that the derived allele—present in almost 100% of present-day Native Americans and East Asians—was not yet fixed by the early Holocene thus implying convergent evolution on both sides of the Pacific.

10.2 Osteological Markers: Health, Lifestyle and Beyond[17]

Osteological markers can provide information on relevant aspects of past populations: diet, health, mobility, growth, interpersonal violence, demographic density, activity patterns and infectious diseases. The frequency of periosteal infection—expressed in the form of superficial bone remodelling (Fig. 22a)—is high in Lagoa Santa. This is suggestive of demographic aggregation, which facilitated the transmission of infectious diseases (Da-Gloria 2012). Parasite eggs—or at least their molecules—can be preserved in archaeological sediments. However, in spite of systematic sampling in Lagoa Santa, results have so far been negative, which is most likely a reflection of preservation issues (Iñiguez et al. 2006). The presence of generalized dental hypoplasia and conoid canines in children from Lapa do Santo (Burial 20) most likely represents an early Holocene case of treponematosis in the New World.

The incidence of cribra orbitalia in Lagoa Santa—a proxy of nutritional deficiency resulting from a poor diet or chronic parasitic infection—is within the normal range for American hunter-gatherers (Da-Gloria 2012). Porotic hyperostosis—another indicator of developmental stress—is absent from Lagoa Santa skeletons (Da-Gloria 2012). Lagoa Santa

[17]For a discussion of caries, please refer to the 'Diet and subsistence' section.

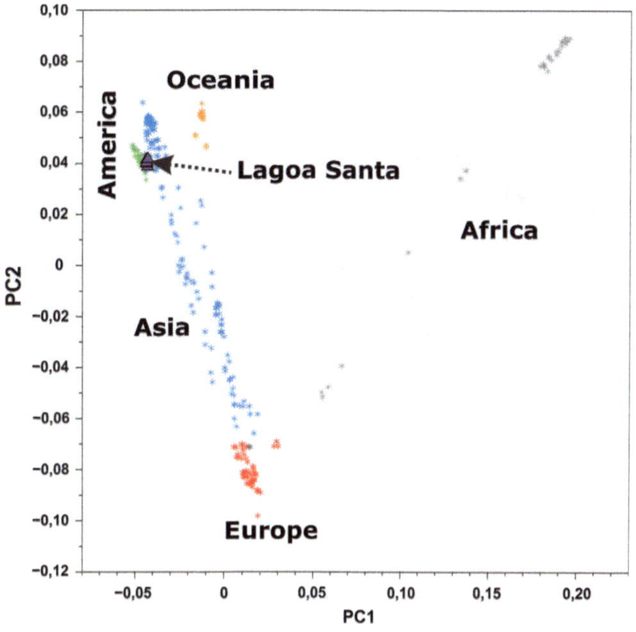

Fig. 21 Genomics affinities of Lagoa Santa individuals (early Holocene). Principal Component Analysis based on 567,000 SNPs with Lapa do Santo (purple triangle) individuals plotted in conjunction with worldwide reference series '(analysis conducted by Cosimo Posth and Tiago Ferraz)'. Note that Lagoa Santa individualsare entirely within Native American genomic variation

individuals were relatively short with an estimated height of 1.60 and 1.50 m for males and females, respectively (Reinhardt 1888; Mello e Alvim 1977; Da-Gloria 2012). Although this could result from nutritional impairment during growth, it might also reflect a genetically short population. Nitrogen isotopic dating is suggestive of a late weaning time and, therefore, long interbirth intervals during the early Holocene in the region (Hermenegildo 2009; Strauss et al. 2016). However, these inferences are based on a sample size of two and need further confirmation.

A high incidence of *antemortem* trauma—particularly to the head (Fig. 22b)—in individuals of all ages and both sexes is compatible with frequent interpersonal domestic violence (Walter 1958; Da-Gloria 2012). In a few instances, non-healed blunt trauma to the head was the cause of death (Da-Gloria 2012). Preliminary studies show that the femurs from Lagoa Santa have a relatively rounded diaphyseal cross section, suggesting reduced levels of mobility. A preliminary analysis carried out on 23 individuals from Lapa do Santo shows that, with one exception, they present a homogenous strontium isotopic signature that is close to—but slightly higher than—the local bioavailability of the Lagoa Santa region (see Fig. 5 in Strauss et al. 2016). This pattern likely reflects a mobility strategy focused in the karst, but with some sporadic time spent in the nearby region of Serra do Espinhaço, a geologically older locality expected to present higher $^{87}Sr/^{86}Sr$ ratios. One important note is that, due to small sample sizes, most of the aforementioned patterns are not statistically significant and must be interpreted with caution (Da-Gloria 2012).

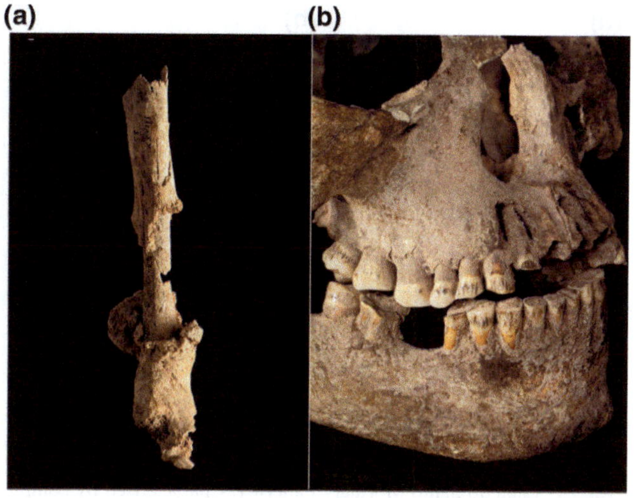

Fig. 22 Example of osteopathologies from early Holocene Lagoa Santa. **a** Burial 28 from Lapa do Santo. Long bone of a child showing a severe periosteal reaction; **b** Burial 1 from Lapa do Santo. *Antemortem* loss of the left lower first molar. Credits: Maurício de Paiva. Housing institution: Institute of Biosciences of the University of São Paulo

10.3 Funerary Practices

In the task of reconstructing the life of past populations, no other aspect of the archaeological record has as much potential to reveal symbolic and ritual aspects as human burials. The burials in Lagoa Santa were very diverse, pointing to cultural diversity in the region during the early Holocene (Strauss 2010, 2011, 2012, 2016a, b, 2017; Strauss et al. 2016). Most of this evidence comes from Lapa do Santo, where dozens of human burials were divided into different mortuary patterns based on their chronology and shared features. Lapa do Santo was occupied from 11.7 to 12.7 kyr cal BP onwards, but its use as an interment site only started from 9.7 to 10.6 kyr cal BP with Lapa do Santo Mortuary Pattern 1 (LSMP-1), which is characterized by primary single burials in a flexed position (Burials 1, 27, 34 and 37) (Fig. 23).

Lapa do Santo Mortuary Pattern 2 (LSMP-2) dates from 9.4 to 9.6 kyr cal BP, and can be further subdivided into three categories: LSMP-2a (Burials 21 and 26), LSMP-2b (Burials 9, 14, 17, 18 and 23) and LSMP-2c (isolated bones). LSMP-2a is characterized by fully articulated partial skeletons with cutting and chopping marks. In Burial 21, the midshafts of both tibiae and fibulae were chopped and removed while soft tissue was still present (Fig. 24). Burial 26 is a decapitated head with the first six cervical vertebrae articulated in anatomical position (Strauss et al. 2015b)

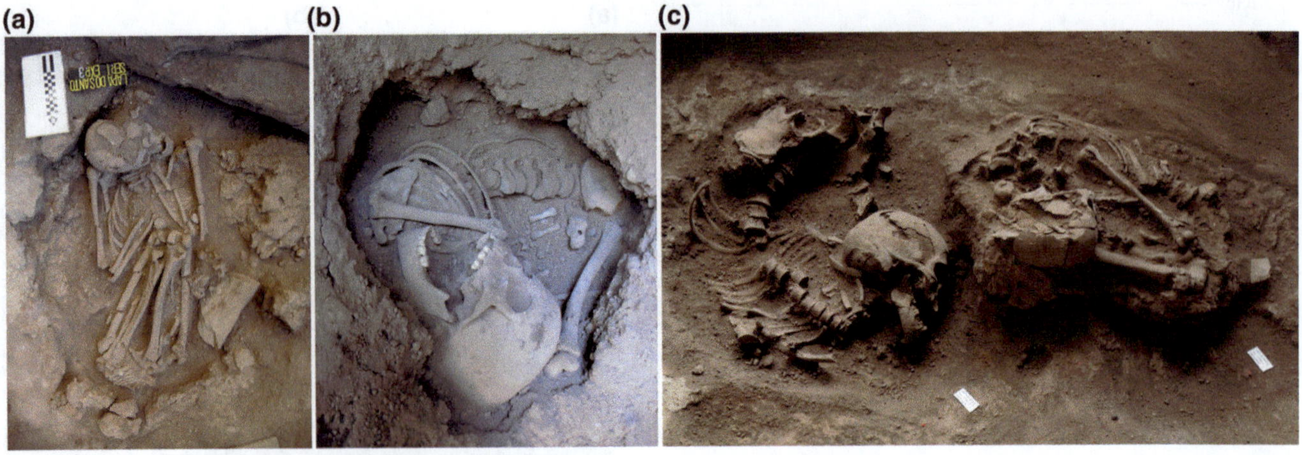

Fig. 23 Lapa do Santo Mortuary Pattern 1. Characterised by complete fully articulated skeletons, such as in burials 1 (**a**), 27 (**b**), 34 (**c**, left) and 37 (**c**, right). Pictures taken in 2008 (**a**, **b**) and 2014 (**c**). Credits and housing institution:Institute of Biosciences of the University of São Paulo

(Fig. 25). Furthermore, the evidence shows that LSMP-2a involved the reduction of fresh corpses (Strauss and Oliveira 2018a, b).

LSMP-2b is characterised by graves filled with the fully disarticulated bones of up to three individuals, presenting a strong selection of anatomical parts (Fig. 26a, b). Some bones show evidence of exposure to fire, the application of a red pigment, defleshing, cutting, chopping and the removal of teeth (Fig. 26c–i). Burials 14, 17 and 18 were composed of a bundle of long bones from one or two individuals, deposited with the individualized cranium and/or mandible of a different individual. Bundles comprising infant postcranial bones were found next to adult crania (Burials 14 and 17) (Fig. 26c), while bundles comprising adult postcranial bones were found next to an infant cranium (Burial 18). The long bones in the bundles had been chopped and segregated into extremities and midshafts and, in some cases, the latter were further chopped into smaller sections (Fig. 26c–e). The cranium of Burial 17 was used as a funerary receptacle and was filled with chopped burnt bones, some of which present defleshing cut marks. Black burn marks, which are limited to the anterior portion of the external maxillary alveolar margin, indicate exposure to fire while soft tissue was still present (Fig. 26h).

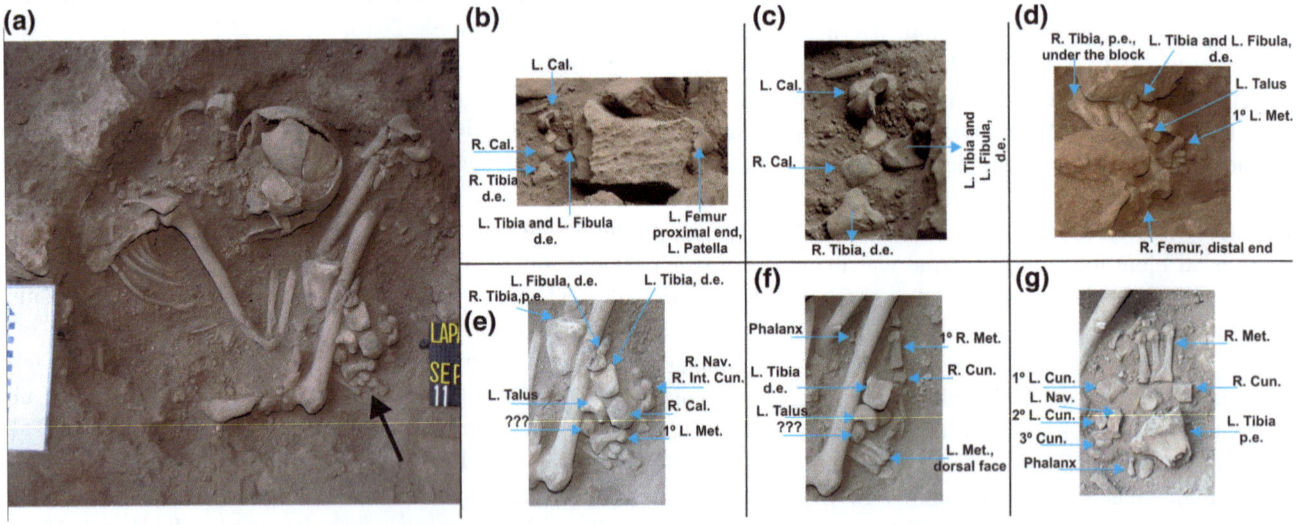

Fig. 24 Lapa do Santo Mortuary Pattern 2A. Burial 21 from Lapa do Santo. The diaphysis of both tibiae and fibulae (black arrow in **a**) were removed (*perimortem*) from this otherwise complete and articulated skeleton. The sequential progress of the exhumation allows for a detailed visualization of the spatial distribution of the affected anatomical parts (**b**–**g**). The following abbreviations apply: left (L.), right (R.), distal extremity (d.e.), proximal extremity (p.e.), calcaneus (Cal.), intermediate cuneiform (Int. Cun.) and metatarsal (Met.). Credits: André Strauss. Housing institution: Institute of Biosciences of the University of São Paulo

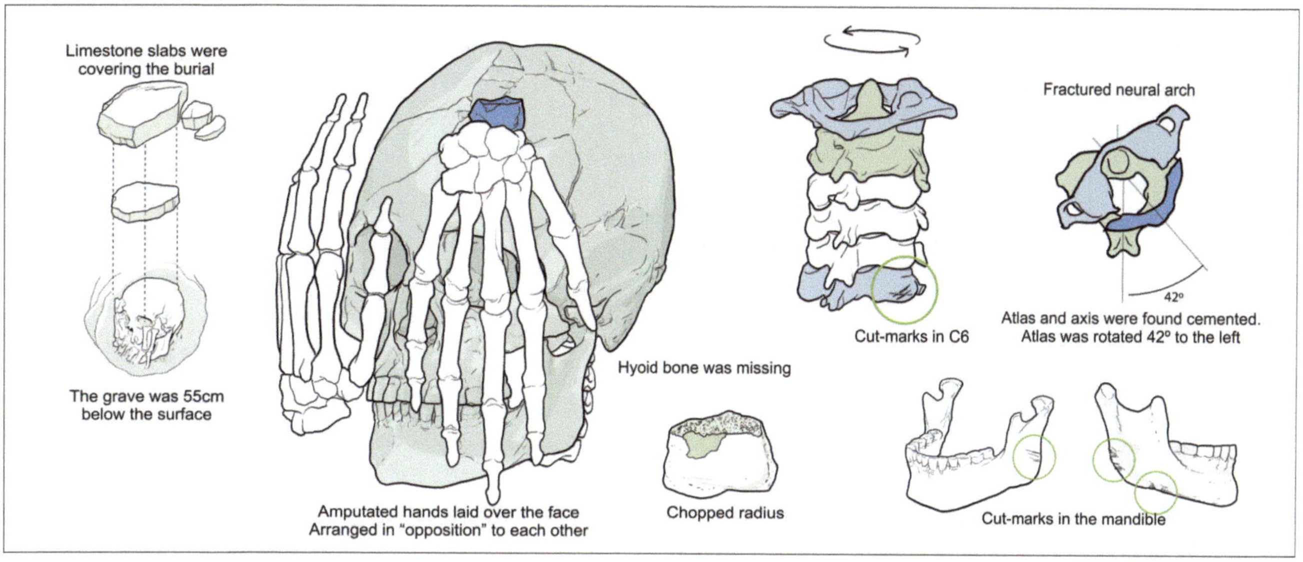

Fig. 25 Lapa do Santo Mortuary Pattern 2A. Burial 26 from Lapa do Santo, directly dated to 9.1-9.4 kycal BP (95.4% interval). The interment was composed of an articulated cranium, mandible and the first six cervical vertebrae. Cut marks with a v-shaped profile were observed in the mandible and the sixth cervical vertebra. The right hand was amputated and laid over the left side of the face with distal phalanges pointing to the chin, and the left hand was amputated and laid over the right side of the face with distal phalanges pointing to the forehead. Schematic representation by André Strauss with drawings by Gil Tokyo. Housing institution: Institute of Biosciences of the University of São Paulo

In Burials 17 and 18, all teeth were intentionally removed and the coronoid processes of Burial 18's mandible were drilled (Fig. 26i). Red pigment was abundantly applied to the bones of Burials 14 and 18. Burial 23 was composed of a cranium *calotte* filled with 54 permanent and 30 deciduous teeth, some of which belonged to the skull of Burial 17. Burial 9 was an individualized child skull placed near the pelvis of an individual of similar age. The deciduous dentition was removed and an assemblage of human teeth and chopped midshafts (accession code: LSt-2253) were deposited next to Burial 9. LSMP-2c is defined by isolated, burnt and chopped bones that were not part of any formal burial; the presence of rodent gnaw marks could indicate they were subject to scavenging and were not immediately buried. The femur presents burn marks, which are indicative of exposure to fire while soft tissues were still present.

LSMP-3 is poorly dated to 8.2–8.6 kyr cal BP, and includes nine burials: 6, 7, 10, 11, 12, 13, 15, 19 and 22. These burials are characterised by shallow circular pits, completely filled with the mostly disarticulated bones of single individuals of various ages and sexes (Fig. 27a, b). Circular stone structures covered some of the burials, but also occurred independently of them (Fig. 27c). Anatomical selection was not observed and, with the exception of some small bones, most elements of the skeleton were present. The midshafts of long bones of adult individuals were, in some cases, intentionally broken in the central region before deposition, resulting in butterfly fractures with impact points indicating the use of some kind of percussion instrument (Fig. 27d, e). LSMP-3 burials are very similar to each other, in contrast with the larger variability observed within LSMP-2 burials. Furthermore, the characteristic elements of the latter, such as cut marks, chop marks, the absence of dentition, red pigment and burn marks, are not present in the former.

In Lapa das Boleiras (Neves et al. 2002; Araujo and Neves 2010) Burial 3 showed signs that red pigment was applied after the *perimortem* fracturing of long bones and the chopping of the extremities (Strauss 2010:660) (Fig. 28). Therefore, it shows a funerary pattern which has not been observed in Lapa do Santo, which implies that there were even higher levels of cultural diversity in the region during the early Holocene. Nevertheless, when compared to the closest sites outside the karst, it seems that Lagoa Santa did constitute a coherent and localized cultural unity. Santana do Riacho (around 150 km away from Lapa do Santo) (Prous 1992/1993) is well known for burials that are assumed to date back to ca. 10.0 kyr cal BP; they are therefore contemporaneous with LSMP-1 burials. The former burials are characterised by fully articulated skeletons in a flexed position.

Further away from Lagoa Santa, the sites of Capelinha (Vale do Ribeira—São Paulo) and Toca dos Coqueiros (Serra da Capivara—Piaui, but see Strauss et al. 2018) also have burials that are roughly contemporaneous with LSMP-1 (see Strauss 2015 for a review). In both cases, fully articulated flexed skeletons were found together with projectile points. In Justino (Xingó—Sergipe), five burials were associated with a piece of charcoal dating back to 8.9 kyr non-cal BP. Four of them are fully articulated flexed

Fig. 26 Lapa do Santo Mortuary Pattern 2B. **a** In Burial 14, a bundle composed of the postcranial bones of sub-adult individuals was deposited above the individualized cranium of an adult; **b** In Burial 17, the cranium was used as a funerary receptacle, within which the chopped and burnt bones of the same individual were deposited. Next to it, there was a bundle with the post-cranial bones of one sub-adult individual; **c** Detailed view of the bundle of long bones from a child that was deposited next to the adult cranium from Burial 17; **d** Medial view of the chopped proximal extremity of the right humerus that was found within the adult cranium from Burial 17 (arrow in **b**); **e** Chopped right humerus (the same one as in **d**) and chopped right ulna articulating with each other; **f** Detail of the cut marks in the medial part of the femoral neck of a chopped proximal extremity of a left femur. It was found above the adult individualized cranium from Burial 14; **g** Fragmented (bite marks?) hand bones found within the adult individualized cranium from Burial 17; **h** The maxilla and mandible from Burial 17 have burn marks concentrated in the anterior portion of the external alveoli margin, and the teeth were intentionally removed; **i** The mandible from Burial 18 had its teeth removed and holes drilled in its coronoid processes (see detail). Credits: Institute of Biosciences of the University of São Paulo (**a**, **b**), Maurício de Paiva (**h**) and André Strauss (**c**–**g**, **i**). Housing institution: Institute of Biosciences of the University of São Paulo

skeletons and one is 'secondary' (Vergne and Amâncio 1992; Vergne 2002). These sites are much too far away from each other and from Lagoa Santa to allow any kind of direct connection to be postulated. After 8.0 kyr BP, several sites outside of Lagoa Santa do have evidence of rituals involving some kind of manipulation and reduction of the bodies. Loca do Suin (Pains—Minas Gerais) has two burials directly dated to 7.4 and 7.5 kyr non-cal BP, including a case of cremation (Strauss et al. 2011). In Burial 1 from Lapa do Malhador (Vale do Peruaçu—Minas Gerais), the right arm

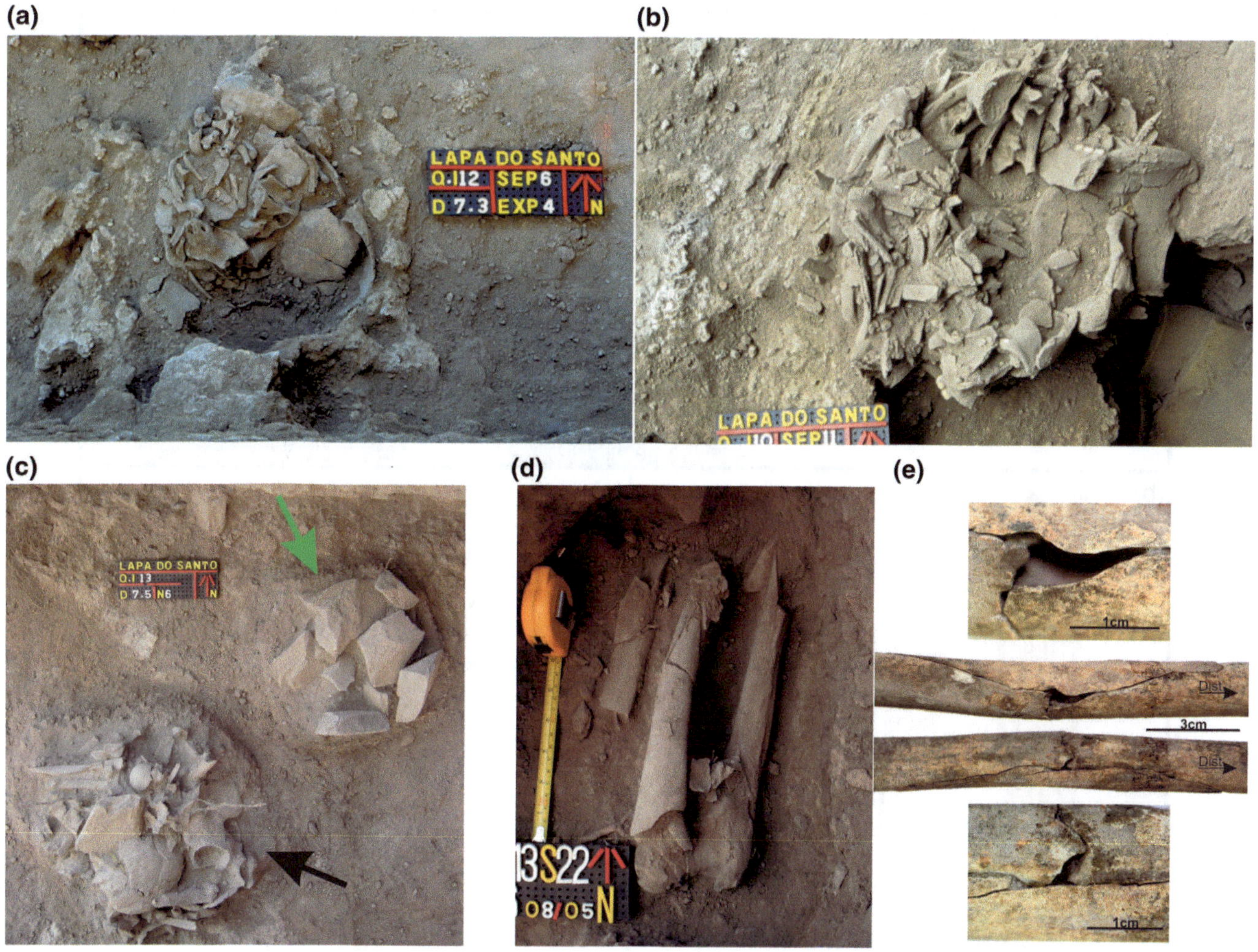

Fig. 27 Lapa do Santo Mortuary Pattern 3. **a** Burial 6 was composed of a small circular grave filled with the chaotically distributed bones of a child; **b** Burial 11 was composed of a circular grave with the chaotically distributed bones of an adult; **c** Burial 15 (black arrow) is typical of LSMP-3. The picture was taken after the circular structure composed of stones covering the grave was removed. The green arrow points to a circular structure of stones that was not directly associated with any burial; **d** The long bones at the bottom of the grave of Burial 22 exemplify the practice of breaking the midshafts; **e** The humerus of Burial 22, exemplifying a butterfly fracture and impact points (details above and below) Credits and housing institution: Institute of Biosciences of the University of São Paulo

of a fully articulated skeleton was amputated, leaving a profusion of cut marks on the bones. Another example of *perimortem* manipulation comes from the site at Pedra do Cachorro, in Pernambuco, dating back to 760 ± 60 years non-cal BP (Solari and DaSilva 2017).

In a broader continental perspective, the evidence from Lagoa Santa demands a re-evaluation of the chronological depth and geographical dispersion of the practice of body manipulation in the New World. The Chinchorro, for example, can no longer be considered '*the oldest known rituals in America*' (Llagostera 2003:21). Before Lapa do Santo and Lapa das Boleiras, all sites containing analogous rituals were located in the Andean region. Pampa de los Fossiles, Cerro Mangote, OGSE-80, Palli Aike, Patapatane, Nemocón, Guavio, Tequendama, Aguazuque, CA0927, CA0952 and CA0928 are some examples where the mortuary prescriptions mainly focused on reducing the body, either by removing parts of the anatomy or exposing them to fire (Santoro et al. 2005). Taking into account the geographic dispersion of these cases and the paucity of information coming from the lowlands, a consensus was established that the funerary ritual which placed emphasis on the manipulation of the body was a phenomenon whose origins and further development were mostly restricted to the Andean region (Ferguson 1990:247; Cordy-Collins 1992:208; Verano 1999:59; Jandial et al. 2004:1215; Santoro et al. 2005:330; Conlee 2007:438).

The mortuary records from Lagoa Santa show remarkable similarities to sites in the Sabana de Bogotá region in Colombia, especially Aguazuque (Correal 1990, 2001). At this site, funerary rituals were also characterised by the strong selection and reallocation of different parts of the

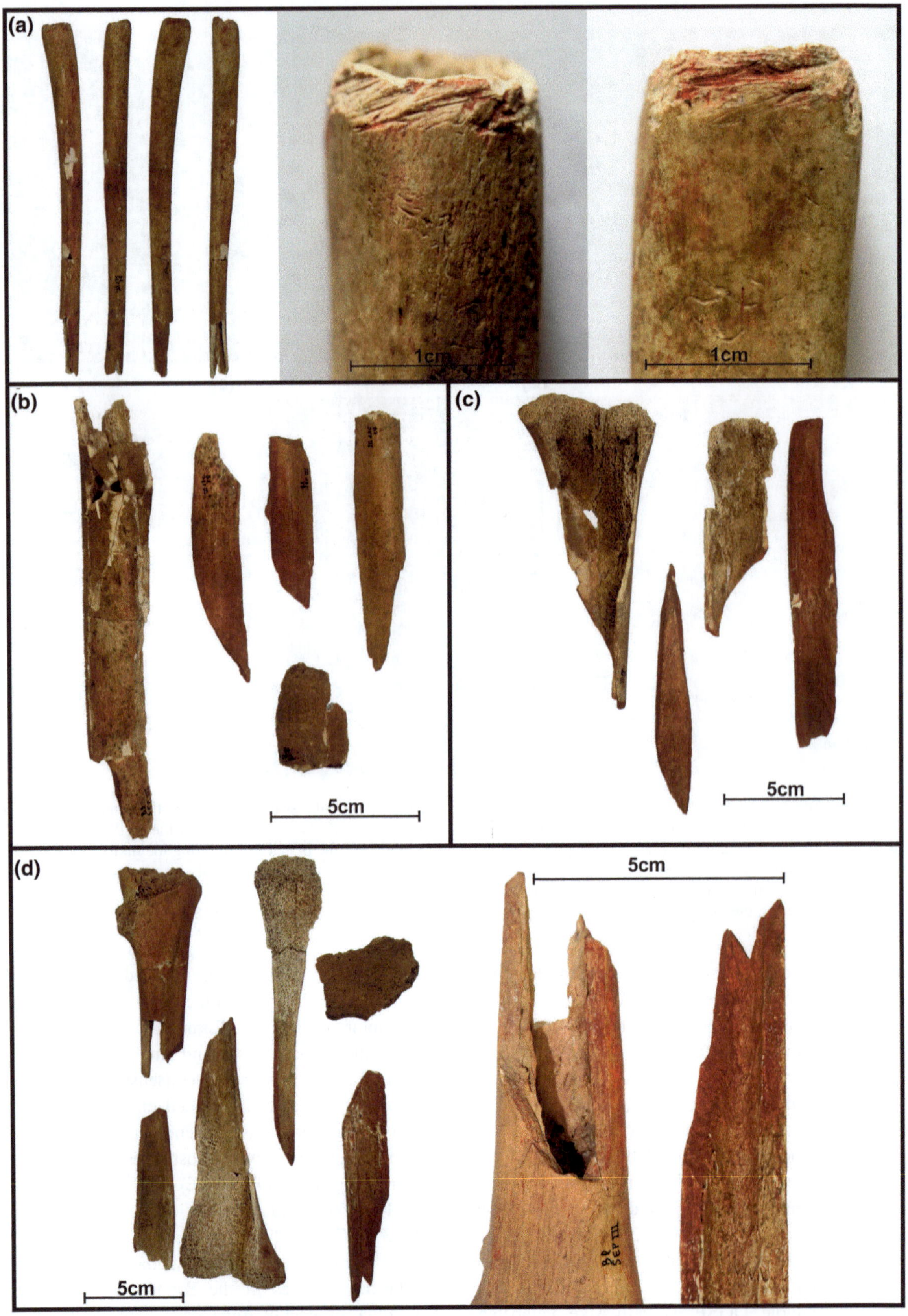

Fig. 28 Burial 3 from Lapa das Boleiras. **a** The left ulna had its proximal extremity removed. Note the presence of red pigment in the chopped margin. The right tibia (**b**), left femur (**c**) and right femur (**d**) all present *perimortem* fractures and the application of red ochre Credits: André Strauss. Housing institution: Natural History Museum and Botanical Garden of the Federal University of Minas Gerais

human body, including the individualization of human skulls and the sectioning of long bones to separate midshafts from the extremities. It is hard to appreciate the extent to which such similarities are the result of random factors and the extent to which they reflect a common cultural *milieu* or background. This similarity between very specific types of mortuary procedure is by no means an exception on a broad continental scale during the Archaic period. The high level of similarity between burials at the Las Vegas 80 and Cerro Mangote sites, located thousands of kilometres away from each other, has long been noticed by archaeologists, although the implications of such similarities have never been fully discussed.

11 Geoarchaeology and Site Formation Processes

One of the most remarkable aspects of archaeological sites in the Lagoa Santa karst is a diverse range of formation processes (Araujo and Piló 2017) which always combine two main sedimentary sources. In-depth geoarchaeological studies were done at the cave sites of Lapa do Sumidouro (Piló et al. 2004, 2005) and Lapa Vermelha IV (Feathers et al. 2010), at the rock shelter sites of Lapa das Boleiras (Araujo et al. 2008), Lapa Grande de Taquaraçu (da Silva and Prous 2014) and Lapa do Santo (Villagran et al. 2017), as well as at the open air site of Sumidouro (Araujo et al. 2013). The mixing of geogenic sediments (clay aggregates) and anthropogenic sediments (ashes) is a recurrent feature of rock shelters and cave sites (Fig. 29). Variation exists in the ratio of geogenic to anthropogenic sediments, as well as in the frequency of depositional agents adding geogenic material to the sites. The anthropogenic source, when present, is persistently associated with human-made fires, both from intact or reworked combustion features.

The geogenic source is associated with soil material which developed from the rock unit overlying the limestones of the Sete Lagoas Formation (metacalcareous unit). Yellow goethite-rich oxisols and red hematite-rich oxisols developed from the weathering of the metapellites of the Serra de Santa Helena Formation via fast and low hydrolysis, respectively (Piló 1998 and see chapter "Soils of the Lagoa Santa Karst", this volume). The soil material is transported inside the karstic cavities as a result of diverse processes, including runoff and colluviation, among others. Both the red and yellow soils are easily identified, not only by their distinct colours and mineralogy but also by their chemical and micromorphological signatures. Micromorphological differences between the two soils include the granular microstructure with compound packing voids of the red oxisol (Fig. 29a, b) and the angular blocky to massive microstructures, with fissures and channel voids, of the yellow oxisol (Fig. 29c) (Piló 1998; Araujo et al. 2013; Villagran et al. 2017).

Cave sites frequently contain a higher amount of geogenic sediments embedding the archaeological finds (e.g., human remains and artefacts). Such is the case at Lapa do Sumidouro, which exclusively contains geogenic sediments, and Lapa Vermelha IV, where very little anthropogenic material is integrated into the mostly geogenic matrix. Piló et al. (2004, 2005) studied the sedimentation processes at Lapa do Sumidouro, and identified the role of yellow goethite-rich sediments and red hematite-rich sediments inside the cave. The sediments were transported from the outside through fissures and by seasonal flood events. Most of the human remains are contained within a layer of brownish-yellow massive clayey sand around 1 m thick, with a minimum age of 8150 ± 450 years (the U-series age was obtained from interstitial calcite). Although sediments come from the yellow goethite-rich soil above the limestone, the human remains were originally positioned somewhere near the southern entrance of the cave and were then transported further inside by fluctuations in the water table.

Feathers et al. (2010) studied the formation processes at the key site of Lapa Vermelha IV, where the skeleton of Luzia was found (Cunha and Guimarães 1978). The authors describe contents of the 6–7 m stratigraphic succession from

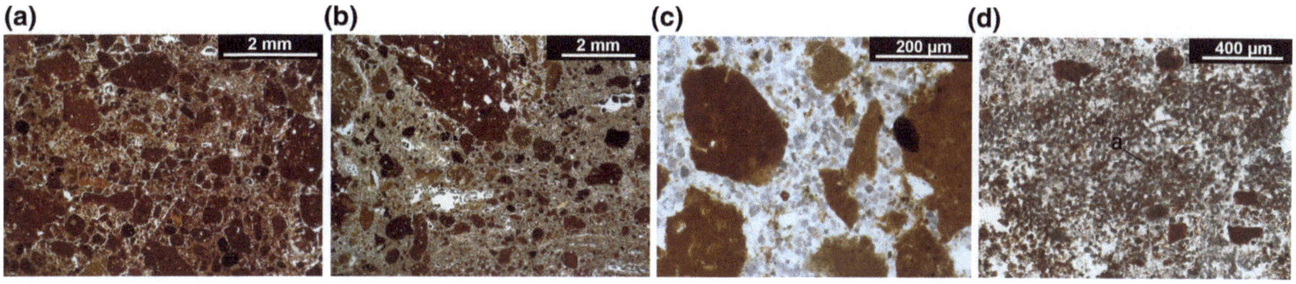

Fig. 29 Photomicrographs from the site of Lapa do Santo. **a** A layer consisting of rounded and sub-rounded clay aggregates associated with the red hematite-rich oxisol developed above the limestone; **b** Mixed deposit consisting of red and yellow clay aggregates embedded in a greyish matrix made of ash crystals; **c** More detailed view of rhombohedral ash crystals with sharp edges, indicating good preservation. **d** Accumulation of well-preserved ash crystals in a light brown, possibly phosphatic micromass Credits: Ximena Villagran. Housing institution: Museum of Archaeology and Ethnology of the University of São Paulo

bottom to top: laminated, reddish-yellow sandy mud with gravel, combining rock fall from the cave roof and yellow goethite-rich sediments from the soil outside; heterogeneous red sandy mud, sometimes embedded in ash crystals, containing a few aggregates of yellow goethite-rich sediment and rare charcoal and plant tissue (possibly reworked from the cave entrance); and laminated water-lain red sandy clay, with rare ash crystals and sand-sized bone fragments, made of red hematite-rich soil transported by gravitation through dejection cones. The skeleton of Luzia was found at the interface between the upper and lower sediments, which were dated by OSL to between 12.7 and 16 kyr BP, respectively. The middle stratum, which was rich in anthropogenic sediments, was dated to about 9 kyr BP.

Sites located in rock shelters are rather different from cave sites. They usually include a larger amount of anthropogenic sediments embedding the archaeological finds; they can also contain intact combustion features. At Lapa das Boleiras, the 1 m deep sedimentary succession contains: dark reddish-brown sediments at the top, dated to 0.745–0.680 kyr cal BP; red clayey sediments interbedded with greyish ash, dated to 9.9–9.6 kyr cal BP; and ashy sediments at the base interbedded with red clayey sediments, aged between 12.1 and 11.4 kyr cal BP. The red sediments are always associated with the red hematite-rich oxisols above the limestone, episodically entering the rock shelter through a colluvial fan in the southern part of the site. The laminations of coarse and fine red clay aggregates are interpreted as signalling depositional hiatuses of high vs. low energy colluviation. The density of red soil aggregates is low in the ashy layers, which contain common phytoliths and rare plant tissue, charcoal and bone fragments. The ash layers are interpreted as having been mobilized from ancient hearths through site maintenance activities, human trampling or by local air currents (Araujo et al. 2008).

The sediments at Lapa do Santo show a slight resemblance those at Lapa das Boleiras; however, the former has higher amounts of red clay aggregates (Fig. 29b), as well as several intact combustion features. Using a micro-contextual approach, including archaeological, modern reference and experimental samples, Villagran et al. (2017) demonstrated that the early Holocene deposit (dated to 7.9–12.7 kyr cal BP) formed after the mixed accumulation of ashes (Fig. 29c, d) and red hematite-rich oxisol aggregates fell continuously from above the limestone cliff into the rock shelter. A combination of intact and remobilized combustion features make up the archaeological deposit. Most of the red soil aggregates have been heated to temperatures of above 500–600 °C, possibly as a result of falling into fires or through being in the substrate of the ash-rich sediments where hearths were constantly lit. Fragments from heated termite mounds have also been described, proving the use of this local resource by early Holocene populations. Decayed wood and forest litter have been identified as the most common fuel sources. The conspicuous differences the thickness of archaeological deposits at Lapa do Santo (almost 5 m) and Lapa das Boleiras (about 1 m) may be better explained by local geogenic/anthropogenic sedimentation processes than by variations in site occupation intensity.

Finally, the open air site of Sumidouro, located on a lakeshore, shows a completely different set of sedimentary dynamics. At least three archaeological layers, containing charcoal and stone artefacts, are described within the red hematite-rich oxisol. Araujo et al. (2013) interpreted the three archaeological layers as indicative of temporally distinct moments of human activity: one at a depth of 137 cm, dated to 5.92 and 5.6 kyr cal BP; another at a depth of 160 cm, dated to between 9.45 and 9.24 kyr cal BP; and one at a depth of 180 cm, possibly from the Pleistocene/Holocene boundary (no charcoal was found in this layer). The authors interpret the deposit as having been formed by discrete episodes of soil accretion and sediment deposition by colluviation from the top of the slope, intercalated by human settlements. Besides the expected differences between sedimentation processes, depositional rates also differ between open air archaeological sites and those situated in caves/rock shelters. The highest values, calculated by contrasting the calibrated radiocarbon and OSL ages with the depths of the samples, come from Lapa do Santo, with an estimated sediment input of 4.14 mm per year. The lowest values were estimated for the open air site of Sumidouro (0.2 mm per year), and intermediate values were calculated for two archaeological sites in rock shelters (Lapa das Boleiras: 1.2 mm per year; Lapa Vermelha: 0.35–3.4 mm per year) (Strauss and Araujo 2012).

In summary, the existing geoarchaeological work carried out in two cave sites and two rock shelters of the Lagoa Santa region provides evidence for the long-term input of local oxisols into the cave systems, whether this occurred through gravitational transport, runoff, colluviation or soil fall (Piló et al. 2004, 2005; Araujo et al. 2008; Feathers et al. 2010; Villagran et al. 2017). The present-day seasonal climate, which is characterised by intercalating wet and dry seasons, favours soil erosion in the karst. Drier climates in the past are believed to have been responsible for the introduction of goethite or hematite-rich sediments into the karst, while the present climate and vegetation conditions favour their removal. Since the soil cover originates from the weathering of pellites overlying the karst, ongoing soil erosion will eventually lead to the subaerial exposure of the limestone (Auler 1999; Auler et al. 2009). Each archaeological site reveals a unique scenario, according to which geogenic sediments are incorporated and mixed with anthropogenic sediments consisting of ashes from human-made fires (Araujo et al. 2008; Feathers et al. 2010; Villagran et al. 2017) and/or discrete archaeological finds (human remains, artefacts, charcoal, etc.) (Piló et al. 2004, 2005; Feathers et al. 2010). Moreover, a recent study of virtual micromorphology suggests the occurrence of termite mounds' fragments in Lapa do Santo—likely resulting from human use (Villagran et al. 2019). This proves the need for unambiguous site formation studies that account for the specificities of each archaeological context. It also shows that estimations of the intensity of site occupation can only be done after a full assessment of the relative input of geogenic, vs., anthropogenic sediments at each site.

11.1 Palynological Analysis in Lapa do Santo

Palynological analyses of materials from archaeological contexts can provide a great deal of information concerning both dietary and medicinal plant use as well as environmental data. In a preliminary study at Lapa do Santo (unit P11), a total of 9 samples of sediment were extracted from an early Holocene wall at 6 cm intervals starting at the surface (vertical) (Table 5). Pollen grains were only recovered from the upper 18 cm of soil and are considered to be recent pollen types rather than archaeologically relevant pollen types. The results of these analyses demonstrate the low preservation potential of pollen grains in alkaline soils with ash deposits. These analyses also stress that samples collected near the surface in similar environments are not suitable for investigations of archaeological pollen grains.

12 Concluding Remarks

The archaeological record of the Lagoa Santa region in east-central Brazil documents more than 12,000 years of human presence. Traditionally, archaeological studies have focused on the early Holocene component of the region, resulting in the most comprehensive and multidisciplinary depiction of life and death during this period in eastern South America. Accordingly, between 12.7–11.7 kyr cal BP and 8.3–8.0 kyr cal BP, Lagoa Santa was occupied by culturally diverse, early Archaic generalized foragers who underwent a constant transformation through time. The use of rock shelters for interment purposes started with primary burials between 10.3 and 10.6 kyr cal BP. Between 9.4 and 9.6 kyr cal BP, body reduction—whether by mutilation, defleshing, tooth removal and exposure to fire followed by the secondary burial of the remains according to strict rules—became a central element in the treatment of the dead. In the absence of monumental architecture or grave goods, parts of fresh corpses were used to develop rituals; this is evidence that this practice was not restricted to the Andean region at this time. Between 8.2 and 8.6 kyr cal BP, another change occurred when the pits were instead filled with the disarticulated bones—often presenting with pervasive *perimortem* fractures—of a single individual, without signs of body manipulation. In a regional perspective, Lagoa Santa is unique in terms of the presence of these elaborate rituals; there is no other location in Brazil in which similar practices have been found for this time frame.

Lithic technology, zooarchaeology and multi-isotopic analyses from the early Holocene component in Lagoa Santa all corroborate models drawn from evolutionary ecology, which suggest that plant resources complemented by small to medium-sized animals would have been an important part of the dietary and subsistence strategies of the foraging societies of Neotropical South America (Kipnis 2002). The stability of lithic types in Lapa do Santo and elsewhere in the region were once considered as indicative of a low degree of cultural diversity over time. The shared use of a specific techno-functional strategy might not, however, be the best proxy of cultural identity. In Lapa do Santo, mortuary practices underwent transformation within relatively short periods of time, showing how a purely technologically oriented interpretation of the past can obliterate the elevated cultural diversity that had already been established in the region by the beginning of the Holocene.

Lagoa Santa also has a rich history of rock art, including one of the oldest securely dated instances on the continent. The ample diversity of forms and themes combines with the funerary record of the region to testify to the great cultural diversity of those groups. Due to the wide variety of motifs/styles and the lack of absolute chronology, the identification of synchronic and diachronic patterns is extremely challenging. Nevertheless, the motifs found in

Table 4 Results of pollen analysis for sediments collected from an early Holocene wall from the site Lapa do Santo (unit P11). Samples were sequentially collected from the surface at 6 cm intervals (vertical)

	Sample Number								
	1	2	3	4	5	6	7	8	9
Total Pollen Counted	196	254	215	38	2	2	1	0	0
Lycopodium spores counted	96	24	147	50	66	51	62	77	79
Lycopodium spores added	37,500	37,500	37,500	37,500	37,500	37,500	37,500	37,500	37,500
Sample weight (g)	27.3	23.0	29.2	24.8	27.8	29.9	31.0	33.1	28.6

Lagoa Santa clearly correspond with those found in other regions of Brazil, sometimes more than one thousand kilometres away, possibly indicating the existence of broad cultural networks during the Holocene.

Although pottery from this region has traditionally been neglected, sites such as Vereda III show the potential of Lagoa Santa for such studies. The ritual feast recorded in Vereda III has made significant advancements to knowledge of vessel function among groups that produce ASU pottery. Moreover, the presence of a great variety of pottery types and sizes makes Lagoa Santa relevant in the discussion surrounding the possibility that current categories of pottery are artificially partitioning the material culture produced by the same human group. Finally, with its rich postcolonial history, Lagoa Santa can shed light on the fate of local Amerindians, the local expression of the African Diaspora, as well as the more recent and intricate economic flux of foodstuffs and gold in the New World. For this, studies of historical archaeology in the region must expand and intensify to encompass an investigation of farms, slave houses, gold mines and ancient churches.

Acknowledgements We are grateful to Augusto Auler and Paulo Pessoa for inviting us to contribute this chapter and for their careful editorial revision. André Prous provided valuable comments on an earlier version of the manuscript. We thank all of the individuals who participated in the Lagoa Santa studies, José Hein for supporting our research on his farm, the IEF for providing us with accommodation and logistical support, the Brazilian authorities (IPHAN and IBAMA) for swiftly providing the necessary documentation, João Bárbara Filho for logistical assistance in the field and the cities of Matozinhos, Lagoa Santa and Pedro Leopoldo for facilitating our work. We would like to thank FAPESP (04/01321-6, 08/51747-0, 15/19405-6, 17/16451-2), CNPq (409474/2016-9, 152582/2018-5 and 435980/2018-1), the Alexander Von Humboldt Foundation, the Eberhard-Karls Universität Tübingen, the Konrad Lorenz Institute, Ciências Sem Fronteiras and the Max Planck Society for their financial support.

References

Agostini C (1997) Cachimbos de escravos e a reconstrução de identidades africanas no Rio de Janeiro. Bachelor thesis, Universidade Estácio de Sá

Angeles Flores R (2016) Before Lagoa Santa: micro-remain and technological analysis in a lithic artifact from the Itaparica industry. J Lithic Stud 3:6–29

Araujo AGM (2015) On vastness and variability: cultural transmission, historicity and the paleoindian records in eastern South America. An Acad Bras Ciênc 87:1239–1258

Araujo AGM, Neves WA, Piló LB (2002) Paleoindian subsistence and technology in central Brazil: results from new excavations at Boleiras rockshelter, Lagoa Santa. Curr Res Pleistocene 19:4–6

Araujo AGM, Neves WA, Piló LB, Atui JPV (2005) Holocene dryness and human occupation in Brazil during the 'Archaic Gap'. Quat Res 64:298–307

Araujo AGM, Feathers JK (2008) First notice of open-air Paleoamerican sites at Lagoa Santa: some geomorphological and paleoenvironmental aspects, and implications for future research. Curr Res Pleistocene 25:27–29

Araujo AGM, Feathers JK, Arroyo-Kalin M, Tizuka MM (2008) Lapa das Boleiras rockshelter: stratigraphy and formation processes at a paleoamerican site in Central Brazil. J Archaeol Sci 35:3186–3202

Araujo AGM, Pugliese F (2009) The use of non-flint raw materials by paleoindians in eastern South America: a Brazilian perspective. BAR International series

Araujo AGM, Neves WA (2010) Lapa das Boleiras - um sítio Paleoíndio do Carste de Lagoa Santa, MG, Brasil. Anna Blume / FAPESP, São Paulo

Araujo AGM, Pugliese F Jr (2010) A indústria lítica. In: Araujo AGM, Neves WA (eds) Lapa das Boleiras - um sítio Paleoíndio do Carste de Lagoa Santa, MG, Brasil. São Paulo, Anna Blume / FAPESP, pp 75–106

Araujo AGM, Neves WA, Kipnis R (2012) Lagoa Santa revisited: an overview of the chronology, subsistence, and material culture of paleoindian sites in eastern central Brazil. Lat Am Antiq 23:533–550

Araujo AGM, Strauss AM, Feathers JK, Paisani JC, Schrage TJ (2013) Paleoindian open-air sites in tropical settings: a case study in formation processes, dating methods, and paleoenvironmental models in central Brazil. Geoarchaeology 28:195–220

Araujo AGM, Pugliese FA, Santos RO, Okumura M (2017) Extreme cultural persistence in eastern-central Brazil: the case of Lagoa Santa Paleaeoindians. Anais da Academia Brasileira de Ciências 1–21

Araujo AGM, Piló LB (2017) Towards the development of a tropical geoarchaeology: Lagoa Santa as an emblematic case study. In: Da-Gloria P, Neves WA, Hubbe M (eds) Archaeological and paleontological research in Lagoa Santa: the quest for the first Americans. Springer

Auler AS, Smart PL, Wang X, Piló LB, Edwards RL, Cheng H (2009) Cyclic sedimentation in Brazilian caves: mechanisms and palaeoenvironmental significance. Geomorphology 106:142–153

Auler AS, Piló LB (2017) Peter Wilhelm Lund's scientific contributions. In: Da-Gloria P, Neves WA, Hubbe M (Eds.) Archaeological and paleontological research in Lagoa Santa: the quest for the first Americans. Springer

Auler AS (1999) Karst evolution and paleoclimate of eastern Brazil. PhD thesis, University of Bristol

Baeta A (1991) Arte rupestre do centro mineiro: a região arqueológica de Lagoa Santa. Rev História da UFOP 2:196–201

Baeta A (2011) Os grafismos rupestres e suas unidades estilísticas no carste de Lagoa Santa e Serra do Cipó - MG. PhD thesis, Universidade de São Paulo

Baeta A, Prous A (2017) The history of the studies of prehistoric rock paintings in the Lagoa Santa karst. In: Da-Gloria P, Neves WA, Hubbe M (eds) Archaeological and paleontological research in Lagoa Santa: the quest for the first Americans. Springer

Bányai M (1997) Minhas pesquisas arqueológicas na região de Lagoa Santa. Symbiose, Belo Horizonte

Bernardo DV, Strauss AM, Neves WA, Okumura M (2011) Measuring skulls, getting into the biological realm of the settlement of the New World. In: Vialou D (ed) Peuplements et préhistoire en Amériques. Éditions du Comité des travaux historiques et scientifiques, Paris, pp 31–41

Bernardo DV, Neves WA, Kipnis R (2017) The origins project and the first Americans's controversy. In: Da-Gloria P, Neves WA, Hubbe M (eds) Archaeological and paleontological research in Lagoa Santa: the quest for the first Americans. Springer

Binford LR (1982) The archaeology of place. J Anthropol Archaeol 1:5–31
Bissaro Jr MC (2008) Tafonomia como ferramenta zooarqueológica de interpretação: viés de representatividade óssea em sítios arqueológicos, paleontológico e etnográfico. MSc Dissertation, Universidade de São Paulo
Blake CC (1864) On human remains from a bone cave in Brazil. J Anthropol Soc London 2:cclxv–cclxvii
Boëda E, Clemente-Conte I, Fontugne M, Lahaye C, Pino M, Felice GD, Douville E (2014) A new late Pleistocene archaeological sequence in South America: the Vale da Pedra Furada (Piauí, Brazil). Antiquity 88:927–941
Brancante EF (1981) O Brasil e a cerâmica antiga. Companhia Lithográfica Ypiranga, São Paulo
Brochado JP (1989) A expansão dos Tupi e da cerâmica de tradição policrômica amazônica. Dédalo 27:65–82
Bueno L (2010) Tecnologia lítica, cronologia e sequência de ocupação: o estudo de um sítio a céu aberto na região de Lagoa Santa, MG. Rev do Mus Arqueol e Etnol 20:91–108
Bueno L (2012) Entre abrigos e lagoas: tecnologia lítica e territorialidade em Lagoa Santa (Minas Gerais, Brasil). Rev de Arqueol 25:62–83
Bueno L, Dias A, Steele J (2013) The late Pleistocene/early Holocene archaeological record in Brazil: a geo-referenced database. Quatern Int 30:74–93
Bueno L, Isnardis A (2017) Lithic technology in Lagoa Santa in the early Holocene. In: Da-Gloria P, Neves WA, Hubbe M (eds) Archaeological and paleontological research in Lagoa Santa: the quest for the first Americans. Springer
Burmeister H (1853) Reise nach Brasilien durch die Provinzen von Rio de Janeiro und Minas Geraes. Mit besonderer Rücksicht auf die Naturgeschichte der Gold und Diamantendistricte. Drick von Georg Reimer, Berlin
Burton R (1869) Explorations of the highlands of Brazil, with a full account of the gold and diamond mines. Tinsley Brothers, London
Cartelle C (1994) Tempo passado: mamíferos fósseis em Minas Gerais. Editora Palco, Belo Horizonte
Chim EN (2018) Zooarqueologia da Lapa Grande de Taquaraçu. Master Dissertation—Museum of Archaeology and Ethnology of University of São Paulo
Conlee CA (2007) Decapitation and rebirth – A headless burial from Nasca, Peru. Curr Anthropol 48:438–445
Cordy-Collins A (1992) Archaism or tradition?The decapitation theme in Cupisnique and Moche iconography. Lat Am Antiq 3:206–220
Cornero S (2005) Biologia esqueletal de los primeros americanos: un caso de estudio en Santana do Riacho I, Brasil central. PhD thesis, Universidad Nacional de La Plata
Corrêa AA (2014) Pindorama de Mboîa e Îakaré. Continuidade e mudança na trajetória das populações Tupi. PhD Thesis, Universidade de São Paulo
Correal UG (1990) Aguazuque evidencias de cazadores, recolectores y plantadores em la altiplanice de la cordillera oriental. Fundación de Investigaciones Arqueológicas Nacionales. Banco de la República, Bogotá
Correal UG (2001) Patrones mortuorios em cazadores recolectores del Pleistoceno y Holoceno en Colombia. Revista Chungará (Arica) 33
Costa FWS (2017) The Minas Gerais Academy of Sciences: Lund's inheritors. In: Da-Gloria P, Neves WA, Hubbe M (eds) Archaeological and paleontological research in Lagoa Santa: the quest for the first Americans. Springer
Chim E (2018) Zooarqueologia da Lapa Grande de Taquaraçu. MSc thesis, Universidade de São Paulo
Cunha FLS, Guimarães ML (1978) Posição geológicado homem de Lagoa Santa no grande abrigo de Lapa Vermelha Pedro Leopoldo – Estado de Minas Gerais. In: Coletânea de Estudos em Homenagem a Annette Laming-Emperaire. Museu Paulista da Universidade de São Paulo
Da-Gloria P (2012) Health and lifestyle in the Paleoamericans: early Holocene biocultural adaptation at Lagoa Santa, central Brazil. PhD thesis, Ohio State University
Da-Gloria P, Larsen C (2014) Oral health of the Paleoamericans of Lagoa Santa, central Brazil. Am J Phys Anthropol 154:11–26
Da-Gloria P, Oliveira RE (2017) History of research into health and lifestyle in Lagoa Santa. In: Da-Gloria P, Neves WA, Hubbe M (eds) Archaeological and paleontological research in Lagoa Santa: the quest for the first Americans. Springer
Da-Gloria P, Oliveira RE, Neves WA (2017a) Dental caries at Lapa do Santo, central-eastern Brazil: an early Holocene archaeological site. An Acad Bras Ciênc 89:307–316
Da-Gloria P, Larsen CS (2017) Subsisting at the Pleistocene/Holocene boundary in the New World: a view from the Paleoamerican mouths of central Brazil. PaleoAmerica 3:101–121
Da-Gloria P, Neves WA, Hubbe M (2017) Archaeological and paleontological research in Lagoa Santa: the quest for the first Americans. Springer
Da-Gloria P, Hubbe M, Neves WA (2018) Lagoa Santa's contribution to the origins and life of early Americans. Evol Anthropol 27:121–133
Da-Gloria P, Bueno L (2018) Resilience among hunter-gatherers of Lagoa Santa, Central-Eastern Brazil. In Temple DH, Stojanowski CM (Orgs) Hunter-gatherer adaptation and resilience: a bioarchaeological perspective 141–167
Da Silva LV, Prous A (2014) O papel dos resíduos de combustão na formação dos estratos sedimentares na Lapa do Niactor: o elemento antrópico como agente dominante na sedimentação em abrigos. Arquivos do Museu de História Natural e Jardim Botânico da UFMG 23:105–139
Delibrias G, Guillier M-T, Labeyrie J (1986) GIF natural radiocarbon measurements X. Radiocarbon 28:9–68
Damasceno C (2011) Arraiais e vilas d'el rei - Espaço e poder nas Minas setecentistas. Editora UFMG, Belo Horizonte
Debret JB (1839) Voyage pittoresque et historique au Brésil. Martins e Edusp, São Paulo
Dias AS, Bueno L (2013) The initial colonization of South America eastern lowlands: Brazilian archaeological contributions to settlement of America. In Graf K, Ketron CV, Waters MR (Orgs) The Paleoamerican Odyssey. Santa Fe: Center for the Study of the First Americans 339–357
Feathers J, Kipnis R, Piló LB, Arroyo-Kalin M, Coblentz D (2010) How old is Luzia? Luminescence dating and stratigraphic integrity at Lapa Vermelha, Lagoa Santa, Brazil. Geoarchaeology 25:395–436
Ferguson RB (1990) Blood of the Leviathan: Western contact and warfare in Amazonia. Am Ethnol 17:237–257
Fernandes HLA (2003) Os sepultamentos do sítio Aratu de Piragiba. MSc thesis, Universidade Federal da Bahia
Fernandes HLA (2011) As lâminas de machado lascadas Aratu de Piragiba-BA. PhD thesis, Universidade Federal da Bahia
Forline LC (1997) The persistence and cultural transformation of the Guajá Indians: foragers of Maranhão State. University of California, Brazil
Galland M, Friess M (2016) A three-dimensional geometric morphometrics view of the cranial variation and population history in the New World. Am J Hum Biol 28:646–661
Gardiman GG, Rodrigues IMM, Cascon LM, Isnardis A (2014) Vereda III e a preparação do cauim. Arquivos do Museu de História Natural 23:66–106

Gardiman GG, Rodrigues IMM, Cascon LM, Isnardis A (2016) A morfologia dos vasos Jê na produção de cauim de milho em Vereda III: uma proposição. Rev Mus Arqueologia Etnologia USP 27:111–120

Gaspar MD (2009) Arqueologia, cultura material e patrimônio. Sambaquis e cachimbos. In: Granato M, Rangel MF (eds) Cultura material e patrimônio. Museu de Astronomia e Ciências Afins, Rio de Janeiro, pp 39–52

Haddad EA (2004) Evolução geo-histórica da Fazenda Jaguara - Matozinhos - MG. BSc monograph, Universidade Federal de Minas Gerais

Hansen S (1888) En anthropologisk undersogelse af jordfundne menneskelevninger fra Brasilianske huler. Med et Tillaeg om det Jordfundne Menneske fra Pontimelo Rio de Arrecifes, La Plata. Museo Lundii 1:1–34

Henriques Jr G (2006) Arqueologia regional da província cárstica do Alto São Francisco: um estudo das tradições ceramistas Una e Sapucaí. MSc thesis, Universidade de São Paulo

Hermenegildo T (2009) Reconstituição da dieta e dos padrões de subsistência das populações pré-históricas de caçadores-coletores do Brasil central através da ecologia isotópica. MSc thesis, Universidade de São Paulo

Holten B, Sterll M (2017) Peter Wilhelm Lund: life and work. In: Da-Gloria P, Neves WA, Hubbe M (eds) Archaeological and paleontological research in Lagoa Santa: the quest for the first Americans. Springer

Hrdlička A (1912) Early man in South America. Bur Am Ethnol Bull 52. Government Printing Office, Washington

Hubbe A, Haddad-Martim PM, Hubbe M, Mayer EL, Strauss A, Auler AS, Piló LB, Neves WA (2011) Identification and importance of critical depositional gaps in pitfall cave environments: the fossiliferous deposit of Cuvieri Cave, eastern Brasil. Palaeogeogr Palaeoclimatol Palaeoecol 312:66–78

Hubbe A, Hubbe M, Karmann I, Cruz FW, Neves WA (2013a) Insights into Holocene megafauna survival and extinction in southeastern Brazil from new AMS 14C dates. Quat Res 79:152–157

Hubbe A, Hubbe M, Neves WA (2013b) The Brazilian megamastofauna of the Pleistocene/Holocene transition and its relationship with the early human settlement of the continent. Earth Sci Rev 118:1–10

Hubbe M, Strauss A, Hubbe A, Neves WA (2015) Early South Americans cranial morphological variation and the origin of American biological diversity. PLoS ONE 10:1–17

Humphrey LT, De Groote I, Morales J, Barton N, Collcutt S, Bronk Ramsey C, Bouzouggar A (2014) Earliest evidence for caries and exploitation of starchy plant foods in Pleistocene hunter-gatherers from Morocco. Proc Natl Acad Sci USA 111:954–959

Hurt W, Blasi O (1969) O projeto arqueológico Lagoa Santa – Minas Gerais, Brasil (nota final). Arq do Mus Parana 4:1–63

Imbelloni J (1938) Tabla classificatoria de los indios. Regiones biológicas y grupos raciales humanos de América. Physis 12:229–249

Jandial R, Hughes SA, Aryan HE, Marshall LF, Levy ML (2004) The science of shrinking human heads: tribal warfare and revenge among the South American Jivaro-Shuar. Neurosurgery 5:1215–1222

Junqueira PA, Malta IM (1978) Sítios cerâmicos da região de Lagoa Santa. Arquivos do Museu de História Natural da UFMG 3:117–162

Keuller, ATAM (2017) The Anthropological Studies of Lagoa Santa in the National Museum of Rio de Janeiro: Insertion, Debates, and Scientific Controversies at the Turn of the Nineteenth to the Twentieth Century. In: Da-Gloria P, Neves WA, Hubbe M (eds) Archaeological and paleontological research in Lagoa Santa: the quest for the first Americans. Springer

Kipnis R (2002) Foraging societies of eastern central Brazil: an evolutionary ecology study of subsistence strategies during the terminal Pleistocene and early/middle Holocene. PhD thesis, University of Michigan

Kipnis R, Bissaro Jr MC, Prado HM (2008) Os restos faunísticos. In: Araujo AGM, Neves WA (Eds.) Lapa das Boleiras - Um sítio paleoíndio do carste de Lagoa Santa, MG, Brasil. Anna Blume, São Paulo

Kollman J (1884) Schadeln von Lagoa Santa. Z Ethnol 16:194–199

Koole E (2015) Entre as tradições planálticas e meridionais: caracterização arqueológica dos grupos caçadores coletores a partir da análise de sete elementos e suas implicações para a ocupação pré-cerâmica da Região Cárstica do Alto São Francisco, Minas Gerais, Brasil. PhD thesis, Universidade de São Paulo

Lacerda F, Peixoto R (1876) Contribuições para o estudo antropológico das raças indígenas do Brazil. Arch do Mus Nac do Rio Janeiro 1:47–79

Laming-Emperaire A (1979a) Missions archéologiques franco-brésiliennes de Lagoa Santa, Minas Gerais, Brésil – Le Grand abri de Lapa Vermelha. Revista de Pré-história 1:53–89

Laming-Emperaire A (1979b) Missions archéologiques franco-brésiliennes de Lagoa Santa, Minas Gerais, Brésil – Le Grand abri de Lapa Vermelha. Rev Pré-história 1:53–89

Laming-Emperaire A, Prous A, Moraes ÁV, Beltrão MC (1975) Grottes et abris de la region de Lagoa Santa, Minas Gerais, Bresil - Premier rapport de la mission archéologique Franco-Brésilienne de Lagoa Santa. Cahiers D'Archeólogie d'Amerique du Sud, 1. École Pratique des Hautes Etudes Vie Section: Sciences Economiques et Sociales, Paris

Llagostera AM (2003) Patrones de momificación Chinchorro em las colecciones Uhle y Nielsen. Revista Chungará (Arica) 35:5–22

Lourdeau A (2015) Lithic technology and prehistoric settlement in Central and Northeast Brazil: definition and spatial distribution of the Itaparica technocomplex. PaleoAmerica 1:52–67

Luna P (2007) Peter Wilhelm Lund: o auge das suas investigações científicas e a razão para o término das suas pesquisas. PhD thesis, Universidade de São Paulo

Lund PW (1844) Notícia sobre ossadas humanas fósseis achadas numa caverna do Brasil. In: Paula Couto P (ed) Memórias sobre a paleontologia brasileira. Instituto Nacional do Livro, Rio de Janeiro

Lyman RL, Savelle JM, Whitridge P (1992) Derivation and application of a meat utility index for phocid seals. J Archaeol Sci 19:531–555

Machado JR (2011) Panelas de barro: aspectos técnicos e sociais de um fazer tradicional. Undergraduate thesis. Universidade Federal de Minas Gerais

Mello e Alvim MC (1977) Os antigos habitantes da área arqueológica de Lagoa Santa, Minas Gerais - estudo morfológico. Arq do Mus História Nat 2:119–174

Mingatos GS (2017) Caça ou deixa passar? A dieta dos grupos humanos do sítio Lapa do Santo, Lagoa Santa, Minas Gerais. Master Dissertation Universidade Federal de Minas Gerais

Mingatos GS, Okumura M (2016) Modelo de amplitude de dieta aplicada a restos faunísticos do sítio Lapa do Santo (MG) e suas implicações para o entendimento da dieta em grupos paleoíndios do Brasil central. Paleoindian Archaeology 1

Monteiro J (1994) Negros da terra: índios e bandeirantes nas origens de São Paulo. Companhia das Letras, São Paulo

Moreno-Mayar et al (2018) Early human dispersal within the Americas. Science 2621:1–27

Moreno de Sousa JC (2014) Cognição e cultura no mundo material: os Itaparicas, os Umbus e os "Lagoassantenses". MSc thesis, Universidade de São Paulo

Moreno de Sousa JC, Araujo AGM (2018) Microliths and polished stone tools during the Pleistocene-Holocene transition and early

Holocene in South America: the Lagoa Santa lithic industry. PaleoAmerica 4:219–238

Moura W (2016) Método de localização de sítios arqueológicos de mineração colonial de ouro nas margens do Rio das Velhas com uso de imagem de satélite. Rev Arqueol 29:202–214

Neto VVG, Rodrigues-Carvalho C (2017) The physical anthropology archives of the National Museum of Rio de Janeiro: Lagoa Santa in the first half of the twentieth century. In: Da-Gloria P, Neves WA, Hubbe M (eds) Archaeological and paleontological research in Lagoa Santa: the quest for the first Americans. Springer

Neves WA, Cornero S (1997) What did South American Paleoindians eat? Curr Res Pleistocene 14:93–96

Neves WA, Kipnis R (2002) A late-paleoindian secondary burial from Lagoa Santa, Minas Gerais, Brazil. Curr Res Pleistocene 19:83–85

Neves WA, Piló LB (2003) Solving Lund's dilemma: new AMS dates confirm that humans and megafauna coexisted at Lagoa Santa. Curr Res Pleistocene 20:57–60

Neves WA, Kipnis R (2004) Further evidence of a highly cariogenic diet among late Paleoindians of Central Brazil. Curr Res Pleistocene 21:81–83

Neves WA, Hubbe M (2005) Cranial morphology of early Americans from Lagoa Santa, Brazil: implications for the settlement of the New World. Proc Natl Acad Sci USA 102:18309–14

Neves WA, Araujo AGM, Bernardo DV, Kipnis R, Feathers JK (2012) Rock art at the Pleistocene/Holocene boundary in Eastern South America. PLoS ONE 7:e32228

Neves WA, Hubbe M, Bernardo D, Strauss AM, Araujo AGM, Kipnis R (2013) Early human occupation of Lagoa Santa, eastern-central Brazil: craniometric variation of the initial settlers of South America. In: Graf K, Ketron C, Water M (eds) Paleoamerican Odyssey. Texas A&M University Press, Austin, pp 397–412

Noelli FS (1996) As hipóteses sobre o centro de origem e rotas de expansão dos Tupi. Rev Antropol 39:7–53

Noelli FS (1998) The Tupi: explaining origin and expansions in terms of archaeology and of historical linguistics. Antiquity 72:648–663

Oliveira R, Da-Gloria PJT, Neves WA (2018) A percepção do meio ambiente por parte da população atual de Lagoa Santa e suas implicações para a arqueologia regional. Revista de Arqueologia 31:104–130

Paiva ZC (2015) 'Uma baforada sim sinhô': cachimbos de escravos e as dinâmicas socioculturais da Diamantina oitocentista. MSc thesis, Universidade Federal do Piauí

Panja S (2003) Mobility strategies and site structure: a case study of Inamgaon. J Anthropol Archaeol 22:105–125

Parry W, Kelly RL (1987) Expedient core technology and sedentism. In: Johnson JK, Morrow M (eds) The organization of core technology. Westview Press, Boulder, pp 285–304

Perego UA, Achilli A, Angerhofer N, Accetturo M, Pala M, Olivieri A, Hooshiar Kashani B, Ritchie KH, Scozzari R, Kong QP et al (2009) Distinctive Paleo-Indian migration routes from Beringia marked by two rare mtDNA haplogroups. Curr Biol 19:1–8

Perez SI, Lema V, Diniz-Filho JAF, Bernal V, Gonzalez PN, Gobbo D, Pucciarelli HM (2011) The role of diet and temperature in shaping cranial diversification of South American human populations: an approach based on spatial regression and divergence rate tests. J Biogeogr 38:148–163

Perez SI, Monteiro LR (2009) Nonrandom factors in modern human morphological diversification: a study of craniofacial variation in southern South American populations. Evolution 63:978–993

Piló LB, Auler AS, Neves WA, Wang X, Cheng H, Edwards RL (2004) Revisitando a Lapa do Sumidouro: marco paleoantropológio do quaternário americano. Rev Bras Paleontol 7:337–348

Piló LB, Auler AS, Neves WA, Wang X, Cheng H, Edwards RL (2005) Geochronology, sediment provenance, and fossil emplacement at Sumidouro Cave, a classic late Pleistocene/early Holocene Paleoanthropological site in eastern Brazil. Geoarchaeology 20:751–764

Piló LB, Auler AS (2002) Apresentação: bicentenário de Peter Wilhelm Lund (1801–1880). O Carste 14:4–7

Piló LB, Neves WA (2003) Novas datações ^{14}C (AMS) confirmam a tese da coexistência do homem com a megamastofauna pleistocênica na região cárstica de Lagoa Santa, MG. In Annals of the XXVIIIth Brazilian Speleology Congress, pp 100–104. Sociedade Brasileira de Espeleologia (SBE), Januária

Piló LB (1998) Morfologia cárstica e materiais constituintes: dinâmica e evolução da depressão poligonal Macacos-Baú - Carste de Lagoa Santa, MG. PhD Thesis. Universidade de São Paulo

Politis GG, Angriazani RC, Hernando A, Ruibal AG, Coelho EB (2013) Quarenta e oito saídas: etno-arqueologia das saidas diárias de forrageio dos Awá. In: Hernando A, Coelho BEM (Eds) Estudos sobre os Awá – Caçadores-coletores em transição: IWGIA y EDUFMA 155–185

Posth C et al (2018) Reconstructing the deep population history of Central and South America. Cell 175:1185–1197

Prado HM (2009) O impacto da caça versus a conservação de primatas numa comunidade indígena Guajá. Master thesis. Universidade de São Paulo

Prous A, Baeta A, Rubbioli E (2003) O patrimônio arqueológico da região de Matozinhos: conhecer para proteger. Ed. do autor, Belo Horizonte

Prous A, Fogaça E (1999) Archaeology of the Pleistocene-Holocene boundary in Brazil. Quat Int 53–54:21–41

Prous A, Fogaça, Ribeiro L (1998) Patrimônio arqueológico. APA de Lagoa Santa - MG. (IBAMA/CPRM), Belo Horizonte

Prous A, Lima TA (2008) Os ceramistas tupiguarani. Volume 1: sínteses regionais. Sigma, Belo Horizonte

Prous A (1977) Missão de estudo da arte rupestre de Lagoa Santa. Arquivos do Museude História Natural da UFMG 2:51–65

Prous A (1986) L'Archéologie au Brésil. 300 Siécles d'occupation humaine. L'anthropologie 90:257–306

Prous A (1991) Artefatos de cerâmica, argila, osso, chifre, dente, vegetal e concha. Arquivos do Museu de História Natural e Jardim Botânico

Prous A (1992) Arqueologia brasileira. Brasilia: Univesidade de Brasilia, Brasilia

Prous A (1992/1993) As estruturas aparentes: os sepultamentos do Grande Abrigo de Santana do Riacho, os sepultamentos da escavação Nº1. Arquivos do Museu de História Natural e Jardim Botânico

Prous A (2017) Archaeological missions to the Lagoa Santa region in the second half of the twentieth century. In: Da-Gloria P, Neves WA, Hubbe M (eds) Archaeological and paleontological research in Lagoa Santa: the quest for the first Americans. Springer

Pugliese F (2007) Os líticos de Lagoa Santa: um estudo sobre organização tecnológica de caçadores-coletores do Brasil Central. MSc thesis, Universidade de São Paulo

Raczka MF (2017) Timining and consequences of Pleistocene megafaunal population collapse in South America. Doctoral Thesis, Florida Institute of Technology

Raczka MF, Bush MB, Folcik AM, McMichael CH (2016) *Sporormiella* as a tool for detecting the presence of large herbivores in the Neotropics. Biota Neotrop 16:e20150090

Raczka MF, Bush MB, Oliveira PE (2017) The collpase of megafaunal populations in southeastern Brazil. Quaternary International 1–16

Raghavan et al. (2015) Genomic evidence for the Pleistocene and recent population history of Native American. Science 349: aab3884

Reinhardt J (1888) De brasilianske Knoglehuler og de I dem forekommende Dyrelevninger. En Samling af Afhandlinger. E Mus Lundii 4:1–56

Reitz EJ, Wing ES (2008) Zooarchaeology. Cambridge University Press, New York

Rivet P (1942) Les Origines de l'Homme Américain. Les Éditions l'Arbre, Montreal

Rodrigues IMM (2011) Fora das grandes aldeias: a ocupação do recôndito sítio arqueológico Vereda III. MSc thesis, Universidade Federal de Minas Gerais

Rodrigues IMM (2014) O Sítio Vereda III: uma ocupação de grupos ceramistas e horticultores fora das grandes aldeias. Arquivos do Museu de História Natural da UFMG 23:16–65

Rodrigues IMM, Volkmer-Ribeiro C, Machado VS (2017) Cauixi em cerâmica arqueológica da região de Lagoa Santa, Minas Gerais: Inclusão de esponjas processadas ou exploração de depósitos sedimentares com espículas? Bol Mus Para Emílio Goeldi Ciências Humanas 12:85–100

Santoro CM, Standen VG, Arriaza BT, Dillehay TD (2005) Archaic funerary pattern or postdepositional alteration? The Patapatane burial in the higlands of South Central Andes. Lat Am Antiq 16:329–346

Santos RO (2011) As tecnologias esqueletais: uma investigação sobre o uso de matérias-primas de origem esqueletal por meio de análise comparativa entre coleções arqueológicas e etnográficas. MSc thesis, Universidade de São Paulo

Schiffer MB (1972) Archaeological context and systemic context. Am Antiq 37:156–165

Snow C, Abreu J (1977) A cerâmica neobrasileira em regiões vizinhas a Belo Horizonte - MG: um estudo da produção atual. Arq do Mus História Nat Belo Horiz 175–191

Solari A, daSilva SFSM (2017) Sepultamentos secundários com manipulações intencionais no Brasil: um estudo de caso no sítio arqueológico Pedra do Cachorro Buíque, Pernambuco Brasil. Bol do Mus Para Emilio Goeldi Ciencias Humanas 12:135–155

Souza SM, Liryo A (2017) The National Museum's Contributions to Lagoa Santa Research in the second half of the twentieth century. In: Da-Gloria P, Neves WA, Hubbe M (eds) Archaeological and paleontological research in Lagoa Santa: the quest for the first Americans. Springer

Souza MAT, Agostini C (2012) Body marks, pots and pipes: some correlations between African scarifications and pottery decoration in eigtheenth- and nineteenth-century Brazil. Historical Archaeology 46:102–123

Strauss A (2008) Early Holocene human occupation around a karstic lake of the Lagoa Santa region – MG: an interdisciplinary approach. Bachelor thesis, Universidade de São Paulo

Strauss A (2010) As práticas mortuárias dos caçadores-coletores pré-históricos da região de Lagoa Santa (MG): um estudo de caso do sítio arqueológico 'Lapa do Santo'. MSc thesis. Universidade de São Paulo

Strauss A (2011) As práticas mortuárias dos caçadores coletores pré-históricos da região de Lagoa Santa (MG): um estudo de caso do sítio arqueológico Lapa do Santo. Revista Brasileira de Arqueologia da SAB 24:136–139

Strauss A (2012) Interpretative possibilities and limitations of the Saxe/Goldstein hypothesis. Bol do Mus Para Emilio Goeldi Ciencias Humanas 7:525–546

Strauss A (2015) As práticas mortuárias dos primeiros sul-americanos. Arquivos do Museu de História Natural e Jardim Botânico da UFMG 23:89–154

Strauss A (2016a) Os padrões de sepultamento do sítio arqueológico Lapa do Santo (Holoceno Inicial, Brasil). Bol do Mus Para Emilio Goeldi Ciencias Humanas 11:243–276

Strauss A (2016b) As práticas mortuárias da região de Lagoa Santa: história das pesquisas. In: Da-Gloria P, Hubbe M, Neves WA (eds) A história das pesquisas na região de Lagoa Santa. Anna-Blume

Strauss A (2017) Burial practices in the Lagoa Santa region: history of research. In: Da-Gloria P, Hubbe M, Neves WA (eds) Archaeological and paleontological research in Lagoa Santa. Springer

Strauss A, Hubbe M (2010) Craniometric similarities within and among human populations. J Hum Biol 82:1–29

Strauss A, Koole Edward, de-Oliveira R, Da-Gloria PT, Nunes T, Robazzini A, Walter F, Neves W (2011) Two directly dated Early Archaic burials from Pains, State of Minas Gerais, Brazil. Curr Res Pleistocene 28:123–125

Strauss A, Hubbe M, Bernardo DV, Neves WA, Atui JPV (2015a) The cranial morphology of the Botocudo indian, Brazil. Am J Phys Anthropol 157:202–216

Strauss A, Oliveira RE, Bernardo DV, Salazar-García DC, Talamo S, Jaouen K, Hubbe M, Black S, Wilkinson C, Richards MP, Araujo AGM, Kipnis R, Neves WA (2015b) The oldest case of decapitation in the New World (Lapa do Santo east-central Brazil). PlosOne 10:e0137456

Strauss A, Oliveira RE, Villagran XS, Bernardo DV, Salazar-García DC, Bissaro MC, Pugliese F, Hermenegildo T, Santos R, Barioni A, de Oliveira EC, Moreno de Sousa JC, Jaouen K, Ernani M, Hubbe M, Inglez M, Gratão M, Rockwell H, Machado M, de Souza G, Chemale F, Kawashita K, O'Connell TC, Israde I, Feathers J, Campi C, Richards M, Wahl J, Kipnis R, Araujo A, Neves W (2016) Early Holocene ritual complexity in South America: the archaeological record of Lapa do Santo (east-central Brazil). Antiquity 90:1454–1473

Strauss A, Oliveira RE (2018a) A prática de individualização de crânios e de decapitação na região de Lagoa Santa durante o Holoceno inicial (Brasil). Revista do Museu de Arqueologia e Etnologia da USP 28:86–108

Strauss A, Oliveira RE (2018b) Early Holocene mortuary practices in Lagoa Santa region: a 9600-9495 years cal BP case of perimortem body manipulation from Lapa do Santo (east-central Brazil). Clio Arqueológica 33:138–183

Strauss A, Oliveira RE, Gratão M, daCosta A, Fogaça E, Boëda E (2018) Human skeletal remains from Serra da Capivara, Brasil: Review of the available evidence and report on new findings. In: Harvati K, Jäger G, Rezes-Centeno H (Orgs). New perspectives on the peopling of the New World. Kerns Verlag

Ströhle A, Hahn A (2011) Diets of modern hunter-gatherers vary substantially in their carbohydrate content depending on coenvironments: results from an ethnographic analysis. Nutrition Research 31: 429-435

Ten Kate H (1885) Sur les Crânes de Lagoa Santa. Bull Soc Anthr 8:240–244

Verano JW, Uceda S, Chapdelaine C, Telllo R, Paredes MI, Pimentel V (1999) Modified human skulls from the urban sector of the pyramids of Moche, northern Peru. Latin American Antiquity 10:59–70

Vergne C, Amâncio S (1992) A necrópole pré-histórica do Justino/Xingó-Sergipe: nota prévia. Clio 1:171–182

Vergne C (2002) Estruturas funerárias do sítio Justino: distribuição no espaço e no tempo. Revista Canindé 2(251):273

Villagran X, Strauss A, Miller C, Ligouis B, Oliveira R (2017) Buried in the ashes: formation processes of an early South American's site (Lapa do Santo, Brazil). J Archaeol Sci 77:10–34

Villagran X, Strauss A, Alves M, Oliveira R (2019) Virtual micromorphology: the application of micro-CT scanning for the identification of termite mounds in archaeological sediments. J Archaeol Scie Reports 24:785–795

Von Cramon-Taubadel N, Strauss A, Hubbe M (2017) Paleoamerican cranial variation in global microevolutionary perspective: Implications for the settlement of the Americas. Sci Adv 3:e1602289

Walter HV, Cathoud A, Mattos A (1937) The Confins Man. A contribution to the study of Early Man in South America. In: MacCurdy GG, editor. Early man—as depicted by leading authorities at the international symposium the Academy of Natural Sciences Phyladelphia. J. B. Lippincott Company, London

Walter HV (1958) Arqueologia da região de Lagoa Santa. SEDEGRA, Rio de Janeiro. Sedegra, Rio de Janeiro

Wied-Neuwied M (1820) Reise nach Brasilien in den jahren 1815 bis 1817. Henry Colburn & Co, London

Environmental Problems in the Lagoa Santa Karst

Paulo Pessoa, Dora Atman, and Gisele Kimura

Abstract

Evolution of any karst environment will inevitably undergo different levels of changes related to two main factors: natural evolution processes and anthropogenic interventions. Numerous and recurring environmental problems have been identified in the Lagoa Santa Karst, such as deforestation, water pollution, damage to paleontological and archeological records, uncontrolled use of groundwater causing modifications to water dynamics, and geotechnical problems. The majority of these problems are consequences of human occupation characterized in the region by deficient sanitation infrastructure, unplanned land occupation, lack of institutional articulation and empowerment, and industrial and mining activities. In karst regions, these issues become even more critical since natural fragility of the environment is higher than that in any other system due to the inherent dissolution of carbonate rocks and its associated hydrological processes. Anthropogenic occupation processes doubled the urbanized area in the karst region of Lagoa Santa within the last 20 years, maintaining and increasing some of the environmental problems that have been identified since the 1990s. These issues have prevailed during the urbanization history of the Lagoa Santa Karst and seem to have no endpoint because the stakeholders have not provided any straightforward plans to establish specific environmental directives. All of these factors highlight the need to intensify social and educational actions at all levels to emphasize the importance of the karst environment and its associated fragility as the root of all the environmental problems. The population must understand their importance and role in preserving natural resources and how significant their presence is for the overall sustainability of the karst environment and its natural attribute; otherwise, no other action will be efficient enough for preservation of the karst environment.

Keywords

Karst environment • Land occupation • Environmental impacts • Groundwater management

1 Introduction

The Lagoa Santa Karst (LSK) exhibits a set of natural and historical attributes of great scientific and cultural interest. This portion of the territory of Minas Gerais State (MG) is characterized as one of the most important Brazilian regions in terms of the karst landscape and natural history (Berbert-Born 2002). Archeological sites found in the Lagoa Santa Karst record the occupation of man for the last ten thousand years, where human groups found not only shelter but also water, mineral resources, and diverse fauna, allowing survival and cultural development (Neves and Piló 2008; Cartelle 2012; Prous 2012). The archeological and paleontological findings in this region have worldwide relevance, contributing to discussions about the mode of human occupation of the Americas (Da-Gloria et al. 2017). In the nineteenth to twenty-first centuries, it is also evident that the geological and geomorphological context was the main driver for the occupation of the Lagoa Santa Karst, which was suitable for mining (especially limestone, sand, and clay exploitation) and farming activities with water and accessible facilities.

Currently, the natural conditions and the strategic location in the northern region of Belo Horizonte City (Minas Gerais State) continue to establish a favorable context for

P. Pessoa (✉) · D. Atman · G. Kimura
Hidrovia Hidrogeologia e Meio Ambiente, Rua Albita 131/301, Belo Horizonte, MG CEP 30310-160, Brazil
e-mail: ppessoa@hidrovia.com.br

D. Atman
e-mail: dora@hidrovia.com.br

G. Kimura
e-mail: gisele@hidrovia.com.br

A. S. Auler and P. Pessoa (eds.), *Lagoa Santa Karst: Brazil's Iconic Karst Region*, Cave and Karst Systems of the World, https://doi.org/10.1007/978-3-030-35940-9_13

occupation. Belo Horizonte is the sixth-largest city in Brazil (IBGE 2010) with a metropolitan region that encompasses 34 municipalities, forming the third-largest urban complex in the country, with an estimated population of 5,313,480 (IBGE 2018). Within recent decades, the northern portion of the metropolitan region of Belo Horizonte has undergone transformations that have driven urban growth, such as the expansion of Confins International Airport, improvements in the road network, new industrial district, and construction of the Administrative Center of the state of Minas Gerais. Altogether, the nine municipalities in the Lagoa Santa Karst area have an estimated population of 570,000 inhabitants.

Human activities developed on a karst substrate related to urban and rural environments affect, in some way, the conditions that maintain the equilibrium of the system (Fig. 1). The main results of karst system degradation are landform and cave destruction, impoverished ecology, air, soil and water pollution, and damage to paleontological and archeological sites. Karst landscape is formed by the interaction of geological and climatic factors over thousands of years, and water is an essential agent for its development. In karst development, there is a strong interaction between surface and groundwater, with features such as dolines, sinks, caves, and karst valleys serving as recharge or discharge areas for water flowing over the surface or through underground conduits carved into limestone. A higher velocity of flows is typical in karstified limestone aquifers, and water movement may occur in different directions. Karst systems are sensitive to changes in the runoff and infiltration patterns. Adjustments to flow can naturally occur in events of concentrated rainfall in a short period. However, persistent changes are common because of alterations in the drainage system, reservoir construction, aquifer exploitation, mine dewatering, irrigation, soil layer reduction, and vegetation removal, among others. Therefore, subsidence, collapse, and cracks in the ground can be induced by these activities.

Impacts on water dynamics may increase the risk of water contamination, mainly when it is followed by improper waste disposal and pollution of soil, and indirect environmental impacts can be seen on plants, animals, and human health. Similarly, mining activities can interfere with water availability and landscape, which often manifests itself as

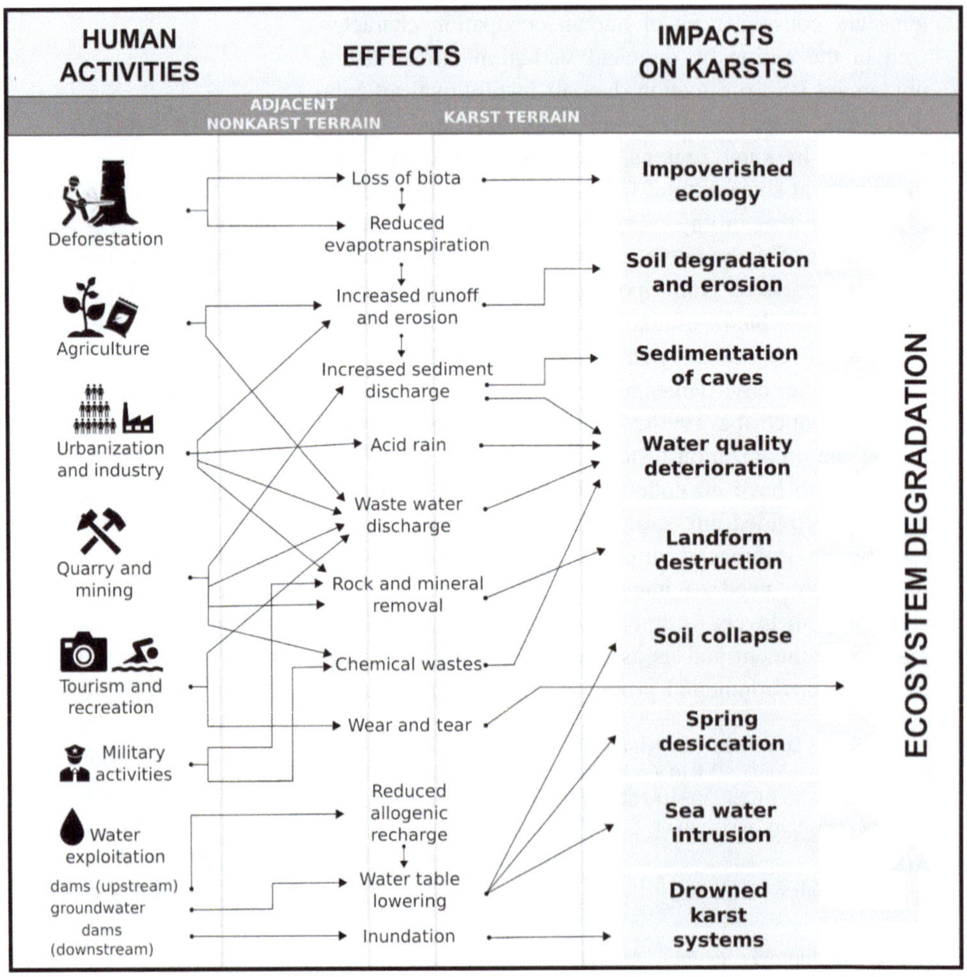

Fig. 1 Human activities, effects, and impacts on karst environments. From Williams (1993)

irreparable losses to caving structures, ecosystems and paleontological and archeological records.

Ford and Williams (2007) pointed out that the environmental changes induced by the human presence in karst ecosystems can be transmitted over large distances in relation to the initial point of the impacts due to the hydrological processes that advance widely through the system of underground flows. In this way, the effects are not obvious or perceived until changes are observed at an advanced stage. Activities in adjacent, upstream nonkarst areas often have an impact on karst as a result of transmission by allogenic drainage into karst (Ford and Williams 2007).

The Lagoa Santa Karst has experienced many of the environmental problems mentioned in Fig. 1, although they occur at different intensities depending on the land use and geological and geomorphological conditions. Several studies have emphasized environmental problems of the Lagoa Santa region and Environmental Protection Area (APA) Lagoa Santa Karst (Pessoa 1996; Souza 1997; Herrmann et al. 1998; Piló 1999; Meneses 2003; Queiroz et al. 2007; Alt 2008; Garcia 2011; Tayer 2016; Auler 2016 as well as several others). Historical records show that after the beginning of the processes of deforestation, intensification of agriculture and anthropization of physical space by urbanization, the first impacts on karst ecosystems are triggered (Williams 1993).

The Lagoa Santa Karst Environmental Protection Area (APA) was created by the Federal Government in 1990, through Decree 98,881, to protect one of the most important archeological and speleological sites in the country and its scientific wealth and culture, regulate human activities and ensure the sustainable use of natural resources without the need of land expropriation. The main product was the Environmental Zoning of the Lagoa Santa Karst APA (Souza 1997), which aims to regulate activities that are compatible with the maintenance of the karst ecosystem and promote the population's well-being.

Over the past twenty years, the need to review the planning and development bases of the region has been revealed because of the current occupation scenario, and this knowledge must be expanded to neighboring areas that present a similar natural context, such as the municipalities of Sete Lagoas, Vespasiano, and São José da Lapa. During this time, academic works were published and they provided insights into the dynamics of the karst areas and their surroundings (Galvão 2015; De Paula and Velásquez 2019; Vieira 2018), as well as the effectiveness of the karstic environment protection actions (Alt 2008; Auler 2016; Pereira 2018).

The aim of this chapter is to describe how the occupation of the Lagoa Santa Karst modifies the natural dynamics of karst and how such interferences become environmental problems. It presents a brief description of the natural features of the Lagoa Santa karstic system, how these features are distributed in space and are related to each other forming distinct domains, and how these domains can be understood in the set represented by the Lagoa Santa Karst. This knowledge is essential for planning and implementing karst protection programs, which need to be comprehensive enough to guarantee the energy flow that maintains the surface and subsurface processes of the karst system.

2 Lagoa Santa Karst Features and Environmental Fragility

The Lagoa Santa Karst encompasses, in total or part, nine municipalities in the central part of Minas Gerais State, the southeastern region of Brazil (Figs. 2 and 3). The study area is located north of Belo Horizonte, on the left bank of the Velhas River, which is, by length, the major tributary of the São Francisco River Basin.

The original distribution of vegetational formations can be determined from an analysis of the structure and composition of remnant fragments (Herrmann et al. 1998). The Cerrado biome predominates, and deciduous and semi-deciduous forest cores, commonly associated with karst massifs, and transitional vegetation are between these typologies (Fig. 2). In the floodplains, swamps and ponds occur with associated aquatic vegetation. Native vegetation has been largely modified through conversion to pasture and agricultural fields, silviculture, road construction, and the urbanization process (Fig. 2).

The landscape is composed of rounded hills and rocky massifs (elevations between 590 and 1350 m.a.s.l.) in the context of the Brazilian Atlantic Plateau in the geological domain of the Bambuí Group, which is composed by an intercalation of pelites, rhythmites, and limestones, with siliciclastic rocks at the base and at the top (Pinto and Martins-Neto 2001; Reis et al. 2017).

The Lagoa Santa Karst was developed on the oldest units of the Bambuí Group: Sete Lagoas Formation and Serra de Santa Helena Formation (Fig. 3). The Sete Lagoas Formation can be divided into two different units (Ribeiro et al. 2003): the Pedro Leopoldo Member, with a thickness of approximately 80 m (Campos 1994; Piló 1998), composed of fine limestones, dolomites, marls and pelites; and the Lagoa Santa Member, consisting of predominantly medium to thick, black, carbonaceous, and laminated limestones whose thickness can reach 200 m (Tuller et al. 1991).

Deposited above the Sete Lagoas Formation is the Serra de Santa Helena Formation, with an estimated regional

Fig. 2 **a** Aerial view of Lagoa Santa Karst in Matozinhos region: limestone massif with forest remnants (1) surrounded by pastures (2), crops (3), silviculture (4) and urban area (5). The massif in the center constitutes a State Natural Monument (MONAE—Experiência da Jaguara), with cavities and a karst spring associated. **b** Mine in a limestone massif. At the top of the pit slopes, there is a thin orange soil that partially covers the limestone

average thickness of 200 m (Costa and Branco 1961). In this formation, fine siliciclastic lithotypes (argillites and siltstones with sandstone levels) predominate with subordinate bodies of lenticular geometry composed of loam and limestone of fine texture.

These units are now partially covered by unconsolidated deposits, resulting from in situ alterations of rock massifs and/or colluvial transport. The alluvial deposits occur along the Velhas River valley and the valley of its right bank tributary Mata Creek, whose slopes expose the weathering mantle of the basement migmatite gneiss complex that outcrops in the west portion of the Lagoa Santa Karst region.

Caves are preferably developed in areas where limestone is pure (high content of calcium carbonate) with structures capable of conducting water flow and favoring the dissolution process. The Sete Lagoas Formation is the main aquifer in the Lagoa Santa Karst and where cavities and most of archeological and paleontological sites occur.

Landscape evolution in the Lagoa Santa Karst generated areas where the karst is covered and areas where it is exposed. The differentiated exposition of geological units and discontinuities has a strong relationship with the development of relief and karst hydrodynamics. According to Pessoa (2005), in some places, the carbonate rocks are covered by thick soils produced by weathering of the metapelitic rocks of the Serra de Santa Helena Formation, which characterize the covered karst zones. Aquifer recharge in these areas tends to be reduced, which decreases the vulnerability of the aquifer to groundwater pollution. On the other hand, the gentle relief and the existence of perennial water bodies favor human occupation, construction of roads and accesses, and rural activities.

In the zones of outcropping limestones—bare karst, a great number of karst features are developed, with multiple points of the capture of surface waters as well as large linear walls, canyons, blind valleys, and sinking dolines characterizing fluviokarst segments (see chapter "Karst Landforms in the Lagoa Santa Area", this volume). In the plateaus, there are "fields of dolines", with small lakes often associated with the dolines, while larger seasonal lakes occur in extensive flattened plains (Berbert-Born 2002). The bare karst corresponds to the recharge zone "*stricto* sensu", where the control of the underground flow depends directly on the catchment area that feeds the cracks and fractures of the subsurface. In exposed karst areas, the vulnerability of aquifers is higher, and the existence of many karst features on the surface leaves vulnerable the speleological heritage, caves, and lakes ecosystem, as well the archeological and paleontological sites.

Due to the geological and geomorphological configuration, bare karst areas and covered karst areas can be connected in two main ways. It is very common that watercourses developed in the covered karst areas, usually in higher topography, flow towards the exposed karst areas. In this case, it is common for some of these waters to seep into sinks, recharging the Sete Lagoas aquifer, and resurface downstream in a karst spring or pond. The second form is established by the condition of the Sete Lagoas aquifer itself, where conduits are distributed in depth along both domains. The superficial and groundwater flow in LSK are carried to the Velhas River.

The use and occupation of the catchment areas of the recharge zone (in bare karst or covered karst areas) can trigger both surface and subsurface environmental problems. The contamination of surface water, for example, can spread through the underground environment because water communicates via aquifer conduits and cavities. In turn, such conduits may be in connection with the outcrops and ponds

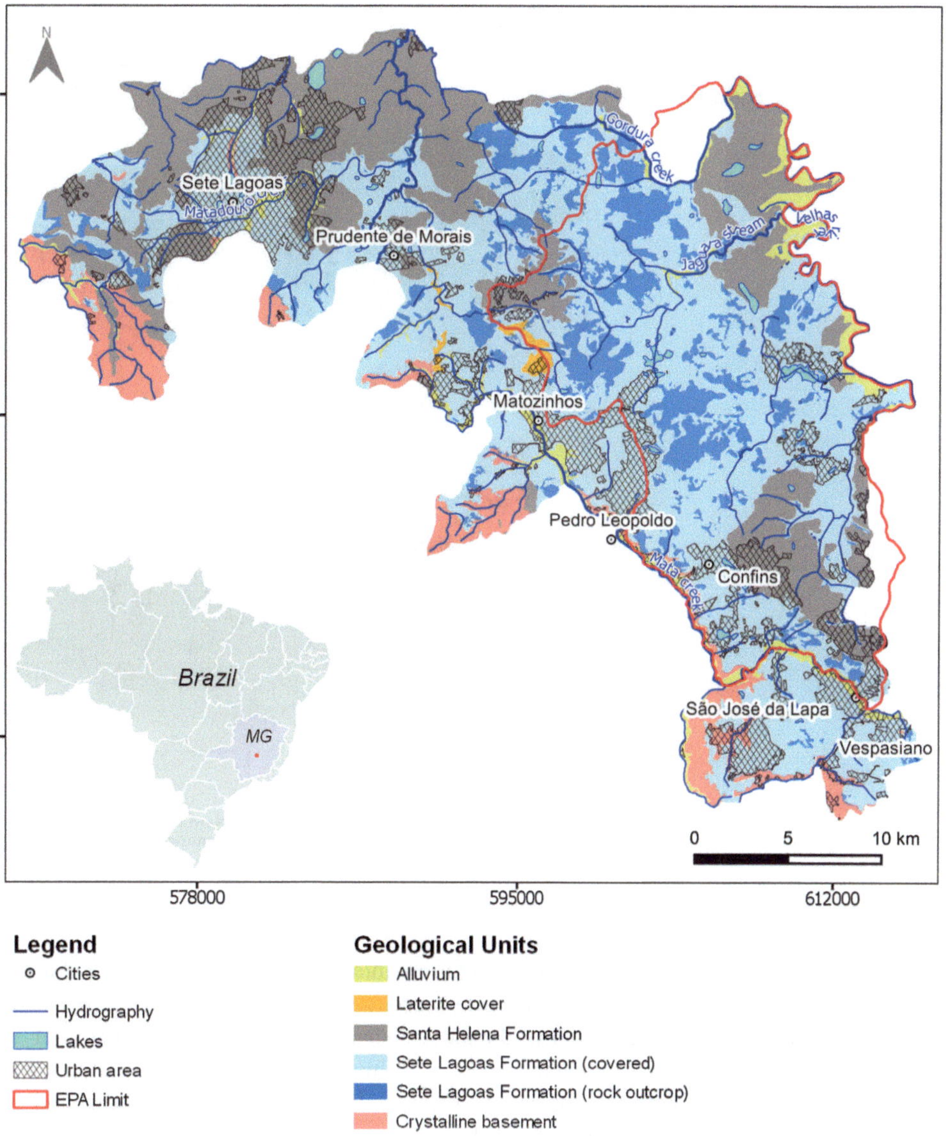

Fig. 3 Lagoa Santa Karst location and geological units

located in watersheds distinct from the polluting sources and harm the flora and fauna of those environments.

A series of lakes are particular features of Lagoa Santa Karst, and they occur in both covered and bare karst areas. These ponds generally have temporary characteristics with annual or multiannual cycles, and they are conditioned by the pluviometric regime or associated with the water table and the flow systems of the karst aquifer. Some lakes are more stable, such as those developed on phyllite, whereas others present a floating water level and may dry under certain climatic conditions (see chapter "Karst Landforms in the Lagoa Santa Area", this volume).

Lagoa Santa Karst lakes also represent areas of greater environmental fragility. More than 50 species of water birds have been encountered (see chapter "The Waterbirds and Dynamics of Lagoa Santa Karst Temporary Lakes", this volume). Nóbrega (2015) showed that the varieties of bird species and their abundances are directly related to the size (water surface area) and duration of the ponds throughout the year, which control the microhabitats and the quantity and type of food available.

Figure 4 shows the main landscape features of the Lagoa Santa Karst and some important aspects of the water dynamics. Underground routes are integrated with lakes,

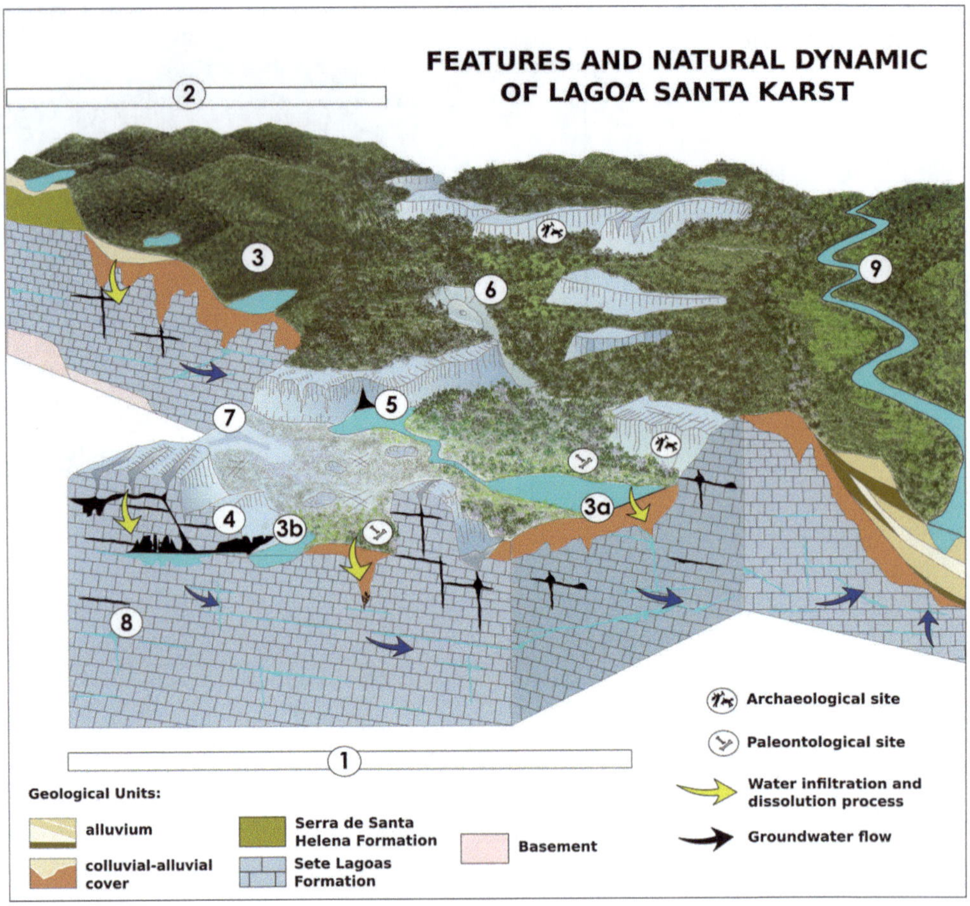

Fig. 4 Sketch showing natural features and hydrodynamics of exposed (1) and covered karst (2) areas in Lagoa Santa Karst System

sinkholes, and dolines and they are important to maintaining the ecological balance of surface and subsurface ecosystems. Therefore, the hydrogeological view of the karst system can improve the analysis of environmental aspects and impacts. Table 1 describes the elements of the LSK indicated in Fig. 4, and relates the natural relevance of these features and aspects that bring environmental concerns.

3 Human Activities in the Lagoa Santa Karst and Associated Environmental Problems

The features of Lagoa Santa Karst, sensitive environmental aspects and the human occupation should be evaluated to conduct a conceptual study and highlight major environmental impacts. Some environmental problems were confirmed in academic papers and/or denounced by the media. However, due to the lack of specific monitoring programs and public data, the loss of environmental quality cannot be properly measured. The analysis of Lagoa Santa Karst occupation and related environmental problems was carried out based on literature revision, analyses of land use and occupation map of the Lagoa Santa Karst region (PROAPA 2018), study of aerial images, and fieldwork.

In the Lagoa Santa Karst, 42% of the area (Fig. 5, Table 2) has experienced some impact on the vegetal cover (areas of agricultural crops, silviculture, altered seasonal forest areas, pastures with isolated trees). Urban areas cover 19% of the land, and mines cover 2%. Natural vegetation (Cerrado, seasonal forest and vegetation on limestone outcrops) and humid areas correspond to 37% of the study area. In these lands, the vegetation, relief and soil characteristics are preserved; however, a loss of environmental quality is expected due to the impacts on the surrounding areas. Native vegetation is impoverished because of many years of selective withdrawal of valuable woods, coal production, and extensive deforestation (Herrmann et al. 1998).

In the areas corresponding to improvements, urban areas, accesses, and mines (21% of the area), vegetation was suppressed, relief features were altered, and soil properties were modified. Therefore, in these areas, the natural water dynamics have already been considerably altered, and soils and groundwater are exposed to polluting sources. The disorderly occupation is usually accompanied by open sewage, unpaved roads susceptible to erosion, occupation of steep slopes and deforestation of areas near springs. Only 2% of the Lagoa Santa Karst area corresponds to the use of mining, however, there is a great pressure to increase these

Table 1 Features of the Lagoa Santa Karst, relevance and sensitive environmental aspects

Natural feature		Relevance	Sensitive environment aspects										
			Sewage discharge	Improper solid waste disposal	Removal of vegetation cover	Changes in natural surface flow and groundwater flow	Erosion	Vibration and noise generation	Dust generation	Trampling	Overload	Surface impermeabilization	Water abstraction
1	Exposed Karst	Contains the largest number of caves and archeological and paleontological sites and holds preferred recharge areas and higher outflow springs	x	x	x	x	x	x	x	x	x	x	x
2	Covered Karst	Permits shallow drainages that contribute to karst water flows and refuge areas of native vegetation and functions as a filter for seepage waters	x		x		x						
3	Lakes	Function as aquifer recharge or discharge points and maintain special ecosystems; surrounded by archaeological and paleontological sites	x	x	x	x	x	x			x	x	x
3A	Sinkhole Lake												
3B	Epikarstic Lake												
4	Cave	Sensitive ecosystem, important paleontological and archaeological records; cultural and religious value and speleological interest; water infiltration zones and vadose circulation into the aquifer	x	x	x	x	x	x	x	x	x	x	x
5	Karstic Spring	Maintains the base flow of watercourses and increase water quality of the streams	x	x	x	x	x					x	x
6	Sinkhole	Direct communication between surface and subsurface dissolution features; allows the flow of water and air and promotes the passage of animals	x	x	x						x	x	x

(continued)

Table 1 (continued)

	Natural feature	Relevance	Sensitive environment aspects										
			Sewage discharge	Improper solid waste disposal	Removal of vegetation cover	Changes in natural surface flow and groundwater flow	Erosion	Vibration and noise generation	Dust generation	Trampling	Overload	Surface impermeabilization	Water abstraction
7	Doline and Uvala	Potential infiltration points and areas, classic karst landforms related to karst aquifer evolution and hydrogeologic mapping indications	x	x	x		x				x	x	x
8	Sete Lagoas Aquifer	Main aquifer system, permitting groundwater storage and recharge/discharge	x	x	x	x					x	x	x
9	Velhas River	Regional base level, main stream dictating karst aquifer potentials, hydraulic heads and gradient for groundwater flows in the Lagoa Santa Karst aquifer system	x	x	x	x						x	x

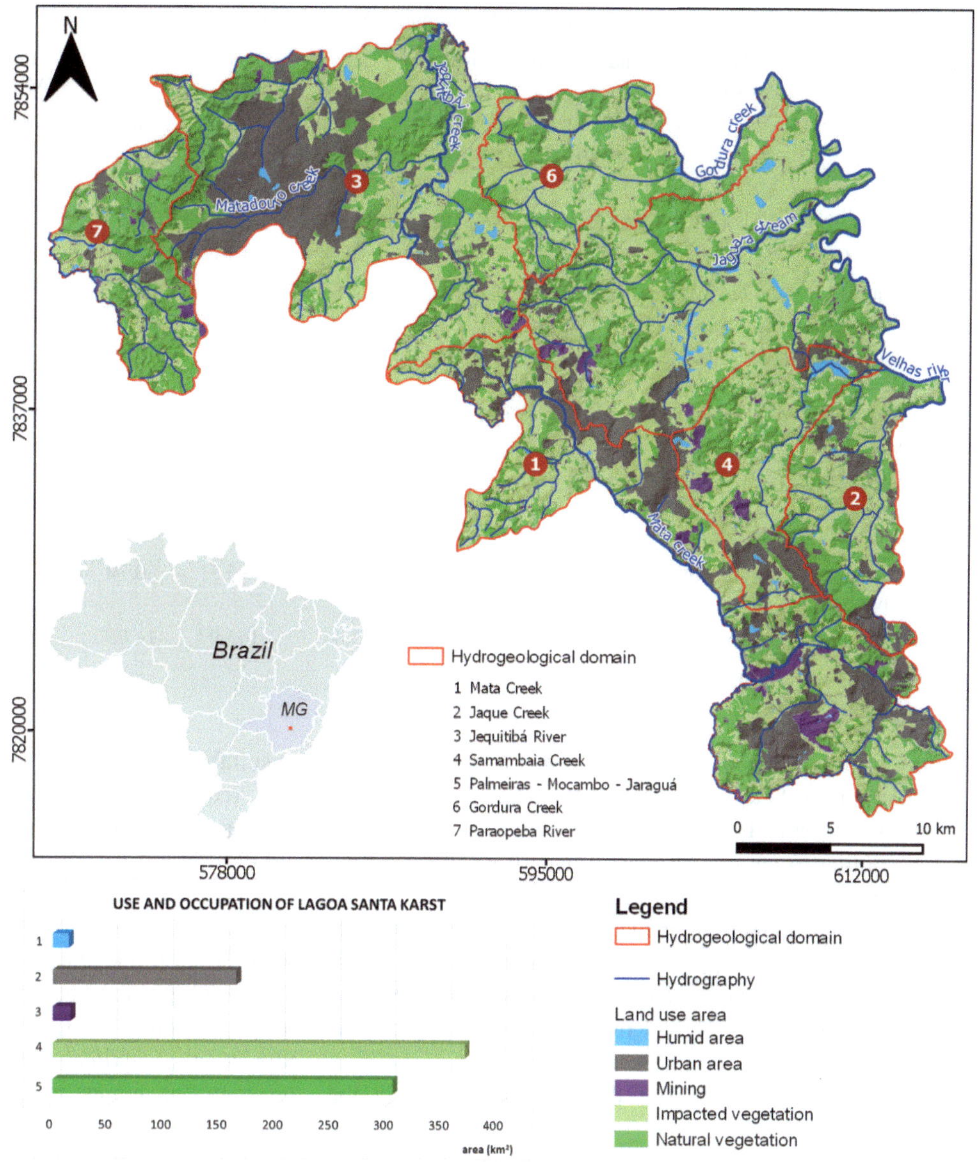

Fig. 5 Land use and occupation of hydrogeological domains in the Lagoa Santa Karst Source: PROAPA (2018)

areas. There are 335 mining titles, in different stages of the mining process, registered in the National Mining Agency (ANM), with the total permitted areas being approximately 504 km^2, 58% of the Lagoa Santa Karst (Fig. 6, Table 2).

Due to the importance of water circulation for karst system maintenance, hydrogeological domains were selected as reference units for the description of the environment and related problems. Seven hydrogeological domains (Figs. 5 and 6) were established based on the results of tracer studies carried out in recent years (Auler 1994; Velásquez et al. 2017). Each domain contains areas of recharge and discharge of surface and groundwater flows and encompasses the locations of sinking streams and relevant karst features in the maintenance of the main routes of water circulation.

- Environmental problems in the Mata Creek Domain

This domain presents less developed karst landforms, although limestone cliffs and sparse outcrops occur. Some caves, as well as Cauê and Retiro springs, are located between the cities of Pedro Leopoldo and Vespasiano. Mares Lake, a lake situated in an area adjacent to Mata Creek valley, is one of the few perennial lakes in the region, which waters drain in the subsurface to a resurgence near Mata Creek (Patrus 1998). Although karst features are not abundant on the surface, the geological map shows the occurrence of the Sete Lagoas Formation over a wide area of this domain, and karst features even occurring, relates to a very shallow depths at the domain near Mata Creek. The main

Table 2 Land use and occupation of Lagoa Santa Karst and ANM information

Domain		Area (km^2)	Land use and occupation					Mining processes and required areas at ANM		
			Natural vegetation (km^2)	Impacted vegetation (km^2)	Mining (km^2)	Urban area (km^2)	Humid area (km^2)	Number of mining titles*	Required area (km^2)	Required area/domain area (%)
1	Mata Creek	181.85	49.64	72.77	7.29	50.96	1.19	154	139.58	77
2	Jaque Creek	67.91	26.05	27.63	0	13.18	1.05	17	5.67	8
3	Jequitibá River	214.73	73.61	73.01	1.57	63.82	2.66	58	66.35	31
4	Samambaia Creek	71.69	22.65	32.5	2.82	11.54	2.16	40	44.67	62
5	Palmeiras—Mocambo—Jaguara	186.49	61.45	100.77	2.8	16.23	5.24	73	130.35	70
6	Gordura Creek	83.42	31.79	47.73	0	2.93	0.96	33	78.2	94
7	Paraopeba River	62.65	39.71	15.49	1.36	5.31	0.79	43	39.6	63
Total		868.74	304.9	369.9	15.84	163.97	14.05	335	504.43	58

* Some of the mining titles belong to more than one domain

cities in this domain are developed on the carbonate substrate of the Sete Lagoas Formation (São José da Lapa, Vespasiano, Pedro Leopoldo and Matozinhos).

According to an environmental diagnosis carried out by the Comitê de Bacia Hidrográfica do Rio das Velhas (2016a), poor sanitation, inadequate solid waste disposal (Fig. 7) and water supply were considered priority environmental problems in the context of the strategic territorial unit of Mata Creek (UTE—Ribeirão da Mata), where the urban population predominates.

Vespasiano and São José da Lapa stand out in the southern portion. These cities are expanding and based on the updated maps of the region, allotments and condominiums are growing around the urban centers. The strategic position in the north vector of Belo Horizonte and the proximity to the administrative city and to Tancredo Neves International Airport in Confins favor the occupation, especially in the subbasins of the Sujo and Carrancas creeks, which drain the land around these urban centers to Mata Creek. Around Pedro Leopoldo and Matozinhos, condominiums and some industries were built.

With growing industries and population, effluent and domestic sewage disposal to Mata Creek are increasing. According to the water quality monitoring program carried out by IGAM (Minas Gerais Institute for Water Management), Mata Creek, at the Pedro Leopoldo monitoring station, had a tendency to worsen water quality, evaluating the period from 2010 to 2017. In general, in the monitored stations in Mata Creek, the parameters showed poor quality waters (with indicative of fecal contamination and organic enrichment) unsuitable for conventional treatment aiming at public supply, requiring more advanced treatments (IGAM 2018). Increased pollutant loads can also compromise the soil and aquifer through leaks from sewage piping, septic tanks, and fuel and chemical tanks and incorrect disposal of solid waste.

The whole scenario shows a tendency of decreasing rates of infiltration and increasing runoff. The increase in urban areas multiply the amount of impermeable surfaces, which reduce the entry of water into the soil and prevent percolation to the aquifers. This change tends to reduce the groundwater inputs to the Mata Creek, which leads to smaller long-term drought flows and higher floods due to the rapid drainage of rainwater in urban areas and roads during heavy rains. Mata Creek already has silted portions and periods of floods and erosion along its banks. In the Mata Creek Domain mining activities occur, and 154 mining concessions processes were granted by the ANM, especially for the exploitation of sand, limestone, and clay.

Population growth and disordered urbanization generate impacts on water quality and water dynamics, thereby decreasing the water availability of the basin and impacting river, lake and cave ecosystems. There is a concern about the expansion of mining activity, especially in relation to the water resource, air pollution (emission of fine particulate matter), noise and shaking caused by detonation.

- Environmental problems in the Jaque Creek Domain

The Jaque Creek Domain corresponds to the Jaque stream basin, which is northwest of the city of Lagoa Santa, in the

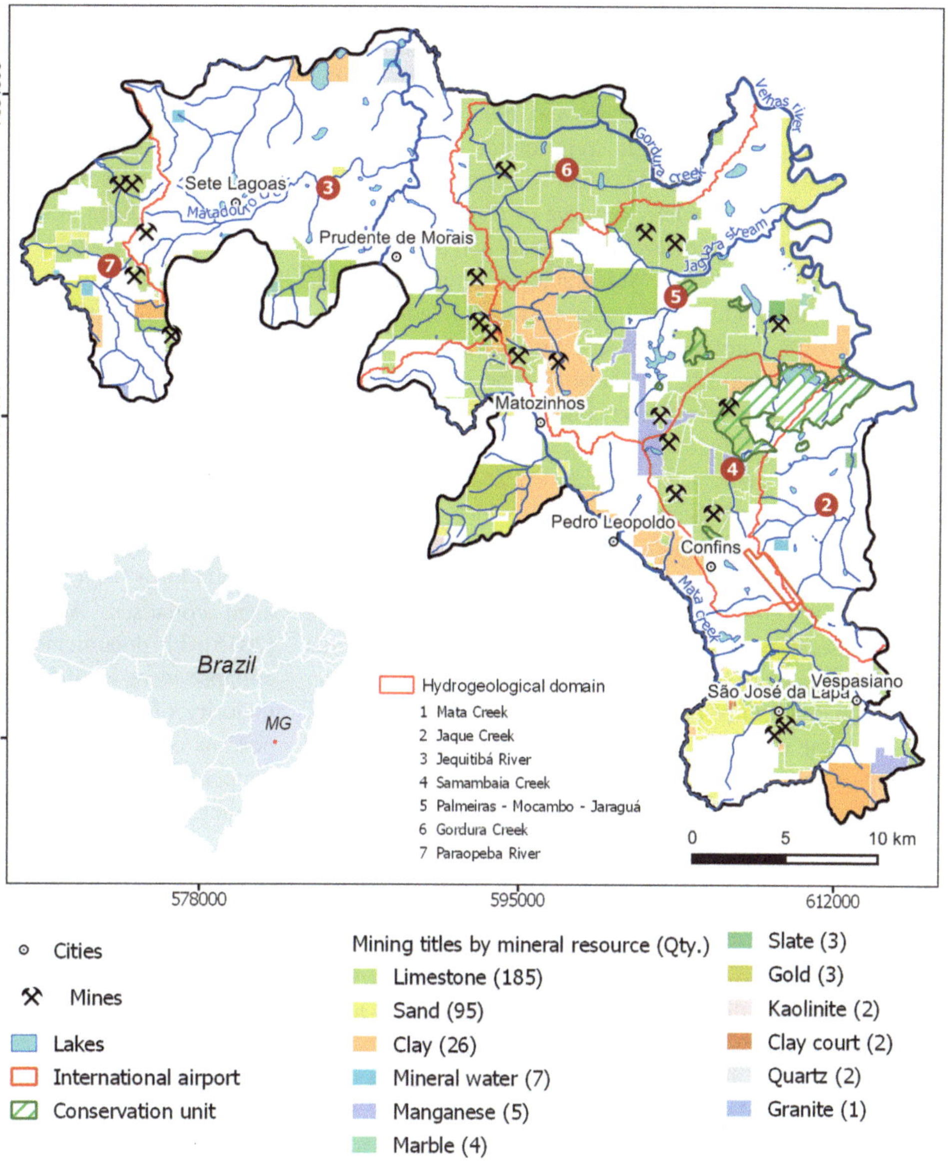

Fig. 6 Mining concession processes granted by National ANM in Lagoa Santa Karst

Fig. 7 Inadequate solid waste disposal in the Mata Creek Domain

south-central portion of the Lagoa Santa Karst. The headwaters are in Confins region, above the Serra de Santa Helena Formation, and the main tributaries are the Fidalgo and Capão de Santana streams. The drainages gather in the valley of the Jaque creek, are excavated over the Sete Lagoas Formation, and continue along this unit until its confluence with the Velhas River.

There are few (17) mining titles registered in the ANM in this domain. Along the valley, there are records of some interventions, such as water dams and pumping wells. The waters are used in rural properties mainly for livestock and irrigation of small temporary crops. The International Airport of Confins, as well as residential condominiums, are located in the headwaters of Jaque Creek. Capão de Santana Creek is a receiver of effluents from the sewage treatment plant of the International Airport of Confins. According to evaluations performed by Patrus (1998) in the Jaque stream near Lapinha district, no heavy metals and toxic substances were detected at relevant concentrations in none of the samples. However, oils and greases and indicators of organic contamination of animal origin were found.

The quality of the surface water and groundwater is a cause for concern, especially because of leaks from sewage pipes and cesspits and the use of fertilizers and pesticides for agriculture. The area most vulnerable to aquifer contamination is in the middle and lower reaches of Jaque basin, where the karst aquifer is covered by unconsolidated sediments. In these areas, the number of dolines and karst lakes stand out and can present direct connections with the aquifer through karst conduits.

Outstanding lakes in the karst landscape include the Fidalgo, Velhas, and Sangrador lakes, which occur in the Lapinha District. Caves and shelters in limestone are located close to the lakes such as Várzea da Lapa Cave, which is a State Natural Monument. The urban area of the Lapinha District is very close to these karst features, which highlights the importance of maintaining a system to control urban expansion, especially interferences in the drainage network and disposal of solid waste and sewage. As a result of vegetation removal and occupation, lakes, dolines, and cavities may be silted, impairing the natural karstic dynamics.

- Environmental problems in the Jequitibá River Domain

The largest urbanized area within the Lagoa Santa Karst area is located in the hydrogeological domain of the Jequitibá River. An industrial, mineral and metallurgical center is located along the axis of the BR-040 federal highway and around the city of Sete Lagoas. Among the various environmental problems that result from these processes, the following can be considered priorities: surface and groundwater pollution, lowering of the aquifer water level and soil collapse. In the context of the strategic territorial unit of Jequitibá River, Comitê de Bacia Hidrográfica do Rio das Velhas (2016b) corroborates this analysis, since it has indicated that the most critical environmental problems include those related to urbanization and the quality and quantity of water resources.

In Sete Lagoas, 95% of sewage is collected, but only 10.8% of the urban population is served by sewage treatment (FEAM 2018). The sewage from Sete Lagoas, a city of more than 210,000 inhabitants, contributes to the pollution of Jequitibá River and Velhas River. The water quality points monitored by IGAM (2018) in the Jequitibá river basin showed medium to low water quality indexes. Fecal contamination, organic enrichment and toxic substances (free cyanide in the Jequitibá River) were identified. In addition, some wells in the central portion of the urbanized area of Sete Lagoas have water with nitrate (as NO^{-3}) that exceeds 10 mg/L, and it could be interpreted as a local contamination by superficial infiltration, probably by domestic sewage or septic tank (Galvão 2015).

The hydrogeological dynamics in the region of Sete Lagoas have been influenced by the groundwater withdrawal due to pumping wells' continuous operation, which is often destined to supply population in the city and the industries installed in the surroundings. The Water Supply and Sewage Service (SAAE) of Sete Lagoas has more than 100 public wells located in urbanized areas, where the highest discharge rates may occur (Galvão 2015). A significant drawdown ($\sim$40 m) was observed in some areas due to pumping wells, which led to voids of dissolution in the vadose zone (Galvão 2015). In this region, along with the decrease in the water level of the aquifer, problems related to the instability of the massifs have been triggered, thus reflecting new doline and karst collapses. According to data collected by Galvão (2015) 17 documented occurrences of subsidence or/and collapse were identified in Sete Lagoas since 1988, and the majority of these occurrences were clustered in the central portion of the urbanized area. The most important event happened in 1988 (Silva 1988) when a collapse with a diameter of approximately 20 m occurred (see chapter "Karst Landforms in the Lagoa Santa Area", this volume). Failures in water plumbing distribution are also factors that increase the risk of subsidence and collapse and waste of water resources.

Moreover, there are many clandestine wells in the region. There is an urgent need to regulate users so that the real values of demand can be known, and thus, strategies for the development of the city can be established, while maintaining basic and elementary principles for the management and conservation of groundwater resources. An important issue related to the water management and supply in Sete Lagoas is determining how the continuous increases in the groundwater pumping affect the dynamics of the lakes and rivers of the region.

- Environmental problems in the Samambaia Creek Domain

The hydrogeological domain of Samambaia Creek is developed in an essentially rural area that presents mining activities and irregular allotments. Abundant karst features with intense water circulation are observed, and they have several points of connection between surface and groundwater, and protected areas with relevant archeological and paleontological sites are also located in this region. Dolines, lakes, springs, caves and limestone cliffs are some of the abundant karst features.

Sumidouro Lake is the main lacustrine body of the entire Lagoa Santa Karst region and it is located in north part of the Samambaia Creek Domain. The waters of Sumidouro Lake mostly come from Samambaia Creek, and the lake is not perennial, but rather presents a multiyear cycle. In 2018, the lake almost dried, with only a small water surface area remaining, and small channels forming that drained to a sinkhole near the limestone massif. Therefore, Sumidouro Lake is an important point for maintaining the water quality of the aquifer. According to Patrus (1998), although the discharge of domestic sewage has not been observed in the lake, this hypothesis should not be discarded due to the urban occupation along its border. The waters of Sumidouro Lake are used for irrigation of crops in adjacent farms and for livestock, which also reinforces the importance of maintenance and conservation of this water body. Occupation in the surrounding area may also indirectly contribute to the lake siltation due to rain drainage in exposed soil areas and the discharge of sawmill effluents (Garcia 2011), which is common in this region for the processing of calcium phyllite rocks, i.e., "Lagoa Santa" stone.

Sumidouro State Park aims to protect the diverse karst features of the paleontological and archeological patrimony. However, the city of Confins and some limestone exploitation mines are in the highest areas of the Samambaia Domain, which correspond to the recharge zones of the karst aquifer, whose groundwater flow in the Sumidouro Lake direction. The mines in operation are points of greater vulnerability since the opening and deepening of the pits expose the massifs, allow direct recharge of rainwater in fractures and conduits, and in some cases expose the water table forming a small lake in the bottom of the pits. In cases where dewatering is necessary, there may be interferences in the functioning of springs and lakes in the surroundings, changing the natural dynamics of the karst aquifer.

Santo Antônio Lake is the upstream portion of the Samambaia domain, near the border with the Mata Creek and Palmeiras-Mocambo-Jaguara domains. In this region, the limits of the groundwater flow are not yet well known. The lake is eutrophic (Fig. 8), and proximity to houses and roads indicates a risk of water contamination by rain drainage, sewage discharge, and improper waste disposal.

Fig. 8 Eutrophic waters of Santo Antônio Lake, Pedro Leopoldo municipality

Based on an analysis of ANM data, a conflict of interest is evident, based on the total mining titles in the area (40). Many mining titles have been registered for limestone (30) throughout this domain and some titles for clay extraction (2) have been registered in the north region of Sumidouro Lake. The extraction of calcium phyllite, which is used for flooring and walls, is an important activity concentrated in this domain, and it generates large amounts of rock waste without adequate disposal. After intensive action by the environmental prosecutor and state environmental agency, the most harmful calcium phyllite mining operations were banned.

In some areas of Confins, Matozinhos and Pedro Leopoldo, in this domain, urban occupation is very close to caves, in the catchment zone of dolines, or near springs and swallets. According to Pereira (2018), Confins municipality has many areas that conflict with Lagoa Santa Karst APA guidelines and state conservation units and has several local pressures, such as lack of sewage collection and treatment, urban and rural drainage, burning, deforestation, and real estate expansion. Activities in the Confins region, especially around the lakes, manifest risks of contamination and silting. Although it has not yet been confirmed, there is the possibility of connecting the waters of the Confins lakes with the Lapa Vermelha Cave. It is known that the waters that sink in this cave discharge to Samambaia Spring, and waters continue flowing to Sumidouro lake and, finally, reaching the aquifer.

Another point of great concern in this area is the Lapinha Cave, a touristic cave with many halls and speleothems. As pointed out by Dutra et al. (1998) and Xavier et al. (2017), tourist exploration in Lapinha Cave developed with some failures that caused damage to the cave environment and the local community, such as artificial lighting, cementation of sections of the floor for walkways or staircases, poor planning of the number and frequency of visitors, and lack of involvement with surrounding residents.

- Environmental problems in the Palmeiras-Mocambo-Jaguara Domain

The Palmeiras-Mocambo-Jaguara Domain drains a predominantly rural region of great scenic beauty and encompasses special karstic springs that form an interconnected system. The Palmeiras basin is the upstream and the

Mocambo basin is in the middle portion. Approximately 3 km downstream of the Mocambo spring is the great Jaguara spring, whose waters starting at the confluence with the Mocambo stream form the Jaguara Creek. The groundwater flow is generally directed to the Jaguara Creek and afterward to the Velhas River.

The conditions in the Palmeiras-Mocambo-Jaguara hydrogeological domain are similar to Samambaia Creek Domain, both in terms of their natural, paleontological, and archeological attributes and in the type of pressures, such as mining, rural activities and some condominiums and small urban centers.

In the upstream area, industrial and mining activities increase the risk of contamination and water dynamics alterations in the karst system. The connection between the Bom Jardim sinkhole, which is in the Palmeiras Creek watershed, and the Mocambo Spring shows the importance of maintaining the quality of waters from the Palmeiras basin and managing the water resources for the continuity of the downstream system in the subbasin of the Mocambo and Jaguara Creeks.

The waters of the Mocambo and Jaguara creeks are used for crop irrigations including central pivot systems, as well as for livestock and water supply to some rural properties. In the region of Mocambeiro, there are large farms for horse and cattle breeding (Fig. 9). Agricultural activity and groundwater pumping wells coexist side by side with natural monuments and areas of high tourist and scientific interest, such as Cerca Grande State Park and Vargem da Pedra State Natural Monument. Large dolines in the Mocambeiro depression have a connection with the underground karst, functioning as recharge points during periods of low water level and discharge points during rainy years. However, the lakes have dried in recent years, indicating that there has been a noticeable decrease in water levels in this region. Dolines that are connected to cavities that functioned as discharge areas have now the potential to function as sinks,

Fig. 9 Water use for pastures

Fig. 10 **a** View of karst massif (arrow) and the Vargem da Pedra Natural State Monument (MONAE Vargem da Pedra) (circle). Note the proximity of these features to the urban area of Mocambeiro, district of Matozinhos. **b** MONAE Vargem da Pedra is an important archeological site where rock paintings are depredated

thus capturing runoff water and recharging the aquifer. These karst features are near the urban areas of Mocambeiro and Matozinhos and thus are vulnerable to depredation and pollution (Fig. 10a). Notable rock art is located on the limestone walls in this area, and some of them are already depredated (Fig. 10b).

- Environmental problems in the Gordura Creek Domain

This domain also encompasses a complex underground flow system and rich karst features and it is marked by the occurrence of springs, caves, and sinks. Limestone outcrops in large areas of this domain, and it favors aquifer recharge. In the middle portion of the basin, the waters captured in the upstream karst structures emerge, forming a karst pond (Gordura Spring, which is the largest in the region according to Pessoa and Mourão 1998) whose groundwater flows form Gordura Creek which afterward joins the Velhas River. The Gordura Spring has a high discharge flow, and studies with tracers conducted by UFMG (Velásquez et al. 2017) showed that part of these waters come from the Flor stream watershed, where is located the Goiabeiras sink and Veredas sink regions. The space delimited by the two river basins may present other connecting routes that interact in this space, which is why these regions were generally considered a single hydrogeological domain. However, this domain is likely to extend north and northwest beyond the area defined for Lagoa Santa Karst.

The hydrogeological domain of the Gordura Creek is mainly occupied by rural and tourist activities. The waters of the Gordura Creek are used for domestic supply, livestock, irrigation of agricultural crops, including central pivot systems, and recreational activities of primary contact and recreational fishing. Tourist activities, especially when karst features are exposed in vast areas, are associated with graffiti and cave depredation, soot from fires or other corrosive substances in cave walls and garbage accumulation. Rural activities around lakes and rivers impact these environments via the discharge of effluent rich in phosphorus and nitrogen compounds, vegetation burning and interference in riparian forest. Although there is no significant mining activity in this domain, there are 28 mining titles for limestone, and the titles areas represent 94% of the area of this domain.

- Environmental problems in the Paraopeba River Domain

The Paraopeba River Domain represents the only area whose surface waters do not drain to the Velhas River, which most of these waters flowing to the São João stream basin, whose substrate corresponds to the crystalline basement (non karstic terrain), as well as the Taboa, Barreiro, Riachinho, and Saquinho streams basins, whose relief is supported by the rocks of the Serra de Santa Helena and Sete Lagoas formations. There are many areas of preserved vegetation in the headwaters and rural areas on the slopes and valleys.

In this domain, two features of special interest for preservation stand out: the Serra de Santa Helena, which extends towards the Jequitibá domain, and the Rei do Mato Cave State Natural Monument. The Serra de Santa Helena hosts the headwaters of several streams and important areas of preserved vegetation, which promote the regional fauna, the climate, and the conservation of soil and water bodies. However, as pointed out by Santos (2011), this area is impacted by burning, irregular occupation, and lack of sanitary sewage. Moreover, soil compaction and ravines and gullies created from the practice of trail, rally, and motocross also occur.

The Rei do Mato Cave, which is located on BR-040 in Sete Lagoas, received more than 22,000 visitors per year between 2005 and 2009, and approximately 30,000 in 2010 and 2011 (Morato et al. 2012). According to the authors, the implementation of internal and external infrastructure for cave tourism (lighting, walkways, and stairs), the large number of visitors, religious use, mining, and high vehicle traffics on the highway affected the environmental quality of this karst environment. Speleothem destruction, physical and chemical changes to the rock surface, loss of fauna microhabitat, alteration of water flows, accumulation of solid waste (including glass bottles, candles, and ceramic plates), sedimentation by trampling, graffiti on ceilings and walls, rock paintings, and dust accumulation are some of the impacts observed.

4 Discussion

An overlap of several political, economic, and natural factors has increased human occupation in the Lagoa Santa Karst region within the last several decades, constituting the main driver for most of the activities and impacts on the karst environment, as described in each hydrogeological domain. Figure 11 summarizes the main pressures and resulted impacts on physical, biotic, and socioeconomic environment of Lagoa Santa karst.

Mining, rural activities, tourism, and recreation all have an important influence on the impacts assessed in the domains that correspond to the karst core of the study area: Samambaia creek, Palmeiras-Mocambo-Jaguara and Gordura creek domains. According to Berbert-Born (2002) common damage in cave environments of Lagoa Santa Karst APA are: speleothems breakage, scribblings and rubbish accumulations, impregnation of the surfaces by grime from fires or other corrosive substances, the algae growing induced by artificial lightening, the covering of areas by mobilized sediments, flooding, drying, alterations on water physicochemical and biological composition and modifications on the natural drainage that alter the cycle of activity of the speleothems, the natural evolution of the forms and the development of the fauna.

In the Paraopeba and Jaque Creek domains, the existence of non-carbonate rocks decreases the concern of some activities. Urbanization and industrialization were mapped as the main pressures in the domains of the Jequitibá River and Mata Creek. However, the progress of urban occupation of this space also reflects other types of pressure, such as increased demand for water and sanitation, which are relevant in all domains assessed.

While almost the totality of the water for human supply in the region is provided by groundwater, little or no effort is applied to the conservation of this resource, with most of the municipalities still disposing solid wastes and domestic sewage inadequately in a vulnerable environment (Comitê

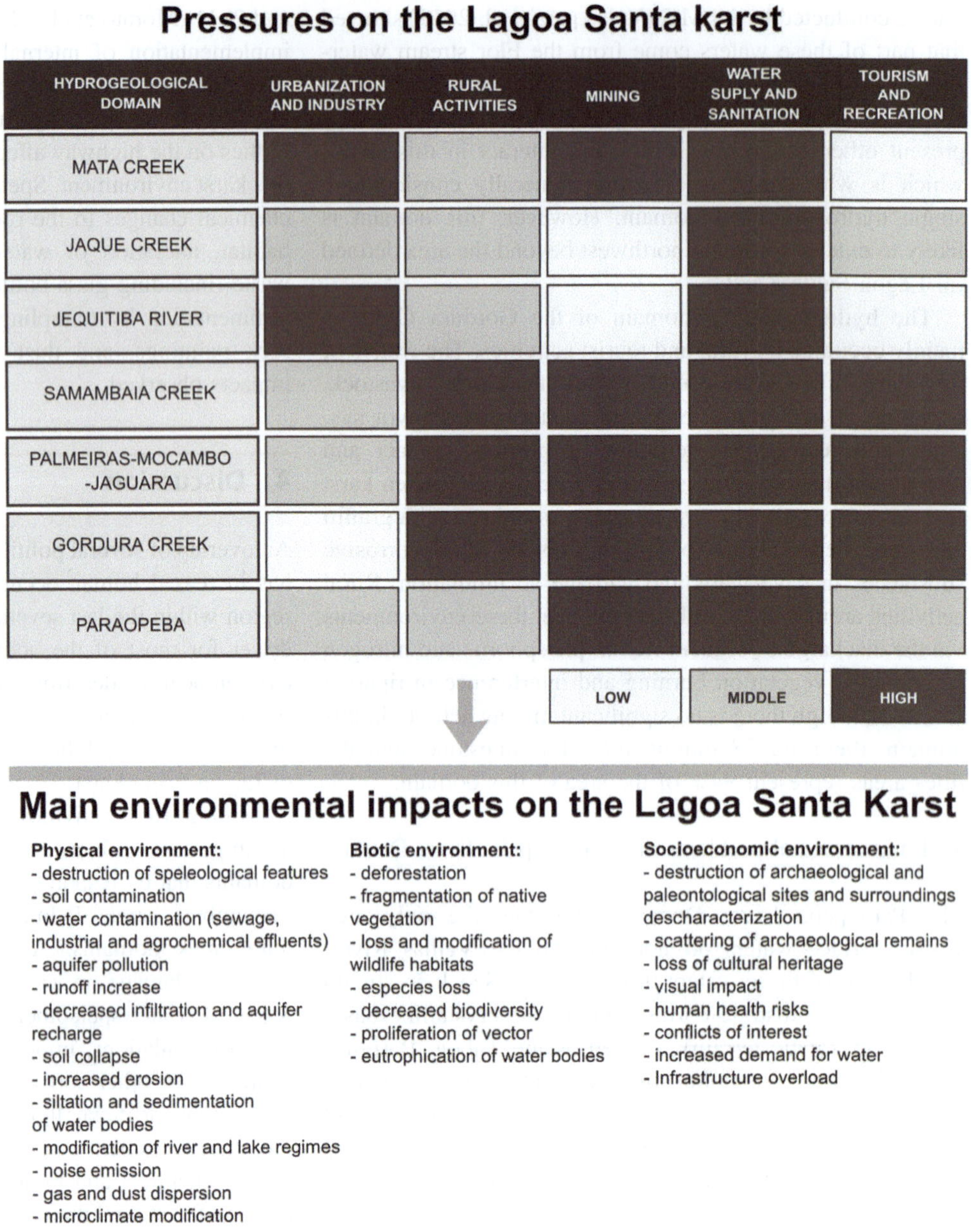

Fig. 11 Pressures and main environmental impacts on Lagoa Santa Karst

de Bacia Hidrográfica do Rio das Velhas 2016c). In rural areas, although groundwater exploitation does not occur as intensely as in urban areas, there is concern about the impact on aquifer dynamics by increasing discharge rates of the pumping wells. Moreover, the lack of effective control of the number of active pumping wells and the actual discharged volumes endanger the use of groundwater, especially near cavities with drainages and areas where the groundwater level is associated with a shallow epikarst profile.

According to the information presented by Pessoa and Mourão (1998) for the Lagoa Santa Karst APA, there were 141 water catchment points, including pumping wells, springs, and manual drilling wells. Among these, 95 were pumping wells were for urban, rural, and industrial supply. The pumping rate of the wells at that time amounted to 2,220 m^3/h. At present, in the whole area of Lagoa Santa Karst, there are 54 points totaling 2,435 m^3/h in the water concession records. These values, however, do not correspond to the reality of the pumped discharges since there are many users who do not yet have a water use permit. Population expansion in urban areas and the intensification of agricultural and mining activities corroborate this understanding. According to Galvão (2015), just in the urban area of Sete Lagoas the pumping rate in wells is estimated to reach 3,125 m^3/h. Velásquez et al. (2018), in an area equivalent to the domains of Mata creek, Jaque creek, Samambaia creek, Palmeiras-Mocambo-Jaguara, and Gordura creek, estimated an exploitation flow of 6.857 m^3/h

associated to 380 pumping wells. In the same area, Velásquez et al. (2018) also accounted for 385 m^3/h in surface water uses.

As a critical point for the effective management of the waters and karst environment of the Lagoa Santa karst, it is evident the need of water users registers accompanied by a plan for monitoring quantity and quality in springs, watercourses, and points of high vulnerability of Sete Lagoas aquifer. In this sense, valuable contributions for understanding the karst system were made by Velásquez et al. (2018), Amaral (2018), Vieira (2018) and De Paula (2019). De Paula (2019) showed that in some basins in Lagoa Santa Karst, groundwater withdrawal is greater than recharge. However, studies have also shown that the flow rates of springs observed in this work and measurements taken more than 25 years ago show a similarity of over 90%, demonstrating that there were no significant impacts on the water dynamics of these springs. The result indicates the existence of flow compartments within the aquifer, showing that there may be areas of greater potential for safe groundwater exploitation without causing damage to the flow of karst springs. Studies like this are strategic for the planning of water supply systems.

However, in all karst systems, it is crucial that the main sources of groundwater that already are known in the field as, at least, the main karst springs, be included in a wide monitoring program. The quantity and distribution of groundwater that emerges at the surface in many places around Lagoa Santa Karst System (see chapter "Karst Hydrogeology of the Lagoa Santa", this volume) need to be well monitored for quality and quantity evaluation of its natural behavior, starting a consistent groundwater databases.

It is noteworthy that, after an evaluation of the characteristics of the water collection and distribution system in Sete Lagoas, Botelho (2008) identified lacking water supply infrastructure (networks and abstraction), difficulties of management related to human resource and lack of databases and strategies for building support systems for planning and decision making.

Tayer (2016) carried out a vulnerability study focused on Lagoa Santa Karst APA, and concluded that 11% of the studied area was included in zones of very high and high vulnerability associated with features with high water absorption characteristics, such as sinks, blind valleys, and limestone outcrops. In addition, the relationship between the coverage of each class of vulnerability and the limits of the environmental zoning of the Lagoa Santa Karst APA indicates an association of less restrictive zones with areas considered to have high and very high vulnerability, which reinforces the need for an update of the territorial and environmental management instruments implemented in the region.

In some situations, such as the case of Sete Lagoas city center, coupling monitoring of groundwater uses with geotechnical monitoring is extremely recommended. Events of doline appearance and collapse can occur in urban and industrial areas based on karst limestones, and it is important to incorporate more detailed investigations of the subsoil and to control the leakage of pipes and drainages.

Looking at the current situation of the mines, as well as the number of DNPM-registered mining titles within the Lagoa Santa Karst area, mining represents a historical interest conflict between the necessity of protection of karst environment and economic development. Limestone is the most important mineral product in the Lagoa Santa Karst region (Ribeiro et al. 2003). The extraction of mineral assets can impact the karst environment by both the mining process and the economic and social effects of these developments. In mining, the massifs are consumed, and the physiography and the natural hydrological dynamics are altered, leading to changes in both surface runoff and groundwater, especially if there is a need to lower the water table. Air particulates dispersed primarily from lime plants have covered the walls and floors of many caves, a type of environmental impact that is difficult to restore and with unknown effects on cave fauna (Auler 2016). Not only speleological features but also ecosystems, archeological, and fossil records can be threatened by mining activities, as well as the knowledge that the entire environment contains. Other activities such as the creation of accesses and drainage systems for the roads, intensification of the flow of vehicles, and increases in the population of the surroundings happen during the mining process and are linked to the industries and companies that use the rocks as inputs.

On the other hand, exploration and development of mineral resources can bring richness and visibility to the region and foster research development, environmental education, creation, and maintenance of protected areas such as Private Natural Heritage Reserves (RPPNs), and recovery and compensation of areas specifically impacted by each undertaking, formally instituted by means of constraints by the licensing body and other intervening agencies.

However, the possibility of protected areas removal through the adoption of compensatory measures, in many types of environmental licensing process, may result in environmental setbacks, especially under pressure of projects with high economic impact. This point raises an important issue related to environmental management. Pereira (2018) argues that parks and protected areas created as compensatory measures related to the installation of Tancredo Neves International Airport, implementation of the Green Line (Linha Verde) and the Administrative City contributed to the existence of multiple environmental planning instruments, surpassing the Lagoa Santa Karst APA. However, in a situation of expansion of the urban fabric of Belo Horizonte and facing the lack of coherence between the Municipal Master Plans and the zoning guidelines of Lagoa Santa Karst APA (which were prepared including karst vulnerability), the multiplicity of instruments has become an obstacle to management. As Pereira (2018) points out, this resulted in flexibility of Lagoa Santa Karst APA in the face of anthropic pressures already installed in its territory and those economic and State Government's interest in infrastructure, weakening conservation efforts in the region.

Alt (2008) points out that the set of environmental legislation that regulates mining and industrial activities, which protect the caves, archeological sites, water resources, and others, have contributed more to the preservation of local heritage than Lagoa Santa Karst APA management instruments (see chapter "Protection of the Environmental and Cultural Heritage in the Lagoa Santa Karst: History and Challenges"). The author makes an important critique of some aspects of the APA environmental zoning that need to be re-evaluated to clearly incorporate geotechnical constraints and risks of aquifer contamination.

Pereira (2018), in the context of the Lagoa Santa Karst APA, concludes that immediate management practices prevail over planning-based management practices. Given that anthropogenic pressures are often situated in areas of high karst vulnerability, such immediacy can surpass planning and cause irreversible human and material damage. This view finds historical support through the words of the Danish naturalist Peter Wilhelm Lund in 1841, about saltpeter extraction, the earliest record of mining activities in Lagoa Santa Karst:

> The same neglect, which is such an important trait in the character of Brazilians, is also announced in the elaboration of this branch of industry. If the emptied cave were supplied with a new layer of loose soil, it would be (experience proved it) sooner or later, soaked again with saltpeter; but, just as the system of Brazilian agricultural culture results in each year a part of the most beautiful and fertile tracts of land being turned into desert, so it works, unconcerned with the future, here too, with a system that over time, it will cause the source of this industry branch, so important for the country, to dry up.
>
> No one who has a feeling for lofty natural beauty can witness this planned annihilation of the greatest wonders of the tropical world, the majestic virgin forest, and perhaps botanists will have to mourn the forever loss of many of tropical nature's most beautiful productions. However, what is this loss compared to what zoology suffered in the millions of remains of an extinct animal creation, which was annihilated and disappeared without a trace, due to the extraction and preparation of saltpeter land from the caves? (Holten and Sterll 2011: 177).

Despite centuries of degradation and immediate management practices, cave destruction was made difficult by the 2008 Federal Decree, which imposed highly restrictive criteria for the determination of cave significance. According to Auler (2016), in Lagoa Santa Karst, approximately 95% of all caves are of either maximum (no impact allowed) or high significance (two high significance caves must be permanently preserved for each high significance cave to be removed). The author points out that, although caves are the only karst landform officially protected according to Brazilian law, the superposition of buffer zones results in the protection of vast portions of the karst (law dictates that a protection buffer zone—preliminarily of 0.2 km^2—must also be taken into account).

5 Conclusion and Final Considerations

The literature review about the Lagoa Santa Karst environmental problems revealed that most of the issues identified since the 1990s, some of them since the nineteenth century, are still present in the region. An overlap of several political, economic, and natural factors has increased human occupation in the Lagoa Santa Karst region within the last several decades, constituting the main driver for most of the impacts on the karst environment. However, society's lack of knowledge of its natural riches, their finitude, vulnerability and interdependence, and the immediate and predatory decisions common in all spheres of interaction between man and nature, manifest the urgent need for environmental education. A new ecological and ethical awareness must be created, and shared by the whole society, from the residents of Lagoa Santa Karst to politicians and businessmen.

Karst systems are complex and vulnerable by nature, demanding a specific approach to mitigate negative human impacts and allow sustainable development. As Brinkmann and Parise (2012) pointed out, humans must make a strong effort to learn to live "in harmony with" karst rather than to live "on" karst. Many problems faced when interacting with karst environments are best solved through changes in human systems rather than through alterations to karst environments.

Despite karst system fragility, there are already case studies (Ford and Williams 2007; Kresic and Stevanovic 2010; Brinkmann and Parise 2012) that can be replicated, with adaptations, to guide the forms of occupation and use of karst terrains in a harmonious and viable way. Successful solutions have sound scientific knowledge of the natural system and a strong concern for interferences in the water circulation in the karst system and in its neighboring areas, combined with continual social engagement.

When some of the negative effects begin to be noticed, the desirable solution is a combination of actions that are consistent with existing regulations, institutional capacity, economic balance, social acceptance, stakeholder involvement, and political will. Adaptability is essential in each case. The goal is sustainable development that considers the whole set of resources rather than anyone in particular.

However, a team integrated with various local stakeholders should be formed to strengthen surveillance through a shared monitoring and management system based on representatives of the communities such as residents and municipalities.

In this sense, some proposals for actions of great value for the preservation of the Lagoa Santa karst are highlighted below.

- The revision of the Lagoa Santa Karst APA management plan and environmental zoning, incorporating new hydrogeological studies that delimit domains and the most vulnerable areas of karst and prioritizing conservation areas of important environmental and historical–cultural heritage of the region. Integrated analyses that show the relationship of these environments and the surrounding areas must be considered.
- Revision of Municipal Master Plans making them consistent with protected area management plans.
- Integration between federal, state, and municipal levels of power and integration of existing management tools and databases.
- Environmental education programs, technical training programs for stakeholders, and massive regional campaign to promote the Lagoa Santa Karst evaluating the possibility of including the theme in basic education.
- Implementation of monitoring programs: qualitative and quantitative surface and groundwater, climatological, air quality, and karstic lakes ecosystems.
- Restoration of riparian forests and degraded areas around springs and lakes.
- Establishment of technical and legal parameters to control construction and sanitation aspects in areas that are already undergoing urbanization on the karst under conditions of greater vulnerability or geotechnical risks.
- Funding research and science communication material that gives technical support to Water Management Institute of Water (IGAM), river basin committees, entrepreneurs, and the population itself to discuss environmental issues of relevance to the preservation of the Lagoa Santa Karst and alternative activities for local development. Relevant topics are: water supply alternatives, improved irrigation practices, climate changes, sewage treatment, karst aquifer characterization, urban planning, local economy, life quality of local community, organic agriculture, conservation area management, karst lakes ecology, environmental education, degraded land reclamation, eco-tourism, expansion of knowledge, and preservation of archeological and paleontological heritage.
- Strengthen the IBAMA of Lagoa Santa, the organ responsible for the management of Lagoa Santa Karst APA, increasing and qualifying its technical staff, infrastructure conditions and available resources.

Acknowledgements The authors would like to thank the colleagues from Hidrovia company, with whom we had many meaningful discussions, and especially Giancarlo Ramos, Tamires Nogueira, Rogério Lucas, Lucas Souza, and Anna Luiza Matos for figure elaboration and providing technical support for developing this chapter. The authors thank Instituto PROAPA in the name of Paulo de Tarso for providing complementary information.

References

Alt LR (2008) Efetividade sócio-ambiental da APA Carste de Lagoa Santa-MG: uma avaliação a partir de suas ferramentas de planejamento e gestão. Dissertation, Universidade de Minas Gerais

Amaral DGP (2018) Análise do comportamento e desempenho hídrico das depressões cársticas da região da APA Carste Lagoa Santa (MG). Dissertation, Universidade Federal de Minas Gerais

Auler AS (1994) Hydrogeological and hydrochemical characterization of the Matozinhos-Pedro Leopoldo Karst, Brazil. MSc thesis, Western Kentucky University

Auler AS (2016) Cave protection as a karst conservation tool in the environmentally sensitive Lagoa Santa Karst, southeastern Brazil. Acta Carsologica 45:131–145

Berbert-Born M (2002) Carste de Lagoa Santa, MG – Berço da Paleontologia e da Espeleologia Brasileira. In: Berbert-Born M, Campos DA, Queiroz ET, Schobbenhaus C, Winge M (eds) Sítios Geológicos e Paleontológicos do Brasil. Comissão Brasileira de Sítio Geológico e Paleobiológicos, Brasília

Botelho LALA (2008) Gestão dos recursos hídricos em Sete Lagoas/MG: uma abordagem a partir da evolução espaço-temporal da demanda e da captação de água. Dissertation, Universidade Federal de Minas Gerais

Brinkmann R, Parise M (2012) Karst environments: problems, management, human impacts, and sustainability an introduction to the special issue. J Cave Karst Stud 74(2):135–136

Campos AB (1994) Relações entre as características faciológicas e estruturais das unidades do Grupo Bambuí e a morfologia cárstica na região de Lagoa Santa-Pedro Leopoldo (Minas Gerais). Universidade Federal de Minas Gerais/Fundação de Amparo à Pesquisa do Estado de Minas Gerais, Belo Horizonte

Cartelle C (2012) Das grutas à luz: os mamíferos pleistocênicos de Minas Gerais. Bicho do Mato, Belo Horizonte

Comitê de Bacia Hidrográfica do Rio das Velhas (2016a) Cartilha Plano Diretor de Recursos Hídricos Unidade Territorial Estratégica Ribeiro da Mata. http://cbhvelhas.org.br/wp-content/uploads/2016/04/09_cartilha_ribeiraodamata_225x27cm_2016_04_13_issuuu.pdf. Accessed 24 May 2019

Comitê de Bacia Hidrográfica do Rio das Velhas (2016b) Cartilha Plano Diretor de Recursos Hídricos Unidade Territorial Estratégica Ribeirão Jequitibá. http://cbhvelhas.org.br/wp-content/uploads/2016/04/13_cartilha_ribeiraojequitiba_225x27cm_2016_04_13_issuuu.pdf. Accessed 24 May 2019

Comitê de Bacia Hidrográfica do Rio das Velhas (2016c) Cartilha Plano Diretor de Recursos Hídricos Unidade Territorial Estratégica Carste. http://cbhvelhas.org.br/wp-content/uploads/2016/04/11_cartilha_carste_225x27cm_2016_04_13_issuuu.pdf. Accessed 24 May 2019

Costa MT, Branco JJR (1961) Roteiro para a excursão Belo Horizonte - Brasília. In: 14th Congresso Brasileiro de Geologia, Universidade de Brasília, Brasília, November 1960

Da-Gloria P, Hubbe M, Neves WA (eds) (2017) Archaeological and paleontological research in Lagoa Santa: the quest for the first Americans. Springer, New York

De Paula RS, Velásquez LNM (2019) Balanço hídrico em sistema hidrogeológico cárstico, região de Lagoa Santa. Minas Gerais. Águas Subterrâneas 33(2):119–133

Dutra GM, Horta LS, Berbert-Born, ML (1998) Levantamento Espeleológico. In: Herrmann G, Kohler HC, Duarte JC, Carvalho PGS (1998) Gestão Ambiental APA Carste de Lagoa Santa. IBAMA – Instituto Brasileiro do Meio Ambiente e dos Recursos Naturais Renováveis/ Fundação BIODIVERSITAS/ CPRM – Companhia de Pesquisa de Recursos Minerais, Belo Horizonte

Feam – Fundação Estadual do Meio Ambiente (2018) Minas Trata Esgoto. Plano para Incremento do Percentual de Tratamento de Esgotos Sanitários na Bacia Hidrográfica do Rio Paraopeba. Belo Horizonte: FEAM, 2018. 499 pp

Ford D, Williams P (eds) (2007) Karst hydrogeology and geomorphology. Wiley, West Sussex

Galvão PHF (2015) Hydrogeological conceptual model of Sete Lagoas (MG) and associated implications of urban development in Karst region. Universidade de São Paulo, Thesis

Garcia VC (2011) Propostas de medidas de controle ambiental para a revitalização da área urbana do distrito de Fidalgo em Pedro Leopoldo – MG. Universidade Federal de Ouro Preto, Thesis

Herrmann G, Kohler HC, Duarte JC, Carvalho PGS (1998) Gestão Ambiental APA Carste de Lagoa Santa. IBAMA – Instituto Brasileiro do Meio Ambiente e dos recursos naturais renováveis/ Fundação BIODIVERSITAS/ CPRM – Companhia de Pesquisa de Recursos Minerais, Belo Horizonte

Holten B, Sterll M (2011) Peter Lund e as grutas com ossos em Lagoa Santa. UFMG, Belo Horizonte

Instituto Mineiro de Gestão das Águas—IGAM (2018) Avaliação da qualidade das águas superficiais de Minas Gerais em 2017: resumo executivo anual. Belo Horizonte: Instituto Mineiro de Gestão das Águas

Instituto Brasileiro de Geografia e Estatística (2010) População residente em 2010. https://cidades.ibge.gov.br/brasil/mg/belohorizonte/pesquisa/23/25207?tipo=ranking. Accessed 14 May 2019

Instituto Brasileiro de Geografia e Estatística (2018) Estimativas da população residente no Brasil e unidades da federação com data de referência em 1° de julho de 2018. https://www.ibge.gov.br/estatisticas-novoportal/sociais/populacao/9103-estimativas-de-populacao.html?=&t=downloads. Accessed 14 May 2019

Kresic N, Stevanović Z (2010) Groundwater hydrology of springs: engineering, theory, management, and sustainability. Butterworth-Heinemann (Elsevier), Amsterdam

Meneses ICRRC (2003) Análise Geossistêmica na Área de Proteção Ambiental (APA) Carste de Lagoa Santa. Thesis, Potifícia Universidade Católica de Minas Gerais, MG

Morato JB et al (2012) Avaliação de impacto ambiental no "Monumento Natural Estadual Gruta Rei do Mato"/MG. Paper presented at the 9° Simposio Nacional de Geomorfologia, Rio de Janeiro, 21–24 October 2012

Neves WA, Piló LB (2008) O povo de Luzia: Em busca dos primeiros americanos. Editora Globo, São Paulo

Nóbrega PFA (2015) Aves aquáticas da Área de Proteção Ambiental Carste de Lagos Santa: ecologia e conservação. Universidade Federal de Minas Gerais, Thesis

Patrus MLRA (1998) Estudos hidrológicos e qualidade das águas de superfície. In: CPRM – Companhia de Pesquisa de Recursos Minerais/IBAMA – Instituto Brasileiro do Meio Ambiente e dos Recursos Naturais Renováveis (ed) Área de Proteção Ambiental Carste de Lagoa Santa – Meio Físico. CPRM/IBAMA, Belo Horizonte, p 58

Pereira MC (2018) O carste sob os olhares da superfície: fragilidade, antropismo e planejamentos urbano e ambiental na região cárstica de Lagoa Santa – MG. Universidade Federal de Minas Gerais, Thesis

Pessoa PFP (1996) Caracterização Hidrogeológica da Região Cárstica de Sete Lagoas, MG: Potencialidades e riscos. Dissertation, Universidade de São Paulo

Pessoa PFP (2005) Hidrogeologia dos Aquíferos Cársticos Cobertos de Lagoa Santa. Thesis, Universidade Federal de Minas Gerais, MG

Pessoa PFP, Mourão MAA Levantamento Hidrogeológico (1998) In: CPRM – Companhia de Pesquisa de Recursos Minerais/IBAMA – Instituto Brasileiro do Meio Ambiente e dos Recursos Naturais Renováveis APA Carste de Lagoa Santa – Meio Físico/Volume I. CPRM/IBAMA, Belo Horizonte, pp 1–39

Piló LB (1998) Morfologia cárstica e materiais constituintes: Dinâmica e evolução da Depressão Poligonal Macacos-Baú - Carste de Lagoa Santa. Thesis, Universidade de São Paulo, Minas Gerais

Piló LB (1999) Ambientes cársticos de Minas Gerais: Valor, fragilidade e impactos ambientais decorrentes da atividade humana. O Carste 11(3):50–58

Pinto CP, Martins-Neto MA (eds) (2001) Bacia do São Francisco: Geologia e Recursos Naturais. Sociedade Brasileira de Geologia, Belo Horizonte

PROAPA (2018) Instituto PROAPA- Meio Ambiente, Desenvolvimento, Sustentabilidade. MAPA de Usos do Solo e Cobertura Vegetal. Unpublished report

Prous A (2012) Le plus ancien art rupestre du Brésil central: état de la question. In: Clottes J (ed) L'art pléistocène dans le monde (ed) Actes du Congrès IFRAO, Tarascon-sur-Ariège, Septembre 2010

Queiroz PGM, Jacomino VMF, Menezes MA de BC (2007) Composição elementar do material particulado presente no aerossol atmosférico do município de Sete Lagoas, Minas Gerais. Química Nova 30(5):1233–1239

Reis HLS, Alkmin FF, Fonseca RCS, Nascimento TC, Suss JF, Prevatti LD. In: Heilbron M, Cordani UG, Alkmim FF (eds) (2017) São Francisco Craton, Eastern Brazil – Tectonic Genealogy of a Miniature Continent. Springer, Cham, pp 117–143

Ribeiro JH, Tuller MP, Danderfer Filho A (2003) Mapeamento geológico da região de Sete lagoas, Pedro Leopoldo, Matozinhos, Lagoa Santa, Vespasiano, Capim Branco, Prudente de Morais, Confins e Funilândia, Minas Gerais (1:50.000). In: Projeto Vida. CPRM, Belo Horizonte

Santos SKC (2011) A Paisagem Cultural da Serra Santa Helena. Dissertation, Escola de Arquitetura da Universidade Federal de Minas Gerais

Silva AB (1988) Abatimento de solo na cidade de Sete Lagoas, Minas Gerais, Brasil. Revista Água Subterrânea 12:57–66

Souza H (1997) Zoneamento ambiental da APA Carste de Lagoa Santa. CPRM/IBAMA, Belo Horizonte

Tayer TC (2016) Avaliação da vulnerabilidade intrínseca do aquífero cárstico da APA de Lagoa Santa, MG, utilizando o método COP. Dissertation, Universidade Federal de Minas Gerais

Tuller MP, Ribeiro JH, Danderfer Filho A (1991) Geologia da região cárstica de Sete Lagoas-Lagoa Santa (MG). Projeto Vida – Programa Gestão e Administração territorial. Companhia de Pesquisa de Recursos Minerais, Belo Horizonte

Williams PW (1993) Karst Terrains: environmental changes and human impacts. Catena Supplement, Reiskirchen

Velásquez LNM, Sampaio JLS, Campello M, Dussin TM, Rodrigues PCH, Minardi PSP (2017) 2º Relatório Parcial de 04 do Projeto de Adequação e Implantação de uma Rede de Monitoramento de Águas Subterrâneas em Áreas com Cavidades Cársticas da Bacia do Rio São Francisco Aplicado à Área Piloto da APA Carste de Lagoa Santa. Minas Gerais, Belo Horizonte

Velásquez LNM et al (2018) Projeto de adequação e implantação de uma rede de monitoramento de águas subterrâneas em áreas com cavidades cársticas da Bacia do Rio São Francisco aplicado à Área Piloto da APA Carste de Lagoa Santa, Minas Gerais. Relatório parcial, PROCESSO FUNDEP/GERDAU/UFMG n22.317/Plano de ação nacional para a conservação do patrimônio espeleológico nas áreas cársticas da bacia do São Francisco. Pan Cavernas do São Francisco

Vieira LCM (2018) Hidrogeoquímica dos Aquíferos da Região da APA Carste de Lagoa Santa, MG. Dissertation, Universidade Federal de Minas Gerais

Xavier FV, Ribas RD, Braz AM (2017) Avaliação de impactos socioambientais na Gruta da Lapinha, Lagoa Santa/MG e seu entorno pela atividade turística. Geografia 26(1):19–33

Protection of the Environmental and Cultural Heritage in the Lagoa Santa Karst: History and Challenges

Luciana Alt and Vitor Moura

Abstract

The anthropic pressure over the use of Lagoa Santa Karst's natural resources increased exponentially in recent decades. Simultaneously, a great effort to protect this complex and vulnerable environment sought to counter the surge in human presence and the impacts of economic activities in the area. The Lagoa Santa Karst located in the Belo Horizonte Metropolitan Region houses archaeological and paleontological sites of international interest as well as caves that are relevant for their multifaceted role, i.e., in the history of South American Anthropology and in the complex account of local conservation and sustainability efforts. The Lagoa Santa Karst, among all other Brazilian karst sets, is the most often researched and the best-documented example. However, this is the Brazilian karst region currently under the heaviest impact of anthropic activities, i.e., mining, industrial processing, and urban occupation. In recent decades, several attempts at regulation and sustainable development were partially successful. Other analogous efforts were never totally implemented despite legal planning. This article recounts and contextualizes the escalation of threats to heritage sites in the target area to the overall historical development of Brazilian Environmental Legislation. It underscores the existence of gaps and challenges for karst protection and recommends actions that may contribute to enhance the region's environmental quality. The political and economic crisis since the late 2010's may jeopardize the future of conservation in the Lagoa Santa Karst and in the rest of the country, as it poses yet another complication to an already defective set of legal practices that subsists despite the very advanced—in spirit—nature of the Brazilian environmental legal corpus.

Keywords

Karst conservation • Protected areas • Environmental policy • Lagoa Santa Karst • Cave protection

1 Introduction

Brazil in its continental dimensions presents karstic systems in different climatic environments, biomes, and lithologies. Within Brazilian karst areas, there are numerous examples of unique archaeological and paleontological sites, often found in caves where, for millennia, they rest undisturbed by weather or human activity (Cartelle 2012). The environmental stability found in caves also makes them prone to harboring endemic fauna, highly specialized and sensitive to environmental changes (Gnaspini and Trajano 2000). Furthermore, karst areas house aquifers, some of which are large and dynamic, with faster water flows and increased structural complexity when compared to aquifers found in other types of terrain (Veni et al. 2001). The aquifers located within karst regions are remarkably vulnerable to contamination and changes in water quality that can affect large areas relatively quickly (Ford and Williams 1989; Hötzl 1999).

The protection of Brazilian karsts is increasingly challenging, as the growing pressure for economic development imposes the continued expansion of agricultural, urban, and industrial boundaries over this environment. In many cases these activities happen without proper planning or formal control, generating, historically, direct and indirect impacts over the terrain.

The Lagoa Santa Karst preserves forest remnants associated with limestone massifs and wetlands. Within its boundaries remains the largest concentration of officially cataloged caves in Brazil (Auler 2016 and chapter "Caves and Speleogenesis in the Lagoa Santa Karst", this volume). Moreover, the region has

L. Alt (✉) · V. Moura
Instituto Do Carste, Rua Barcelona 240/302, Belo Horizonte, MG 30360-260, Brazil
e-mail: luciana.alt@gmail.com

V. Moura
e-mail: vmamoura@gmail.com

A. S. Auler and P. Pessoa (eds.), *Lagoa Santa Karst: Brazil's Iconic Karst Region*, Cave and Karst Systems of the World, https://doi.org/10.1007/978-3-030-35940-9_14

several archaeological sites, some of which have historically contributed to the debate on the origin and evolution of hominids in the Americas (Neves et al. 2012 and chapter "The Archaeological Record of Lagoa Santa (East-Central Brazil): From the Late Pleistocene to Historical Times", this volume). There are also paleontological sites where invaluable finds in Pleistocene megafauna occurred (Neves and Piló 2008 and chapter "Cave Paleontology in the Lagoa Santa Karst", this volume). The region's natural and cultural heritage first became known to the world in the nineteenth century through the work of the Danish scientist Peter Wilhelm Lund (Cartelle 2012). From then on, the Lagoa Santa Karst is considered extremely relevant at the national and international levels both for research and for conservation purposes.

This karst environment is located at the northern boundary of the third-largest urban complex in Brazil, known as the Belo Horizonte Metropolitan Region—BHMR. This agglomeration houses approximately 5.8 million inhabitants (IBGE 2014). Despite its rich environmental heritage, some key activities contributed to polarize the recent anthropic expansion over the Lagoa Santa Karst. For instance, the establishment of quarries and limestone processing industries since the late 1950s; the installation of an international airport wedged amid important speleological and archaeological sites in the early 1980s; and in the beginning of the new millennium, the implementation of the new administrative infrastructure for the State Government stimulates the economic growth in the north part of BHMR, and promoted a rapid urban sprawl toward this karst area (UFMG 2011).

The demand for land and for environmental resources generated by these activities historically resulted in harmful and lasting changes on karst. In recent decades, many landscape changes took place such as deforestation (Alt 2008) and destruction of karst features, including caves, archaeological and paleontological sites (Prous et al. 2003), moreover, different impact was registered, such as surface and underground water contamination, induced sinkhole development (Galvão et al. 2015), potential loss of biodiversity on surface and underground systems (Herrmann et al. 1998), air pollution (Queiroz et al. 2007), accumulation of particulate matter over sensitive features (Alt 2010), actions of vandalism—the breakage of speleothems, graffiti, accumulation of garbage—in caves and archaeological sites (Alt and Moura 2016) among others (see chapter "Environmental Problems in the Lagoa Santa Karst", this volume).

The pursuit of a balance between economic development and the Lagoa Santa Karst protection is complex and challenging. This is expressed by several attempts to create and manage protected areas and to generate specific protective legal instruments, since the 1960s. Historically, however, combined factors meant that the first protected areas created by public initiative in the region, between 1980 and 1990, had low effectiveness (Alt 2008). Meanwhile, natural heritage has been gradually fragmented. Virtual islands of life with low connectivity were consolidated. These were vulnerable to environmental problems, i.e., aquifer contamination, air pollution, and biodiversity loss.

At the beginning of the present millennium, the protection of the Lagoa Santa Karst had drawn a more positive outlook, though. Some initiatives searched to promote the sustainable development of the region. For instance, a few "protected areas" that were abandoned or functioned defectively since the early or mid-1980 were finally implemented. New "smaller in size" protected areas, i.e., Natural Heritage Private Reserves NHPRs and Natural Monuments were created to protect karst features or loci of interest within the Lagoa Santa Karst. Along with that, environmental education initiatives sought to expand the local population's knowledge about Lagoa Santa Karst's natural functioning and its fragilities. Most of these conservation efforts became possible through the Environmental Compensation demanded from potentially impact-causing enterprises. The empowerment of local environmental agencies and the development of Brazilian legal apparatus since 1980 was relevant to achieve this. Nowadays the process of scientific evaluated Environmental Licensing is vital for providing environmental protection. It articulates Compensatory Measures favoring the rational use of local environmental resources.

This chapter provides an overview of the development of Brazilian environmental laws along with a panorama of local economic, social progresses, and environment actions aiming protection of the Lagoa Santa Karst. This chapter has named the challenges related to the protection of this fragile environment to contribute to the discussion of actions needed to protect this heritage complex.

2 Study Area

In the south-central part of the State of Minas Gerais, about 30 km north of the state capital, Belo Horizonte, lays the Lagoa Santa Karst. The limits of this karst area comprise the municipalities of Vespasiano, São José da Lapa, Confins, Lagoa Santa, Pedro Leopoldo, Matozinhos, Prudente de Morais, and Sete Lagoas. All but the latter two belong to the BHMR (Fig. 1). Nowadays this karst area is home to about 500,000 people (Auler 2016).

The Lagoa Santa Karst is primarily composed of carbonate rocks of the Bambuí Group. It has two distinct morphological domains. First, the exokarst presents a high concentration of large limestone outcrops as well as large areas of partially covered limestone. The surface common features include vast doline plains, the presence of lakes associated with dolines and blind valleys (see chapter "Karst Landforms in the Lagoa Santa Area", this volume). Second, the endokarst presents a complex underground drainage

Fig. 14.1 Lagoa Santa Karst location in relation to BHMR

system and it also contains numerous caves (CPRM 1994, chapter "Caves and Speleogenesis in the Lagoa Santa Karst", this volume).

The local climate is warm with two well-defined seasons. Warm–wet summer and a cool–dry winter. Temperature varies only 5 °C year-round. The mean annual rainfall stands around 1,381 mm. Regarding its flora, the Lagoa Santa Karst has remaining patches of savanna and semi-deciduous forest, the latter being associated with doline fields and limestone outcrops (see chapter "The Vegetation of Lagoa Santa Karst", this volume, Herrmann et al. 1998).

The Lagoa Santa Karst is the cradle of Brazilian paleontology, archaeology, and speleology. Its assets, recognized as highly relevant at the international level, became known to the world in the second quarter of the nineteenth century through the works of the Danish naturalist Peter Wilhelm Lund. His research set the spotlight on the region due to the debate over the origin of hominids in America, and the possible coexistence of man with extinct megafauna (Neves and Piló 2008, chapter "The Archaeological Record of Lagoa Santa (East-Central Brazil): From the Late Pleistocene to Historical Times", this volume). Eugen Warming also held relevant botanical studies in the region. These were published in 1892 (Gomes 2006). Also, many nineteenth century naturalist travelers who followed the course of Velhas River, such as Sir Richard Burton, also noted in their travelogues some of the peculiarities of what would later be known as the Lagoa Santa Karst (chapter "History of Research in the Lagoa Santa Karst", this volume).

In the early twentieth century, the area housed several archaeological excavations promoted by scientists such as Jorge Padberg-Drenkpohl, Wesley Hurt, Harold Walter, Carlos de Paula Couto (Da Gloria et al. 2014). National and international institutions funded fieldwork and publications. In the 1970s, the Franco-Brazilian Archaeological Mission, coordinated by Annette Laming-Emperaire delved into the Lagoa Santa hominid, furthering Dr. Lund's pioneer finds. In later decades, the region received several local university research teams. Federal University of Minas Gerais' Archaeology Sector, coordinated by Dr. André Prous, was

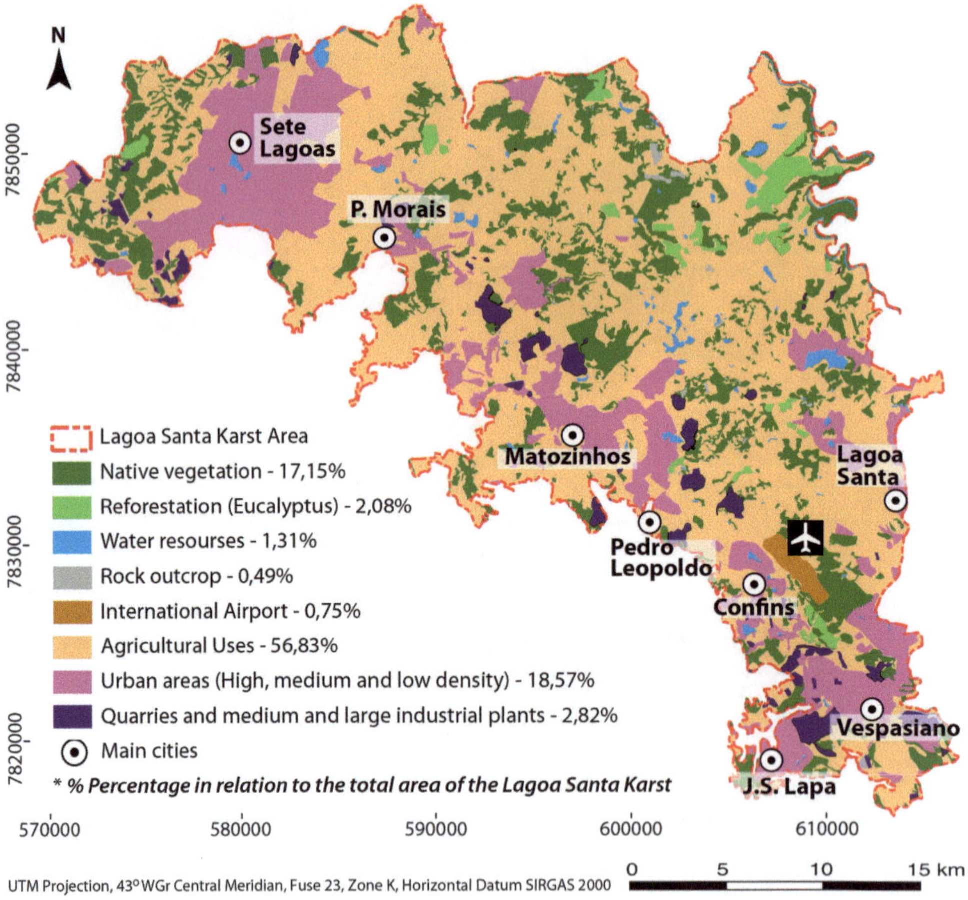

Fig. 14.2 Map of the main land uses in the Lagoa Santa Karst. Adapted from Brandt Meio Ambiente (2008)

particularly prominent. From 2000 on, University of São Paulo's Biosciences Institute team (IB-USP), coordinated by Dr. Walter A. Neves, lay emphasis on biological anthropology. The discussion of the origin and evolution man in America was expanded opening new conceptual possibilities concerning the settlement of the continent (Neves and Piló 2008, chapter "The Archaeological Record of Lagoa Santa (East-Central Brazil): From the Late Pleistocene to Historical Times", this volume).

In the field of speleology, particularly during the 1980s and 1990s, caving groups voluntarily set forth several prospection and cave mapping efforts. These works were later followed by "Projeto Vida" and Lagoa Santa Karst Environmental Protected Area Project, both carried out in the 1990s by the Geological Survey of Brazil—CPRM. From the year 2000 on, several companies specialized in environmental licensing ran intensive speleological surveys. An increase in the number of scientifically described caves was achieved since specific legislation conditioned the feasibility of economic endeavors to the technical inventorying of impacted areas. Nowadays, more than 2000 caves are documented throughout the Lagoa Santa Karst (Auler 2016).

Besides its importance as a unique geomorphological and biological domain, the Lagoa Santa Karst also stands out nationwide thanks to the significance of human activities carried out upon it from prehistorical to modern times. As far as modern endeavors, the region has a long history in mining and industry. The state of Minas Gerais owes a great deal to the Lagoa Santa Karst as its mines churned out ton upon ton of raw material to produce cement, lime, gravel, and sand since the real estate construction boom and population growth from the 1960s (Herrmann et al. 1998). This karst region also stands out, at the national level, for its expressive anthropic occupation.

Currently, human activities predominate over the Lagoa Santa Karst surface. The agricultural uses occupy 56.83% of the area, followed by urban uses (18.57%), industrial and mining activities (2.82%), and eucalyptus plantation (2.08%). These islands of native vegetation correspond only to 17.15% of the area (Fig. 2).

3 Methods

To evaluate the strategies and challenges related to the management of the protected areas, a series of interviews was conducted with regional coordinators of the State Forest Institute—IEF-MG and the Chico Mendes Institute for Biodiversity Protection—ICMBio. For the assessment of specific issues regarding protected areas that figure in Table 1, the managers of the following areas were consulted: International Airport Special Protected Area—SPA,

Table 1 Protected areas in the Lagoa Santa Karst

Protected area	Year established	Instance	Current total area (ha)	Corresponding category in IUCN	Land expropriation regime	Management plan?	Technical staff?	Management councils?	Infrastructure?
International Airport SPA	1980	State	37.715,6	VI	NM	No	Yes[a]	No	No
Sumidouro State Park	1980	State	2001,9	II	Partial	Yes	Yes	Yes	Yes
Lagoa Santa Karst EPA	1990	Federal	35.600,0	VI	NM	Yes	Yes	Yes	Yes
Fazenda Bom Jardim NHPR	1997	Federal	172,8	III	Yes	Yes	Yes	NM	Yes
Municipal SPA Serra de Santa Helena	1997	Municipal	4928,0	VI	NM	Yes	NE	NE	NE
Fazenda Campinho NHPR	2001	State	43,0	III	Yes	Yes	NE	NM	NE
Fazenda Vargem Alegre NHPR	2001	State	9,7	III	Yes	No	NE	NM	NE
Sol Nascente NHPR	2004	State	60,3	III	Yes	Yes	NE	NM	NE
Rei do Mato Cave SNM	2009	State	141,4	III	Partial. NM	Yes	Yes	Yes	Yes
Cerca Grande State Park	2010	State	134,2	II	No	No	Yes	Yes	No
Vargem da Pedra SNM	2010	State	10,1	III	NM	No	Yes	Yes	No
Experiência da Jaguara SNM	2010	State	38,5	III	NM	No	Yes[a]	Yes	No
Várzea da Lapa SNM	2010	State	23,5	III	NM	No	Yes[a]	Yes	No
Santo Antônio SNM	2010	State	31,1	III	NM	No	Yes[a]	Yes	No
Lapa Vermelha SNM	2010	State	33,7	III	NM	In process	Yes[a]	Yes	No
Serra de Santa Helena Municipal Park	2011	Municipal	Not delimited	II	No	No	No	No	No

[a]It has only one manager
NE Not evaluated
NM Not mandatory

Lagoa Santa Karst Environmental Protected Area—EPA, Sumidouro State Park, Cerca Grande State Park, and Lapa Vermelha State Natural Monument. Seeking further clarification on the social mobilization processes and challenges related to the management of the region was interviewed the Chairman of the Subcommittee on Mata Creek Hydrographic Basin. All interviews were conducted in October 2016 and followed a basic scope of 10 questions, qualitative and semi-structured in nature. The average interview lasts about an hour and a half.

The maps produced for this article were made using Quantum GIS.—Geographic Information System, version

QGIS 2:18. The final layout of the maps was done in Adobe Illustrator CS5. The perimeters of protected areas were provided in digital media by IEF-MG. The most complete database on the caves registered in the region, which was used in this article, is CANIE (National Register for Speleological Information), and it was provided by CECAV (2016). A voluntary caving association "Grupo Bambuí de Pesquisas Espeleológicas" provided the maps of the Lapa Vermelha caves.

4 Results

4.1 Scenario of the Protected Areas in the Lagoa Santa Karst

The last 40 years saw environmental and cultural heritage initiatives that underscored Lagoa Santa Karst's preservation. Some "protected areas" were created which in Brazil are called "conservation units" (Fig. 3). Civil society was organized into task forces of interest/action, i.e., providing education and expanding knowledge about existing heritage, contributing to improving water quality, protecting caves, and more. At the institutional (state and federal) level, structuring efforts were strengthened in some conservation units already established. It includes the elaboration of Management Plans, the planning and installation of supporting infrastructure, and the oversight of potentially conflicting activities. At the municipal level, land usage Master Plans were created or reviewed in accordance to some use restrictions imposed by the protected areas' zoning assessment. At the private initiative level, it was sought to balance/rationalize the exploitation of environmental resources through the realization of environmental licensing processes and through application of approved Environmental Compensation measures.

Starting in 1980, 16 protected areas were created in the Lagoa Santa Karst. The State Forest Service, IEF-MG, is responsible for managing 75% of those conservation units. The federal agency ICMBio is responsible for 12.5%, and the municipality of Sete Lagoas covers the remaining 12.5% (Table 1).

These protected areas created in Lagoa Santa Karst are inserted in the National System of Conservation Units—NSCU (Brasil 2000). The NSCU was established in Brazil,

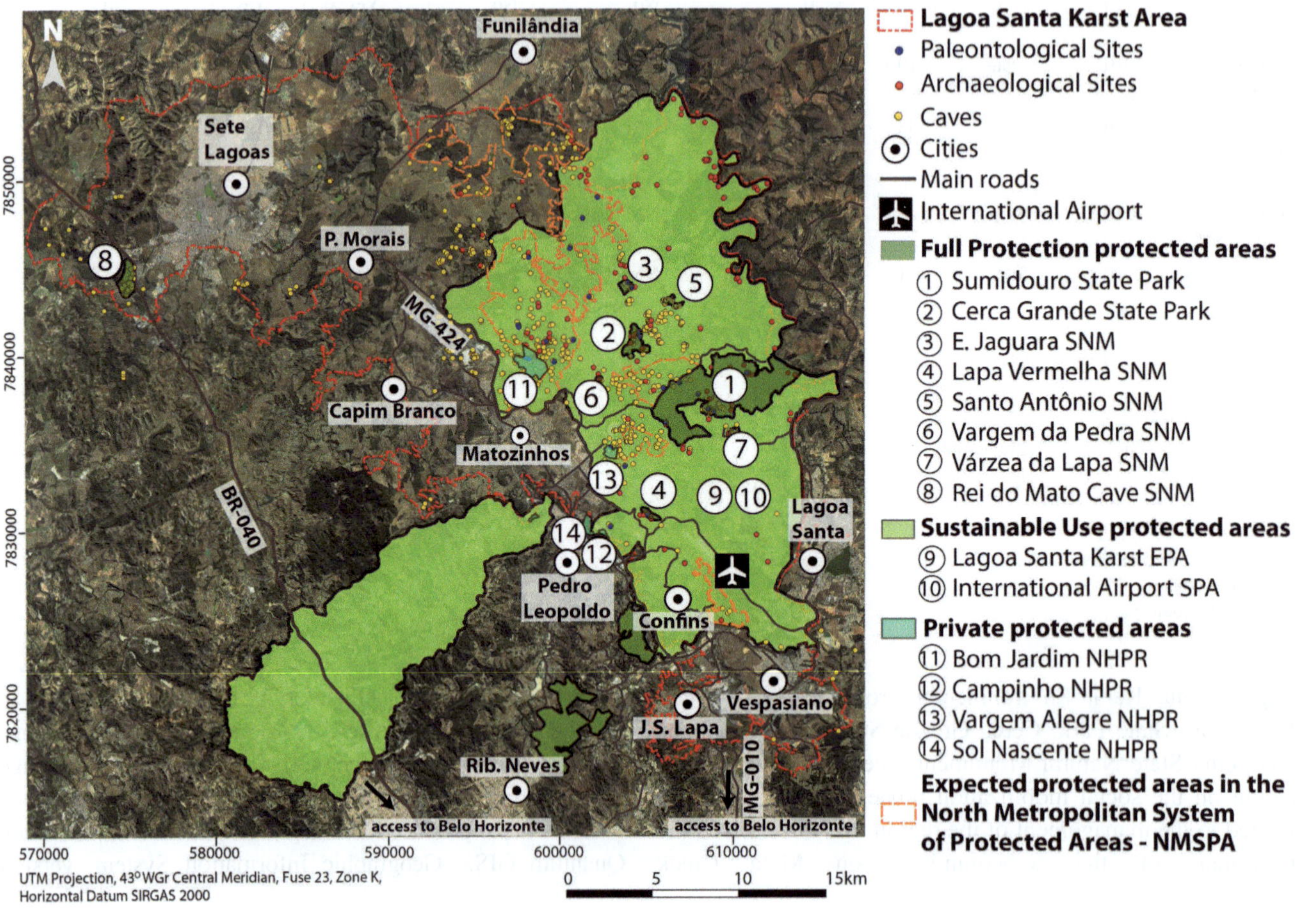

Fig. 14.3 Map of the protected areas in the Lagoa Santa Karst

in the year 2000, determining norms for the creation, implementation, and management of the protected areas. The NSCU categories of protected areas have high level of correspondence to the international system of classification of protected natural areas, established by the International Union for Conservation of Nature—IUCN, although there are differences of nomenclature between both systems (Table 14.1). The Brazilian protected areas are divided into two groups: Full Protection and Sustainable Use (Brasil 2000). The first group aims to keep ecosystems protected of changes caused by human interference, and only allow the indirect use of their natural resources. In this group, there are five categories of protected areas in Brazil, which corresponds to IUCN categories I to III: Ecological Station (IUCN—subcategory Ia), Biological Reserve (IUCN—subcategory Ia), National Park (IUCN—category II), Natural Monument (IUCN—category III) and Wildlife Refuge (IUCN—category Ib). The Sustainable Use protected areas group aims to protect natural ecosystems but, at the same time, allows different levels of use of natural resources, in a sustainable approach. This group contain seven categories of protected areas which corresponds to IUCN categories IV to VI: Environmental Protected Area (IUCN—category VI), Area of Relevant Ecological Interest (IUCN—category IV), National Forest (IUCN—category IV), Extractive Reserve (IUCN—category VI), Fauna Reserve (category V), Sustainable Development Reserve (category VI). The Natural Heritage Private Reserve, despite being included in the Sustainable Use group, has management objectives similar to IUCN—category III, related with Full Protection group.

The effectiveness of Lagoa Santa Karst protected areas depends on varied factors, i.e., land tenure issue resolution, Management Plan implementation, existence of ground staff and management council, infrastructure installation, etc., (IBAMA-WWF 2007). Information on those was volunteered by managers of certain conservation units.

As for land expropriation, it was not mandatory in little more than half of the conservation units created (Table 1). In these areas, the Management Plans mandate land-use restrictions to private owners. This happens, for example, to safeguard certain features present within the still private parcel (i.e., caves) that are deemed "of public interest/constitutionally protected." Since its inception, these type conservation unit entails only a very small initial expenditure on the part of the federal or state government. For that they produce attractive conservation statistics. However, the management constantly brushes against the owners' pursuit of private interests, thus, the effectiveness of heritage protection tends to be low (Alt 2008). Among these protected areas without mandatory expropriation of land figure Lagoa Santa Karst EPA and International Airport SPA. These superposed conservation units include almost half of Lagoa Santa Karst's total area. Another example that does not necessarily involve mandatory expropriation is the State Natural Monuments (SNM). Compared to the EPA and SPA, their extension is small as they target specific heritage protection, such as caves and archaeological sites. Natural Monuments may demand destitution of ownership, i.e., particularly in the case that a private owner's activities put in danger the protected area's management goals or heritage integrity. Only the Natural Heritage Private Reserve (NHPR), created by owner's initiative, present a stable status on the issue of land ownership. Among State Parks, where is mandatory expropriation, only Sumidouro State Park presents partial land regularization, while the others remain without this implemented action. Overall, it is possible to notice that the level of governmental investment on land regularization of protected areas is still small within the Lagoa Santa Karst (Table 1).

Almost half of the protected areas, especially those created before 2010, present Management Plans. So, almost all NHPR areas already have their Management Plans, or are in the process of drafting one. As for new Natural Monuments, created from 2010, there is still a gap in the usage of the provision. That indicates a long way to go to accomplish effective management especially in the public protected areas.

Most of the protected areas have the technical staff responsible for their management, generating positive conservation statistics. However, about one-third of these areas are "manager-only". These managers lack direct contact with the protected area. The absence of a robust technical staff makes it difficult to perform actions such as oversight, visitation support, forest fire control; thus, the manager becomes figurative. This type of problem is common to new Natural Monuments. In the search for a solution IEF-MG recently adopted the sharing of the technical staff, mainly of the forest fire brigade from Sumidouro State Park with Natural Monuments, located in its immediate surroundings. This may prove to be a cost-effective solution for the future. For instance, Natural Monuments created before 2010 have shared management councils. This new directive may benefit more protected areas since it bypasses red tape promoting systematic integration among different conservation units through social participation.

As for infrastructure, only about a quarter of the protected areas have arrangements in place to support maintenance and/or visitation, a rather low percentage in relation to the heritage value and the educational or recreational potential of these karst resources. This indicates that public use activities need to be valued and made available as an instrument of social inclusion and environmental protection.

The IEF-MG was not able to provide detailed information on municipal protected areas and state run NHPRs. This may indicate problems in the integrated management of protected areas at the state and municipalities' level. Thus, a more fluid integration rests to be developed among all players involved in the conservation efforts.

4.2 Brazilian Legal Evolution and the Protection of the Lagoa Santa Karst

Since the 1930s, a legal corpus regarding environment conservation and the rational use of natural resources was gradually created. A few laws are milestones in the legislative process. Some examples are the Forest, Water, and Mines Code, in the 1930s; the implementation of the National Environmental Policy in 1980; the enactment of the articles concerning the environment in the 1988 New Federal Constitution and, in 2000, the creation of a National System of Conservation Units—NSCU. This set of laws, summarized in Table 2, constitutes the current basis for the management of national environmental resources, for the licensing and compensation for environmental damage, for the creation and management of protected areas, and for the protection of the speleological, archaeological, and paleontological heritage.

Regardless of the challenges met for their implementation and further enforcement, Brazil has now a complex institutional system of environmental management oriented by a comprehensive legislative corpus. A challenge to their application may lay at the institutional fragility of certain public organizations overseeing natural resources. Shortage of financial resources, lack of political interest, and even shortage of technical staff are just a few reasons for failure. Another difficulty met by public organizations is the issue of defining the area of action of each of these organizations.

In the New Federal Constitution of 1988, all Brazilian caves were considered Federation assets. In 1990, following this line, new laws determined the unconditional preservation of all caves in the country (Brasil 1990). This fact nonetheless proved to become in practice a contentious point if one considers the political bias that favors economic development over conservation. For instance, right before the enactment of this law, in 1989, a mining company deliberately brought down the "Lapa do Arco," an archaeological site in the municipality of Matozinhos (Prous et al. 2003). The site contained a significant amount of rock art, all vanished forever. This disrespect for the local natural and cultural heritage underscores the vulnerability of such sites, despite the existence of strict legislation.

In 2004 a new legislation (MMA 2004) softened the preservationist rigor of previous legislation. In this perspective, it was determined that the environmental licensing of potentially damage-causing activities in karst areas should distinguish between relevant or nonrelevant caves. In 2008 and 2009 another set of laws, currently in force, determined the criteria for establishing cave relevance. For the first time, the legalized suppression of caves was authorized by law through environmental compensation.

These new laws, promulgated since 2004, stimulated the discovery and accomplishment of detailed environmental studies in hundreds of caves in Lagoa Santa Karst area increasing the knowledge about the local heritage and making difficult its illegal destruction. In addition, they stimulated the preservation of karstic features not protected by specific legislation, i.e., dolines and outcrops, when inserted in cave influence area. Minimally the legislation requires a perimeter of protection of 250 m around the horizontal projection of the cave until its permanent influence area is established, which can be much larger. As in Brazil, the natural voids of sufficient size to allow the entrance of a human being are considered caves, and that in Lagoa Santa Karst there are more than 2000 known caves, the amalgamation of their influence areas would result in the legal protection of vast karst areas (Auler 2016).

The environmental licensing processes around protected areas should receive consent from their managers. Recently, IEF-MG and ICMBio, responsible for the protected areas management in the Lagoa Santa Karst, intensified their activities, increasing the inspection of potentially damaging activities and in the suggestion of compensatory measures. The empowerment of these institutions and the increase in their influence represents a significant advance in the protection of the regional heritage.

However, fundamental issues are still relevant once heritage preservation is opposed to economic interests. If a cave or an archaeological site is unique and irreplaceable, how could their destruction be legalized and absorbed by society? In case of the authorized destruction of a cave, how the obliteration of a unique structure could be properly quantified and then accordingly compensated? How to determine if the compensatory measures are excessive? Are the current scientific and technical knowledge accurate enough to allow for safe decision-making, which will not be regrettable in the future? These and other issues are currently under discussion, as the implicated agents continuously seek to define the future parameters to ensure the protection of environmental and the cultural heritage in Lagoa Santa Karst and in other Brazilian karst areas.

The pressure for economic development and the current (2017) political instability in Brazil may both account for the weakening of the national environmental legislation, as occurred with the revision of the Forest Code (Brasil 2012), (Table 2). Currently, the karst features are under the same kind of legislative pressures. In a deregulation scenario, the preservation of relevant features would be under constant threat.

Table 2 Key federal laws related directly or indirectly to the management of speleological heritage in Lagoa Santa Karst

Year	Legal instrument	Summary
1934	Forest Code (modified in 1965, 1989 and 2012), (Brasil 2012)	Legally defined "protective forests" and allowed for the creation of national, state, and municipal parks. In 1965, it established as Permanent Preservation Areas (PPA) forest belts situated along rivers, lakes, fresh water springs, and on hill tops
1934	Water Code and the National Water Resources Policy (modified in 1997), (Brasil 1934)	Established criteria for protection of surface and groundwater resources. Water was deemed a limited resource with economic value; its management should be decentralized and combined with environmental management
1934	Mining Code (amended in 1967 and 1996), (Brasil 1967b)	Established criteria for the granting of research permits and of mineral exploration. Established that the subsoil is the property of the Federation, so caves, although not directly mentioned in the legal text, are property of the federal government since 1934
1937	Protection of Historical, Artistic and Archaeological Heritage (Brasil 1937)	Established the legal instrument of "tombamento", the decree of safeguarding of historical or natural heritage sites deemed at risk
1961	Protection Act of archaeological and prehistoric monuments (Brasil 1961)	The archaeological or prehistoric monuments in national territory fall under the care and the protection of the federal government. The economic exploitation, the destruction or alteration of archaeological or prehistoric sites is forbidden.
1967	Fauna Protection Code (Brasil 1967a)	Established that the wildlife (including the cave fauna) as well as their nests, shelters, and natural breeding grounds (which includes caves) are federal property. Their use, poaching or destruction are prohibited.
1977	Law protecting sites of touristic interest (Brasil 1977)	Established criteria for protecting areas of special touristic interest, applicable to caves that allow tourist visitation. Any activity that hinders the site's touristic potential should be restricted
1981	National Environmental Policy (Brasil 1981)	Framework for the management of the Brazilian environment. Established among other instruments, the Impact Assessment; Environmental Licensing; disciplinary or compensatory penalties for non-compliance with conservation measures or environmental degradation correction. Created spaces specially protected, i.e., the Environmental Protection Areas (EPA), Extractive Reserves and others. It created the National Environmental System—SISNAMA and the National Environmental Council—CONAMA, improving coordination between municipal, state and federal governments
1985	Diffuse Interests Law (Brasil 1985)	Expanded the concept of environmental damage through the definition of Diffuse Interests, granting Public Ministry more power to deal with environment relevant cases
1988	Federal Constitution (Brasil 1988)	Recognizes the environment as a public good essential to the maintenance of quality of life. Caves and archaeological sites are declared federal property
1990	Federal Decree No. 99.556 (Brasil 1990) and Decree No. 887/90 (Ibama 1990)	States "any natural cavity penetrable by man and its respective area of influence is unconditionally to be preserved." It mandates the preparation of Environmental Impact Study for activities located in areas of speleological potential States that the area of influence of a cave must be defined by specific technical studies, the minimal length of 250 m surrounding the horizontal projection of the given cave being adopted until those studies are concluded
1998	Environmental Crimes Law (Brasil 1998)	Determines criminal and administrative sanctions to any conduct or activity that proves to be harmful to the environment
2000	National System of Protected Areas (Brasil 2000)	Establishes norms for the creation and management of protected areas
2001	Cities Statute (Brasil 2001)	Included among the instruments of urban policy the environmental impact study and a study of neighborhood impact
2004	CONAMA Resolution No. 347/04 (MMA 2004)	Recognized the need to determine the relevance of caves. Mandates environmental compensation for impacts over the speleological heritage
2008	Decree No. 6640 (Brasil 2008)	Mandates the classification of relevance for caves per ecological, biological, geological, hydrological, paleontological, scenic, historical, cultural, and socioeconomic attributes
2009	Normative Instruction No. 2 (MMA 2009)	Regulates Decree No. 6640. Determines methodology for establishing cave relevance
2012	Forest Code revision (Brasil 2012)	In 2012, developmental pressure prevailed in the modification of this Law, implying losses in previous environmental achievements: PPA reduction and de-characterization; End of the obligation to recover illegally deforested areas, until July 22, 2008, for properties of a specific size; Reduction of the legal reserve in the Amazon; among others

4.3 Case Studies

After outlining Lagoa Santa Karst's current environmental protection status it is important to show in more detail the history of the relationships involved in the protection of this natural and cultural heritage. Therefore, some examples contextualized in time and space are summarized in Fig. 4.

4.3.1 Before the 1980s

In precolonial times the existing evidence allows us to imagine that the probable type of human settlement in Lagoa Santa Karst was scant and sparse, inserted in areas of easier access, such as the plains surrounding karst lagoons, which concentrated environmental resources (Neves and Piló 2008). Several caves and rock shelters near these features show indications of prehistoric use (see chapter "The Archaeological Record of Lagoa Santa (East-Central Brazil): From the Late Pleistocene to Historical Times", this volume).

These same areas of subdued relief were historically occupied with greater intensity from the eighteenth century on. From then on, plains were grounds for the implementation of agricultural and cattle rising, which soon had impacts on vegetation and other natural resources. However, limestone outcrops, due to their difficult access, were naturally avoided for agricultural or urban activities.

In colonial times (1500–1889), and even during the First Republic (1889–1930), due to the search for basic resources in the endokarst, many caves were intensely altered. Their deposits were often prospected for saltpeter, a key ingredient for gunpowder. Saltpeter extraction caused the destruction of unique sedimentary deposits, many containing fossil materials. This activity was described as early as in the nineteenth century by P. W. Lund as a huge scientific loss (Holten et al. 2011). Other extraction activities such as small lime quarries, the removal of fossils for commercial or scientific purposes, or even timber extraction, albeit their small scale left permanent marks in caves and their surroundings.

In the late nineteenth century, a railroad improved access along Lagoa Santa Karst's southwest limit, stimulating economic development in the region. From 1950s on, several mining and limestone processing industries were established along this railroad, that runs parallel to the highway MG-424, and connects the municipalities of Lagoa Santa to Sete Lagoas. Such industrial activities stimulated the improvement of the road network and fostered the economic development of urban areas along the road's path. At that time, loose law enforcement and incipient environmental legislation meant that several caves, archaeological, and paleontological sites were significantly impacted or destroyed by mining. Among these stand out the archaeological sites of the Carrancas Complex, in São José da Lapa; the Lapa Vermelha in Lagoa Santa, researched by Lund; and Trevo Rockshelter, home to rock art in Sete Lagoas (Prous et al. 2003). To judge the state of heritage preservation from the 1950's to the 1980's, it suffices to recall that until the mid 1980s, speleothem fragments were still sold as souvenirs to visitors at Lapinha Cave, inside Sumidouro State Park (Herrmann et al. 1998).

The first formal initiative of specific site protection within the Lagoa Santa Karst, took place in 1962 with the declaration of national heritage ("tombamento") of the Archaeological Site of Cerca Grande. This initiative was carried out at the federal level by SPHAN—National Service for Historic and Artistic Heritage (Santiago 2015). This occurred about a year after the creation of a specific federal law to protect the archaeological heritage (Fig. 4). In Brazil "tombamento" is a key legal protection instrument designed for cultural and natural heritage sites. In effect, it serves as legal proof of value for specific assets, legally protecting those from alterations to their present physical state (Lima and Breviglieri 2017). In practice, if not followed by law enforcement, surveillance or heritage preservation through educational efforts, these same sites become frequent targets to disfigurement or destruction. The archaeological site of Cerca Grande is an example of that. A target of vandalism actions caused by decades of uncontrolled visitation, such as numerous *graffiti* on rock art panels, speleothems breakage, wide disturbance of the superficial sedimentary layers, and degradation of vegetation in its surroundings by agricultural use, among other problems (Prous et al. 2003).

Throughout the 1970s, encouraged by the nascent environmental movement, members of the scientific community and of the civil society exerted political pressure on lawmakers. They demanded the creation of karst protection strategies for Lagoa Santa. Solutions started to be pointed mainly in a specific work group that gathered researchers from the Federal University of Minas Gerais (Herrmann et al. 1998).

4.3.2 The 1980's and the Creation of the First Protected Areas Within the Lagoa Santa Karst

In the early 1980s, a combination of factors led to the creation of the first protected areas in the Lagoa Santa Karst. A specific environmental protection policy was implemented at the national level and laid the foundations for the current processes of Environmental Licensing and Environmental Compensation (Table 2). Concomitantly, the International Airport was also being installed amid important archaeological sites such as the complex of Lapa Vermelha, in Pedro Leopoldo. This fact implied broad landscape alterations, with possible direct and indirect impacts on the speleological and archaeological heritage. Therefore, based on this new national environmental policy and on social pressure for

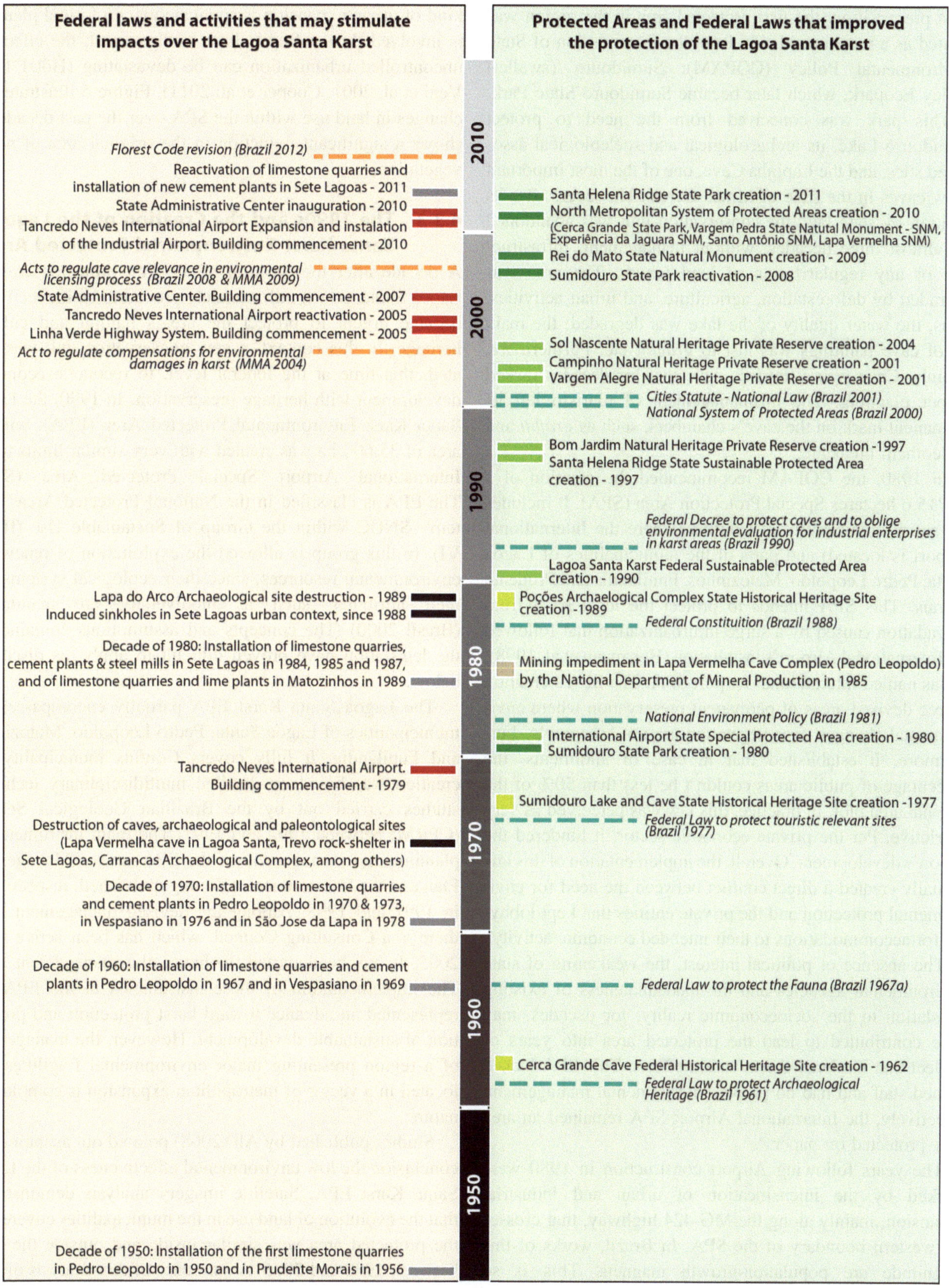

Fig. 14.4 Timeline: environmental protection actions versus main threats to the protection of the Lagoa Santa Karst

karst preservation, the first protected area in the region was created as a recommendation from the Commission of State Environmental Policy (COPAM): Sumidouro (swallet) Valley Ecopark, which later became Sumidouro State Park.

This park was conceived from the need to protect Sumidouro Lake, its archaeological and speleological associated sites, and the Lapinha Cave, one of the most important show caves in the state of Minas Gerais. However, after its creation by state law, the conservation unit was abandoned for almost three decades, without management, infrastructure or any regularization of land tenure. The area was degraded by deforestation, agriculture, and urban activities. Thus, the water quality of the lake was degraded; the main set of cave paintings was lost to *graffiti,* etc. Furthermore, Lapinha Cave was a touristic cave that for decades lacked proper planning. The mismanagement of tourism left its permanent mark on the cave's chambers, such as *graffiti* and speleothem breakage.

In 1980, the COPAM recommended the creation of a 37,715.6 hectares Special Protection Area (SPA). It includes the entire municipality of Confins (where the International Airport is located) and parts of the municipalities of Lagoa Santa, Pedro Leopoldo, Matozinhos, Funilândia and Prudente Morais. This SPA intends to protect the local karst from degradation caused by a surge in urbanization that followed the International Airport's installation (Herrmann et al. 1998). It was named International Airport SPA. The SPA's creation decree devised areas of permanent preservation where environmental changes over vegetation are not acceptable. Furthermore, it established that in case of allotments, the percentage of public areas couldn't be less than 50% of the original area plot. In practice, this law was perceived as very restrictive. Per the private economic sector, it hindered the region's development. Overall, the implementation of this law virtually created a direct conflict between the need for environmental protection and the private entities that kept lobbying for accommodations to their intended economic activity.

The absence of political interest, the weakening of state environmental agencies and the inadequateness of existing legislation to the socioeconomic reality, for decades, may have contributed to lead the protected area into years of neglect that hindered conservation efforts. This SPA lacked trained staff and had no proper environmental management. Effectively, the International Airport SPA remained an area only protected on paper.

The years following Airport construction in 1980 were marked by the intensification of urban and industrial expansion, mainly along the MG-424 highway, that crosses the western boundary of the SPA. In Brazil, works of this magnitude are population-growth magnets. This is so because the intensification of public and private investment generates direct and indirect jobs. That, in turn, stimulates the creation or development of urban centers. Often, this kind of urban expansion is spontaneous, no formal planning is involved. In the fragile karst environment, the effects of uncontrolled urbanization can be devastating (Hötzl 1999; Veni et al. 2001; Cooper et al. 2011). Figure 5 illustrates the changes in land use within the SPA over the past decades. It shows a significant reduction in the original area of native vegetation.

4.3.3 The 1990s and the Creation of the Lagoa Santa Karst Environmental Protected Area

A decade after the creation of the first protected areas in Lagoa Santa, which unfortunately proved to be ineffective, it became urgent to protect the area's natural and cultural heritage. In this context, a new conservation unit was created, this time at the federal level, to reconcile economic development with heritage preservation. In 1990, the Lagoa Santa Karst Environmental Protected Area (EPA), with an area of 35,600 ha was created with very similar limits to the International Airport Special Protected Area (SPA). The EPA is classified in the National Protected Areas System—SNUC within the Group of Sustainable Use (IUCN VI). In this group is allowed the exploitation of renewable environmental resources, since their ecological systems and their attributes—such as biodiversity—are maintained (Brasil 2000). The concepts and assumptions contained in the legislation have proven difficult to apply, as discussed below.

The Lagoa Santa Karst EPA partially encompasses the municipalities of Lagoa Santa, Pedro Leopoldo, Matozinhos and Funilândia. It fully covers Confins municipality. Its creation was based on detailed multidisciplinary technical studies carried out by the Brazilian Geological Service (CPRM) between 1992 and 1997. The main instruments of planning and management of the EPA are the Management Plan and the Environmental Zoning, published, respectively, in 1997 and 1998. Among the devised management tools there is a Consulting Council, which has been active since 2005. It has headquarters to house the management staff. The implementation of these instruments in the EPA has represented an advance toward karst protection and promotion of sustainable development. However, the management of a region presenting major environmental fragilities and located in a vector of metropolitan expansion is complex by nature.

Studies published by Alt (2008) pointed out as their main conclusion the low environmental effectiveness of the Lagoa Santa Karst EPA. Satellite imagery analysis demonstrated that the evolution of land use in the municipalities covered by the protected area was similar inside and outside the EPA between 1980 and 2006. A summary of this analysis of land use, in Fig. 5, shows the significant reduction of endemic vegetation, indicating the low effectiveness to halt deforestation processes. Furthermore, the evaluation of the

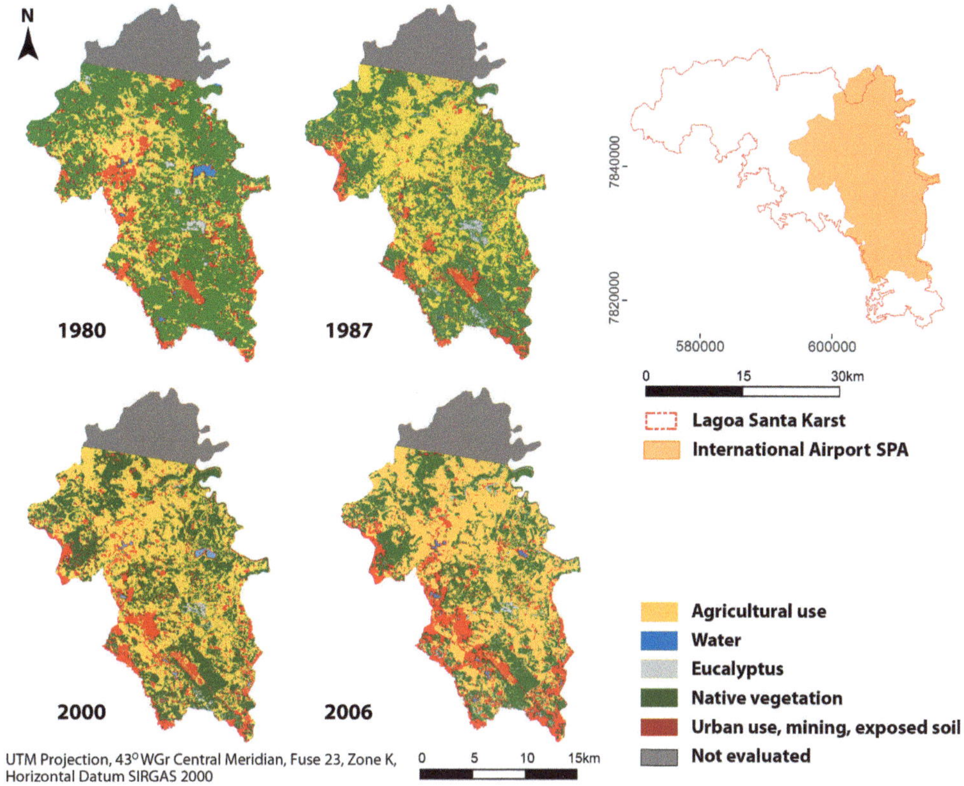

Fig. 14.5 Land use evolution in International Airport SPA, between 1980 and 2006 (Alt 2008)

existing instruments of territorial management indicated technical contradictions between the EPA Environmental Zoning and Geotechnical Zoning. Between these instruments and the Municipal Master Plans there are also technical inconsistencies, a fact that probably led to an urban and industrial growth incompatible with the fragility of the natural environment (Fig. 6). Most of them interviewed by Alt (2008) did not recognize themselves as protected area residents and, in addition, did not distinguish the rules in effect for land use. Most respondents did not have proper sewage treatment at home, using rudimentary septic tanks for that purpose. The environmental problems identified in 1998 by the Environmental Zoning, such as deficiencies in basic sanitation, persisted or even intensified in EPA municipalities in 2008. In the municipality of Matozinhos the disposal of solid waste was done on bare ground, in a doline surrounding the EPA, creating a situation of high risk of contamination to the local underground aquifers (Fig. 7). The Lagoa Santa Karst EPA Management Plan never got off the paper and its contents were not even acknowledged by the members of the Consultant Council, when interviewed by Alt (2008). Consequently, by 2008 the creation of the protected area had little influence on the region's environmental protection.

After 2008, the EPA management team was renewed. Nowadays the new team improved the inspection of the existent land uses and implement the restrictions imposed by the EPA Environmental Zoning. These actions increased the environmental relevance of this protected area, as discussed in the following items.

4.3.4 Present Times: Compensation for Environmental Damages Funds Specific Initiatives for the Lagoa Santa Karst Protection

Private reserves: a dynamic model of environmental protection

At the beginning of the new millennium, in a context of inefficiency of existing protected areas, the intervention of certain government sectors proved decisive. Guided by regulations such as the Law of Diffuse Interests/Law of Public Civil Action (Brasil 1985) and the Environmental Crimes Act (Brasil 1998), the Public Ministry, both in its state and in federal instances, reviewed and resolved some pending issues related to past environmental liabilities in the Lagoa Santa Karst. It demanded the intensification of the activities of the environmental agencies responsible for the management of the protected areas and for the environmental licensing processes. In the case of the destruction of the archaeological site of Lapa do Arco, the agency required from the offenders Compensatory Measures involving local environmental education in different forms, i.e., lectures in schools, book publishing, and others (Prous et al. 2003).

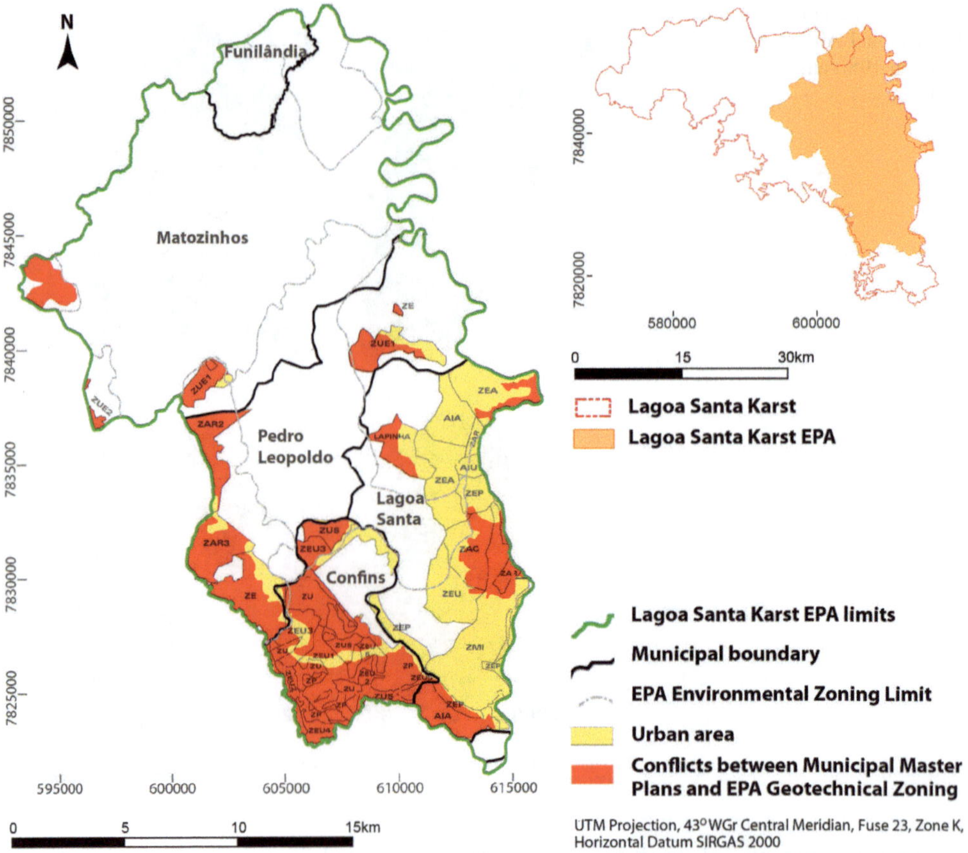

Fig. 14.6 Areas with detected conflicts between the EPA geotechnical zoning and the municipal master plans in the Lagoa Santa Karst EPA (Adapted from Alt 2008)

Fig. 14.7 Matozinhos landfill, located near the northwest boarder of the Lagoa Santa Karst EPA, in 2007 (Photo: Luciana Alt)

Actions that underscore accountability positively influenced the protection of natural and cultural heritage in the region from then on.

Between 1997 and 2004, four Natural Heritage Private Reserves (NHPR) were created by the private initiative, under the influence of environmental agencies and the Public Ministry, which tied the implementation or the expansion of enterprises to environmental licensing conditions. These NHPRs overlap the Lagoa Santa Karst EPA as more restrictive protected areas and are named: "Sol Nascente", "Fazenda Bom Jardim", "Fazenda Campinho," and "Fazenda Vargem Alegre," These NHPRs are near mining and/or real estate developments, and are important initiatives for the protection of specific features of karst, such as the Ballet Cave. The owners are responsible for the protection and management of these protected areas, where only indirect uses of natural resources are allowed, in activities such as scientific research and public use (Brasil 2000). These NHPRs totaling 313 ha, still a small amount, represent less than 1% to the total area of the EPA.

Among the protected areas created by the private sector, a highlight is the NHPR Fazenda Bom Jardim, owned by a cement factory, and established in 1997 for the protection of Ballet Cave and adjacent karst features. This cave has approximately 150 m on horizontal development. It has rock art panels with their own stylistic peculiarity, which place Ballet Cave as one of the most important archaeological sites in Minas Gerais, as attested by a relative dating, made in a cave nearby, with a minimum age of about 10,500 yr AP (Neves et al. 2012) for this style of rock art. The cave also figures as a State Historical Heritage Site, since 1989. Unfortunately, before the creation of the NHPR the cave was altered by decades of uncontrolled visitation. It was even used as seasonal shelter for livestock. At the beginning of the new millennium, the Ballet Cave's conservation status was deplorable. *Graffiti* covered practically every smooth and accessible surface including the rock art panels and speleothems. Furthermore, there was a significant amount of accumulated garbage, several speleothems were broken and cattle feces covered large areas of the cave's floor, as described by David and Moura (2002).

Also as an Environmental Compensation measure, in 2002, the first restoration of the Ballet cave was conceived and coordinated by Helena David (Fig. 8). This was a pioneer initiative in cave preservation in Brazil. The second phase of the project comprised the installation of walkways and signs to guide visitors and preserve the integrity of underlying archaeological sediments. After all this effort, only in 2010, the cave was, in fact, open to controlled educational visitation due to the agreement reached between the Public Ministry and the land owner. Currently, the cave is the centerpiece of a larger scripted visit, lasting about 2 hours, which includes other attractions of the NHPR. The visitors are presented with a booklet that contextualizes the Ballet cave in relation to other archaeological and environmental heritage features of Lagoa Santa's Karst. However, visitation is still low in relation to the educational potential of the cave, averaging 600 visitors/year, that can only enter the cave after a bureaucratic authorization process by the company, including mandatory safety training (Alt and Moura 2016).

Private initiative, if well conducted during the processes of licensing and environmental compensation, can be an important partner for the preservation of the Lagoa Santa Karst. However, without the active involvement of regulatory and environmental agencies, it is difficult to move toward conservation and educational projects, since compensation implies additional expenditures that are beyond the primary objectives of most companies, which is profit.

The expansion of the Belo Horizonte Metropolitan Region and the environmental protection alternatives

In the beginning of the new millennium, great economic and social disparities are present in the Belo Horizonte Metropolitan Region (BHMR). The municipalities contained in the Northern Metropolitan Expansion Vector (NMEV) are characterized by complex and heterogeneous urban occupation patterns. Large areas in NMEV lack in infrastructure and services. They are also home to a population that besides being of low income is subject to high levels of violence and social environmental problems (Gomes 2011). Aiming at promoting better social and economic equity within the metropolitan network, the State Government has chosen to endorse, through the implementation of strategic public works, the economic and social development of the municipalities of the NMEV (UFMG 2011). Among these initiatives stands out a project called Green Line, concluded in 2008. Its goal was the road restructuring of the connecting axis between the municipality of Belo Horizonte and the International Airport, just inside Lagoa Santa Karst area. The International Airport was expanded including the creation of Industrial and International terminals, the latter being completed in 2016. Another important project was the implementation in NMEV of new administrative headquarter for the State Government. This became known as the Administrative City of Minas Gerais and was inaugurated in 2010. A metropolitan ring highway was designed, but not yet implemented. In addition to those, the State Government has granted tax incentives for the implementation of industrial and service activities in NMEV. Obviously, these strategic works were and are still followed by significant population growth through the division of land and increasing of urban density, both of which in many cases are carried out irregularly, imposing strong pressure for the occupation of the Lagoa Santa Karst.

However, karst areas have weaknesses that limit the intensive use of their natural resources, such as sites subject to high risk of contamination of groundwater, high propensity for erosion, and/or mass gravitational movements, areas subject to flooding, areas where it can be induced doline collapses (Galvão et al. 2015; Herrmann et al. 1998). Such factors can generate risks for the resident population and rapid degradation of regional environmental quality. In addition, there is a risk of irreversible loss of specimens and clusters of speleological, archaeological, and paleontological heritage including major alterations in the karst landscape associated with these unique sites (Prous et al. 2003; Alt 2008).

In an overwhelming context of lack of effectiveness of existing protected areas and pressing regional development in an area of great environmental fragility, there was a need to design and implement complementary means to protect the Lagoa Santa Karst, under penalty of irreversible loss or

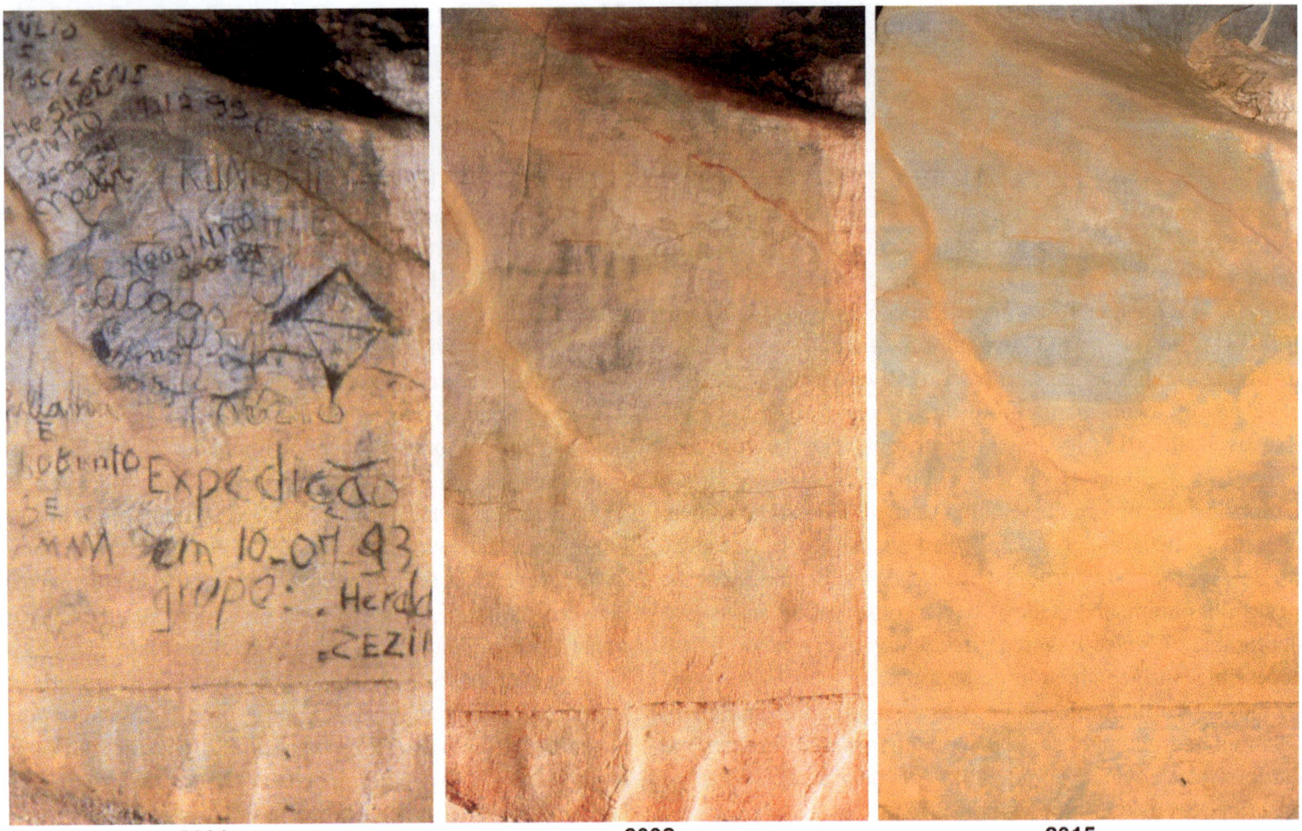

2001 2002 2015

Fig. 14.8 Surface with *graffiti* in Ballet cave in 2001, in 2002 after restoration project and in 2015 (Alt and Moura 2016)

alteration of the environmental heritage (Alt 2009). Under pressure from civil society, with the support of the Public Ministry and environmental agencies, the environmental licensing processes demanded conditions. For the implementation of the Administrative City of Minas Gerais and the expansion of the International Airport the following measures were required as conditions, i.e., effective implementation of the Sumidouro State Park, the creation of a North Metropolitan System of Protected Areas—NMSPA, among other actions.

The reactivation of Sumidouro State Park

In Lagoa Santa and Pedro Leopoldo municipalities, in 2008, after 28 years of oblivion, the effective implementation of the Sumidouro State Park finally started to happen. Funding came from the Green Line road system's Environmental Licensing process. The park's 1980's original area was expanded, totaling 2,004 hectares, and its boundaries were modified to accommodate the socio-environmental context (Fig. 9). A Management Plan was drafted, land expropriation was initiated and staff was hired. Much of the infrastructure to support visitation was completed in 2010. Today the park has guarded entrances, a parking area, hiking trails, a museum, accommodation for researchers, etc. Currently, it is the state conservation unit with the largest number of employees.

However, there are still many challenges to its full implementation. In October 2016, only about 60% of the Park area had been expropriated according to an interview with its manager. Over the years, the price of land has risen making it increasingly difficult to realize the land expropriation needed for the Park consolidation. This incomplete regularization creates conflicts. The owners have their land with restricted use and sale value without receiving financial compensation until the situation is regularized. According to the park manager, criminal fires and deforestation are frequent in non-regularized areas, mainly near Sumidouro Lake. Moreover, according to Faria et al. (2015) neighboring communities feel limited in their leisure activities, since the Park reactivation they no longer have free access to this lake, which historically was their main leisure venue.

Along with the implementation of Sumidouro State Park traditional yet illegal small-scale mining stopped along park borders. And in the urban areas, the processing of calcium phyllite was also banned. These activities, although illicit,

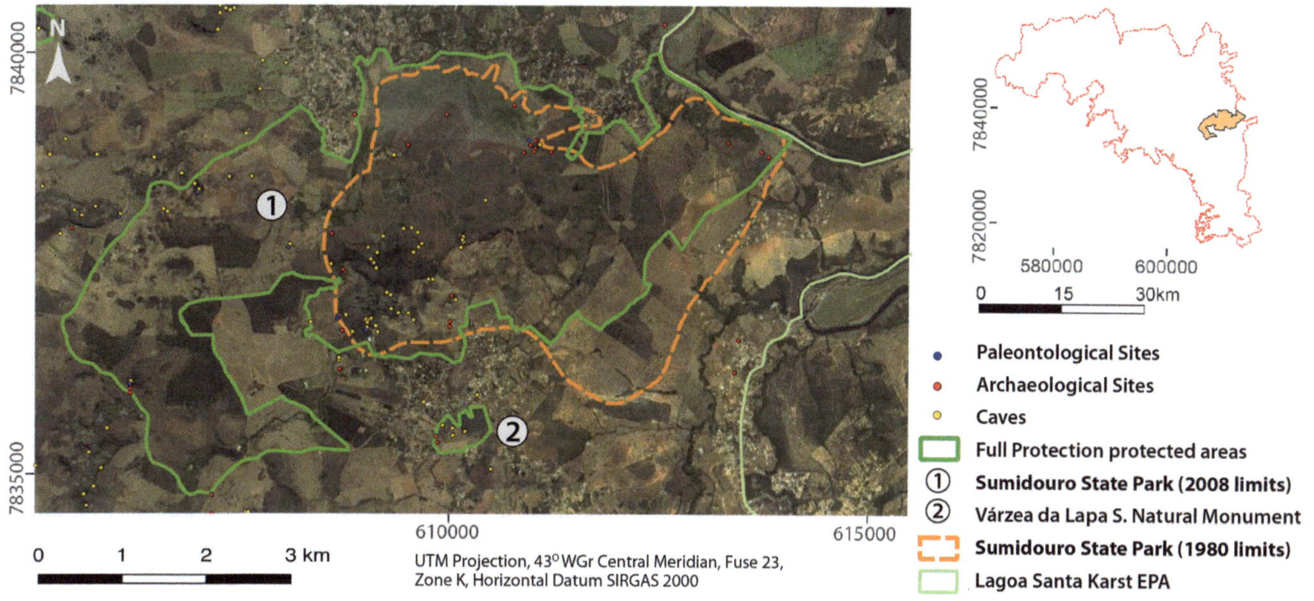

Fig. 14.9 Sumidouro state park borders over 2016 satellite image

were important sources of income for the local population. With the park creation, an increase in the tourist flow was expected. That would create new opportunities for income generation, however, with the slight increase in visitation, these promises have not materialized. The local population felt as if they were let down. According the park's manager, actions are being carried out to improve the park relationship with surrounding communities, such as activities involving environmental education and the encouragement to local leaderships to participate in the Consulting Council of the Park.

Actions such as the inclusion of local people in the Park decision-making processes, the regulation and facilitation of access to Sumidouro Lake, the stimuli to an increase in tourist visitation, and the proposition of alternative means of income generation to the local population can all contribute to a healthy balance between the existence of the park and the sustainability of its surrounding communities. This may bring long-term benefits related to environmental education and heritage appreciation, which will be constructive both to the community and to the Lagoa Santa Karst.

Recently, the State Government implemented a tour itinerary called Peter Lund Caves Route. In this project, some protected areas were included, i.e., Sumidouro State Park, Rei do Mato Cave SNM and Peter Lund SNM, as well as other places of cultural interest. This project aims to value and protect the existing heritage in the Route and to extend the public use and the education possibilities (Cartelle 2012). To improve the management of conservation units included in this tourist route, the Government intends to implement a Public Private Partnerships (PPP). Different from the system of concession of public use activities, common in several protected areas around the world, this proposal may present greater risk. In the PPP model, the State would delegate to a private company its attributions in relation to the direct management of conservation units—conservation and monitoring activities, tourism management—being only responsible for overseeing the process. The payment for services would be conditional on the fulfillment of measurable socio-environmental conditions. However, this management model raises several questions as follows. How will the protection and monitoring of specific karst environmental resources be guaranteed, which sometimes involves complex and costly monitoring? What will be the role of outsourced employees and staff who once worked in these protected areas? How to assure the carrying capacity of caves, in case the visitation be largely increased to generate profit? How will the socio-environmental benefits be guaranteed for the neighbor communities? Will the final cost to be paid by the State for the maintenance of protected areas be smaller than in existent model? How to make the PPP economically viable? With so many uncertainties, unfortunately the PPP does not seem to be a viable solution.

Reactivation of the International Airport SPA

In 2009, the legislation governing the International Airport SPA management was revised and the protected area was reactivated, three decades after its creation. This protective measure was also conducted by the Public Ministry. The SPA's perimeter was modified to coincide with that of Lagoa Santa Karst EPA. The figure of the "Permanent Protection Area" was introduced for certain karst features, i.e., dolines and for areas necessary for the protection of archaeological, paleontological, and speleological sites. In cases of damage to endemic vegetation, the requirement for compensatory reforestation within the SPA itself became mandatory. In the last 2 years, the State Forestry Institute (IEF-MG), through SPE's management staff, has been working more intensively with ICMBio, the federal managing body of the Lagoa Santa Karst EPA. According to interviews performed in October 2016, these agencies expanded their activities in the inspection of potentially impacting activities, in the concession of permits related to the environmental licensing process, and in the suggestion of environmental compensation actions. However, in these actions, because it had well trained and equipped staff, the federal agency could lead more initiatives than the State level institution.

As pointed out in interviews held in October 2016, the mining and industrial sectors and the real estate sector are exerting political pressure over the state government. This pressure is aimed at the idea of APE International Airport being incorporated into the Lagoa Santa Karst EPA, which has less restriction on economic activities. In turn, the environmental agencies would indicate how to include the restrictions imposed by the APE on the future environmental zoning of the Lagoa Santa Karst EPA. This still undefined scenario casts a shadow over the few advances in preservation already consolidated.

The implementation of the North Metropolitan System of Protected Areas—NMSPA and its shortcomings

The NMSPA was designed for the protection of important heritage areas, vulnerable to metropolitan development, located in the Northern Metropolitan Expansion Vector (NMEV), inside and outside of the Lagoa Santa Karst area. The protection of these areas would be made possible by means of: (1) creation of new protected areas of the Integral Protection Group (Brasil 2000), more restrictive character; (2) stimulate the creation of NHPR, or the establishment of legal forest reserves adjoining to the existing protected areas or to be created, contributing to the formation of ecological corridors, which potentially would increase environmental connectivity in the region. With these actions, would be formed as a mosaic of protected areas, that could result in the protection of environmental resources and specific points of interest, such as caves, paleontological, and archaeological sites. Conceptually the NMSPA constitutes a decisive contribution to the strengthening of asset protection and necessary counterpoint to the breakneck economic development. Its implementation is one of the environmental constraints imposed during the environmental licensing of public works carried out by the State Government in the NMEV (IEF-MG 2016).

The NMSPA creation parameters were established in 2007, through the Environmental and Urban Governance Plan of the Belo Horizonte Metropolitan Region, and are regulated by specific law decrees, published later. These decrees establish basic parameters for activities in the area covered by the NMSPA. It is important to note that Environmental Licensing and other types of authorization have been denied in areas where NMSPA conservation units are foreseen (Decree nº 45. 233, of December 3, 2009, Art. 4). This legal resource may already contribute to the protection of karst specific features, even if all areas provided for in the NMSPA have not yet been created.

As of 2010, nine of the twenty conservation units planned for the NMSPA were finally created (IEF-MG 2016). Six of these protected areas have been specifically designed to extend a higher protection standard within the Lagoa Santa Karst EPA, they are: Cerca Grande State Park (134.19 ha); Vargem da Pedra State Natural Monument (SNM) (10.09 ha); Experiência da Jaguara SNM (38.48 ha); Várzea da Lapa SNM (23.53 ha); Santo Antonio SNM (31.10 ha), and Lapa Vermelha SNM (33.71 ha). These new protected areas, belonging to the Integral Protection Group (SNUC 2000), sum up 271.1 ha overlapping the Lagoa Santa Karst EPA. This meant the deployment of larger potential restrictions, although the total area of protection established in this effort is equivalent to less than 1% of the total area of the EPA.

The creation of the NMSPA first conservation units was widely publicized in the media trying to highlight the environmentally positive character related to strategic political ventures installed in the North Metropolitan Vector. However, the State Natural Monuments, which constitute the great majority of protected areas created, do not imply the expropriation of lands, and their creation is facilitated. Despite this, it imposes strict rules on owners, which makes their management difficult and creates conflicts. Another aggravate is the fact that the natural monuments created have extremely small areas, which in some cases are insufficient to protect the cave influence area, i.e., the area needed to protect cave dynamics and evolution, as well as the landscape to their surroundings. This problem will be exemplified in the Lapa Vermelha SNM, but it is common in most of the new protected areas created after 2010 in the Lagoa Santa Karst area.

The Lapa Vermelha SNM, with 33.71 ha, houses four caves, one of them being where a prehistoric human skull was excavated in 1970s by a Franco Brazilian Archeological Mission, and described by Neves and Piló (2008). This anthropological remains was named "Luzia" and became iconic in Brazilian context contributing to expand the debate on the America's occupation. These caves are inserted in a limestone outcrop, located in a wide dissolution doline, with long and smooth slopes. Flanking the outcrop, at the bottom of the doline, is a temporary karst lagoon, with drainage associated with the largest of the caves, which has 1,870 m

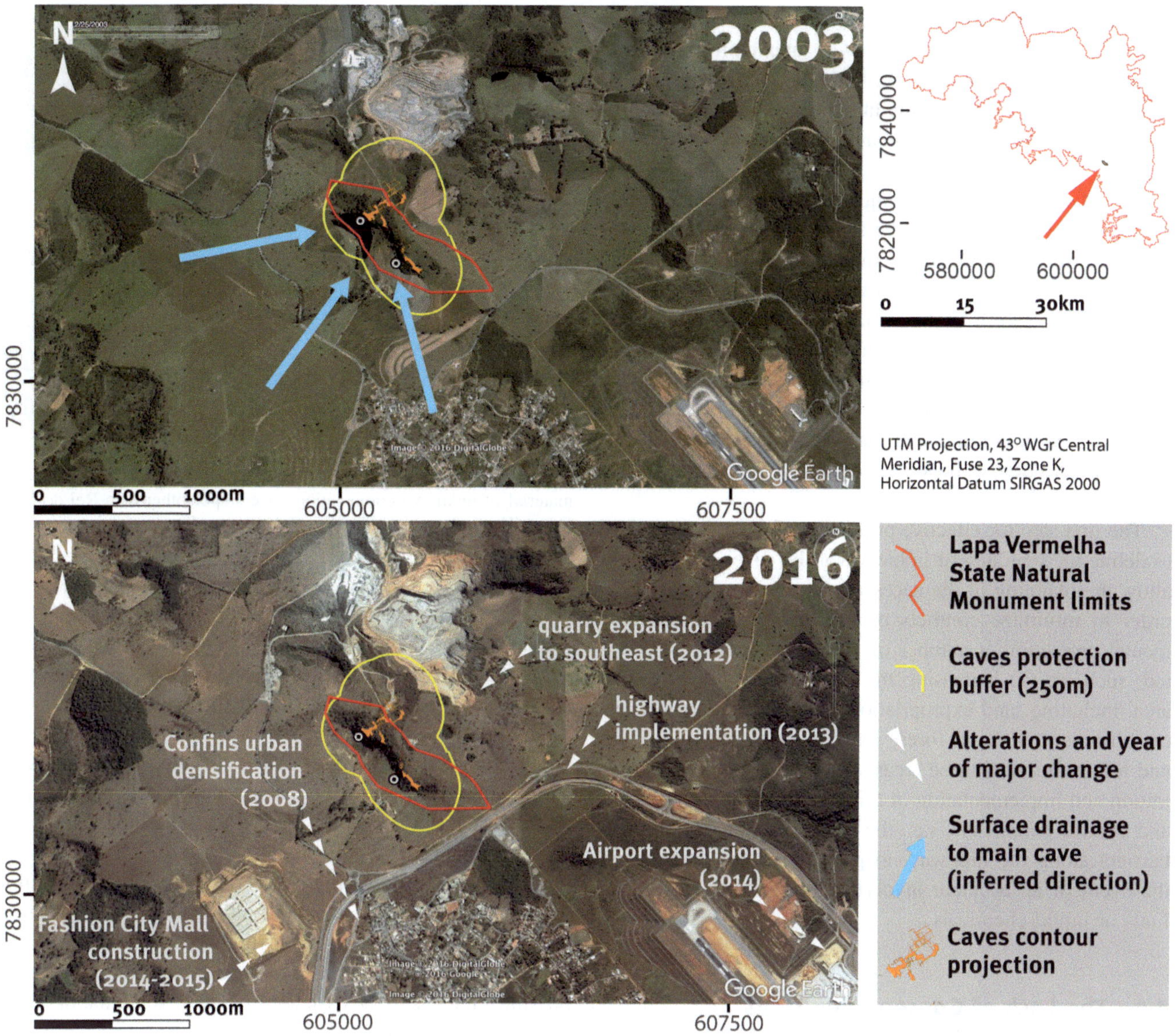

Fig. 14.10 Environmental situation of Lapa Vermelha SNM in 2003 and 2016, showing some alterations that took place in this time interval. The main cave map is currently under review, which may give rise to small inaccuracies in the cave contour

in length. The doline is home to agricultural use, a large commercial center, still under construction, and a stretch of the access road to the International Airport. However, according to EPA Geotechnical Zoning it is an area of high to very high vulnerability to the pollution of the karst aquifer and moderate to high propensity to the occurrence of collapses (Cabral 1998).

The perimeter defined for the Lapa Vermelha SNM forms a polygon of about three hundred meters wide by a kilometer long, which covers the outcrop where these four caves are inserted. It partially encloses the temporary lagoon at its maximum flood level (Fig. 10). Considering that the surface drainage of the doline converges to the temporary lagoon, and enters the largest cave of the SNM, it is coherent to assume that the whole doline is part of the area of direct influence of the main cave. That is, potentially impacting activities located there can directly affect the cave environment. Despite this, only a small strip of the doline, close to the outcrop, was inserted in the protected area.

In addition, the main cave Lapa Vermelha I surpasses the perimeter of the SNM, with cave passages directed toward a limestone quarry. Possibly this cave will be classified, according to Brazilian law (MMA 2009), as a maximum relevant cave. Currently, the mining structures could have reached the government enforced 250-m perimeter buffer zone that aims to safeguard a cave's ecosystem from external impact. This buffer is valid until scientific studies are performed to delineate the real area of influence of the cave environment. This same legislation defines that the area of influence of caves classified as of maximum relevance cannot be the target of irreversible negative impacts. That is, there is a conflict between the land uses already installed in the area, the current legislation and the conservation unit are created.

Such discrepancies lead the question about the real role of this protected area. Would it be a mere statistic to attest and publicize, in beautiful advertising campaigns, the sustainability of the State Government in the development oriented project of the North Metropolitan Expansion Vector? The fact is that an important environmental and cultural heritage is at high risk with the land uses that are being gradually developed in their surroundings, and that can, unfortunately, be intensified in the future. It may compromise the capacity of this area to deliver benefits that may be crucial for future generations.

The first step for effective protection of the area would be to define the real area of influence of the caves. This would allow the planning of the necessary conservation actions. In order to minimize conflicts of use and increase environmental protection, a number of alternatives could be adopted, such as: (1) increasing the perimeter of the protected area, including land expropriation; (2) in private properties, establishment of legal forest reserves bordering the SNM and reforestation of these areas with native species; (3) definition and implementation of minimum impact construction and control parameters, i.e., the installation of containment barriers along the highway, to possibly protect aquifers in the event of an accident involving the possible leakage of fuels or liquid chemicals.

Fig. 14.11 A dark gray color layer composed by fine particulate material of unknown composition covers speleothems in Rei do Mato Cave. The red arrow indicates speleothem cleanup (white rectangle) during the International Course on Cave Restauration and Conservation, 2014 (Hildreth-Werker et al. 2016). Photo: Luciana Alt

4.4 The Lack of Conservation Units in the Northern Lagoa Santa Karst

The nine conservation units created under the NMSPA are concentrated in the southern portion of the Lagoa Santa Karst, or in its surroundings. Most of the northern sector of this karst area is currently devoid of the environmental protection afforded by protected areas.

The municipality of Sete Lagoas is located at the northwest end of the Lagoa Santa Karst. There the institutionalized protection of the natural heritage is lower. Sete Lagoas is today a medium-size city that in the last three decades has developed fastly and without proper planning. Its population 144,014 in 1991, increased to 234,221 inhabitants in 2016 (IBGE 2016). In time Sete Lagoas became an important regional center. Nowadays the most important local industries focus on food processing, textiles, automobiles, metallurgy, and limestone processing.

In the municipality of Sete Lagoas, the karst aquifers are the main source of water for urban and industrial activities (Pessoa 1996). Different authors quote risk or problems of contamination of surface and groundwater (Galvão et al. 2015; Souza 2015). Others quote the geotechnical risk of doline collapse induction associated with the overexploitation of karstic aquifers. This risk is stronger in the urbanized areas, where several cases of subsidence or collapse occurred in the last three decades, offering a risk to the population (Galvão et al. 2015).

Studies have shown high levels of atmospheric pollution in Sete Lagoas, mainly derived from metallurgical industries and limestone transformation processes (Queiroz et al. 2007). This fact may be related to loss of attractiveness and possible changes in the dynamics of the cave fauna, in one of the main tourist caves of the State, the Rei do Mato Cave, and in other caves in its surroundings. In some areas of this cave a thick layer composed by fine particulate matter cover speleothems and other surfaces. This layer of exogenous origin and unknown composition was described on Rei do Mato Speleological Management Plan (Alt 2010) and illustrated by Fig. 11. A State Natural Monument was created, in 2010, for protection of the Rei do Mato cave. However, just like in the Lapa Vermelha SNM, the conservation unit is probably not sufficient to enclose the cave's influence area.

There is an environmental protection gap in the northern Lagoa Santa Karst. To the southeast of the city of Sete Lagoas lays one of the best preserved and most important heritage sets of this karst area. In this region, NMSPA original planning includes the creation of the following protected areas: Poções State Park, Cauaia State Wildlife Refuge, Escrivânia SNM, and Veredas SNM. Currently, due to the national economic and political crisis, there is no prospect for the implementation of these protected areas,

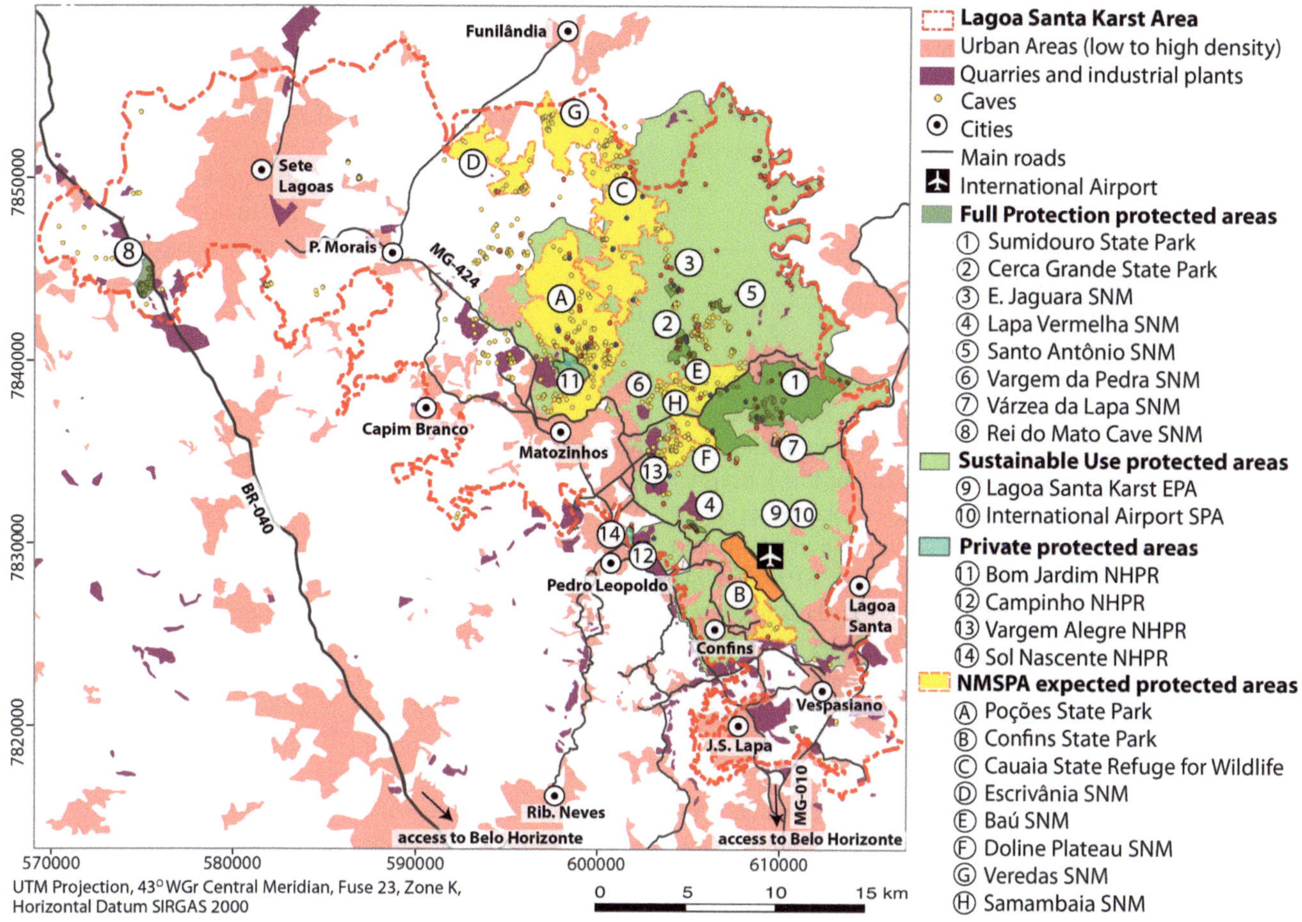

Fig. 14.12 Map of the protected areas and potential harmful activities in Lagoa Santa Karst

according to interviews conducted in October 2016. However, this region suffers a strong trend of conurbation along MG-424 state highway, which connects Sete Lagoas to the International Airport (Fig. 12).

In the southern portion of the Lagoa Santa Karst, where several conservation units are superimposed, it is noted that environmental agencies are struggling to manage the complex existing conflicts and protect the heritage. With conservation units of a very local character, and sparse, the northern portion of this karst area becomes extremely vulnerable to all kinds of threats, coming from different economic sectors. Even with all the problems presented here related to the management of the existing protected areas, the simple existence of this legal protection is still helpful and may have long-term effects as was exemplified by Sumidouro State Park.

4.5 New Alternatives for the Protection of Lagoa Santa Karst

In 2013, a team coordinated by the Biological Sciences Institute of the Federal University of Minas Gerais applied Lagoa Santa Karst EPA, as a site in the Convention on Wetlands, called the Ramsar Convention. This initiative was justified by the existence of a vast system of temporary karst lagoons, which undergoes annual cycles of floods and droughts, attracting countless bird species, including migratory species, as well as mammals threatened with extinction (see chapter "The Waterbirds and Dynamics of Lagoa Santa Karst Temporary Lakes", this volume, Nóbrega et al. 2013). The Ramsar Convention is an intergovernmental treaty that provides opportunities for international cooperation for the environmental conservation and wise use of wetlands and their resources, an additional attempt to protect the region. In 2017, this initiative was approved by the Federal Government, being an opportunity to reinforce the protection needs of Lagoa Santa Karst EPA. It will probably increase the opportunities to access international funds to finance projects and researches.

5 Conclusion: Challenges and Future Opportunities

The set of conservation units created to protect the Lagoa Santa Karst is still incomplete and fragile. This effort still does not correspond to the size and value of the natural and cultural heritage of this area. Less than 2% of this karst area is protected in more restrictive management conservation units. The large and less restrictive sustainable use protected areas made up of the International Airport SPA and the Lagoa Santa Karst EPA, overlap, covering the same territorial portion, and present serious deficiencies in its effectiveness. In addition, protected areas are rare in the entire northern portion of the Lagoa Santa Karst. Despite these problems, the region still holds an extensive heritage to be protected.

The question remains: what factors allowed the perpetuation of this heritage to the present days? As possible responses, could be listed: the creation of conservation units, the action of some parts of civil society and government aimed at environmental protection, increased environmental awareness in Brazil in the last decades. Could be also mentioned: lack of economic interest for urbanization or agricultural uses in some areas with difficult access, protection provided by environmental licensing of enterprises or speculation through land ownership. In fact, the most logical answer may be a combination of all these factors and even the presence of other unlisted ones.

In the last two decades, the main protection actions of the Lagoa Santa Karst were directly or indirectly derived from environmental licensing processes, from public or private enterprises. However, they are currently being experienced a weakening of the environmental licensing process and cave protection legislation in Brazil, which are considered obstacles to national development. The set of Brazilian environmental laws, a model for the world, is now threatened and has difficulties in practical application. This threat comes from current factors such as the widespread weakening of government environmental agencies and increasing pressure for economic development. As Brazil continues to be a major exporter of primary products, the population lives a sad reality in which the exploitation of environmental resources generates the profit of a few. Profit justified by the creation of precarious jobs in a context of widespread impoverishment and overpopulation. In this process, the common interest represented by the natural and cultural heritage is one of the vulnerable points. The right to get an ecologically balanced environment, guaranteed in theory by the New Federal Constitution of 1988, ends up in second place.

The prospects for environmental protection in the Lagoa Santa Karst stand out in three main necessary fronts: (1) strengthening of the Environmental Licensing system, (2) adoption of minimum impact parameters for the use of the karst areas, associated with territorial planning instruments, (3) creation and effective management of conservation units.

Strengthening of the Environmental Licensing System

With the Brazilian nation in serious economic and political crisis (2017), the trend for the next decades is that most environmental conservation actions will be financed by the private sector, mainly through environmental licensing. Therefore, to increase the environmental protection in the Lagoa Santa Karst, it is necessary to invest in the strengthening and technical qualification of the governmental agencies responsible for environmental licensing and inspection, so that they can work more efficiently and rigorously. It is also necessary to increase popular participation in decision-making processes related to environmental licensing, enabling the mediation of social and environmental conflicts, and aiming the collective interests. But before that, broad and continuous education campaigns are needed regarding the importance and fragility of the karst environment, to generate instruments for popular participation.

In environmental licensing processes, it is fundamental that the karstic environment is understood as a system with interconnected components. For that, it would be crucial to create a unified georeferenced database, containing all the territorial planning instruments and information layers available, i.e., natural resources maps, geotechnical zoning, municipal master plans, protected areas zoning, enterprises influence areas, and others. This could be compulsory and for common use by all the agencies that work in the area avoiding the granting of undue and conflicting authorizations in relation to environmental protection and patrimony. In this context, it is also important that the enterprises aiming for the environmental licenses should present complete and integrated data sets for the environmental agencies. Nowadays some enterprises data are presented as fragments, what made the impact analyzes more complex and less precise for the licensing body. These would also contribute to the proposition of more effective mitigation and control measures actions in environmental licensing processes.

Special attention should be given to the prevention of contamination of water resources in the environmental licensing processes since the dynamics of karst aquifers are still little known and there is no wide network for monitoring their quality in the region. In the next decades should happen a massive investment in environmental monitoring in the region, aiming to prevent and control environmental problems and to ensure collective health.

In the case of the strategic development implemented by the State Government in the Northern Metropolitan Expansion Vector, the environmental constraints required in the environmental licensing process were insignificant in relation to the possible direct and indirect damages to the Lagoa Santa Karst heritage. These conditions were not fulfilled in their entirety, even after the complete installation of the planned projects. It is believed that full compliance with these conditions will only be made possible through the action of the Public Ministry, which is the great guardian of collective interests in the region. But, regrettably, the upper echelons of the government have tried to weaken this important body, reallocating the public prosecutors involved with the Lagoa Santa Karst to other municipalities.

Very often, the entrepreneurs make political pressure to reduce as much as possible the financial amounts spent on compensatory measures in the context of environmental licensing. However, they use the results obtained, i.e., the creation of "green areas," as an advertisement to make a sustainable impression on their enterprise. In many cases this adoption increases the aggregate economic value, as in the case of real state enterprises. It is concluded that this process, if well conducted, has the potential to generate direct benefits for the entrepreneur, the environment and the society welfare, being an instrument for the viability of sustainable development. And it would allow the implementation of the model of Socio-environmental and Democratic State of Law advocated by the New Federal Constitution of 1988.

Adoption of minimum impact guidelines for karst areas, associated with territorial planning instruments

The existence of mining enterprises, urban areas, and agricultural activities along with a large concentration of caves associated with archaeological and paleontological sites is a reality that demands constant research, analysis, and environmental monitoring to avoid impacts on the Lagoa Santa Karst heritage.

The municipal Master Plans in the Lagoa Santa Karst and the Governance Plan for Belo Horizonte Metropolitan Region should be based on technical aspects such as geotechnical risks and specific environmental vulnerabilities of karst areas. This would undoubtedly lead to a less damaging occupation on karst environment, and less risk to the population living there. However, the integration between technical knowledge and planning actions is still poor. Currently, some urban areas are inserted or will expand over areas of high geotechnical vulnerability (Alt 2008). In the Lagoa Santa Karst, there are still vast areas occupied by pastures, which will be probably converted into urban areas. Thus, to avoid serious future socio-environmental problems and high remediation costs, long-term territorial and environmental planning actions are urgently needed. In this context, it is urgent to consider the current best practices for a harmonious coexistence with karst (Veni et al. 2001).

There is a demand for the creation of norms and specific guidelines for the Lagoa Santa Karst. These standards could be developed under the umbrella of an environmental agency resulting in products such as a manual and a set of standards for land use in the region. In this context, it could be specified, for example, the type of sanitary system most suitable for each area, guidelines for its construction and maintenance, norms for opening new pumping wells, for hazardous waste disposal, among other guidelines. Following this trend, the municipal Master Plans over this karst area should specify technical criteria to be adopted in public and private works, considering these orientations. This is already done in other karst regions of the Planet (Hötzl 1999; Cooper et al. 2011).

Effectively protecting the Lagoa Santa Karst will be impossible without the deep involvement of the local population, who will need to invest in changing its own rudimentary sanitation system to proper sanitation solutions, for example. To do so, it is necessary that the population, in fact, understand what karst is and its importance and weaknesses, the main environmental risks involved, the implications of living on this type of terrain. It is essential that the population has the means to contribute and live in a more harmonious way in this complex environment. This requires vast environmental education campaigns that must last for generations. A good way to achieve this would be including this technical content in the region's school's curriculum.

Creation and effective management of conservation units

To improve environmental protection over the Lagoa Santa Karst it is fundamental to: (1) increase the surface area protected by more restrictive conservation units; (2) create the protect areas foreseen in North Metropolitan System of Protected Areas (NMSPA); (3) adjust the limit of the small Natural Monuments created under the NMSPA according to the cave influence areas that it is intended to protect (Alt 2009); (4) carry out land expropriation in NMSPA protected areas and in Sumidouro State Park; (5) strength the Lagoa Santa Karst EPA, as an instrument for land use management around the most restrictive protected areas. For that its Management Plan should be revised considering the geotechnical risks for the definition of the new environmental zones; and (6) legally establish the Lagoa Santa Karst mosaic of protected areas, as an instrument to facilitate the joint management of this protected areas and to facilitate conservation fund raising. For all this be possible, should be created political interest in solving conflicts between economic and environmental aims, in parallel the technical staff responsible for the protected areas management should be strengthened.

The Sumidouro State Park, neglected for three decades, has recently been reactivated, demonstrating the importance of creating conservation units to protect the region's environmental and cultural heritage. The protected areas are perpetual under the national legislation, and cannot be easily challenged. With the potential rapid economic development of the region, the creation of new protected areas is fundamental to guarantee the perpetuation of environmental quality. However, the progressive increase in land value makes this increasingly difficult.

With the current economic and political crisis, at the state and national levels, the most viable option to increase the protection of the Lagoa Santa Karst, which has been effective in the last two decades, is the establishment of Natural Heritage Private Reserves (NHPR). Created by initiatives of land owners or as a compensatory measure in the context of environmental licensing, this area becomes perpetually protected by law. In this process, the Government will be decisive as an agent for development and regulation. Through territorial planning actions, more adequate limits can be suggested to the NHPRs to allow greater environmental connectivity with other elements of environmental importance, sometimes located in neighboring properties, i.e., forest reserves, waterfront protection areas, or even to other NHPRs and protected areas. This measure is fundamental for biodiversity conservation. And it would also prevent the creation of isolated islands of protection amid private undertakings, where public use of collective heritage could be hampered, as observed in Ballet Cave, located inside the NHPR Fazenda Bom Jardim.

Currently, there are several ways to make a protected area economically viable. This can be done by environmental compensation, payment for environmental services, tax exemption, concessions for public use activities, and international funds for environmental conservation. Adding economic value to environmental protection can help preserve the Lagoa Santa Karst heritage. These initiatives are still little explored in the area and could even stimulate the creation of new protected areas.

Initiatives are urgently needed to make inhabitants of the Lagoa Santa Karst social partners in the conservation units. To this aim, recreation and education activities in contact with natural environments should be encouraged and facilitated. A broad action in this direction has the potential to increase the quality of life and contribute to raise awareness of the importance of the region's natural and cultural heritage. Fortunately, some of these actions are already underway in the region, driven by River Basin Committees, NGOs, Universities and other agents.

Whatever the alternatives considered for the environmental protection of the Lagoa Santa Karst, the task is complex and urgent. And it will demand the commitment of all sectors of society.

Acknowledgements Our deep gratitude to Andre Alt for organizing the English version and for his valued contributions. For the valued information's during the interview processes we thank to Allan Gonzaga and Mario Douglas Fortini, from ICMBio, to Cecilia Vilhena, Rogério Tavares, Mariana Moura, Rinaldo Souza, Paulo Scheid, from IEF-MG, and to Procópio de Castro, from the Subcommittee on Mata Creek Hydrographic Basin. We also thank Cristiane Froes, Cláudia Rocha, and Mariangela Araújo, from IEF-MG, for providing information about the state protected areas. Special thanks to Dr Roberto Célio Valadão, from Geosciences Institute—UFMG, for his essential contributions as Luciana Alt master thesis supervisor.

References

Alt L (2008) Efetividade sócio-ambiental da APA Carste de Lagoa Santa-MG: Uma avaliação a partir de suas ferramentas de planejamento e gestão. http://hdl.handle.net/1843/MPBB-7YKK6F Accessed 3 Oct 2016

Alt L (2009) Estudos técnicos de espeleologia como suporte para criação de unidades de conservação do Sistema de Áreas Protegidas do Vetor Norte de Belo Horizonte. Instituto Estadual de Floresta de Minas Gerais, Belo Horizonte

Alt L (2010) Diagnóstico e Planejamento: Patrimônio Espeleológico do Monumento Natural Estadual Gruta Rei do Mato. Belo Horizonte: Ambiente Brasil Centro de Estudos, Instituto Estadual de Floresta de Minas Gerais

Alt L, Moura V (2016) Ballet cave: rock art conservation in Brazil. NSS News 74:6–7

Auler AS (2016) Cave protection as a Karst conservation tool in the environmentally sensitive Lagoa Santa Karst, southeastern Brazil. Acta Carsologica 45:131–145

Brandt Meio Ambiente (2008) Mapa de uso de solo e ocupação do solo. In: Contribuição do Setor Produtivo relativa às áreas propostas para unidades de conservação. Sindicalge e Snic, Belo Horizonte

Brasil (1934) Water Code. Decreto nº 24.643, de 10 de julho de 1934. http://www.planalto.gov.br/ccivil_03/decreto/d24643.htm. Accessed 29 Sept 2016

Brasil (1937) Organizes the protection of national historical and artistic heritage. Decreto-lei nº 25, de 30 de novembro de 1937. https://www.planalto.gov.br/CCIVIL_03/Decreto-Lei/Del0025.htm. Accessed 29 Sept 2016

Brasil (1961) It disposes of archaeological and prehistoric monuments. Lei no 3.924, de 26 de julho de 1961. http://www.planalto.gov.br/Ccivil_03/leis/1950-1969/L3924.htm. Accessed 29 Sept 2016

Brasil (1967a) Fauna Protection. Lei n° 5.197, de 3 de janeiro de 1967. http://www.planalto.gov.br/ccivil_03/leis/L5197.htm. Accessed 29 Sept 2016

Brasil (1967b) Mining Code. Decreto-lei nº 227, de 28 de fevereiro de 1967. http://www.planalto.gov.br/ccivil_03/Decreto-lei/Del0227.htm. Accessed 29 September 2016

Brasil (1977) Law on protection of sites of tourist interest. Lei no 6.513, de 20 de dezembro de 1977. http://www.planalto.gov.br/CCivil_03/leis/L6513.htm. Accessed 29 Sept 2016

Brasil (1981) National Environment Policy. Lei nº 6.938, de 31 de agosto de 1981. http://www.planalto.gov.br/ccivil_03/LEIS/L6938.htm. Accessed 29 September 2016

Brasil (1985) Law of Diffuse Interests. Lei no 7.347, de 24 de julho de 1985. http://www.planalto.gov.br/ccivil_03/leis/L7347orig.htm. Accessed 29 Sept 2016

Brasil (1988) Federal Constitution. Constituição da República Federativa do Brasil de 1988. http://www.planalto.gov.br/ccivil_03/constituicao/constituicao.htm. Accessed 29 September 2016

Brasil (1990) It disposes about cave protection. Decreto no 99.556, de 1º de outubro de 1990. http://www.planalto.gov.br/ccivil_03/decreto/1990-1994/D99556.htm. Accessed 29 Sept 2016

Brasil (1998) Law of environmental crimes. Lei nº 9.605, de 12 de fevereiro de 1998. http://www.planalto.gov.br/ccivil_03/leis/L9605.htm. Accessed 29 Sept 2016

Brasil (2000) Creates the national system of conservation units. Lei no 9.985, de 18 de julho de 2000. http://www.planalto.gov.br/ccivil_03/leis/L9985.htm. Accessed 29 Sept 2016

Brasil (2001) City statute. Lei no 10.257, de 10 de julho de 2001. http://www.planalto.gov.br/ccivil_03/leis/LEIS_2001/L10257.htm. Accessed 29 Sept 2016

Brasil (2008) Cave protection. Decreto nº 6.640, de 7 de novembro de 2008. http://www.planalto.gov.br/ccivil_03/_Ato2007-2010/2008/Decreto/D6640.htm Accessed 29 Sept 2016

Brasil (2012) Forest code. Lei nº 12.651, de 25 de maio de 2012. http://www.planalto.gov.br/ccivil_03/_ato2011-2014/2012/lei/l12651.htm Accessed 29 Sept 2016

Cabral JL (1998) Zoneamento Geotécnico e Aptidão dos Terrenos. CPRM, Belo Horizonte

Cartelle C (2012) Das grutas à luz: os mamíferos pleistocênicos de Minas Gerais = From caves to light: the Pleistocene mammals of Minas Gerais. Bicho do Mato Editora, Belo Horizonte

CECAV (2016) Centro Nacional de Pesquisa e Conservação de Cavernas - CECAV - Canie. http://www.icmbio.gov.br/cecav/canie.html. Accessed 7 Dec 2016

Cooper AH, Farrant AR, Price SJ (2011) The use of karst geomorphology for planning, hazard avoidance and development in Great Britain. Geomorphology 134:118–131

CPRM (1994) Projeto Vida: Dinâmica do processo erosivo. CPRM, Belo Horizonte

Da-Gloria P, Neves WA, Hubbe M (2014) História das pesquisas em Lagoa Santa: museus, crânios e antropologia biológica no Brasil. In: Anais 29ª Reunião Brasileira de Antropologia, Natal, pp 1–12

David H, Moura V (2002) The Ballet Cave, Minas Gerais, Brasil, Conservation and Visitation. 10es journées de la Section Française de l'Intitut International de Conservation, pp 164–171

Faria S, Pereira A Costa, M, Santos G (2015) Análise do conflito socioambiental no Parque Estadual do Sumidouro em Minas Gerais: A Importância da relação Comunidade- Parque para gestão ambiental efetiva. http://www.ibeas.org.br/congresso/Trabalhos2015/VI-023.pdf. Accessed 7 Oct 2016

Ford DC, Williams PW (1989) Karst geomorphology and hydrology. Chapman & Hall, London

Galvão P, Halihan T, Hirata R (2015) Evaluating karst geotechnical risk in the urbanized area of Sete Lagoas, Minas Gerais, Brazil. Hydrogeol J 23:1499–1513

Gnaspini P, Trajano E (2000) Guano communities in tropical. In: Wilkens H, Culver DC, Humphreys WF (eds) Ecosystems of the world, subterranean biota. Amsterdam, pp 251–268

Gomes MC (ed) (2006) A Canção das Palmeiras: Eugenius Warming, um jovem botânico no Brasil. Fundação João Pinheiro, Belo Horizonte

Gomes P (2011) Entre o planejamento "estratégico" e a regulação urbanístico-ambiental: um olhar sobre a expansão urbana recente no Vetor Norte da metrópole belo-horizontina. PUC Minas, Belo Horizonte

Herrmann G, Kohler C, Duarte J, Carvalho P (1998) Gestão ambiental. CPRM, Belo Horizonte

Hildreth-Werker V, Werker J, Alt L, Moura V (2016) First International cave conservation and restoration course in Brazil. NSS News 74:8–10

Holten B, Sterll M, Vaz LPR (2011) Peter Lund e as grutas com ossos em Lagoa Santa. Editora UFMG, Belo Horizonte

Hötzl H (1999) Karst hydrogeology and human activities. In: Drew D, Hötzl H (eds) IAH international contributions to hydrogeology. AA Balkema Publishers, Rotterdam

IBAMA (1990) Portaria Nº 887 de 15 de junho de 1990. Diário Oficial 117:11844

IBAMA and WWF (2007) Efetividade de Gestão das Unidades de Conservação Federais do Brasil: Implementação do Método RAPPAM - Avaliação Rápida e Priorização da Gestão de Unidades de Conservação. IBAMA, Brasília

IBGE (2014) Nota Técnica - Estimativas da População dos Municípios Brasileiros com data de referência em 1º de julho de 2014. http://www.ibge.gov.br/home/presidencia/noticias/pdf/analise_estimativas_2014.pdf. Accessed 22 Nov 2016

IBGE (2016) https://cidades.ibge.gov.br/brasil/mg/sete-lagoas/panorama. Accessed 4 Dec 2016

IEF-MG (2016) Unidades de Conservação do SAP Vetor Norte. http://www.ief.mg.gov.br/areas-protegidas/sap-rmbh/1645-unidades-de-conservacao-do-sap-vetor-norte. Accessed 3 Nov 2016

IUCN (2017). https://www.iucn.org/theme/protected-areas/about/protected-areas-categories. Accessed 24 May 2017

Lima MCA, Breviglieri EMB (2017) Proteção do patrimônio histórico, cultural e artístico e o instituto do Tombamento, Revista Quaestio Iuris 10(1). https://doi.org/10.12957/rqi.2017.23446

MMA (2004) Protection of speleological heritage. Resolução no 347, de 10 de setembro de 2004, Brazil, Ministério do Meio Ambiente. http://www.mma.gov.br/port/conama/legiabre.cfm?codlegi=452. Accessed 27 Sept 2016

MMA (2009) Instrução Normativa Nº2, Brazil, Ministério do Meio Ambiente. http://www.icmbio.gov.br/cecav/images/download/IN%2002_MMA_criterios_210809.pdf. Accessed 27 Sept 2016

Neves WA, Araujo AGM, Bernardo DV, Kipnis R, Feathers JK (2012) Rock art at the Pleistocene/Holocene boundary in eastern south America. PLoS ONE 7(2):e32228. https://doi.org/10.1371/journal.pone.0032228

Neves WAA, Piló LB (2008) O povo de Luzia: Em busca dos primeiros americanos. Editora Globo, São Paulo

Nóbrega P, Aguiar J, Dornas T, Drumond M, Figueira J (2013) Proposta de Inclusão da APA Carste de Lagoa Santa, MG, Brasil, como Sítio RAMSAR. In: Information Sheet on Ramsar Wetlands

Pessoa PFP (1996) Caracterização Hidrogeológica da Região de Sete Lagoas – MG: Potencialidades e Riscos, Dissertation, Universidade de São Paulo

Prous A, Baeta A, Rubbioli E (2003) O Patrimônio arqueológico da região de Matozinhos: conhecer para, proteger edn. Do Autor, Belo Horizonte

Queiroz PGM, Jacomino VMF, Menezes MÂ de BC (2007) Composição elementar do material particulado presente no aerossol atmosférico do município de Sete Lagoas, Minas Gerais. Química Nova 30(5):1233–1239. https://doi.org/10.1590/s0100-40422007000500035

Santiago DGA (2015) Proteção do patrimônio arqueológico: motivações, critérios e diretrizes no tombamento de sítios arqueológicos pelo IPHAN. Instituto do Patrimônio Histórico e Artístico Nacional, Rio de Janeiro

Souza T (2015) Avaliação da qualidade higiênico-sanitária da água de poços rasos utilizada para consumo humano no município de Sete Lagoas-MG. Dissertation, Universidade Federal de São João del-Rei

UFMG (2011) Plano Diretor de Desenvolvimento Integrado da Região Metropolitana de Belo Horizonte – PDDI-RMBH, V.1. http://www.rmbh.org.br/pddi/index.php?option=com_docman&task=cat_view&gid=41&Itemid=30&lang=pt-br. Accessed 7 Nov 2016

Veni G, LaMoreaux PE, DuChene H, Crawford NC (2001) Living with karst: a fragile foundation. American Geological Institute, Alexandria

Correction to: The Archaeological Record of Lagoa Santa (East-Central Brazil): From the Late Pleistocene to Historical Times

André Strauss, Igor M. Mariano Rodrigues, Alenice Baeta, Ximena S. Villagran, Marcony Alves, Francisco Pugliese Jr., Marcos Bissaro Jr., Rodrigo Elias de Oliveira, Gustavo Neves de Souza, Lucas Bueno, João Carlos Moreno de Sousa, Johnica J. Morrow, Karl J. Reinhard, Tiago Hermenegildo, Glauco Constantino Perez, Eliane Nunes Chim, Rafael de Oliveira dos Santos, Maurício de Paiva, Renato Kipnis, and Walter Neves

Correction to:
Chapter "The Archaeological Record of Lagoa Santa (East-Central Brazil): From the Late Pleistocene to Historical Times" in: A. S. Auler and P. Pessoa (eds.), *Lagoa Santa Karst: Brazil's Iconic Karst Region*, Cave and Karst Systems of the World, https://doi.org/10.1007/978-3-030-35940-9_12

The original version of the chapter "The Archaeological Record of Lagoa Santa (East-Central Brazil): From the Late Pleistocene to Historical Times" was inadvertently published with errors in the author names, which were incorrectly listed as João Carlos Moreno de Souza and Glauco Constantino instead of João Carlos Moreno de Sousa and Glauco Constantino Perez, respectively. Further, on page 267, the incorrect name "Ferreira" was replaced with the correct name "Ferraz" in the caption for Fig. 21.

The chapter and book have been updated with the changes.

The updated version of this chapter can be found at https://doi.org/10.1007/978-3-030-35940-9_12

A. S. Auler and P. Pessoa (eds.), *Lagoa Santa Karst: Brazil's Iconic Karst Region*, Cave and Karst Systems of the World, https://doi.org/10.1007/978-3-030-35940-9_15

MIX
Papier aus verantwortungsvollen Quellen
Paper from responsible sources
FSC® C105338

If you have any concerns about our products,
you can contact us on
ProductSafety@springernature.com

In case Publisher is established outside the EU,
the EU authorized representative is:
Springer Nature Customer Service Center GmbH
Europaplatz 3, 69115 Heidelberg, Germany

Printed by Libri Plureos GmbH
in Hamburg, Germany